WHISPERS BETWEEN PARTICLES

A Narrative of Quantum Entanglement Told Through Everyday Metaphors

LIOR BENNATHORNE

Whispers Between Particles: A Narrative of Quantum Entanglement Told Through Everyday Metaphors

This book is a work of narrative nonfiction. It draws on the published scientific record, including peer reviewed research papers, official materials from the Nobel Prize committee, and reporting from established scientific publications. Sources are identified in the Notes on Sources and Further Reading section near the end of this book. Every reasonable effort has been made to ensure the accuracy of the scientific and historical information presented. The recurring illustrative characters and metaphors used throughout are original to this work and are offered as aids to understanding rather than as literal descriptions of quantum phenomena.

Cover and interior design by the author.

CONTENTS

A Note Before We Begin

There is a particular kind of silence that settles over a physics lecture hall the moment someone says the words quantum entanglement out loud. It is not the silence of boredom. It is the silence of people trying to decide whether they misheard something impossible.

I have watched that silence happen more times than I can count, first as a student who did not understand what I was being told, and later as someone who spent years trying to explain it to people who, like me, had no interest in becoming physicists but every interest in understanding what the physicists were so excited about. Somewhere in that process, I stopped trying to explain the mathematics and started trying to explain the feeling. The feeling of two things, separated by an ocean or a solar system, behaving as though they were still holding hands.

This book does not require you to know any physics. It does not require you to remember anything from a school science class, and if that class left you certain you were not a science person, I would like to gently suggest that you were lied to. Nobody is born a science person or a non science person. Some of us were simply handed the subject in a way that made it sound like a foreign language read aloud from a phone book, all vocabulary and no story. Entanglement, properly told, is one of the great stories of the twentieth century, and it belongs to everyone who is curious enough to sit with it for an afternoon.

What follows is an attempt to tell that story using the objects and relationships already sitting in your life. Coins in pockets. Twins who seem to think the same thought at the same moment. A pair of gloves mailed to opposite sides of the world. None of these metaphors are perfect, and I will tell you exactly where each one breaks down, because a metaphor that pretends to be flawless is a metaphor that is lying to you. But used honestly, these ordinary images can carry you further into quantum mechanics than most textbooks manage in a full semester.

You will meet real people in these pages. Albert Einstein, who found entanglement so disturbing that he spent the last decades of his life trying to prove it wrong. Niels Bohr, who disagreed with him in public and in private for thirty years. John Bell, a quiet Irish physicist who wrote an inequality on a piece of paper in 1964 that would take other scientists decades to test properly. John Clauser, Alain Aspect, and Anton Zeilinger, who eventually did test it, so convincingly that the Royal Swedish Academy of Sciences awarded them the Nobel Prize in Physics in 2022. You will read about a satellite named after an ancient Chinese philosopher, sent into orbit to prove that entanglement survives the vacuum of space. You will read about diamonds in a Dutch laboratory, cooled and measured a kilometer and a third apart, closing an argument that had lasted half a century.

None of it is science fiction. All of it happened, and continues to happen, in laboratories that you could visit if you had the right credentials and enough patience with airport security.

I wrote this book because I believe the story of entanglement deserves to be told the way any good story deserves to be told, with tension, with characters who disagree passionately, with turns that surprise you, and with an ending that leaves you looking at the world slightly differently than you did before you opened the cover. Read it in one sitting if you can. It was built for that kind of reading, chapters sized to survive a long flight or a rainy Sunday, each one capable of standing on its own but each one also quietly setting up the next.

Somewhere in the universe, right now, there may be a pair of photons behaving as though the distance between them means nothing at all. This book is an attempt to explain why that should astonish you, and why, once you understand it properly, it will.

Lior Bennathorne

CHAPTER ONE

The Two Coins in Separate Pockets

Imagine a woman named Farah standing in an airport in Lisbon, about to board a flight to Tokyo. Before she leaves, her brother Idris hands her a small wooden box and asks her not to open it until she lands. Inside the box, though neither of them has looked, is a single coin. Idris keeps an identical box for himself, and he promises that his box also contains a single coin. He tells her one more thing before she walks toward the gate: the two coins were prepared together, as a pair, and whatever her coin shows when she opens the box, his coin will show the opposite.

Twelve hours later, somewhere over Siberia, Farah opens her box and finds heads. She does not know it yet, but at that exact moment, on the other side of the planet, Idris could open his box and find tails, guaranteed, every single time, no matter how many times they repeat this experiment with new coins.

If this were the whole story, it would be mildly interesting and easily explained. Idris and a friend simply flipped two coins before Farah left, noted the results, and packed the boxes accordingly. The coins were decided in advance. Farah's box always contained whatever it contained, from the moment it was sealed in Lisbon, and Tokyo had nothing to do with it. This is how ordinary correlation works, and it is how the overwhelming majority of correlated things in your life actually behave. Two shoes from the same pair match because they were manufactured to match, not because the left shoe has any awareness of what the right shoe is doing.

Quantum entanglement asks you to accept something considerably stranger, and it took the most brilliant physicists of the twentieth century decades of argument to accept it themselves.

In the quantum version of this story, the coins do not have a result until the moment they are observed. This is not a philosophical flourish or a rhetorical trick. It is, according to the mathematics of quantum mechanics and according to every experiment ever performed to test it, a description of how nature actually behaves at the smallest scales. Before Farah opens her box, her coin is not secretly heads or secretly tails, waiting to be revealed. It exists in a strange mathematical condition called superposition, a genuine mixture of both possibilities at once, and only the act of looking forces it to settle into one outcome or the other. The moment it settles, faster than any signal could possibly travel between Lisbon and Tokyo, Idris's coin settles into the opposite outcome.

Read that again, because it is worth sitting with. The coin does not have an answer until it is asked the question. And when it is asked, its distant partner instantly knows how to answer too, even though nothing, no radio wave, no light beam, no possible signal, had time to travel between them.

This is entanglement, and it is one of the most rigorously tested phenomena in the history of physics. It has been demonstrated with photons sent through optical fiber under city streets, with atoms trapped in laboratory chambers a few feet apart, with electrons embedded in diamond crystals over a kilometer apart, and, in one of the more theatrical demonstrations, with photons beamed from a satellite orbiting five hundred kilometers above the Earth to ground stations more than a thousand kilometers apart from each other. Every serious attempt to explain it away using ordinary, common sense ideas, coins decided in advance, secret signals, hidden instructions folded into the particles at creation, has failed under experimental scrutiny. We will spend a great deal of this book examining exactly how those experiments were designed and why they were convincing, because the credibility of an extraordinary claim rests entirely on the quality of the evidence behind it, and in this case the evidence is extraordinary too.

For now, hold onto the coin story loosely, the way you might hold a paper map you know you will need to redraw later. It gets you moving in the right direction even though, as we will see in a few pages, it is not quite telling the truth about how quantum mechanics works. The honest version is stranger than the coins, and stranger, I think, is more interesting.

A word about the phrase spooky action at a distance. Albert Einstein used a version of this phrase, spukhafte Fernwirkung in his native German, in a letter to his friend and fellow physicist Max Born in the 1940s, and he did not mean it as a compliment. Einstein found the implications of entanglement deeply troubling, not because the mathematics was wrong, he never disputed the mathematics, but because he believed the universe ought to behave in a way that respected locality, the common sense idea that an object can only be directly influenced by its immediate surroundings, and that no influence can travel faster than the speed of light. Entanglement, taken at face value, seemed to violate that principle, and Einstein spent a significant portion of his later career trying to show that quantum mechanics, despite its predictive success, was an incomplete description of reality, one that would eventually be replaced by a deeper theory restoring the common sense rules he believed the universe must obey.

He was, on this particular point, wrong. Not about the mathematics, which he understood better than almost anyone alive, but about the conclusion. Entanglement is real, it has been measured directly, and it does not require any signal to travel between the particles faster than light. This last part surprises people every time I explain it, so let me be precise about it now, early, before the coin metaphor does any more damage to your intuition.

Nothing travels between Farah's coin and Idris's coin. No signal, no information, no whisper carried on some undiscovered channel. What happens instead is that the two coins were never, in the deepest sense, two separate objects to begin with. From the moment they were entangled, they formed a single mathematical system, described by one set of equations rather than two. Measuring one part of that system tells you something about the other part not because a message passed between them, but because they were, in a way that has no good analogy in the world of coins and boxes, one thing wearing two disguises.

This is the part where every metaphor, including mine, starts to strain at the seams. Coins are separate objects. You can hold one in each hand and feel their separateness. Entangled particles are not separate in that sense, even when they are light years apart. It is a little like asking whether the left half of a soap bubble is a different object from the right half. The question assumes a kind of separateness that the bubble itself does not respect.

Physicists have a precise way of describing this using something called a wave function, a mathematical object that captures everything that can be known about a quantum system. For two entangled particles, there is a single wave function describing both of them together, and it cannot be split into two separate wave functions, one for each particle, no matter how far apart they travel. This inability to split the description into independent pieces is the actual definition of entanglement, and it is worth memorizing because everything else in this book builds from it. Entanglement is not a force. It is not a signal. It is a statement about how completely, and how irreducibly, two or more particles can share a single description, even when they are nowhere near each other in space.

Why does this matter to anyone who is not a physicist.

It matters first because it is true, and true things about the structure of reality are worth knowing regardless of their practical use, in the same way that it is worth knowing the age of the universe or the distance to the nearest star even if that knowledge will never help you fix a leaking faucet. But entanglement also happens to be enormously practical, in ways that would have seemed like fantasy to the physicists who first argued about it in the 1930s. It is the working principle behind quantum cryptography systems already deployed to protect financial transactions and government communications in several countries. It is the resource that quantum computers, built by companies including IBM, Google, and IonQ, rely on to perform certain calculations in ways that ordinary computers cannot match. It underlies proposals for a future quantum internet, a network that would use entangled particles to link computers together with a form of security that eavesdropping cannot defeat even in principle, because any attempt to intercept the entangled signal necessarily disturbs it in a detectable way. And it has already improved the precision of atomic clocks, the devices that define the length of a second and that make possible everything from GPS navigation to the synchronization of global financial markets.

None of this was the goal when entanglement was first discovered. Nobody in 1935 was thinking about GPS satellites or unhackable communication networks. The physicists who stumbled onto entanglement were trying to answer a much older and, in some ways, much more human question: what can we actually know about the world, and is there a limit to that knowledge built into the fabric of reality itself, or only into the limits of our instruments.

This question had occupied philosophers for centuries, but by the 1930s it had become an unavoidable, practical problem for physicists, because the equations of quantum mechanics, first written down in the 1920s by Werner Heisenberg, Erwin Schrodinger, and others, kept insisting on a strange kind of fuzziness at the heart of nature. A particle, according to these equations, does not have a definite position until you measure its position. It does not have a definite momentum until you measure its momentum. And, as Heisenberg showed in 1927 with what became known as the uncertainty principle, you cannot even measure both of these properties with unlimited precision at the same time. The more precisely you pin down where a particle is, the less precisely you can know how fast it is moving, and this is not a limitation of our measuring instruments. It appears to be a limitation built into the structure of reality itself.

Einstein found this deeply unsatisfying. He believed, as most physicists before him had believed, that a complete physical theory should describe reality as it actually is, independent of anyone measuring it. A moon, he liked to say in conversations with colleagues, should be there whether or not anyone is looking at it. Quantum mechanics, in its most literal reading, seemed to suggest otherwise, at least for the smallest constituents of matter, and Einstein spent years searching for a way to prove that the theory, whatever its practical successes, could not be the final word.

In 1935, working with two younger colleagues, Boris Podolsky and Nathan Rosen, at Princeton,

Einstein published a paper that he believed would expose a fatal flaw in quantum mechanics. The paper is usually referred to today by the initials of its three authors, EPR, and it introduced, almost as a side effect of trying to defeat the theory, the very phenomenon that would come to be called entanglement. Einstein and his colleagues described a thought experiment involving exactly the kind of correlated pair we have been discussing with Farah and Idris, though dressed in the mathematical language of position and momentum rather than coins in wooden boxes. Their argument, in essence, was this: if measuring one particle instantly tells you the result of measuring its distant partner, then either some influence is traveling between them faster than light, which Einstein considered physically impossible, or the particles must have possessed definite properties all along, hidden variables that quantum mechanics simply fails to describe. Either way, they concluded, quantum mechanics as it stood was incomplete.

It was a serious argument, constructed by serious people, and it could not be dismissed. It also, without anyone quite intending it, gave the phenomenon its name. Reading the EPR paper shortly after its publication, the Austrian physicist Erwin Schrodinger, already famous for his own contributions to quantum theory, coined a German word to describe the strange connection between the particles: Verschrankung, which translates into English as entanglement. Schrodinger recognized immediately that this was not a minor technical curiosity. He called entanglement not one but the characteristic trait of quantum mechanics, the feature that most sharply distinguished it from every physical theory that had come before.

Schrodinger was right about its importance, though it would take another three decades before anyone figured out how to settle the argument between Einstein and the defenders of quantum mechanics experimentally, rather than merely philosophically. That settling required a physicist named John Stewart Bell, working largely alone at the CERN particle physics laboratory in Switzerland, to find a way of turning a metaphysical disagreement into a testable, numerical prediction. We will meet Bell properly in a later chapter, because his contribution deserves the space of its own full telling. For now, it is enough to know that the disagreement between Einstein and the quantum physicists was not settled by argument. It was settled, eventually, by measurement, in laboratories, with real photons and real detectors and real data, and the measurement came down decisively on the side Einstein did not want to be true.

Let us return, briefly, to Farah and Idris, because their story, imperfect as it is, has done its job. It has given you a felt sense of correlation across distance, the eerie experience of one event seeming to know about another event it could not possibly have communicated with. Keep that feeling. Distrust the mechanism, coins prepared in advance, but keep the feeling, because the feeling is closer to the truth than the mechanism is. Something is shared between entangled particles that is not explained by advance preparation, and it is not explained by any signal passing between them either. It is explained by a mathematical unity that our everyday language, built for a world of separate coins and separate boxes and separate people standing in separate airports, was never designed to describe.

That inadequacy of language is not a failure on your part. It is not a sign that you are missing some crucial piece of common knowledge that would make all of this click into place if only someone explained it a little more clearly. The physicists who built quantum mechanics faced the same inadequacy, and some of the greatest minds among them, Einstein prominent among them, wrestled with it for the rest of their careers without fully making peace with it. Richard Feynman, one of the most celebrated physicists of the twentieth century and a man not easily accused of failing to understand his own field, was fond of telling students that nobody understands quantum mechanics, not in the sense of an intuitive, gut level comprehension the way you understand why a dropped glass breaks. You can learn to use the mathematics. You can learn to predict, with extraordinary precision, what an experiment will show. But the feeling of true, intuitive understanding, the feeling you have when you understand why ice floats or why a rainbow appears after rain, may simply not be available for a phenomenon like entanglement, because that feeling was built by a brain evolved to survive on the savanna, not to track particles that share a single description across a thousand kilometers of empty space.

This should not discourage you. It should, if anything, be liberating. You are not failing to understand entanglement in some special, personal way that marks you as insufficiently scientific. You are encountering the same wall that stopped Einstein, and the honest response to that wall is not embarrassment. It is curiosity, and perhaps a certain amount of awe.

In the chapters ahead, we will build up the picture piece by piece. We will look more closely at superposition, the strange quantum habit of being several things at once before measurement forces a choice. We will follow Einstein and Bohr through three decades of respectful, sometimes furious disagreement, staged at scientific conferences in Belgium in the late 1920s and continuing in letters and papers for the rest of their lives. We will meet Schrodinger's cat, one of the most famous thought experiments in the history of science and one of the most widely misunderstood. We will watch John Bell turn philosophy into mathematics, and we will watch three generations of experimental physicists, working in California, in France, in Austria, in the Netherlands, and eventually in orbit above the Earth, turn that mathematics into an unambiguous verdict, one convincing enough to earn the Nobel Prize in Physics in 2022 for Alain Aspect, John Clauser, and Anton Zeilinger.

By the end of this book, you will understand not only what entanglement is, but why it took so long to prove, why the proof required such ingenious experiments, and why, having been proved, it is already quietly reshaping the technology of communication, computation, and measurement in ways that will likely touch your life within the next decade, whether or not you ever think about coins in wooden boxes again.

Somewhere, right now, a laboratory technician in Vienna or Delft or Hefei is preparing a pair of particles that will, in a matter of microseconds, share an outcome across a distance no signal could have crossed in time. It has become, in the truest sense, a routine event. It is also, if you let yourself sit with it honestly, one of the strangest things that has ever been shown to be true.

CHAPTER TWO

The Language Particles Speak

What does it mean for something to be in two places at once.

Not metaphorically. Not in the sense that your attention is divided between a phone call and a boiling pot of pasta. Genuinely, physically, mathematically in two places, or two states, or two conditions, at the same time, until the moment someone checks.

This is the question that superposition forces you to confront, and it is worth confronting slowly, because almost every other idea in this book depends on your having some working grip on it. Entanglement is, at its heart, a story about superposition shared between two or more particles. Without superposition first, entanglement has nothing to stand on.

Let us start smaller than a coin. Let us start with a single particle, an electron, say, sitting inside an atom. Classical physics, the physics that describes cannonballs and planets and the trajectory of a thrown baseball, insists that the electron has a definite position at every instant, even if we do not happen to know what it is. If we had a good enough microscope, classical physics assumes, we could simply look and find out. The not knowing, in this view, is a fact about us and our limited instruments, not a fact about the electron.

Quantum mechanics denies this, and it does not deny it gently. According to the mathematics that has now survived a full century of experimental testing without a single confirmed failure, the electron genuinely does not have a definite position before it is measured. It exists as a spread of possibilities, described by that wave function we mentioned in the previous chapter, a mathematical object that assigns a probability to every place the electron might be found. When you measure the electron's position, this spread of possibilities collapses, instantly and unpredictably, into a single definite outcome. Before the measurement, asking where the electron actually was is a bit like asking what number a die actually was showing while it was still tumbling through the air. The question has no answer, not because you have not looked closely enough, but because the tumbling die has not yet become any particular number.

Erwin Schrodinger, the same physicist who would later name entanglement, gave physicists their most famous illustration of superposition in 1935, in the same year as the EPR paper, and in direct response to it. He imagined, purely as a thought experiment, never intended to be carried out, a cat sealed inside a steel chamber along with a small amount of radioactive material, a Geiger counter, a hammer, and a flask of poison. If the radioactive material decays, an event governed by genuine quantum randomness, the Geiger counter detects it, triggers the hammer, breaks the flask, and kills the cat. If the material does not decay, nothing happens, and the cat continues about its business, presumably unaware of the elaborate machinery built around it.

Radioactive decay is a quantum event, meaning that before we open the box and check, the radioactive atom exists in a superposition of having decayed and not having decayed, both possibilities present at once in the mathematics. Schrodinger's point, and it is a point people frequently miss, was not that this describes what actually happens to cats. His point was to expose how uncomfortable it becomes when you take the quantum description seriously and scale it up to something as large and familiar as a cat. If the atom is in superposition, and the fate of the cat is chained directly to the atom's decay, then a strict, unadjusted reading of quantum mechanics seems to require that the cat itself, until observed, exists in superposition too, simultaneously alive and dead, a single coherent mixture of both possibilities.

Schrodinger did not propose this as a fact about cats. He proposed it as a challenge, almost a taunt, aimed at physicists who were too comfortable applying quantum weirdness to electrons and atoms without asking what happens when that weirdness is allowed to grow. Something, he was suggesting, must prevent superposition from surviving all the way up to the scale of cats and chairs and human beings, and the theory as it stood in 1935 did not clearly explain what that something was.

Modern physics has a partial answer, though it took decades to work out properly, and the answer goes by the name decoherence. A single isolated particle, shielded carefully from its surroundings, can maintain a superposition for a surprisingly long time, long enough that scientists routinely demonstrate it in laboratories today. But a cat is not isolated. A cat is constantly bumping into air molecules, exchanging heat with its surroundings, reflecting stray photons in every direction. Each of these countless interactions acts, in effect, as a tiny measurement, and the accumulated weight of trillions of such interactions per second forces the superposition to collapse almost instantaneously, long before any human observer opens the box. Decoherence explains why you have never seen a superposed cat, a superposed coffee cup, or a superposed friend, without needing to invoke consciousness or human observation as some magical trigger, a misreading of quantum mechanics that unfortunately still circulates in popular culture. It is not the human eye that collapses a superposition. It is the sheer, overwhelming density of physical interaction that any large object experiences with its environment, interaction that a carefully isolated single atom or photon in a laboratory can, for a little while, be shielded from.

This matters enormously for entanglement, because the particles physicists entangle in real experiments, photons especially, are chosen precisely because they interact so weakly with their surroundings that they can preserve their delicate quantum state over meaningful distances and meaningful stretches of time. A photon traveling through the near vacuum of space, as we will see when we get to China's Micius satellite, barely interacts with anything at all along its path, which is exactly why space turned out to be such a promising medium for extending the distance over which entanglement could be reliably demonstrated.

Let us make superposition slightly more concrete using an everyday object, while being honest, as promised, about where the comparison breaks down.

Picture a spinning coin, not yet landed, whirring on a tabletop, moving too fast for your eye to resolve whether it currently shows heads or tails. This is sometimes offered as an image of superposition, and it captures a sliver of the idea, the sense of an outcome not yet settled. But it is a misleading sliver, because the spinning coin, at every instant during its spin, actually does have a definite orientation. It is simply changing too quickly for you to track. A high speed camera could, in principle, catch it at any frozen moment and read off exactly which face was showing. The coin's indefiniteness is a fact about your perception, not a fact about the coin.

Quantum superposition offers no such escape hatch. There is no hypothetical high speed camera, however advanced, that could catch the electron in the act of secretly being in one position rather than the other before you measure it. Multiple lines of experimental evidence, developed over the past several decades, have closed off this possibility with increasing rigor. It is not that we lack a fast enough camera. It is that the electron, in a very real sense that unsettled Einstein for the rest of his life, does not have a single position to be caught in until the measurement forces one into existence.

If the spinning coin will not quite do the job, try this instead, offered with the same caveat that every metaphor eventually strains. Imagine a musical chord, three notes played together on a piano, blending into a single sound. Ask a listener with no musical training which single note they are hearing, and the question itself is malformed. They are hearing a chord, a genuine blend, not merely three individual notes hidden inside a single sound that a sharper ear could separate out one by one, though a sharper ear certainly can identify the components. The chord is, in a loose sense, several things simultaneously present as one coherent whole. Push the analogy further, though, and it breaks the same way all the others do: play the chord loudly enough, next to a piano wire, and you can get one string to resonate at just one of those component frequencies, isolating a single note. Quantum superposition does not offer even that consolation. Measuring an electron's position does not politely isolate one component from a blend that was there all along. It forces the entire system into a single, freshly determined outcome, with the other possibilities not merely unheard but, in the standard interpretation of quantum mechanics, gone.

This raises the obvious and entirely fair question: gone where. What happened to the possibilities that were not selected. This question has produced one of the longest running and most passionate arguments in the philosophy of physics, and the honest answer is that physicists do not agree. The most widely taught position, called the Copenhagen interpretation because it was largely developed by Niels Bohr and his colleagues in Copenhagen during the 1920s, holds that asking what happened to the unselected possibilities is itself a misguided question. The wave function, in this view, is not a description of a hidden physical reality waiting to be uncovered. It is a description of our knowledge, our predictive capacity, and once a measurement occurs, the other possibilities simply were never realized, full stop, with no further story to tell about where they went.

A rival view, called the many worlds interpretation, first proposed by the American physicist Hugh Everett in 1957, offers a strikingly different answer. According to Everett, every possibility contained within a superposition is realized, but each in its own separate, non communicating branch of reality. When you measure the electron and find it here rather than there, in this view, another version of you, in another branch of the universe, is at that very moment finding it there rather than here, and neither version of you will ever know the other exists. It is a dramatic proposal, and it remains a minority position among working physicists, though a respected one, actively defended by serious researchers and taken seriously in academic physics departments around the world.

There are other interpretations too, including pilot wave theory, developed initially by Louis de Broglie and later extended by David Bohm, which attempts to restore a hidden, definite reality beneath the quantum fuzziness, at the cost of allowing certain influences to act instantaneously across arbitrary distances, precisely the kind of nonlocal influence that made Einstein so uneasy in the first place.

I raise these competing interpretations not to resolve them, since nobody has resolved them, and perhaps nobody ever will using experiments alone, since all of these interpretations, remarkably, make the exact same testable predictions about the outcomes of experiments. They differ only in the story they tell about what is happening underneath those outcomes, a story that no experiment yet devised can distinguish between. I raise them because it is important, this early in the book, to be honest about the difference between the mathematics of quantum mechanics, which is not seriously disputed by any working physicist, and the interpretation of that mathematics, which remains a genuinely open, and genuinely fascinating, philosophical question.

What is not in dispute, and what we can build on with total confidence, is this: particles genuinely exist in superposition before measurement, this superposition genuinely collapses into a single definite outcome upon measurement, and the outcome of that collapse is genuinely, fundamentally random, not merely unpredictable due to our ignorance but random in the deepest sense the word can carry. This last point deserves particular emphasis, because it is where quantum mechanics parts ways most decisively with everything that came before it in the history of physics.

Classical physics, whatever its complexities, is deterministic in principle. Given complete information about a system, its equations tell you exactly what will happen next, with no room left over for genuine chance. A flipped coin looks random to us only because we lack complete information about the force of the flip, the air currents in the room, and the precise starting orientation. A physicist with a sufficiently powerful supercomputer and sufficiently precise measurements of the initial conditions could, in principle, predict the coin's landing with certainty. The randomness is, as physicists say, epistemic, a matter of our incomplete knowledge rather than a fact about the coin itself.

Quantum randomness is different in kind, not merely in degree. When a radioactive atom decays at an unpredictable moment, or when a photon passes through a polarizing filter and is either transmitted or absorbed, no amount of additional information about the initial conditions, none whatsoever, allows you to predict the specific outcome in advance. The theory gives you probabilities, sometimes extraordinarily precise probabilities, but never certainties about individual events. This is not a temporary gap in our knowledge waiting to be filled by a better theory. Decades of careful experimental work, much of it directly related to the entanglement experiments we will examine in later chapters, have shown that no hidden information exists to be uncovered. The randomness appears to be a genuine, irreducible feature of how the universe operates at its smallest scales.

This is precisely the feature that so troubled Einstein, expressed in his famous remark, made in a 1926 letter to Max Born, that he found it hard to believe God would choose to play dice with the universe. It was a poetic way of expressing a serious scientific objection: Einstein believed, as a matter of deep conviction about how a rational universe ought to be structured, that underlying the apparent randomness of quantum mechanics there must be some deeper, deterministic layer of reality, some set of hidden variables that, if only we could access them, would restore the orderly, predictable universe that classical physics had always promised.

It is this very conviction, the belief in hidden variables lurking beneath quantum randomness, that the EPR paper was designed to defend, and it is this same conviction that John Bell would eventually find a way to test directly, with results that came down firmly against Einstein's hopes. We are building, chapter by chapter, toward that test, and I want you to feel the full weight of what was at stake in it. This was not an abstract, academic dispute between colleagues who enjoyed arguing for its own sake. It was a genuine disagreement about the deepest nature of physical reality, conducted by some of the most capable minds of the century, and it would take until 1964 for anyone to figure out how to settle it with numbers rather than words, and until the 1970s, 1980s, and beyond for anyone to actually perform the experiments capable of reading those numbers off from nature itself.

Superposition, then, is the strange but well established fact that a quantum system can genuinely exist as a blend of multiple possibilities before measurement, with the outcome of that measurement being irreducibly random. Entanglement is what happens when two or more particles share a single superposition, a single blended description, such that measuring one instantly determines information about the others, no matter how far apart they have traveled. In the next chapter, we turn to the argument that broke out over exactly what this sharing implies, staged across three decades of conferences, letters, and increasingly pointed disagreements between two of the towering figures of twentieth century physics, a disagreement that would eventually pull the entire subject of quantum foundations out of philosophy and into the experimental laboratory, where it has remained, productively and fascinatingly, ever since.

CHAPTER THREE

A Quarrel Between Two Geniuses

In October of 1927, in a hotel in Brussels, a group of the most accomplished physicists alive gathered for the fifth Solvay Conference, an invitation only meeting that has since become legendary in the history of science largely because of a single photograph taken on its final day. Seventeen of the twenty nine attendees pictured in that photograph would go on to win or had already won a Nobel Prize. Marie Curie sits near the center, the only woman in the room and, at that point, the only person ever to have won the Nobel Prize twice. Albert Einstein sits beside her, arms folded, looking, in most reproductions of the photograph, faintly displeased.

He had reason to be. The conference had been dominated by discussion of quantum mechanics, a theory that had matured with startling speed over the preceding three years, largely through the work of younger physicists including Werner Heisenberg and Erwin Schrodinger, and Einstein, though he had been one of the theory's founding contributors decades earlier through his work on the photoelectric effect, found himself increasingly uneasy with where the theory was heading under the leadership of Niels Bohr, the Danish physicist who had become its most forceful and philosophically committed defender.

The disagreement between Einstein and Bohr at that 1927 conference, and at the Solvay Conference that followed in 1930, has become one of the most celebrated intellectual duels in the history of science, and it deserves to be understood properly, because the entanglement experiments we will examine later in this book are, in a very real sense, the final verdict in an argument that began in those Brussels hotel corridors nearly a century ago.

Bohr's position, which came to be known as the Copenhagen interpretation after the city where his institute was based, held that quantum mechanics was a complete theory, and that the apparent fuzziness of the quantum world, the superpositions, the fundamental randomness, the impossibility of knowing both a particle's exact position and exact momentum simultaneously, was not evidence of an incomplete theory waiting to be improved. It was a genuine, irreducible feature of nature, and physicists needed to accept it and adjust their expectations about what physics could tell us, rather than continuing to search for a deeper, more classical layer of reality hiding beneath it.

Einstein could not accept this, and not out of stubbornness or a failure to grasp the mathematics, which he understood as well as anyone in the room. His objection was philosophical, and in a certain sense, deeply personal. He had built his entire career, including the general theory of relativity that had made him the most famous scientist on the planet, on the conviction that the universe possessed an objective structure, independent of any observer, waiting to be discovered through careful reasoning and mathematical elegance. Quantum mechanics, in Bohr's telling, seemed to deny that an electron had a definite position at all until someone measured it, which struck Einstein as an abdication of the very purpose of physics.

At the 1927 conference, Einstein began raising a series of ingenious thought experiments, designed to expose a contradiction within quantum mechanics or a violation of Heisenberg's own uncertainty principle. Bohr, by all contemporary accounts, would spend sleepless nights working through each challenge, and would appear the following morning with a rebuttal that closed the loophole Einstein had proposed. This pattern repeated at the 1930 Solvay Conference, most famously in an exchange that has come to be called the photon box thought experiment, in which Einstein proposed an elaborate device designed to measure the energy and timing of an emitted photon with a precision that the uncertainty principle should have forbidden. Bohr reportedly spent an entire anguished night on the problem before realizing, the following morning, that Einstein's own general theory of relativity, specifically its prediction that gravity affects the passage of time, provided exactly the correction needed to preserve the uncertainty principle. It was, by most tellings, one of the most elegant rebuttals in the history of physics, and it used Einstein's own great theory to defeat Einstein's own objection to another theory.

These early rounds of the debate concerned the internal consistency of quantum mechanics. Had Einstein found a genuine mathematical contradiction, that would have been the end of the theory altogether. He did not find one, and by the early 1930s, most physicists, including eventually a somewhat chastened Einstein, accepted that quantum mechanics was internally consistent. But Einstein shifted his objection onto different ground, ground that would prove far more durable and far more interesting, because it was no longer a question of whether the theory was mathematically consistent, but whether it was complete, whether it told the whole story about physical reality, or merely a useful, predictive fragment of a larger story still waiting to be told.

This shift produced the 1935 paper we mentioned in chapter one, written with Boris Podolsky and Nathan Rosen, and known ever since by the initials EPR. It is worth walking through the actual argument of that paper carefully, because it is a genuinely beautiful piece of reasoning, even though its conclusion did not survive later experimental testing.

Einstein, Podolsky, and Rosen began by proposing a reasonable sounding standard for what it should mean for a physical theory to be complete. Every element of physical reality, they argued, must have a corresponding element in the complete theory describing it. And they proposed a reasonable sounding criterion for identifying an element of physical reality: if, without in any way disturbing a system, you can predict with certainty the value of some physical quantity, then that quantity corresponds to a real, definite element of physical reality, whether or not you have actually measured it.

They then described a pair of particles that had interacted in the past and then separated, in a manner that, according to the mathematics of quantum mechanics itself, entangled their momenta and positions. Quantum mechanics predicted that if you measured the momentum of the first particle, you could predict, with certainty and without disturbing the second particle in any way, what the momentum of the second particle would be. By their own criterion, this meant the second particle's momentum was an element of physical reality, existing independently of measurement. But equally, if you had instead chosen to measure the position of the first particle, you could predict with certainty the position of the second particle, meaning position, too, must be an element of physical reality for the second particle.

Here was the trap they believed they had built. Quantum mechanics itself, through Heisenberg's uncertainty principle, insisted that a particle cannot simultaneously possess a definite position and a definite momentum. Yet the EPR argument seemed to show that both quantities must simultaneously be elements of physical reality for the second particle, since either one could be predicted with certainty depending on which measurement was performed on the first particle, far away. Quantum mechanics, they concluded, must therefore be an incomplete description of reality. Something was missing from the theory, some additional structure, some hidden variables, that would restore a fuller picture in which the particle really did have both a definite position and a definite momentum all along, values that quantum mechanics simply failed to capture.

The only alternative, they noted almost in passing, would be to accept that measuring the first particle somehow instantaneously affects the second particle, no matter the distance between them, an idea they considered so at odds with relativity and common sense that they treated it as effectively ruled out without much further discussion. This dismissal, offered almost as an aside in a paper focused on a different conclusion, turned out to be the single most consequential sentence in the entire argument, because it was precisely this possibility, instantaneous correlation across distance without any local hidden variables to explain it, that experiments would eventually confirm as the actual state of affairs in nature.

Bohr's response to the EPR paper, published later the same year under the same title Einstein's team had used, has struck many later readers, myself included on first encounter, as somewhat difficult to follow, dense with careful qualifications about the meaning of physical reality and the necessity of specifying the whole experimental arrangement before any measurement can be said to have a definite meaning. Bohr's central point, distilled from the surrounding caution, was that the EPR criterion for an element of physical reality was too permissive. You cannot, Bohr argued, speak of the second particle's momentum or position as an independent physical reality without specifying what measurement was actually performed on the first particle, because the very act of choosing which quantity to measure defines what can meaningfully be said about the whole entangled system. There is no view from nowhere, no way of talking about the properties of one half of an entangled pair without reference to the measurement context of the whole pair.

This is a subtle point, and reasonable physicists at the time, and since, have disagreed about whether Bohr's response fully answered Einstein's challenge or merely reasserted the Copenhagen interpretation's assumptions in more elaborate language. What is certain is that the debate, as brilliant as it was on both sides, remained, for nearly thirty years after the EPR paper, a philosophical dispute rather than a scientific one in the strictest sense, because nobody had proposed a way to test the disagreement experimentally. Both the Copenhagen interpretation and the hidden variables alternative that Einstein favored made, as far as anyone could show at the time, identical predictions for every experiment anyone knew how to perform. A dispute that cannot be settled by experiment is not, strictly speaking, a scientific dispute at all, however sharp and intelligent the reasoning on both sides. It is closer to a disagreement in philosophy, conducted, in this particular case, by two of the sharpest philosophical minds to ever also happen to be physicists.

This state of affairs persisted until 1964, when John Stewart Bell, working quietly and somewhat outside the mainstream of fashionable physics research at CERN, found a way to turn the philosophical dispute into a mathematical inequality whose violation or non violation could actually be measured in a laboratory. Bell's achievement, which we will examine in detail in a later chapter, showed that any theory built on Einstein's preferred foundation, local hidden variables, values fixed in advance and unaffected by distant measurements, must obey a certain mathematical constraint. Quantum mechanics, Bell showed, predicted that this constraint would sometimes be violated. The disagreement between Einstein and Bohr had finally been converted into a number that experimenters could go out and measure.

It is worth pausing here to appreciate something about Einstein's later years that popular accounts of physics sometimes flatten into caricature. Einstein was not a crank clinging to outdated ideas out of vanity or a failure to keep pace with a younger generation. He was arguably the single physicist most responsible for the twentieth century's revolution in how we understand space, time, and gravity, and his objections to quantum mechanics were not casual. They were the product of the same rigor and creativity that had produced special and general relativity. He never disputed that quantum mechanics worked, in the sense of making accurate predictions. He disputed that it was the final, complete story, and he was prepared to spend decades of his diminishing scientific reputation, in a field increasingly dominated by younger physicists who had made their peace with quantum weirdness, defending a position he believed the evidence would eventually vindicate.

The evidence, when it finally arrived, decades after Einstein's death in 1955, did not vindicate him. But the manner in which it settled the question, through the ingenious mathematics of a physicist Einstein never lived to meet, and through experiments performed by physicists not yet born when the EPR paper was written, is itself one of the most satisfying stories in the history of science. A disagreement conducted almost entirely in words and thought experiments for thirty years was, in the following sixty years, converted into an experimental program so rigorous and so repeatedly confirmed that it eventually earned its central figures the Nobel Prize in Physics.

Niels Bohr died in 1962, two years before Bell published the paper that would finally give the debate its experimental teeth. He did not live to see it either. In a certain sense, the argument between Bohr and Einstein was settled by history rather than by either man individually, which feels, in its own way, appropriate for a disagreement that was never really about personal vindication. Both men, whatever their disagreements, shared an unshakable commitment to finding out what was actually true about the universe, and both would likely have been fascinated, whatever their private disappointment or vindication, by the sight of physicists sending entangled photons to satellites in order to finally answer the question they had spent their lives arguing about across a hotel dining table in Brussels.

There is a personal dimension to the Einstein and Bohr relationship worth adding here, because it complicates any simple narrative of two rivals locked in permanent opposition. Despite the intensity of their disagreement, which spanned three decades and touched on what each man considered the deepest questions in physics, Einstein and Bohr maintained a relationship of genuine mutual respect and even warmth throughout their lives. Bohr kept a portrait of Einstein in his study. Einstein, for his part, described Bohr in a 1949 essay as one of the greatest scientific discoverers of our time, even while continuing to insist that Bohr's interpretation of quantum mechanics could not be the final word. When Einstein died in 1955, Bohr was reportedly deeply affected, and he continued, for the remaining seven years of his own life, to discuss and refine his responses to Einstein's objections, as though the conversation, however one sided it had become without its other participant, was not yet finished for him either.

This detail matters for understanding the broader story this book tells, because it establishes a model for scientific disagreement that stands in useful contrast to how public arguments, scientific or otherwise, are sometimes conducted today. Einstein and Bohr disagreed about something they both considered enormously important, for decades, in front of their peers, in published papers, and in private correspondence, without either man's respect for the other's intellect or integrity ever seriously wavering. Their disagreement was never about who was the better physicist, or who deserved more prestige or credit. It was about what was actually true, and both men remained willing, until the end of Einstein's life, to keep engaging with the other's strongest arguments rather than retreating into a comfortable dismissal of the opposing view.

It is also worth noting that the Solvay Conferences themselves, the setting for the most dramatic early exchanges between Einstein and Bohr, were a relatively new institution at the time, having been established in 1911 through the funding of the Belgian industrial chemist Ernest Solvay, specifically intended to bring together the world's leading physicists for concentrated, invitation only discussion of the most pressing open problems in the field. The conference's format, small enough that every attendee could participate directly in discussion rather than merely presenting prepared remarks to a passive audience, created exactly the kind of environment in which the Einstein and Bohr exchanges could unfold with the improvisational intensity that contemporary accounts describe, ideas proposed, challenged, and refined within the same afternoon, rather than across the more distant, formal exchange of published papers that might otherwise have been the primary vehicle for such a disagreement.

We will return to Bohr and Einstein's ideas throughout this book, because their disagreement set the terms for nearly every experiment that followed. But first, we need to spend proper time with the thought experiment that made the stakes of this argument vivid to an entire generation of scientists and, eventually, to the wider public: a sealed box, a small amount of radioactive material, and a cat whose fate would become the most famous unresolved question in the popular understanding of physics.

CHAPTER FOUR

The Cat We Cannot Quite See

Here is a confession that might be useful before we go any further: Erwin Schrodinger did not like his own cat.

Not literally, of course. There was no actual cat, no actual box, no actual vial of poison. But Schrodinger, writing in 1935 in a paper titled Die gegenwartige Situation in der Quantenmechanik, or The Present Situation in Quantum Mechanics, introduced his now famous thought experiment specifically because he found something absurd, almost embarrassing, about where the mathematics of quantum mechanics seemed to lead when followed without flinching. He was not celebrating the strangeness of superposition. He was holding it up, half amused and half appalled, and asking his colleagues to look closely at what they had built.

The setup, as we described briefly in chapter two, involves a cat sealed in a steel chamber with a mechanism connected to the quantum state of a radioactive atom. If the atom decays within a given hour, a Geiger counter triggers a hammer that shatters a flask of poison, killing the cat. If the atom does not decay, nothing happens. Radioactive decay is a genuinely quantum event, meaning that before anyone opens the box, the atom exists, according to the standard mathematics of quantum mechanics, in a superposition of having decayed and not having decayed. And because the cat's fate is mechanically linked to that atom with no room for ambiguity, a strict application of the same mathematics seems to require that the cat itself, sealed inside and unobserved, exists in a superposition of being alive and being dead, not as a fifty percent chance one way and a fifty percent chance the other, in the ordinary sense we use when we say a coin flip has even odds, but as a genuine, simultaneous blend of both possibilities, exactly as an unmeasured electron is genuinely both here and there until observed.

Schrodinger's point, sharpened by decades of subsequent discussion into something clearer than his original German prose, was this: nobody actually believes a macroscopic object like a cat can be in two states at once. It offends every intuition we have about the world, built from a lifetime of interacting with cats and doors and coffee cups that are always in one place, doing one thing, at any given moment. And yet nothing in the mathematics of quantum mechanics, as it stood in 1935, drew a clean line explaining exactly where superposition was supposed to stop. If electrons can be smeared across multiple positions, and if a macroscopic object can, in principle, be coupled to an electron's quantum state through a chain of mechanical triggers, what prevents the smearing from propagating all the way up to the cat, and beyond the cat, to the room, and beyond the room, to the person who eventually opens the box.

This is sometimes called the measurement problem, and it remains, even today, one of the genuinely open questions in the foundations of physics, discussed seriously in academic journals and physics department seminars around the world. The mathematics of quantum mechanics describes two very different kinds of process. There is the smooth, continuous evolution of the wave function when nothing is measuring it, governed by an equation Schrodinger himself derived in 1926, an equation so central to the field that it bears his name. And there is the abrupt, discontinuous collapse of that wave function into a single definite outcome the moment a measurement occurs. The theory tells you how to calculate both processes with tremendous precision. What it does not clearly tell you, at least not in its original 1920s formulation, is exactly what counts as a measurement, and exactly why the smooth evolution should suddenly and violently give way to the abrupt collapse at some particular moment rather than another.

We touched briefly in chapter two on decoherence, the modern physics answer to at least part of this puzzle, and it is worth spending more time with it here, because it genuinely does resolve much of what troubled Schrodinger, even if it does not resolve every philosophical question that his cat raises.

The core insight of decoherence theory, developed substantially in the decades after Schrodinger's original paper by physicists including Heinz Dieter Zeh and Wojciech Zurek, is that superposition is not fragile in some vague, general sense. It is fragile specifically with respect to interactions with a large, complicated environment. A single photon, carefully isolated in a vacuum chamber and shielded from stray light, can maintain a superposition for a meaningful stretch of time, because it has very little to interact with. A cat, by grim contrast, is an almost comically poor candidate for isolation. It exchanges heat with the air around it. It reflects countless photons in every direction. Its constituent atoms are constantly jostling against each other and against the surrounding steel chamber. Each of these interactions, individually microscopic and insignificant, functions as a tiny, informal measurement, extracting a minuscule amount of information about the cat's state into the surrounding environment.

The number of such interactions occurring within a cat, every single second, is almost incomprehensibly large, and the cumulative effect of this constant environmental measurement is to force the cat's quantum state to collapse into one classical outcome, alive or dead, far faster than any human observer could open the box to check. By the time anyone lifts the lid, the cat has already been definitively, classically, one thing or the other for a length of time so short that no meaningful superposition ever had the chance to persist at the scale of the whole animal, whatever might have been true for the single radioactive atom that started the chain.

This is a genuinely satisfying resolution, and it explains, without any need to invoke consciousness or human observation as a special, almost mystical trigger, why we never encounter superposed cats, superposed coffee cups, or superposed people in ordinary life, despite the fact that every cat, coffee cup, and person is, at bottom, made of quantum particles obeying quantum rules. The unfortunate popular idea that consciousness collapses the wave function, that reality only becomes definite once a mind looks at it, is a persistent misunderstanding, one that occasionally surfaces in popular science writing and self help literature with rather more mystical flourish than the physics actually supports. Decoherence shows that collapse, or at least something functionally indistinguishable from collapse for all practical purposes, happens through ordinary physical interaction with a complicated environment, with no observer, conscious or otherwise, required.

I say functionally indistinguishable, and this qualification matters, because decoherence does not fully settle the deeper philosophical question Schrodinger's cat raises, and honest physicists will tell you as much. Decoherence explains why we never observe superpositions at large scales, because the information describing the superposition leaks out into the environment so quickly and so thoroughly that recovering it, or ever detecting any interference effects that would reveal the superposition was genuinely there, becomes practically impossible. But some interpretations of quantum mechanics, the many worlds interpretation prominent among them, hold that the superposition never actually collapses in any final, complete sense. It merely becomes entangled with an enormous environment, in a way so complex and so irreversible that it looks, for all practical purposes, exactly like a definite, single outcome to anyone embedded within that environment, even though, in a stricter mathematical sense, the full superposition of alive cat and dead cat, now entangled with the entire surrounding room and everyone in it, technically persists.

Other interpretations, including certain objective collapse theories developed by physicists such as Giancarlo Ghirardi, Alberto Rimini, and Tullio Weber in the 1980s, propose instead that collapse is a genuine, spontaneous physical process, occurring randomly but with a probability that increases sharply with the size and complexity of the system involved, meaning that a single particle almost never spontaneously collapses on its own, but a system made of trillions of particles, like a cat, collapses essentially instantly, not because of environmental interaction per se, but because of an additional physical process built directly into the fundamental laws of nature, a process not yet confirmed or ruled out by any experiment performed to date.

I raise these competing accounts, again, not because I expect you to adjudicate between them, since the finest minds in the field have not managed to do so definitively after nearly a century of trying, but because Schrodinger's cat deserves to be understood as a genuinely open, genuinely productive puzzle, rather than a solved problem with decoherence as its tidy answer. The cat forced physicists to take seriously a question that might otherwise have been dismissed as merely philosophical: where, precisely, does the quantum world end and the classical world begin, and is that boundary a sharp line, a gradual fade, or, as some interpretations suggest, an illusion created entirely by our limited perspective as observers embedded within a much larger, and possibly still fully superposed, universe.

It is worth clarifying one more common misconception before we move on, because I have encountered it often enough in conversation to know it deserves direct correction. Schrodinger's cat is frequently invoked in popular culture as a general purpose metaphor for any situation involving uncertainty, any circumstance where you will not know the outcome until you check, a job application whose result has not yet arrived, a medical test whose results are still pending. This usage, while understandable and often harmless as a figure of speech, actually inverts Schrodinger's original point. An unopened envelope containing a job rejection or acceptance is not a genuine superposition in the quantum sense. The decision has already been made. The letter already says what it says. Your uncertainty about the outcome is purely a matter of your own incomplete information, exactly the kind of ordinary, classical, epistemic uncertainty we discussed in the previous chapter with the example of a tumbling die. Schrodinger's cat, by contrast, was meant to illustrate a genuinely different kind of uncertainty, one built into the structure of reality itself rather than merely into our limited knowledge of it, at least according to the most literal, unadjusted reading of quantum mechanics that Schrodinger was needling his colleagues about.

Getting this distinction right matters, because it is precisely the difference between these two kinds of uncertainty, genuine quantum indefiniteness on one hand and mere classical ignorance on the other, that sits at the very heart of the entanglement debate we are building toward. Recall the EPR argument from the previous chapter. Einstein, Podolsky, and Rosen wanted to show that quantum mechanics was missing some hidden classical layer, some hidden variables that would convert the apparent quantum indefiniteness of entangled particles into ordinary classical ignorance, the same way a shuffled deck of cards feels random to us only because we do not know the order, even though the order is, in fact, perfectly definite the whole time. If Einstein had been right, then the strangeness of entanglement would reduce, on close inspection, to nothing stranger than Farah's coins from chapter one, decided in advance, merely unknown to the people carrying them.

Schrodinger, interestingly, shared some of Einstein's discomfort with quantum mechanics, and the two men corresponded about it extensively throughout 1935, the same year that produced both the EPR paper and Schrodinger's cat. In fact, Schrodinger's cat thought experiment was written partly in direct response to and partly in sympathy with the EPR paper, both pieces published within months of each other, both authored by physicists who had contributed enormously to building quantum mechanics and who were now, in different ways, expressing discomfort with what they had built. It is a genuinely remarkable historical fact that entanglement, as both a word and a well defined physical concept, and the cat paradox, one of the most enduring images in the popular understanding of science, were born in the very same year, from overlapping conversations between physicists who were, in their own ways, trying to sound an alarm about a theory whose predictions they did not dispute but whose deeper meaning they found genuinely troubling.

The alarm, as we will see, was not entirely misplaced, though its resolution came from a direction neither Einstein nor Schrodinger anticipated. It did not come from discovering hidden variables that restored a comfortable, classical picture of reality. It came from proving, through decades of increasingly rigorous and increasingly clever experiments, that no such hidden variables can exist, at least not of the kind that would preserve locality, the principle that objects can only be directly influenced by their immediate surroundings. The universe, it turns out, really is as strange as the mathematics suggested, and the strangeness is not a temporary embarrassment waiting to be resolved by a deeper theory. It appears to be a permanent, confirmed feature of how reality is built.

It is worth pausing on a question that occurs to almost everyone hearing about Schrodinger's cat for the first time: has anyone ever actually tried to build something like it, not with an actual cat, which would be both unnecessary and unkind, but with some object large enough to test whether superposition genuinely breaks down at a particular size, or whether it can be coaxed further than common sense would predict.

The answer is yes, and the results are among the more delightful, under reported achievements of modern experimental physics. Researchers have not managed to put a cat, or anything resembling a cat, into superposition. But they have pushed the boundary of what can be placed into a genuine quantum superposition considerably further than the single atoms and photons of Schrodinger's era, and the record has been advancing steadily for decades.

In 1996, a team led by Serge Haroche at the Ecole Normale Superieure in Paris created what physicists somewhat playfully call a Schrodinger cat state using a single atom and a small number of photons trapped between two mirrors, demonstrating a genuine superposition of distinguishable electromagnetic field states, a miniature but real echo of Schrodinger's thought experiment realized in an actual laboratory. Haroche would go on to share the 2012 Nobel Prize in Physics with David Wineland for developing methods that allow the measurement and manipulation of individual quantum systems without destroying them, techniques that grew directly out of experiments exploring exactly the boundary Schrodinger had asked about six decades earlier.

Since then, the frontier has moved further still. Physicists have demonstrated superposition in molecules substantially larger and more complex than single atoms, sending them through interference experiments that only make sense if the molecule genuinely traveled multiple paths at once, the same underlying logic as an electron's famous ability to pass through two slits simultaneously. Researchers at the University of Vienna, the same city where Anton Zeilinger would later conduct his entanglement experiments, have demonstrated quantum interference with molecules containing more than two thousand atoms, organic compounds engineered specifically to be large, heavy, and structurally complex, and yet still capable, under sufficiently careful isolation from their environment, of behaving as though they occupy a superposition of positions. No cat has been superposed. No coffee cup either. But the boundary Schrodinger asked about, the line separating the quantum world from the classical one, has proven to be considerably further out than a casual reading of decoherence theory might suggest, and pushing that boundary further remains an active, well funded area of experimental physics, not merely because it satisfies curiosity, but because understanding exactly how and why large systems lose their quantum behavior bears directly on the practical challenge of building large scale quantum computers, machines that must preserve delicate superpositions across many interacting components for long enough to perform a useful calculation, a challenge we will examine in detail in a later chapter.

There is also a quieter, more personal dimension to Schrodinger's story worth including here, because it adds useful texture to a man too often reduced, in popular retellings, to the single image of a cat in a box. Schrodinger was, by the account of most biographers, a restless and unconventional figure, uncomfortable within institutions for much of his career, and his path through the 1930s was shaped as much by the rise of Nazism in Europe as by physics itself. He left his position in Berlin in 1933, the same year he shared the Nobel Prize in Physics with Paul Dirac for the development of wave mechanics, unwilling to remain in a country whose new government he found morally repugnant, and he spent the following years moving between Oxford, Graz, and eventually Dublin, where he helped found the Dublin Institute for Advanced Studies and remained for seventeen years, a period during which he continued to think and write about the philosophical foundations of the very theory he had helped build. His famous 1935 cat paper was written, in other words, not from the settled comfort of a permanent chair at a prestigious institution, but during a genuinely unsettled period in his own life, by a man already accustomed to living with more uncertainty than most of his colleagues, which perhaps makes it fitting that he became the physicist most associated with forcing the field to sit honestly with its own uncertainties.

Before we can appreciate exactly how the argument over those uncertainties was finally settled, we need to meet the physicist who made settling it possible, a quiet, largely self effacing man working at a particle physics laboratory in Switzerland, who in 1964 wrote down a short, elegant piece of mathematics that would take the rest of the twentieth century to fully appreciate, and that would eventually turn the ghost of the Einstein and Bohr debate into a number that could be measured with photon detectors and stopwatches rather than argued about across a conference dinner table.

CHAPTER FIVE

The Man Who Wrote an Inequality on a Blackboard

Picture two weather stations, one on a remote hillside in Wales and one on a remote hillside in Scotland, each staffed by a technician whose only job, once a minute, is to check three separate instruments: a thermometer, a barometer, and an anemometer measuring wind speed, and to record whether each reading is above or below a preset threshold, marking the result as either a plus or a minus. Neither technician can see the other station. Neither technician can communicate with the other during the recording session. And crucially, at the start of each minute, a supervisor at each station randomly and independently chooses which single instrument that particular technician will actually check, keeping the other two instruments unread for that round.

Over the course of many thousands of these one minute rounds, the two stations' recorded results are compared. If the weather across Britain is behaving in the ordinary way weather behaves, shaped by common regional patterns, shared fronts moving across the country, shared atmospheric pressure systems, you would expect the two stations' readings to show a certain, boundable amount of correlation, and no more. There is a mathematical ceiling on how correlated two independently measured, locally determined systems can possibly be, no matter how cleverly you design the underlying weather patterns connecting them, so long as neither station's reading at any given moment can be instantaneously altered by what is happening, or what is being measured, at the other station a few hundred kilometers away.

This is, in a deliberately simplified and physically inexact form, the essential shape of the argument John Stewart Bell constructed in 1964, in a paper titled On the Einstein Podolsky Rosen Paradox, published not in a famous journal but in a short lived periodical called Physics Physique Fizika, a journal whose very name, printed in three languages, reflected its brief and somewhat marginal existence. Bell was working at CERN, the European particle physics laboratory near Geneva, primarily on more conventional topics in particle physics. The 1964 paper on entanglement was, by most accounts, something closer to a side project, pursued during a sabbatical, born from Bell's long standing personal fascination with the EPR paradox and his admiration for the physicist David Bohm, whose 1950s work on hidden variable theories had convinced Bell that the debate between Einstein and Bohr deserved a more rigorous mathematical treatment than it had so far received.

What Bell managed to prove, using mathematics considerably more careful and precise than my weather station sketch, was genuinely remarkable, and it is worth stating its structure clearly, because the structure itself, independent of the physics, is a small masterpiece of logical reasoning.

Bell began by taking Einstein's side of the argument completely seriously, on its own terms, rather than dismissing it. He assumed, exactly as

Einstein had hoped might be true, that entangled particles carry with them some set of hidden variables, definite properties fixed at the moment of their creation, that fully determine the outcome of any measurement performed on them later, regardless of which measurement is chosen. He further assumed locality, the principle Einstein held dear, that the choice of measurement performed on one particle cannot instantaneously influence the outcome of a measurement performed on its distant partner. These two assumptions together, hidden variables plus locality, are often bundled under the term local realism, and they represent, essentially, the common sense picture of the world that most people, including most physicists before the 1960s, simply took for granted.

Bell then asked a precise mathematical question: given these two assumptions, and nothing else, what is the maximum possible correlation that could be observed between measurements performed on two entangled particles, across many repeated trials, where the specific property to be measured on each particle is chosen randomly and independently for each trial. He derived an inequality, a mathematical statement establishing an upper bound on that correlation, a bound that any theory built on local hidden variables, no matter how cleverly constructed, could never exceed. This is now known as Bell's inequality, though in the decades since 1964 physicists have developed several related versions, including a particularly practical form derived in 1969 by John Clauser, Michael Horne, Abner Shimony, and Richard Holt, known as the CHSH inequality after the initials of its four authors, a version specifically designed to be testable using real experimental apparatus, real photon sources, and real detectors, rather than the somewhat more idealized mathematical setup of Bell's original 1964 paper.

Here is the crucial second half of Bell's achievement, the part that turned a piece of pure mathematics into a genuine scientific bombshell. Bell then calculated what quantum mechanics itself predicts for the same experimental setup, the same repeated trials, the same randomly chosen measurements on entangled particles. And he found that quantum mechanics predicts correlations that, under the right conditions, exceed the bound set by his inequality. Not by a small, ambiguous margin that might be explained away by experimental imprecision, but by a clear, calculable amount.

This was the moment the argument between Einstein and Bohr stopped being philosophy and became physics in the fullest, most testable sense. If experiments performed on real entangled particles showed correlations obeying Bell's inequality, staying within the bound, then Einstein's picture of the world, local hidden variables, common sense locality, would be vindicated, and quantum mechanics, whatever its practical successes, would indeed be missing something, exactly as Einstein had long suspected. But if experiments showed correlations violating Bell's inequality, exceeding the bound that any locally realistic theory could produce, then Einstein's picture would be definitively ruled out, not merely disfavored or philosophically awkward, but mathematically incompatible with observed reality, and quantum mechanics's stranger predictions would stand fully vindicated, exactly as Bohr, and Bohr's intellectual descendants, had long insisted.

It is worth sitting with how unusual this achievement was within the broader landscape of theoretical physics. Bell had not discovered a new particle or a new force. He had not proposed a new equation describing some previously unexplained phenomenon. He had, instead, taken an argument that had been conducted for thirty years using the tools of philosophy, thought experiments, careful reasoning about hypothetical scenarios, and shown that it could be converted into an actual, falsifiable, numerical prediction, testable with instruments that, while demanding, were not obviously beyond the reach of experimental physics as it existed even in the mid 1960s. He had, in effect, built a bridge between a philosophical disagreement and an experimental program, and once that bridge existed, it was only a matter of time, engineering, and funding before someone walked across it.

That someone turned out to be John Clauser, a graduate student at Columbia University who encountered Bell's paper in the late 1960s and became, in his own later recollection, obsessed with the idea of actually testing it, despite considerable skepticism from more senior physicists in his field, several of whom regarded the entire foundational question as settled, uninteresting, or a waste of a promising young researcher's career. This skepticism deserves a moment of honest acknowledgment, because it complicates any tidy, triumphant narrative about scientific progress. For much of the mid twentieth century, the physics establishment, having absorbed and successfully applied quantum mechanics to enormously productive effect in fields ranging from nuclear physics to the emerging technology of semiconductors, had largely stopped treating questions about the foundations and interpretation of the theory as live scientific business. The mathematics worked. It predicted experimental outcomes with staggering accuracy. Many senior physicists regarded further inquiry into what the mathematics actually meant, whether it implied genuine nonlocality, whether hidden variables might yet be found, as a distraction, closer to philosophy than to serious physics, and not the kind of project likely to advance a young researcher's career.

Clauser pursued the question anyway, alongside his official doctoral research, eventually connecting with Stuart Freedman, a fellow graduate student with strong experimental skills, at Columbia and later at the University of California, Berkeley, and Lawrence Berkeley National Laboratory. Together, working with limited funding and equipment they had to build and adapt themselves, Clauser and Freedman set out, in the early 1970s, to perform the first real experimental test of a Bell type inequality, using pairs of entangled photons generated by exciting calcium atoms with ultraviolet light.

We will spend the entirety of the next chapter examining what Clauser and Freedman actually did, because their experiment, performed in a converted basement space in Birge Hall at Berkeley, deserves to be understood in real detail, not as a footnote but as a genuine scientific adventure, conducted by two young researchers pursuing a question their field had largely stopped taking seriously. For now, though, it is worth pausing to appreciate John Bell himself a little longer, because his story carries a particular kind of poignancy that I think deserves acknowledgment.

Bell never abandoned a certain personal sympathy for Einstein's discomfort with quantum mechanics, even after his own theorem provided the mathematical machinery that would eventually be used to rule out Einstein's preferred resolution. In interviews conducted later in his life, Bell described himself as hoping, at least at some level, that experiments would eventually vindicate local hidden variables, that some subtle loophole in the existing experiments might yet be found, restoring the kind of orderly, locally causal universe that Einstein had spent his life defending. This is a genuinely humane detail, easy to overlook amid the triumphant retelling of how quantum mechanics was vindicated. The man whose mathematics did more than anyone else's to settle the argument against Einstein was, by his own account, not entirely happy about the direction the evidence was pointing, and he spent considerable energy in his later career scrutinizing the loopholes in existing experiments with real rigor, precisely because he wanted to be certain that the verdict against local realism was airtight before he would fully accept it himself.

Bell died suddenly in 1990, at the age of sixty two, from a cerebral hemorrhage, only a few years before some of the most decisive loophole closing experiments of the 1990s and 2000s would begin to appear, and decades before the fully loophole free experiments of 2015, which we will examine later in this book, and decades before the 2022 Nobel Prize that would formally recognize the experimental program his 1964 paper had made possible. He is widely regarded today, among physicists working on the foundations of quantum mechanics, as one of the most important theoretical physicists never to have won the Nobel Prize himself, a gap explained partly by the Nobel committee's longstanding reluctance to award the physics prize posthumously, and partly, perhaps, by how long it took the wider physics community to fully recognize the significance of a paper published in a journal most physicists of the time had never heard of.

There is a particular kind of intellectual courage in Bell's story that I want to highlight before we move forward, because I think it offers something valuable beyond the physics itself. Bell pursued a question that most of his contemporaries considered closed, unfashionable, or simply not worth a serious physicist's time, and he did so not because he was certain of the answer in advance, but because he recognized that the question had never actually been properly settled, only set aside. The distinction between a question being answered and a question merely being abandoned is one worth remembering, in physics and well beyond it. Bell's insistence on reopening a debate that most of his field considered finished is precisely what allowed that debate to finally, decades later, be closed properly, with evidence rather than exhaustion.

In the chapters ahead, we follow that evidence as it accumulated, starting in a Berkeley basement in the early 1970s, continuing through an ingenious French experiment in the early 1980s that closed one of the most stubborn remaining loopholes, and culminating in a pair of nearly simultaneous, fully loophole free experiments performed in 2015 in the Netherlands, the United States, and Austria, experiments so thorough and so carefully designed that they left no reasonable escape route for anyone still hoping, as Bell himself once quietly hoped, that Einstein might yet be proven right.

Before we leave Bell's paper itself, it is worth spending a little more time with why it took physics so long, nearly three decades after the EPR paper, to arrive at this particular piece of mathematics, because the delay says something important about how scientific progress actually happens, and it is not always a story of steady, inevitable advance.

Part of the delay was cultural, in the specific sense that the physics community of the 1940s and 1950s had, for reasons both practical and intellectual, largely moved on from foundational questions about the meaning of quantum mechanics. The Second World War had redirected enormous scientific talent and funding toward nuclear physics, and the postwar decades saw an explosion of productive, applied work: the development of quantum electrodynamics, the construction of ever more powerful particle accelerators, the beginnings of solid state physics that would eventually give the world the transistor and, later, the entire semiconductor industry. Physicists were busy, and enormously successful, and the question of what quantum mechanics actually meant, as opposed to how to use it, came to be regarded by much of the field as a settled matter, or at least a matter that had ceased to be productive to argue about further. Bohr's Copenhagen interpretation, whatever its philosophical loose ends, provided a working framework that let physicists calculate and predict with confidence, and for many working scientists, that was enough.

David Bohm, whose 1952 papers on hidden variable theories directly inspired Bell, found himself increasingly marginalized within American physics during this period, partly for reasons that had nothing to do with physics at all. Bohm had been called before the House Un-American Activities Committee in 1949 amid the anti communist political climate of the early Cold War, and though he was never convicted of any wrongdoing, the professional fallout was severe enough that he left the United States entirely, eventually settling in Brazil, then Israel, and finally at Birkbeck College in London. His exile from the mainstream of American physics meant that his ideas about hidden variables, though taken seriously by a small number of careful readers including Bell, circulated at the margins of the field for years rather than at its center.

Bell himself, working at CERN on entirely different, more mainstream topics in particle physics for most of his career, treated his foundational work on entanglement as something closer to a private intellectual passion than a professional priority, pursued during a 1964 sabbatical year spent partly at Stanford and partly at the University of Wisconsin, and published, notably, in a journal, Physics Physique Fizika, that folded after only a handful of volumes and that few physicists of the era would have encountered through their ordinary reading habits. It is worth sitting with this fact for a moment: one of the most important theoretical results in the history of quantum mechanics was published somewhere close to the periphery of the field's ordinary attention, by a physicist who was not, at the time, considered a specialist in the subject, working through a question that most of his more prominent contemporaries had already filed away as settled.

This is not, I think, a story about the system failing, exactly. Bell's paper was not suppressed or actively rejected. It was simply, for a number of years, quietly overlooked, the way a great many genuinely important ideas in every field of intellectual life spend some period of time overlooked before the right combination of readers, circumstances, and follow up work brings them to wider attention. What changed the trajectory, in this case, was the appearance of young researchers like John Clauser, willing to take the paper seriously enough to build an actual experiment around it, despite the discouragement of more senior colleagues who considered the whole enterprise a poor use of a promising career. Bell's theorem needed not only to be written. It needed to be noticed, and then it needed to be tested, and each of those steps required a specific person, at a specific moment, willing to treat a supposedly settled question as genuinely open.

There is a broader lesson here about the sociology of science that is easy to miss when the story is told only in its triumphant, final form, Nobel medals and all. Ideas do not advance on their own momentum. They require individual physicists, often working against the prevailing judgment of their field about what counts as a worthwhile use of time and resources, to keep pushing on questions that everyone else has agreed to stop asking. Bell asked. Clauser tested. And the decades that followed, which we turn to next, would show, with increasing and eventually overwhelming rigor, just how right they both were to keep asking.

CHAPTER SIX

Twins Separated at Birth

There is a particular kind of story that circulates every few years in newspapers and daytime television segments, always with slight variations but always following the same basic shape. Twins, separated shortly after birth and raised by different families, sometimes in different countries, are reunited as adults and discover, to everyone's amazement, that they married spouses with the same first name, or worked in the same unusual profession, or independently named their firstborn sons the same slightly uncommon name. The tone of these stories is invariably wonder struck, treating the coincidences as evidence of some mysterious, almost telepathic bond between identical twins, a connection that survived years of separation and completely different upbringings.

There is a more sober explanation, and it is worth walking through carefully before we return to entangled particles, because the contrast between the two cases, twins and particles, illuminates exactly what makes quantum entanglement so different from anything else correlation ordinarily has to offer.

Identical twins share, as their name suggests, essentially identical genetic material, having developed from a single fertilized egg that split early in development. This shared genetic starting point, combined with certain broad similarities that tend to exist across different households in the same general culture, common naming trends, common professional opportunities available at a given time and place, common patterns in what people find attractive in a partner, is entirely sufficient to explain an above average rate of coincidence between separated twins, without requiring any mysterious ongoing connection between them. The twins are correlated because they share a common cause in their past, their shared genetics and their broadly overlapping cultural environment, not because some invisible thread continues to link their present day choices across the distance separating them. If one twin, at this very moment, decides to redecorate a kitchen, nothing about that decision reaches across the ocean to nudge the other twin's decorating choices. The correlation, however striking it might feel in an individual case, was baked in decades ago, at the moment of the shared genetic inheritance, and has simply been unfolding independently, in parallel, ever since.

This is precisely the kind of explanation Einstein, Podolsky, and Rosen hoped would eventually be shown to underlie the correlations observed in quantum entanglement. Two particles, prepared together at a common source, might carry with them some kind of shared genetic inheritance, in the form of hidden variables, definite properties fixed at the moment of their creation, that would fully determine how each particle would respond to any measurement performed on it later, entirely independently of what happened to its distant partner. Under this picture, entangled particles are simply an extreme version of the twins story: correlated because of a shared past, not because of any mysterious ongoing connection operating in the present.

Bell's theorem, examined in the previous chapter, gave physicists a precise mathematical tool for distinguishing between these two very different explanations of correlation: shared past versus ongoing connection. And the experiments built on that theorem, which we begin examining properly in the next chapter, showed, with mounting and eventually overwhelming confidence, that the twins style explanation, shared hidden variables fixed at the moment of creation and nothing more, simply cannot account for the correlations observed between entangled particles. Something else is going on, something that does not reduce to a shared past however cleverly you try to construct it.

It is worth spending real time understanding exactly why the twins explanation fails for entangled particles, because the failure is not obvious, and grasping it properly is, in my experience, the single most important step in genuinely understanding what makes entanglement strange.

Return to Bell's weather station setup from the previous chapter, but let us make it more precise and closer to how real entanglement experiments are actually structured. Instead of thermometers, barometers, and anemometers, imagine that each technician has access to a single measuring device that can be rotated to one of three different angles before each measurement, chosen randomly and independently by each technician's own local coin flip, with neither technician able to know, in advance, which angle their counterpart will choose for that round. Each measurement, at whatever angle it happens to be performed, yields one of two possible outcomes, which we can simply call plus or minus.

If the two particles being measured carry hidden variables, definite predetermined instructions for how to respond to a measurement at every possible angle, laid down back when the particles were created together at a common source, then there exists, in principle, a complete table, fixed in advance, specifying exactly what result each particle would produce at each of the three possible measurement angles. The particle does not know in advance which angle will actually be chosen, since that choice is made randomly, independently, and locally at each station, but the hidden variable theory assumes the particle carries a definite, predetermined answer ready for every possible angle it might be asked about, the way a well prepared twin might have a definite answer ready for any of several possible interview questions, regardless of which one actually gets asked.

Bell's mathematical insight, developed further and made experimentally practical by Clauser, Horne, Shimony, and Holt in 1969, was to show that if such a complete, predetermined table exists for each particle, fixed independently at the source and unaffected by anything happening at the distant measurement station, then the statistical pattern of agreement and disagreement between the two stations' results, averaged across thousands of repeated trials with randomly varying angle combinations, must satisfy a specific mathematical inequality. There is a hard ceiling on how strongly correlated the two stations' results can possibly be, a ceiling imposed purely by the logical structure of predetermined, locally fixed answers, regardless of exactly what values fill in that predetermined table.

Quantum mechanics predicts that entangled particles, measured in precisely this kind of experimental arrangement, will violate that ceiling. Not narrowly, not by an amount plausibly attributable to experimental noise or measurement error, but by a clear, calculable, and repeatedly confirmed margin. And when physicists actually built the apparatus and ran the experiment, starting with Clauser and Freedman in the early 1970s and continuing through increasingly rigorous versions over the following five decades, nature agreed with quantum mechanics, not with the ceiling that any predetermined, locally fixed table of answers could produce.

This result rules out, quite decisively, the twins style explanation for entanglement. It is not merely that we have not yet discovered the hidden variables. It is that no possible set of hidden variables, fixed locally at the source and unaffected by distant events, no matter how cleverly constructed, no matter how many additional hidden properties you propose the particles might secretly carry, can reproduce the statistical pattern that quantum mechanics predicts and that experiments have now measured, again and again, across laboratories on four continents and, as we will see in a later chapter, in orbit above the planet as well.

Something in the correlation between entangled particles genuinely does depend on which measurement is chosen, in real time, at the distant location, in a way that a fixed, predetermined table of answers, however elaborate, cannot capture. This does not mean, and I want to be careful here because this is the single most common misunderstanding that arises at exactly this point in the explanation, that a signal travels from one particle to the other, informing it which measurement was chosen so that it can adjust its response accordingly. If such a signal existed, it would have to travel faster than light to explain the experimental results, since some entanglement experiments have been performed with the two measurements separated by a distance and a time interval that would require exactly that kind of faster than light signal. And a faster than light signal carrying usable information would create severe problems for Einstein's theory of relativity, potentially even allowing, under certain configurations, effects to precede their causes, a violation of causality that essentially no physicist, however comfortable with quantum weirdness, is prepared to accept.

The actual resolution, confirmed by decades of careful theoretical and experimental work, is more subtle than either a shared predetermined past or a faster than light signal, and it requires letting go of an assumption so basic that most people, including most physicists before the mid twentieth century, never thought to question it: the assumption that measurement outcomes are, at bottom, always local, meaning that the result you get when you measure a particle is fully determined by that particle's own local properties, whatever they might be, independent of anything happening anywhere else in the universe.

Entangled particles, it turns out, do not obey this assumption. The measurement outcome obtained at one location genuinely is connected, in the sense that the statistical correlations between the two locations exceed what any locally determined story could produce, to the choice of measurement made at the other location, even though no usable signal passes between them, and even though neither individual measurement outcome, taken on its own, in isolation from its distant partner, shows any pattern at all. This last point deserves emphasis, because it resolves the apparent tension with relativity that troubles many people encountering this material for the first time. If you look only at Farah's results, the sequence of pluses and minuses she records over many trials in Tokyo, that sequence looks, on its own, completely random, exactly fifty percent plus and fifty percent minus, with no detectable pattern whatsoever, and certainly no detectable trace of which measurement angle Idris happened to choose back in Lisbon at each corresponding moment. The strangeness, the violation of Bell's inequality, only becomes visible when you later bring the two separate lists of results together and compare them side by side, calculating the correlation between them. And bringing two separate lists of results together for comparison requires, in every real experiment ever performed, an entirely ordinary, well within the speed of light, communication channel, an email, a phone call, a shared database, something entirely mundane connecting the two laboratories after the fact.

This is precisely why entanglement, remarkable as it is, cannot be used to send a message faster than light, a point we will return to with more technical precision in a later chapter, because it has significant implications for what quantum technologies can and cannot achieve, and because it addresses one of the most persistent misconceptions surrounding entanglement in popular treatments of the subject. Farah, sitting in Tokyo with her list of random looking results, cannot extract any information whatsoever about what Idris chose to measure in Lisbon by examining her own results alone. The correlation is real, and it is genuinely stronger than any locally determined explanation could produce, but it only becomes visible, only becomes usable in any sense, once both lists of results are compared side by side, and that comparison necessarily travels through ordinary, subluminal channels.

What entanglement actually demonstrates, then, is not communication faster than light, and not a shared predetermined past written into the particles at the moment of their creation. It demonstrates something genuinely without precedent in classical physics: a form of correlation between distant events that cannot be fully explained by either mechanism, a correlation that is, in the technical language physicists use, nonlocal, without requiring any nonlocal signal, cause, or influence that would violate relativity's speed limit. This combination, nonlocal correlation without nonlocal signaling, strikes most people, on first encounter, as a kind of contradiction, or at best a semantic trick papering over a deeper problem. It is not a trick. It is a genuinely new category of physical relationship, one that simply did not exist in the conceptual toolkit that physics had developed prior to the 1960s, and reconciling oneself to its existence, as opposed to merely learning to calculate with it, remains, honestly, one of the more demanding intellectual exercises that modern physics has to offer.

I find it helps, at this point in the explanation, to abandon the search for a perfect analogy altogether, and instead to sit briefly with the negative space the analogies leave behind. Entangled particles are not like twins with a shared genetic inheritance, because Bell's theorem rules out any story built purely on a shared, predetermined inheritance. They are not like two people talking on a telephone, because no signal, faster than light or otherwise, actually passes between them at the moment of measurement. They are not quite like anything else you have direct experience of, and the honest, if slightly uncomfortable, conclusion is that the universe, at its most fundamental level, contains a genuine category of relationship for which ordinary human experience, shaped entirely by objects and events unfolding at everyday speeds and everyday distances, simply did not prepare you.

This should not be received as bad news. Consider how many other genuinely novel categories physics has already asked you to accept without complaint: that time runs measurably slower for a fast moving spacecraft than for an observer at rest, confirmed repeatedly using atomic clocks flown on aircraft and satellites, or that light behaves, depending on how you choose to measure it, sometimes like a wave spreading through space and sometimes like a discrete particle striking a single point, a duality confirmed in laboratories for more than a century. Physics has, throughout its history, periodically asked us to expand our sense of what is possible beyond what everyday intuition, evolved for tracking predators and prey across an African savanna, was ever built to anticipate. Entanglement is simply the latest, and in some ways the most vivid, chapter in that long and ongoing expansion.

There is a formal version of this same point, one that physicists take seriously enough to have proven it as a rigorous mathematical theorem rather than leaving it as an informal observation. It is called the no communication theorem, and it was established independently by several physicists during the 1970s and 1980s, building directly on the mathematics of quantum mechanics itself. The theorem shows, with full rigor, that no matter how cleverly you try to exploit entanglement, no matter what measurement you choose to perform on your half of an entangled pair, the statistical results you obtain, examined entirely on their own without reference to your distant partner's results, are always completely random and completely uninformative about what your partner chose to do. This is not a practical limitation that better engineering might one day overcome. It is a mathematical consequence of the structure of quantum mechanics itself, as solid and as unlikely to be overturned by future discoveries as the Pythagorean theorem.

This matters enormously for how we should think about proposals, and there have been many over the decades, some serious and some considerably less so, to use entanglement for instantaneous communication, perhaps between Earth and a distant space probe, sidestepping the delays imposed by the speed of light that currently make conversation with a Mars rover a matter of minutes rather than seconds. The no communication theorem closes this door firmly. Whatever entanglement is good for, and as we will see in later chapters it is good for a great deal, sending usable messages faster than light is not among its talents, and no future refinement of quantum technology is expected to change that, because the limitation is built into the mathematical bones of the theory rather than into any particular piece of laboratory apparatus.

It may help to consider one more everyday scenario, chosen specifically because it captures the shape of the resolution, the way correlation can be real and yet only become visible upon later, ordinary comparison, even though this scenario, like every other one in this book, will eventually need to be set aside.

Picture two ice skating judges seated at opposite ends of a rink, each independently scoring a series of routines without conferring, each recording a single number after every performance. Suppose, remarkably, that whenever judge one gives a score ending in an even digit, judge two, examined afterward, turns out to have given a score ending in an odd digit for that same routine, and vice versa, an exact and unbroken pattern across hundreds of performances, despite the two judges never speaking to each other during the competition and each one's individual sequence of scores looking, on its own, entirely unremarkable and unpatterned. If you interviewed either judge individually and asked them to predict their own next score, they could not do it. Their own sequence of scores carries no signal, no hidden order, nothing that would let them, or anyone else, anticipate what number was coming next. Only when someone collects both scorecards after the competition and lays them side by side does the astonishing pattern of perfect complementary correlation reveal itself. Even then, discovering the pattern requires physically bringing the two scorecards together, an entirely ordinary act of communication, no faster than a courier walking across the rink.

This is close, though of course not identical, to the shape of what happens with entangled particles. Each individual stream of measurement results is, taken alone, indistinguishable from pure randomness. The correlation, the genuinely nonclassical pattern that no locally predetermined table of hidden variables could reproduce, only becomes apparent, only becomes meaningful or useful in any sense, once the two separate records are brought together through an entirely mundane channel operating well within the limits that relativity imposes. The strangeness of entanglement lives entirely in that comparison, not in any single measurement taken in isolation, and this is precisely why the phenomenon, remarkable as it is, poses no threat to the causal structure of the universe that Einstein spent his career mapping out so carefully, even though it denied him, in the end, the fully local, fully deterministic picture of reality he had hoped that structure would ultimately support.

In the next chapter, we leave the realm of mathematics and thought experiments behind and step into an actual laboratory, a converted basement space at the University of California, Berkeley, in the early 1970s, where two young physicists, working with modest funding and equipment they built largely themselves, performed the first real experimental test of Bell's inequality, transforming the argument between Einstein and Bohr, once and for all, from a philosophical dispute into an empirical question with a measurable, numerical answer.

CHAPTER SEVEN
Closing the Back Doors

The sub basement of Birge Hall at the University of California, Berkeley, was not, by any account, a glamorous place to conduct physics in 1972. It was the kind of space universities tend to hand over to projects considered promising enough to fund but not prestigious enough to merit the newer, better ventilated laboratories upstairs, and it suited John Clauser and Stuart Freedman's purposes precisely because nobody was paying very close attention to what they were doing down there.

Clauser had, by his own later account, developed something close to an obsession with John Bell's 1964 paper during his graduate studies, first at Caltech and then at Columbia University, at a time when almost none of his professors considered the foundations of quantum mechanics a serious or career advancing area of research. He wrote to Bell directly in 1969, and Bell, apparently pleased that a young experimentalist was taking the theoretical question seriously enough to attempt an actual test, encouraged him warmly. That same year, Clauser, working with Michael Horne, Abner Shimony, and Richard Holt, published the CHSH inequality, the experimentally practical refinement of Bell's original theorem that we discussed in chapter five, specifically designed to be testable using photon pairs and polarizing filters, apparatus that, while demanding to build with the technology available at the time, was not fundamentally beyond reach.

Clauser moved to Lawrence Berkeley National Laboratory as a postdoctoral researcher in 1969, and it was there that he found a collaborator in Stuart Freedman, then a graduate student with strong hands on experimental skills, exactly the kind of practical expertise Clauser's theoretical enthusiasm needed to become a functioning apparatus. Freedman would go on to a distinguished career in nuclear and particle physics at Berkeley, and his early death in 2012 was mourned by colleagues who remembered him as both a rigorous experimentalist and a generous mentor, but in 1972, he was still relatively early in his career, working alongside Clauser to build, largely from scratch and with a research budget that both men later described as modest, an apparatus capable of testing Bell's inequality for the first time in the history of physics.

The experimental design, at its core, needed to accomplish three things. It needed a reliable source of entangled photon pairs. It needed a way to measure the polarization of each photon at a freely chosen angle. And it needed a way to record, with enough precision and enough repeated trials, whether the resulting correlations between the two photons' measured polarizations obeyed or violated the CHSH inequality.

For their photon source, Clauser and Freedman used calcium atoms, excited using a specific sequence of light at carefully chosen wavelengths, causing the atoms to emit pairs of photons as they cascaded back down through a series of energy levels toward their ground state. This particular atomic cascade had a valuable property: the two emitted photons were entangled in their polarization, meaning that the direction in which each photon's electromagnetic field oscillated was correlated with its partner's in exactly the way the theory predicted, and in exactly the way needed to test Bell's inequality experimentally. Building a stable, sufficiently bright source of these entangled photon pairs, using the vacuum technology, light sources, and detectors available in the early 1970s, decades before the highly refined laser and photon detection equipment that would become standard in later experiments, was itself a significant engineering achievement, one that consumed a substantial portion of Clauser and Freedman's efforts.

On each side of the apparatus, the two researchers placed a polarizing filter, a device that allows light oscillating in one particular direction to pass through while blocking light oscillating in other directions, followed by a photon detector capable of registering the arrival of a single photon, an extremely faint signal requiring careful shielding from any stray light in the surrounding laboratory. The angle of each polarizing filter could be adjusted, and Clauser and Freedman ran their experiment across several different combinations of filter angles on each side, exactly the kind of varied measurement settings that Bell's mathematics required in order to produce a meaningful test.

Over the course of about two hundred hours of actual data collection, spread across many separate runs to manage the practical challenges of keeping such a sensitive apparatus stable, Clauser and Freedman recorded the rate at which photon pairs were detected together at various combinations of filter angles. When they compiled and analyzed the results, published in 1972 in Physical Review Letters under the title Experimental Test of Local Hidden Variable Theories, the correlations they measured violated the inequality that any local hidden variable theory could produce, and they did so by a margin large enough, relative to the experiment's statistical uncertainty, to represent a genuinely meaningful result rather than a marginal, ambiguous one.

This was, by any reasonable measure, a landmark achievement, the first experimental evidence, gathered under carefully controlled conditions and analyzed with appropriate statistical rigor, suggesting that Einstein's preferred picture of the world, local hidden variables fixed at the source and unaffected by distant measurement choices, could not fully account for the behavior of entangled particles. And yet, in the years immediately following its publication, the Clauser and Freedman result did not settle the argument once and for all, for reasons that are themselves instructive about how rigorous experimental science actually works, as opposed to how it is sometimes portrayed in simplified retellings.

The problem was not that anyone seriously doubted the honesty or competence of Clauser and Freedman's work. The problem was that their apparatus, like every apparatus of its era, contained certain unavoidable practical limitations, and these limitations opened what physicists call loopholes: gaps in the logic of the experiment through which a sufficiently determined defender of local hidden variables could, at least in principle, still construct an alternative explanation for the observed results, an explanation that did not require abandoning locality after all.

The most significant of these was what has come to be known as the detection loophole, or sometimes the efficiency loophole. The photon detectors available to Clauser and Freedman in 1972 were far from perfect. They failed to register a substantial fraction of the photons that actually arrived, a limitation of the photomultiplier tube technology of the era rather than any flaw in the experimental design itself. This meant that the data Clauser and Freedman analyzed represented only a subset, a sample, of all the entangled photon pairs actually produced by their source, and it is a subtle but genuine logical possibility that this sample was not perfectly representative of the whole. Imagine, hypothetically, that the entangled photons carried some hidden property that made them simultaneously more likely to violate Bell's inequality and more likely to be successfully detected, while photons that would have obeyed the inequality were, for some hidden reason connected to that same property, disproportionately likely to slip past the detectors undetected. Under this admittedly contrived scenario, the measured violation of Bell's inequality could be, in principle, an artifact of imperfect detection rather than genuine evidence against local hidden variables.

Nobody involved in the field seriously believed this particular scenario reflected what was actually happening in nature. It required an oddly specific and physically unmotivated conspiracy between the hidden variables and the detection apparatus, precisely the kind of contrived, ad hoc rescue that scientists are trained to view with deep suspicion. But the logical possibility, however implausible, remained genuinely open, and closing it required detectors efficient enough to capture a sufficiently high percentage of all photon pairs produced, technology that did not yet exist in a form suitable for this kind of delicate experiment in the early 1970s.

A second significant loophole, called the locality loophole or sometimes the communication loophole, concerned the timing and arrangement of the measurements themselves. In Clauser and Freedman's apparatus, and in most of the early experiments that followed over the subsequent decade, the settings of the polarizing filters on each side were fixed before each run began, rather than being changed randomly and rapidly while the photons were already in flight between the source and the detectors. This left open another admittedly strained but logically genuine possibility: that some conventional, subluminal signal, traveling no faster than light, might have passed between the two measurement stations during the course of the experiment, somehow coordinating the filter settings and the detector responses in a way that produced the appearance of a Bell inequality violation without any genuine nonlocal quantum effect actually being involved. Closing this loophole required an experimental design bold enough to change the measurement settings after the photons had already left their source, and fast enough that no signal traveling even at the speed of light could possibly cross the distance between the two detectors in time to influence the outcome.

Closing this particular loophole required both a more sophisticated experimental design and, frankly, considerable audacity, since it meant deliberately building an apparatus specifically engineered to rule out an explanation that most physicists already considered exceedingly unlikely. The physicist who took on this challenge, working at the Institut d'Optique near Paris in the early 1980s, was a Frenchman named Alain Aspect, and his experiments, which we will examine in full in the next chapter, represent one of the most elegant and consequential refinements in the entire history of this line of research.

It is worth pausing here, before we move to Aspect's work, to reflect on what Clauser and Freedman's 1972 experiment actually accomplished, properly weighed against what it did not yet accomplish, because this kind of honest accounting is, I think, one of the more valuable habits of mind that the history of entanglement research has to offer any reader, scientist or otherwise. Clauser and Freedman did not deliver an absolutely final, loophole free verdict against local hidden variables. What they delivered was something almost as valuable, and in some ways more valuable for the health of the field going forward: they delivered proof that Bell's abstract, mathematical challenge could actually be built, tested, and measured using real apparatus in a real laboratory, and they delivered a first, genuinely suggestive result pointing firmly in quantum mechanics's favor. They also, through the loopholes their own honest, careful work revealed, handed the next generation of experimentalists a precise and well defined engineering challenge: build better detectors, build faster switching mechanisms, close the remaining logical gaps one at a time, until no reasonable escape route remained for anyone still hoping, as Bell himself quietly did, that Einstein might yet be vindicated.

This is, in miniature, how a great deal of careful experimental science actually proceeds, not through a single decisive, all resolving experiment, but through a sustained, multi decade campaign in which each new experiment closes a specific gap left open by the one before it, gradually narrowing the space in which alternative, more conventional explanations could still hide. It is a slower, less cinematic story than popular accounts of scientific breakthroughs sometimes suggest, but it is, I would argue, a considerably more trustworthy one, precisely because it takes seriously, and painstakingly eliminates, every reasonable objection along the way rather than declaring victory prematurely.

Clauser would go on, more than fifty years after his basement experiment with Freedman, to share the 2022 Nobel Prize in Physics with Alain Aspect and Anton Zeilinger, recognition that arrived, as recognition for foundational work in physics often does, only after decades in which the full significance of the original contribution had time to become unmistakably clear to the wider scientific community. Freedman, tragically, did not live to see that recognition, having passed away in 2012, a full decade before the Nobel committee's announcement, though colleagues at Berkeley Lab and elsewhere have made a point, in the years since, of ensuring that his essential contribution to the original 1972 experiment is not forgotten in the retelling of a story that is sometimes, for the sake of narrative simplicity, compressed down to Clauser's name alone.

The story that began in a converted basement in Birge Hall did not end there. It moved, in the following decade, to a laboratory outside Paris, where a physicist with a gift for elegant experimental design set out to close the loophole that had proven hardest of all to eliminate: the possibility that the two measurement stations, however far apart, might still be quietly, conventionally talking to each other.

Before we make that journey, it is worth understanding a little more about how word of the Clauser and Freedman result actually traveled through the physics community, because reception, in science as in publishing, is rarely as simple or as immediate as the phrase published in Physical Review Letters might suggest to a modern reader.

The paper did not, on its release, produce headlines or widespread public attention of any kind. It was received, within the relatively small community of physicists who had been following Bell's work closely, as a genuinely significant result, but that community remained small throughout the 1970s, a niche within a niche, populated by researchers whose senior colleagues frequently regarded their choice of research topic with a mixture of bemusement and mild concern. Abner Shimony, one of the four physicists behind the CHSH inequality, later recalled that foundational research on quantum mechanics carried, for much of this period, something close to a stigma within mainstream physics departments, an association with philosophy rather than serious experimental science that made hiring committees and funding agencies visibly wary.

This is part of why the loopholes left open by Clauser and Freedman's experiment were not closed quickly. It was not a matter of the physics being unclear about what needed to be done next. Physicists in this small community understood exactly what a more decisive experiment would require: higher efficiency detectors, and a way of changing measurement settings while the photons were already in flight. The obstacle was largely one of resources and institutional attention, in a field that had not yet been recognized, by the broader scientific establishment or by the funding bodies that supported it, as central to the future of physics rather than a curious sideline.

It took the specific ambitions of Alain Aspect, a French physicist completing his doctoral work at the Institut d'Optique in the late 1970s, to gather the resources, the technical ingenuity, and the sheer persistence needed to push the experimental program to its next major milestone. Aspect had, notably, written to John Bell directly before beginning his doctoral research, asking whether Bell considered the project of closing the locality loophole worthwhile, given how many senior physicists in France had questioned whether it was a wise use of a young researcher's most productive years. Bell's answer, by Aspect's own later recollection, was warmly encouraging, and it is a small but telling detail that both of the two physicists who would perform the most consequential early tests of Bell's theorem, Clauser in America and Aspect in France, found themselves needing direct reassurance from Bell himself before committing years of their careers to a line of research their own institutions viewed with some skepticism.

What Aspect set out to build was, in its ambition, considerably more demanding than what Clauser and Freedman had attempted a decade earlier, and the following chapter tells that story in full, because it represents one of the most carefully engineered experiments in the history of physics, designed from the ground up to answer, decisively and without escape routes, the single objection that had continued to give even sympathetic physicists some room for doubt.

CHAPTER EIGHT

The Frenchman Who Switched the Locks

Alain Aspect has told, in various interviews over the decades since his most famous experiments, a version of the same origin story: as a young physics student in France in the late 1960s, he came across the EPR paper and Bell's 1964 response to it almost by accident, outside the standard curriculum of his formal training, and found himself captivated by a question that his professors treated as, at best, a philosophical curiosity rather than live experimental physics. By the time he began his doctoral research at the Institut d'Optique, part of the University of Paris Sud in Orsay, just outside Paris, he had already decided that he wanted to spend his career trying to answer it properly.

What set Aspect's approach apart from Clauser and Freedman's pioneering but necessarily more limited 1972 experiment was a specific, precisely targeted ambition: he wanted to close the locality loophole, the possibility, however contrived it might seem to most physicists, that some ordinary, subluminal signal could pass between the two measurement stations during an experiment and produce the appearance of a Bell inequality violation without any genuine quantum nonlocality being involved. Closing this loophole required an apparatus capable of changing the polarization measurement setting at each detector after the entangled photons had already left their shared source, and doing so quickly enough that even a signal traveling at the speed of light, the fastest speed permitted by relativity, could not possibly cross the distance between the two detectors and influence the outcome in time.

This was, in the early 1980s, an extraordinarily demanding engineering requirement. It meant the entire measurement, from the moment the setting was changed to the moment the photon detection was recorded, needed to happen faster than light could travel the separation distance between the two detector stations, a separation Aspect and his team eventually built out to about thirteen meters, meaning the switching needed to occur in a matter of tens of nanoseconds, billionths of a second, using the electronics and optical technology available at the time, technology that, by modern standards, would be considered almost primitive.

Aspect's solution, developed with his collaborators Philippe Grangier, Gerard Roger, and Jean Dalibard, and published across a series of papers between 1981 and 1982, was ingenious in its specifics. Rather than relying on a mechanical device to physically rotate a polarizing filter, far too slow a process to achieve the necessary switching speed, Aspect's team built an optical switch using acousto optic devices, essentially using high frequency sound waves traveling through a specially chosen transparent medium to rapidly redirect each photon toward one of two different polarization analyzers, each already fixed at a different measurement angle. This effectively allowed the experiment to choose, at extremely high speed and in a manner not predetermined in advance, which of two possible polarization measurements would be performed on each photon, changing the choice while the photons were already in flight between the source and the detectors, exactly the condition needed to close the locality loophole.

It is worth taking a moment to appreciate just how tight this timing requirement actually was, because numbers like tens of nanoseconds tend to slide past a reader without registering their true strangeness. Light travels roughly thirty centimeters in a single nanosecond, meaning that across Aspect's thirteen meter separation, a light signal would need approximately forty three nanoseconds to cross from one detector station to the other. For the experiment to genuinely close the locality loophole, the entire process of choosing a measurement setting and completing the corresponding detection needed to happen within a window shorter than that forty three nanosecond crossing time, a demand that pushed the electronics and optics of the early 1980s to their absolute limit. To put this in a scale more familiar to everyday experience, a single blink of a human eye lasts somewhere between one hundred and four hundred milliseconds, which is to say roughly ten million times longer than the entire window Aspect's apparatus had to work within. Building a laboratory instrument capable of operating reliably on a timescale that dwarfs human perception by seven orders of magnitude, using equipment assembled largely by hand in a French university laboratory, remains, considered purely as a feat of applied engineering, as impressive as the physics result it was built to obtain.

The 1982 paper reporting this experiment, titled Experimental Test of Bell's Inequalities Using Time Varying Analyzers and published in Physical Review Letters, remains one of the most cited papers in the history of quantum foundations research, and for good reason. The results, once again, violated Bell's inequality, and did so in a configuration specifically engineered to rule out the locality loophole that had remained open in Clauser and Freedman's earlier work. Taken together with the earlier experiment and a series of related, increasingly refined tests performed by Aspect's team and other groups internationally throughout the 1980s, the case against local hidden variables had become, by the end of that decade, exceptionally strong, strong enough that a genuine consensus began forming across the physics community that Einstein's preferred resolution of the EPR paradox was not going to be vindicated by further experimentation.

Even so, and this is a detail that speaks well of the rigor of the field rather than reflecting any lingering, reasonable doubt about the basic physics, Aspect's own switching mechanism, while extremely fast for its era, did not switch in a manner that was fully and truly random in the strictest sense that later, more demanding versions of the locality loophole would require. The acousto optic switches Aspect used cycled through their settings in a regular, periodic pattern rather than a genuinely unpredictable, random one, leaving open, at least in principle, an extremely narrow and contrived possibility that some conspiracy between the periodic switching pattern and the photon source could still produce a spurious violation. Aspect himself was entirely candid about this limitation in his own published work and subsequent writing, and it became, over the following decades, the next target for experimentalists determined to close every conceivable gap.

I want to dwell on this pattern for a moment, because it recurs throughout the history of Bell test experiments and it says something valuable about how science actually earns its credibility. Each generation of experiment closed one loophole while leaving another, often narrower and more contrived, still technically open. Critics of the emerging consensus, and there were a small number of serious, careful physicists who continued to press these objections well into the 1990s and 2000s, were not being unreasonable or obstructionist in raising them. They were doing exactly what good scientific skepticism requires: refusing to accept a conclusion, however likely it seemed, until every reasonable alternative explanation had been experimentally excluded, not merely argued against on grounds of implausibility.

Aspect's experiments, despite this remaining limitation, represented a watershed moment in the field's history for reasons that went beyond the purely technical. They marked the point at which testing Bell's inequality shifted from being regarded, within the wider physics community, as a slightly eccentric pursuit undertaken by researchers on the margins of the field's mainstream priorities, to being recognized as a serious, well funded, and increasingly prestigious area of experimental physics in its own right. Aspect's results received substantially more attention, both within physics and in the broader scientific press, than Clauser and Freedman's had a decade earlier, partly because of the growing cultural moment quantum mechanics was beginning to enjoy more broadly, and partly because the specific loophole Aspect had closed, the locality loophole, addressed so directly the core of Einstein's original relativistic objection that its resolution felt, to many observers, genuinely climactic.

It is worth pausing to consider what this shift in institutional attitude actually meant in practical terms for the physicists working in this area. Grants became somewhat easier to secure. Positions in quantum foundations research, once considered a poor career choice by cautious doctoral advisors, began to appear, gradually, as legitimate and even desirable specializations at leading physics departments. A new generation of graduate students, inspired by Aspect's results and unburdened by the institutional skepticism that had shadowed Clauser's and even Aspect's own early careers, began entering the field in the 1980s and 1990s, bringing with them steadily improving laser technology, steadily improving photon detectors, and steadily more creative experimental designs. Among this new generation was a young Austrian physicist named Anton Zeilinger, who would go on, over the following decades, to extend entanglement experiments in directions that Clauser, Freedman, and even Aspect had not yet attempted: entangling not just pairs but larger groups of photons, demonstrating quantum teleportation of quantum states across meaningful distances, and eventually helping to lay the groundwork for satellite based quantum communication, the subject of a later chapter in this book.

Aspect himself continued refining his experimental techniques throughout the 1980s and beyond, eventually collaborating on later experiments that used genuinely random number generators, rather than the periodic switching of his original 1982 apparatus, to select measurement settings, closing even the narrow residual gap his own earlier candor had identified. He remained, throughout a long and distinguished career at the Institut d'Optique and later at the elite French research institution known as the College de France, closely associated with the broader project of testing and refining the experimental case against local hidden variables, a project he had committed himself to as a young doctoral student pursuing a question his own professors had once viewed with polite skepticism.

There is a particular satisfaction, I think, in knowing that Aspect wrote to John Bell directly before beginning his most consequential experiments, asking whether Bell believed the project was worth pursuing, given the doubts some of Aspect's own senior colleagues in France had expressed about devoting a promising young researcher's most productive years to what they considered a settled or at least unglamorous question. Bell's response was warmly encouraging, and it is worth noting that this same pattern, a young experimentalist reaching out directly to Bell for validation before committing years of research to testing his theorem, repeats itself across multiple continents and multiple decades in the history we are tracing. Bell, who never won the Nobel Prize himself and who died in 1990 without seeing the full experimental vindication of his 1964 paper, nonetheless served, informally and repeatedly, as a kind of intellectual encouragement for an entire generation of experimentalists willing to take his mathematics as seriously as he had hoped it deserved.

By the mid 1980s, the combination of Clauser and Freedman's original 1972 result and Aspect's more sophisticated 1982 experiments had shifted the balance of informed opinion within physics decisively, though not yet completely unanimously, against local hidden variable theories. A small number of dedicated skeptics, working carefully and in good faith, continued to press on the remaining loopholes, and their persistence, far from being an obstacle to progress, helped drive the field toward the truly decisive experiments that would follow over the subsequent three decades: experiments capable of closing every remaining loophole simultaneously, in a single unified test, leaving no reasonable escape route whatsoever for anyone still hoping to preserve Einstein's preferred picture of a locally real universe.

The next major step in that direction came in 1998, sixteen years after Aspect's original result, from a team led by Gregor Weihs working in Anton Zeilinger's group, by then established at the University of Innsbruck in Austria. Weihs and his colleagues, working alongside Thomas Jennewein, Christoph Simon, and Harald Weinfurter, addressed the specific residual concern that had lingered around Aspect's own experiment, the fact that Aspect's acousto optic switches had cycled through their settings in a predictable, periodic pattern rather than a genuinely unpredictable one. The Innsbruck team built a new apparatus using fast electro optic modulators controlled by independent, physically separate quantum random number generators at each measurement station, positioned four hundred meters apart on the university campus, connected to their respective photon sources by optical fiber. Each random number generator operated entirely locally, with its own independent source of physical randomness, capable of producing a fresh, unpredictable choice roughly every few nanoseconds. Because the two stations were separated by four hundred meters, light itself would need about 1.3 microseconds to cross between them, and the Innsbruck apparatus completed its full cycle of choosing a measurement setting and registering the corresponding detection comfortably within that window, ensuring that not even a light speed signal could travel from one station to the other in time to influence the outcome. The 1998 paper, published in Physical Review Letters under the title Violation of Bell's Inequality Under Strict Einstein Locality Conditions, closed the locality loophole with a rigor that went meaningfully beyond what Aspect's apparatus, remarkable as it was for 1982, had been able to achieve, and it did so using genuinely random, physically independent setting choices rather than a predictable periodic cycle.

The Innsbruck result mattered not because anyone seriously expected it to overturn the accumulating consensus, but because it represented the field's characteristic insistence on tightening every bolt that a careful critic could still point to. Weihs and his colleagues were, like Clauser and Aspect before them, explicit in their published paper about which loopholes their experiment closed and which, including the detection loophole that would not be closed until Hanson's diamonds and their contemporaries seventeen years later, remained technically open. This habit of precise, unglamorous bookkeeping about exactly what has and has not yet been proven is, as much as any single dramatic result, what has made the eventual case against local hidden variables as close to unimpeachable as anything in experimental physics has ever become.

That final, decisive chapter of the experimental story would not arrive until 2015, more than three decades after Aspect's Paris experiments, and it would require technology that did not yet exist when Aspect first switched his acousto optic devices in a basement laboratory outside Paris: diamonds cooled to within a fraction of a degree of absolute zero, electron spins measured across more than a kilometer of Dutch countryside, and a level of engineering precision that would have struck even Aspect, in 1982, as almost fantastical.

There is one more figure from this period worth introducing properly before we move forward, because his name will recur throughout the remainder of this book: Anton Zeilinger, born in 1945 in Ried im Innkreis, Austria, who completed his doctoral studies at the University of Vienna in the early 1970s and who would, over the following four decades, become perhaps the single most prolific and inventive experimentalist in the entire field of quantum information science. Where Clauser had proven that Bell's inequality could be tested at all, and Aspect had proven that the locality loophole could be closed, Zeilinger would spend his career asking a further, more expansive question: given that entanglement is real, what can we actually build with it. That question, and Zeilinger's decades long pursuit of its answer, threads through nearly every chapter that follows, from the loophole free experiments of 2015 to the satellite based quantum communication systems of the 2020s, and it is worth remembering, as his name reappears, that it began with the same simple fascination that had gripped Clauser in New York and Aspect in Paris: a conviction that a question most of the field considered closed deserved, in fact, to be asked again, more carefully, with better instruments, until no reasonable doubt remained.

CHAPTER NINE

Diamonds, Photons, and a Bicycle Path in Delft

Ronald Hanson still remembers, by his own account in later interviews, the particular mixture of excitement and dread that comes from designing an experiment specifically built to fail if the universe turns out to work the way common sense suggests it should. This is an unusual position for a scientist to be in, and it is worth understanding why Hanson and his team at Delft University of Technology in the Netherlands found themselves there in the early 2010s, pursuing what would become, when finally published in 2015, the first genuinely loophole free test of Bell's inequality in the history of physics.

By the early 2010s, four decades of accumulated experimental evidence, beginning with Clauser and Freedman in 1972 and continuing through Aspect's refinements in the 1980s and a long series of increasingly sophisticated follow up experiments in the 1990s and 2000s, had left the case against local hidden variables extremely strong but not, in the strictest logical sense, absolutely complete. No single experiment had yet managed to close both the detection loophole and the locality loophole simultaneously, within one unified test. Various experiments had closed one loophole while leaving the other open, and various experiments had closed the other while leaving the first open, but a small number of careful, technically minded skeptics continued to point out, correctly as a matter of pure logic, that these were, strictly speaking, two separate experiments testing two separate configurations, and that a fully rigorous, unimpeachable test would need to close both loopholes at once, in a single run, with a single set of entangled particles.

This might sound, to a reader encountering it for the first time, like an almost pedantic objection, and in a certain sense it was, given how overwhelmingly the accumulated evidence already pointed in one direction. But it was also, and this is worth taking seriously, exactly the kind of pedantry that good experimental science depends on. Extraordinary claims, and the claim that reality itself does not respect the locally real picture Einstein championed for the final decades of his career is about as extraordinary as claims in physics get, deserve to be tested against every conceivable alternative explanation, no matter how narrow or how many earlier experiments have already pointed the same direction.

Hanson's team chose an entirely different physical system from the photon based experiments of Clauser, Freedman, and Aspect. Rather than entangling pairs of photons directly, they worked with electron spins trapped inside nitrogen vacancy centers, tiny defects within diamond crystals where a nitrogen atom sits adjacent to a missing carbon atom in the diamond's crystal lattice, creating a location where a single electron's magnetic spin state can be manipulated and measured with remarkable precision, and, crucially, held stable for a useful length of time, cooled to extremely low temperatures inside specialized laboratory equipment.

The experimental design, published in Nature in 2015 under the title Loophole Free Bell Inequality Violation Using Electron Spins Separated by 1.3 Kilometres, worked roughly as follows. Two diamonds, each containing a nitrogen vacancy center with a trapped electron, were placed in separate laboratories on the Delft University campus, physically separated by 1.3 kilometers, a distance that required running dedicated fiber optic cable underneath a public bicycle path that cuts across the university grounds, a detail that has since become something of a beloved anecdote among physicists retelling the story, since it meant that ordinary Dutch cyclists were, quite literally, riding their bicycles directly above one of the most important physics experiments of the decade without any awareness of what lay beneath them.

The two electrons were entangled using a clever intermediate step involving photons emitted by each electron, sent toward a station located roughly halfway between the two diamond laboratories, where the photons themselves were subjected to a joint measurement that had the effect, when successful, of entangling the two distant electrons even though the electrons themselves never directly interacted. This technique, known as entanglement swapping, had been developed and refined over the preceding two decades, substantially through the work of Zeilinger's group in Vienna, and it allowed Hanson's team to create genuine, verified entanglement between two electron spins separated by more than a kilometer, a physical system entirely different from the photon pairs used in earlier experiments and one specifically chosen because its detection efficiency could be made close enough to complete, unlike the leaky photon detectors that had left the detection loophole open in Clauser and Freedman's original 1972 apparatus.

Because the electron spin measurements at each end could be performed with extremely high, near complete efficiency, essentially every entangled pair generated could be reliably measured rather than a mere fraction, the detection loophole was closed. And because the 1.3 kilometer separation between the two laboratories was large enough that the entire process, from randomly choosing which measurement to perform to completing that measurement and recording the result, could be finished well before a light speed signal could travel between the two locations and influence the outcome, the locality loophole was closed simultaneously, within the very same experimental run, for the very first time in the history of this line of research.

The randomness used to select each measurement setting deserves particular mention, because Hanson's team, along with the community of physicists reviewing their proposed methodology in advance, took the question of genuine randomness with an almost philosophical seriousness that speaks to just how carefully this final, decisive experiment needed to be constructed. A measurement setting chosen by a predictable or pseudo random process, however sophisticated, could theoretically be anticipated in advance by a sufficiently exotic hidden variable theory, reopening, in a suitably contrived and admittedly far fetched way, exactly the kind of loophole the experiment was designed to close. To guard against even this remote possibility, later refinements of this experimental program, including versions performed in subsequent years by other collaborating groups, incorporated randomness generated by genuinely unpredictable physical processes, including, in one particularly imaginative variant, randomness partly derived from the measured brightness fluctuations of distant stars, chosen specifically because any hidden coordination between starlight emitted centuries or millennia ago and a laboratory experiment on Earth today would require an almost absurdly implausible cosmic conspiracy to explain.

When Hanson's team analyzed their results in 2015, the violation of Bell's inequality was clear, statistically significant, and, for the first time, achieved under conditions that closed every loophole that decades of careful, skeptical scrutiny had identified. The paper's publication in Nature was quickly followed, within the same year, by two additional loophole free experiments performed by independent teams, one led by researchers including Krister Shalm at the National Institute of Standards and Technology in Boulder, Colorado, using photons rather than electron spins but achieving comparably high detection efficiency through improved superconducting photon detectors, and another led by Anton Zeilinger's group in Vienna, also using photons and also achieving the necessary combination of high efficiency detection and rapid, unpredictable setting selection.

The near simultaneous appearance of three independent, loophole free confirmations, using different physical systems, different laboratories, different countries, and different specific experimental techniques, all converging on the same conclusion within a period of months in 2015, represents one of the more remarkable episodes of scientific consensus building in recent physics history. It is not that any single one of these three experiments was, on its own, absolutely unimpeachable in every conceivable respect, since even the most careful experimentalists acknowledge that some vanishingly narrow theoretical loopholes remain difficult to close with complete, mathematical finality, including exotic scenarios involving a lack of what physicists call measurement independence, the assumption that the choice of measurement setting is not itself somehow correlated, through some deeper and currently unknown mechanism, with the properties of the particles being measured. But taken together, three independent experiments, using different technologies and located on different continents, converging on the same violation of Bell's inequality under conditions closing every practically addressable loophole, left remarkably little room for reasonable doubt.

The scientific community's response reflected this. Physics World, the magazine published by the Institute of Physics and widely read across the international physics community, named the 2015 loophole free Bell tests its Breakthrough of the Year, an honor reflecting the widespread sense within the field that this represented, in the most meaningful sense available to experimental science, a genuine and final resolution to the question that Einstein, Podolsky, and Rosen had first raised in 1935, eighty years earlier.

It is worth reflecting, briefly, on what exactly had been established, and what had not, by this point in our story. What had been established, about as firmly as anything in experimental physics can be established, was that local hidden variable theories, theories in which particles carry predetermined properties fixed at their creation and unaffected by distant measurement choices, cannot account for the correlations observed between entangled particles. This is a genuinely profound conclusion about the structure of physical reality, and it took eighty years, from Einstein's original 1935 paper to Hanson's 2015 experiment, of sustained theoretical and experimental effort to establish it beyond reasonable doubt.

What had not been established, and what remains, even now, a subject of active and genuinely open discussion among physicists and philosophers of physics, is exactly what this conclusion means for our broader understanding of reality. As we discussed in chapter two, several different interpretations of quantum mechanics, the Copenhagen interpretation, the many worlds interpretation, pilot wave theories, and others, remain fully consistent with the experimental evidence, because they all, despite their radically different pictures of what is happening beneath the surface, make identical predictions for every measurable outcome. The 2015 experiments closed the door on local hidden variables specifically.

They did not, and could not, single out one remaining interpretation as uniquely correct, and the philosophical work of understanding what nonlocal correlation without nonlocal signaling actually implies about the deep structure of reality continues, productively, to this day.

Ronald Hanson, reflecting later on the years of preparation that led to the 2015 result, described the specific satisfaction of finally closing a scientific question that had remained technically open for eighty years, not through a single dramatic breakthrough but through the patient, decades long accumulation of increasingly rigorous experiments, each one closing a specific gap left by the one before it. It is, I think, one of the finest examples available of how science actually earns its conclusions, not through certainty declared in advance, but through the sustained, collaborative, and often unglamorous work of eliminating alternative explanations one at a time, across generations of researchers, until nothing reasonable remains standing except the conclusion the evidence has been pointing toward all along.

Seven years after Hanson's diamonds settled the argument definitively on the ground, John Clauser, Alain Aspect, and Anton Zeilinger would be honored jointly, by the Royal Swedish Academy of Sciences, for the full arc of experimental work that had made this resolution possible, from the Berkeley basement of 1972 through the Parisian optical switches of 1982 to the broader body of work in quantum information science that Zeilinger, in particular, had continued to build throughout the decades between.

It is worth pausing here, at what feels like a natural resting point in the historical narrative, to take stock of the full journey this book has walked through so far, because the accumulated weight of it is easy to lose amid the individual technical details of any single chapter. In 1935, Einstein, Podolsky, and Rosen raised a philosophical objection to quantum mechanics, an objection so carefully constructed that it took the rest of the century to answer properly. Schrodinger, reading their paper, named the phenomenon at its center and dramatized its strangeness with an imaginary cat. For nearly thirty years, the disagreement between Einstein's camp and Bohr's camp remained, in the strictest sense, untestable, a matter of interpretation and philosophical preference rather than empirical science. Then, in 1964, a physicist working quietly on the margins of mainstream particle physics found a way to convert that philosophical disagreement into an actual number, a mathematical inequality that nature itself could be asked to either satisfy or violate.

It took eight more years before anyone built an apparatus capable of asking nature that question directly, in a Berkeley basement staffed by two young researchers many of their colleagues considered to be wasting their early careers. It took another decade before a French physicist, driven by the same stubborn fascination, closed the most philosophically troubling of the remaining loopholes using switches fast enough to outrun the speed of light across a laboratory bench. And it took more than three decades beyond that, spanning the full arc of an entire scientific career for many of the physicists involved, before a team in the Netherlands, using diamonds and a fiber optic cable buried beneath a bicycle path, finally closed every remaining door at once.

This is not a story of a single flash of insight settling a question overnight. It is a story of a question being asked patiently, carefully, and repeatedly, by successive generations of physicists willing to treat a supposedly settled matter as genuinely worth reopening, each one building directly on the specific limitations that the physicist before them had been honest enough to acknowledge. That pattern of honest acknowledgment, the willingness of Clauser to publish alongside a clear statement of his experiment's detection loophole, the willingness of Aspect to note candidly that his switching mechanism was not yet truly random, is itself part of what makes the eventual 2015 resolution so trustworthy. Nothing was swept under the rug. Every gap was named, targeted, and eventually closed by someone willing to spend years of their career doing the difficult, unglamorous engineering required.

But before we arrive at the formal recognition the Nobel committee eventually offered this decades long collaborative effort, the story of entanglement testing takes one more remarkable turn, leaving the laboratories of Europe and America behind entirely and traveling, for the first time, into orbit above the Earth itself, aboard a satellite named, fittingly for a project this ambitious, after an ancient philosopher who spent his own life arguing that careful, disciplined reasoning could reveal truths the untrained eye would never notice on its own.

CHAPTER TEN

A Satellite Named After a Philosopher

Mozi lived in China roughly twenty five centuries ago, founding a school of philosophy that valued impartial care, careful logical argument, and, intriguingly for a thinker so far removed from modern physics, some of the earliest recorded investigations into the behavior of light, including an early description of the pinhole camera effect that would not be properly explained by optical science for another two thousand years. When Chinese physicist Jian Wei Pan and his colleagues at the University of Science and Technology of China needed a name for the satellite at the center of their ambitious quantum communication program, they chose Micius, the Latinized form of Mozi's name, a deliberate nod to a thinker who had, according to the team's own public statements, anticipated something of the spirit of careful, evidence based inquiry that their own project embodied.

The satellite itself, launched from the Jiuquan Satellite Launch Center in the Gobi Desert on August 16, 2016, represented the centerpiece of a roughly one hundred million dollar research program called Quantum Experiments at Space Scale, generally abbreviated QUESS, funded by the Chinese Academy of Sciences and representing, at the time of its launch, the most ambitious attempt yet undertaken to determine whether quantum entanglement, already confirmed repeatedly across laboratory distances ranging from a few meters to just over a kilometer, could survive being extended across the much greater distances that a satellite in low Earth orbit makes possible.

The scientific motivation for taking entanglement into space was straightforward, even if the engineering required to pursue it was not. Every laboratory based entanglement experiment discussed in the previous chapters relied on sending photons either through open air across a room or a campus, or through optical fiber cables running between buildings or across a city. Both of these methods share a common limitation: photons are gradually absorbed and scattered as they travel through any medium, whether ordinary air or specially engineered optical fiber, and the rate of this loss increases, roughly exponentially, with distance. Sending entangled photons through a standard telecommunications fiber, the same basic technology that carries internet traffic beneath city streets, loses roughly half the photons for every fifteen kilometers or so of cable, a rate of loss that makes maintaining usable entanglement across distances of hundreds or thousands of kilometers essentially impossible using ground based fiber alone, since the signal would be reduced to an unusably faint trickle long before it reached its destination.

Space offered a way around this fundamental limitation. Once a photon leaves Earth's atmosphere, it travels through the near perfect vacuum of space with almost no absorption or scattering at all, meaning that a photon sent from a satellite to a ground station loses relatively little of its intensity compared to the same distance traveled through fiber or through the thicker, more turbulent air near the Earth's surface. The atmosphere itself, which the photon must cross twice, once as it descends from the satellite and, in some experimental configurations, once more depending on the exact path, does introduce some loss and distortion, but nowhere near enough to make the experiment infeasible, particularly if the satellite operates at night, avoiding the overwhelming background light that sunlight would otherwise introduce into the delicate single photon detectors on the ground. This is, in fact, why Micius has always operated exclusively during nighttime passes over its ground stations, a constraint that shapes the entire operational rhythm of the mission and one of many practical details that rarely makes it into popular summaries of the project's headline achievements.

Micius orbits at an altitude of approximately five hundred kilometers, low enough to allow reasonably strong signal strength between the satellite and the ground, and it carries onboard a crystal, potassium titanyl phosphate, chosen for its ability to convert a single incoming laser photon into a pair of lower energy entangled photons through a well established nonlinear optical process. When the satellite passes within range of appropriately equipped ground stations, timed carefully to coincide with clear, dark skies, it generates entangled photon pairs and sends onc photon of each pair down to one ground station while sending the other photon of the same pair down to a second, separate ground station, potentially located a very great distance away on the Earth's surface.

The landmark experiment establishing the feasibility of this approach, led by Jian Wei Pan and a large team of collaborators and published in the journal Science on June 16, 2017, under the title Satellite Based Entanglement Distribution Over 1200 Kilometers, distributed entangled photon pairs to three ground stations located in the mountains of the Chinese province of Qinghai and the autonomous regions of Xinjiang and Yunnan, chosen specifically for their high altitude, which reduces the amount of atmosphere the photons need to cross, and their remoteness, which reduces light pollution and atmospheric turbulence from nearby cities. The two most widely reported ground stations in this particular experiment, located at Delingha in Qinghai province and at the Gaomeigu Observatory near Lijiang in Yunnan province, sit 1203 kilometers apart, more than ten times the previous distance record for entanglement distribution, which had stood at roughly one hundred kilometers using ground based optical fiber and open air links.

The measured violation of the Bell type inequality used in this experiment confirmed that the entanglement between the two widely separated photons had survived the journey from the satellite to both ground stations, a result that had significant implications well beyond simply setting a new distance record for its own sake. It demonstrated, for the first time, that entanglement is not some fragile laboratory curiosity, dependent on the careful, controlled conditions of an enclosed optical bench, but a robust physical phenomenon capable of surviving transmission across the vacuum of space and through substantial stretches of Earth's turbulent atmosphere, opening the door to the possibility of a genuinely global network built around entangled particles, a prospect that had remained purely theoretical before Micius provided the first convincing experimental proof of concept.

Pan's team did not stop at demonstrating long distance entanglement distribution alone. In the years surrounding the 2017 Science paper, the Micius mission achieved a series of related milestones that, taken together, established China as the clear leader in the emerging field of space based quantum communication, a position that has prompted, in the years since, significant investment from other countries, including the United States, several European nations, and Japan, all pursuing their own satellite based quantum communication programs, generally with considerably smaller budgets and later timelines than China's pioneering effort.

Among these additional milestones, only ten days after its August 2016 launch, Micius successfully demonstrated quantum key distribution, a cryptographic technique we will examine in detail in the next chapter, between the satellite and a ground station near Beijing, using a protocol called decoy state BB84, a technique that does not require entangled photon pairs at all but instead relies on sending carefully prepared single photons whose quantum properties make any attempted eavesdropping detectable. Micius also achieved ground to satellite quantum teleportation, a technique, which we will also examine more carefully in a later chapter, for transferring the complete quantum state of one particle onto another distant particle without physically transporting the original particle itself, a feat that sounds, and in some popular accounts is unfortunately described, as something like the transporter technology of science fiction, though the reality, while still remarkable, is considerably more constrained and does not involve moving any matter or energy from one location to another, only information about a quantum state, transferred in a manner that still requires an ordinary, subluminal communication channel to complete, exactly as we discussed regarding entanglement's inability to transmit faster than light information in chapter six.

Perhaps most significantly for the long term practical implications of this research program, Micius has also been used as what physicists call a trusted relay, distributing quantum cryptographic keys between ground stations located in China and, in a landmark 2018 demonstration, a ground station in Graz, Austria, establishing the first intercontinental quantum secured video call, conducted between the Chinese Academy of Sciences and the Austrian Academy of Sciences, using encryption keys generated through the quantum properties of entangled photons distributed via the satellite. The word trusted in trusted relay carries a specific technical meaning worth clarifying, because it represents an important limitation in the current state of the technology rather than a fully realized end to end quantum secured global network. In this configuration, the satellite itself must be trusted not to record or leak the cryptographic keys it helps distribute, since the satellite briefly possesses both halves of each key exchange during the relay process. A fully secure, trust free global quantum network, one where not even the relay satellite itself needs to be trusted, remains an active area of ongoing research, requiring satellites capable of distributing genuinely entangled photon pairs directly between two ground stations without the satellite ever possessing usable information about the resulting key on its own, a capability the original 2017 entanglement distribution experiment demonstrated in principle but that has not yet been fully scaled into an operational, continuously available global network.

It is worth situating the Micius program within the broader context of China's substantial and sustained investment in quantum information science over the past two decades, an investment that reflects a clear, publicly stated strategic assessment that quantum technologies, spanning communication, computation, and sensing, represent a critical frontier for both economic competitiveness and national security in the coming decades. Beyond the Micius satellite itself, China has constructed the Beijing Shanghai quantum backbone, a roughly two thousand kilometer fiber optic quantum key distribution network connecting Beijing, Jinan, Hefei, and Shanghai through a series of trusted relay nodes, providing quantum secured communication links for government and financial institutions along that corridor since its completion in 2017. Jian Wei Pan, the driving scientific force behind both the Micius satellite and much of this ground based infrastructure, trained partly in Europe, completing doctoral work under Anton Zeilinger's supervision at the University of Vienna in the late 1990s before returning to China to build what has become one of the world's leading quantum information science research programs at the University of Science and Technology of China in Hefei, a detail that nicely illustrates how thoroughly international, and how thoroughly collaborative across national boundaries even amid periods of intense geopolitical competition, the foundational science underlying these efforts has actually been.

The strategic and economic dimension of this research deserves honest acknowledgment, because it represents a genuine shift from the earlier decades of this story, when Bell inequality testing was, as we saw in earlier chapters, a niche pursuit struggling for institutional respect and adequate funding. By the 2010s and 2020s, quantum information science had become a recognized priority for national research funding and industrial policy across multiple major world governments, including not only China but the United States, through programs including the National Quantum Initiative Act signed into law in December 2018, the European Union, through its own Quantum Flagship program launched in 2018 with a planned budget exceeding one billion euros over ten years, and other nations including the United Kingdom, Japan, Canada, and Australia, each pursuing substantial national investment in quantum technology research and development. The transformation of Bell test physics from a marginalized academic curiosity in the 1970s to a subject of serious national strategic investment by the 2020s is, in its own right, a remarkable historical arc, one that traces directly back to the patient, often institutionally unsupported work of physicists like Clauser, Aspect, and their early collaborators, whose careers would likely have looked very different had they pursued the same research questions a single generation later, once the practical stakes of understanding and controlling entanglement had become unmistakably clear to funding agencies and governments around the world.

I want to close this chapter by returning briefly to the image with which it began: an ancient Chinese philosopher, dead for well over two thousand years, lending his name to a satellite engineered with technology he could not possibly have imagined, built to test a physical phenomenon that would have struck even the most imaginative thinkers of his era as pure fantasy. Mozi's own philosophical school placed considerable emphasis on the idea that careful, systematic observation, rather than received authority or unexamined intuition, was the proper foundation for genuine knowledge, a principle that resonates, whether or not the connection was fully intended by the scientists who chose his name, with the entire multi generational project this book has been tracing: the patient, repeated insistence, from Einstein's original 1935 challenge through Bell's 1964 mathematics through five decades of increasingly rigorous laboratory experiments and finally into orbit above the Earth itself, that questions about the deep structure of reality deserve to be settled by careful observation and measurement, not by intuition, however deeply held, and not by authority, however distinguished.

There is also a quieter, less frequently discussed dimension of the Micius program worth appreciating, one that speaks to the sheer scale of coordination such a project demands. Getting a satellite traveling at roughly seven and a half kilometers per second relative to the ground to align its photon transmission precisely enough with a ground based telescope, itself only a meter or two in diameter, to successfully deliver single photons across five hundred kilometers of atmosphere, requires a tracking and pointing precision often compared, in scientific press coverage of the mission, to the difficulty of throwing a coin into a moving car from a considerable distance, except that the coin in question is a single particle of light and the moving car is traveling at orbital velocity. Achieving this level of precision required custom built tracking telescopes at each ground station, capable of following the satellite's rapid motion across the sky while simultaneously maintaining the exquisite alignment needed to couple incoming photons into single photon detectors sensitive enough to register individual quanta of light arriving after their long journey from orbit. The engineering achievement embedded in this pointing and tracking system, while it rarely receives the same attention as the headline entanglement results themselves, represents years of dedicated optical and mechanical engineering work by teams whose contributions are easy to overlook amid the more conceptually dramatic story of entanglement surviving the vacuum of space.

It is also worth noting, in the interest of giving a complete and honest picture, that Micius does have real operational limitations that distinguish it clearly from a mature, always available technology. Because the satellite requires clear night skies at both ground stations simultaneously, and because it is only within useful range of any given pair of ground stations for a matter of minutes during each orbital pass, the effective data rates achieved by these experiments, while scientifically decisive, remain far too low to support routine, everyday communication traffic at any meaningful scale. Successor missions and next generation satellite designs, pursued by multiple national space agencies and several private companies since Micius demonstrated the basic feasibility of the approach, aim to address these limitations through larger telescope apertures, improved photon sources, and constellations of multiple satellites working together rather than relying on a single spacecraft, an approach conceptually similar to how modern satellite internet services use large constellations of smaller satellites rather than a handful of larger ones. Whether a genuinely continuous, high bandwidth global quantum network eventually emerges from this line of development remains, honestly, an open engineering question, though the fundamental physics underlying it, thanks to Micius, is no longer in serious doubt.

In the next chapter, we turn from the achievement of proving entanglement real to the increasingly urgent, increasingly practical question of what that proof now allows us to build, beginning with a technology already protecting real communications for real governments and real financial institutions today: quantum cryptography, a security technique that does not merely make eavesdropping difficult, in the way that ordinary encryption does, but makes it, under the right conditions, detectable in principle the moment it happens, a genuinely new category of security guarantee that owes its existence entirely to the strange, whispered connection between particles that this book has spent its opening chapters carefully, patiently establishing as real.

CHAPTER ELEVEN

The Unbreakable Whisper

Ask most people what makes their online banking password secure, and you will likely get an answer involving length, complexity, or the wisdom of avoiding your pet's name followed by the year of your birth. Ask a cryptographer the same question, and you will get a considerably more unsettling answer: the security of nearly every encrypted communication sent across the internet today, from banking transactions to private messages to classified government communications, rests on a mathematical bet, a wager that certain calculations are so difficult to reverse that no realistic computer, given any realistic amount of time, could break them.

This is worth sitting with for a moment, because it is a genuinely different kind of security than most people assume they are relying on. Modern encryption, in its most widely used forms, does not make eavesdropping impossible. It makes eavesdropping impractically slow, based on the current state of computing technology and mathematical understanding. The most common method, called RSA encryption after its inventors Ron Rivest, Adi Shamir, and Leonard Adleman, who developed it in 1977, relies on the fact that multiplying two very large prime numbers together is computationally easy, while factoring the resulting large number back into its original two primes is, for sufficiently large primes, extraordinarily difficult, requiring, for the key sizes currently considered secure, more computing time than is practically available even to a well resourced adversary using the fastest conventional computers that exist today.

The crucial word in that sentence is currently. Mathematicians have not proven that factoring large numbers is fundamentally, permanently hard. They have only observed that nobody has yet found an efficient method for doing it using conventional computers, and the entire security architecture of the modern internet rests on the accumulated confidence built from decades of failed attempts to find such a method. This confidence, while reasonably well founded for the moment, carries an inherent vulnerability: it could, in principle, be undermined either by a clever new mathematical insight or, more concretely and more urgently for anyone thinking seriously about long term data security, by the arrival of sufficiently powerful quantum computers, a technology we will examine in detail in the next chapter, capable of running an algorithm, developed by the mathematician Peter Shor in 1994, that can factor large numbers dramatically faster than any known method available to conventional computers.

This looming vulnerability, often referred to within the security industry as the harvest now decrypt later threat, in which adversaries collect encrypted data today with the intention of decrypting it once sufficiently powerful quantum computers become available, has driven significant recent investment in what is called post quantum cryptography, new mathematical encryption methods, based on problems believed to remain hard even for quantum computers, that the United States National Institute of Standards and Technology formally began standardizing in 2024, after a multi year evaluation process. Post quantum cryptography, however, remains, at its core, the same fundamental kind of security as RSA: a mathematical bet that a certain class of problem remains hard to solve, rather than a security guarantee derived from the fundamental laws of physics themselves.

Quantum key distribution, the technology we touched on briefly in the previous chapter's discussion of the Micius satellite, offers something categorically different, and understanding exactly why requires returning to the strange, foundational property of quantum measurement that has run through every chapter of this book so far: the fact that observing a quantum system generally disturbs it, in a way that cannot be avoided or concealed.

The original and still most widely referenced protocol for quantum key distribution, called BB84 after its inventors Charles Bennett and Gilles Brassard, who proposed it in 1984, works, in simplified outline, as follows. Imagine Farah, back in our recurring example, wants to send Idris a secret cryptographic key, a long string of random ones and zeros that the two of them will later use to encrypt and decrypt ordinary messages using conventional, fast, and otherwise unremarkable encryption methods. Rather than sending this key through an ordinary communication channel where an eavesdropper, traditionally named Eve in cryptography literature, could potentially intercept and copy it without detection, Farah encodes each bit of the key onto the quantum property of a single photon, choosing randomly, for each photon, between two different possible ways of encoding the information, corresponding to two different orientations of the photon's polarization that quantum mechanics tells us cannot both be measured with certainty by the same detector setup.

When Idris receives each photon, he also randomly and independently chooses which of the two possible measurement orientations to use. Roughly half the time, by pure chance, Idris happens to choose the same measurement orientation Farah used to encode the bit, and in those cases, he correctly recovers the bit she intended to send. The other half of the time, his measurement orientation does not match Farah's encoding choice, and the result he obtains is essentially random and unusable. After the transmission is complete, Farah and Idris communicate openly, over an ordinary, unsecured channel that an eavesdropper is welcome to listen in on, comparing only which measurement orientations they each used for each photon, without revealing the actual measurement outcomes themselves. They discard every bit where their orientations did not match, keeping only the roughly half of the bits where they did, and this remaining string of matched bits becomes their shared secret key.

The security of this scheme rests on precisely the strange feature of quantum measurement we have discussed since chapter two: any attempt by an eavesdropper to intercept and measure the photons in transit, in order to learn the key before it reaches its destination, necessarily disturbs the delicate quantum state of at least some of those photons, because the eavesdropper, not knowing in advance which encoding orientation Farah used for any given photon, cannot help but sometimes choose the wrong measurement orientation herself, introducing detectable errors into the transmission in a way that has no equivalent in the world of classical, non quantum communication. After exchanging their key, Farah and Idris can sacrifice a small, randomly chosen portion of their matched bits, comparing them openly to check for unexpectedly high error rates. If an eavesdropper has been listening in, the disturbance her measurements introduced will show up as an elevated error rate in this check, alerting Farah and Idris to the presence of an eavesdropper and allowing them to discard the compromised key entirely and try again, rather than unknowingly using a key that has already been partially compromised.

This is the genuinely novel security guarantee that quantum key distribution offers, and it is worth being precise about exactly what is and is not being claimed. Quantum key distribution does not make it impossible for an eavesdropper to attempt interception. It makes eavesdropping detectable, in principle, every single time it occurs, because the laws of quantum mechanics themselves, not merely the current limitations of an adversary's computing power, guarantee that measuring an unknown quantum state disturbs it. This is a security guarantee rooted in physics rather than mathematics, and it does not degrade over time the way RSA's security might degrade as quantum computers become more powerful, because it does not depend on any computational problem remaining hard. It depends only on the basic structure of quantum measurement remaining true, a proposition that, as the preceding chapters of this book have documented in exhaustive experimental detail, rests on considerably firmer ground than any assumption about the future difficulty of factoring large numbers.

It is worth noting, because I promised at the outset of this book to be honest about the limits of every claim, that quantum key distribution as actually deployed today is not a flawless, unconditionally perfect technology, and serious cryptographers and security researchers have identified genuine practical vulnerabilities in specific real world implementations over the years, generally arising not from any flaw in the underlying quantum physics but from imperfections in the physical devices used to implement it. Real photon detectors, for instance, have sometimes been shown to be vulnerable to a class of attacks called detector blinding, in which an attacker shines a bright, continuous beam of light at a photon detector, forcing it into a classical mode of operation where it can be manipulated in ways that were not anticipated in the original theoretical security proof. These vulnerabilities are real, have been demonstrated in laboratory conditions by security researchers acting in good faith to strengthen the technology, and have generally been addressed through improved detector designs and additional protocol safeguards in subsequent generations of commercial quantum key distribution equipment, in much the same way that vulnerabilities discovered in conventional encryption implementations are routinely patched without undermining the underlying mathematical soundness of the encryption scheme itself.

Quantum key distribution has moved well beyond the laboratory and into genuine, if still limited, commercial and governmental deployment over the past two decades. Beyond the Chinese Beijing Shanghai quantum backbone discussed in the previous chapter, connecting major financial and government institutions along a roughly two thousand kilometer fiber optic corridor since 2017, quantum key distribution networks have also been deployed in several other countries. Switzerland used quantum key distribution to help secure the transmission of vote counting data during a national election in the canton of Geneva as early as 2007, an early and often overlooked milestone in the practical deployment of this technology. Several major banks, including in China and in Europe, have piloted quantum secured links for particularly sensitive financial transactions. And multiple national governments, recognizing the long term threat that future quantum computers pose to conventionally encrypted communications and data, have invested in quantum key distribution infrastructure specifically to protect communications considered sensitive enough to warrant protection lasting decades into the future, well past the point when today's RSA encrypted traffic might reasonably be expected to become vulnerable to a sufficiently advanced quantum computer.

It is worth being clear about one more important limitation, because popular discussions of quantum cryptography sometimes blur past it in ways that overstate the technology's current maturity. Quantum key distribution, in its most widely deployed form, secures only the distribution of the cryptographic key itself, the shared secret string of random bits that Farah and Idris use to encrypt their subsequent messages. The actual encryption and decryption of the messages themselves, once the key has been securely shared, typically still uses conventional, classical encryption algorithms, meaning that quantum key distribution should be understood as strengthening one specific, critical link in the overall security chain, the initial exchange of a shared secret, rather than replacing the entire cryptographic infrastructure of the internet outright. This is not a modest achievement, since the secure distribution of cryptographic keys has historically been one of the most persistent and difficult problems in the entire field of cryptography, a problem sometimes called the key distribution problem, which conventional cryptography has typically solved using the same kind of computational hardness assumptions, and therefore the same long term vulnerability to future quantum computers, that we discussed regarding RSA encryption earlier in this chapter.

There is also a more ambitious, still largely experimental extension of this technology worth mentioning before we close this chapter, because it returns us directly to the entanglement based satellite experiments of the previous chapter rather than the single photon, non entangled BB84 protocol we have focused on so far. Entanglement based quantum key distribution, first proposed theoretically by the physicist Artur Ekert in 1991 in a paper that arrived at a functionally similar security guarantee to BB84 through an entirely different theoretical route, grounded directly in Bell inequality violations rather than the disturbance of single photon measurements, uses entangled photon pairs distributed to two separate parties, who then generate a shared key by measuring their respective photons and comparing measurement orientations in a manner closely analogous to the BB84 protocol, but with an additional and rather elegant security guarantee built directly into the process: if the measured correlations between the two parties' results violate a Bell inequality, exactly the kind of violation this book has spent its opening chapters carefully establishing as a hallmark of genuine, undisturbed entanglement, then the parties can be confident, based on the fundamental physics established by Clauser, Aspect, Zeilinger, and the many other physicists we have discussed, that no eavesdropper has successfully intercepted and measured the photons along the way, since any such interception would necessarily have reduced the strength of the observed correlation below the threshold Bell's mathematics establishes.

This entanglement based approach is precisely the technique underlying the intercontinental quantum secured video call conducted between China and Austria via the Micius satellite in 2018, discussed in the previous chapter, and it represents, in the assessment of many researchers working in the field, the more elegant and in some respects more fundamentally secure long term direction for quantum cryptography, precisely because its security proof connects directly back to the same rigorously tested Bell inequality violations that this entire book has been devoted to explaining, rather than resting on the disturbance of single, non entangled photon measurements as the original BB84 protocol does.

Whichever specific protocol ultimately proves most practical for widespread deployment, the underlying promise is the same, and it is worth stating plainly before we move to the next chapter's exploration of quantum computing: for the first time in the history of cryptography, a security guarantee exists that does not rest on an assumption about the future limits of computational power, an assumption that history has repeatedly shown to be a fragile foundation, but rests instead on the measured, repeatedly confirmed, and by this point in our story thoroughly established behavior of entangled particles themselves. Whether or not quantum key distribution eventually becomes a routine, invisible part of everyday internet infrastructure, the way encryption itself became invisible and routine over the preceding several decades, the physics underlying its promise is no longer speculative. It has already been tested, repeatedly, across university basements, French optical benches, Dutch bicycle paths, and Chinese satellites, and it has, every single time, held.

CHAPTER TWELVE

Machines That Think in Whispers

A conventional computer, whatever its speed or sophistication, ultimately performs its work using bits, tiny physical switches that exist, at any given moment, in exactly one of two states, conventionally labeled zero and one. Every photograph you have ever taken, every message you have ever sent, every calculation a bank ever performs on your account balance, is, at the deepest level of the machine actually doing the work, nothing more than an enormous number of these simple zero or one switches, flipped in carefully orchestrated sequences according to the instructions of a computer program.

A quantum computer replaces the ordinary bit with something called a qubit, and the difference, though it might sound at first like a minor technical variation, turns out to unlock computational possibilities that no conceivable arrangement of ordinary bits could ever match, for certain specific kinds of problems. A qubit, exactly like the electrons and photons we have discussed throughout this book, can exist in superposition, a genuine blend of the zero state and the one state simultaneously, not merely an unknown value waiting to be revealed but a mathematically real mixture of both possibilities at once, in precisely the sense established in chapter two.

This alone would be interesting but not, on its own, transformative. What makes quantum computing genuinely powerful is entanglement, the same phenomenon this entire book has been devoted to explaining, applied not to a pair of particles sent to distant laboratories but to a large collection of qubits held together within a single computing device. When many qubits are entangled with one another, the resulting system can represent and manipulate an enormous number of possible combinations of zeros and ones simultaneously, growing exponentially with the number of qubits involved. A system of just fifty entangled qubits, for instance, can represent more simultaneous combinations than there are stars in the observable universe, a genuinely staggering scale of parallel information processing that no conventional computer, however large, could replicate by simply adding more ordinary bits.

It is worth being precise here, because popular descriptions of quantum computing frequently overstate what this parallelism actually allows a quantum computer to do, in ways that can leave readers with a distorted picture of the technology's real promise. A quantum computer does not simply try every possible answer to a problem simultaneously and then hand you the correct one, the way an overly simplified explanation sometimes suggests. If it worked that way, extracting a useful answer would still require measuring the qubits at the end of the calculation, and measurement, as established throughout this book, forces a quantum superposition to collapse into a single, definite, and essentially random outcome, discarding all the other possibilities that were briefly present in the superposition. A naive quantum computer built without further cleverness would simply return one random answer out of the vast number of possibilities it had briefly explored, with no way of knowing whether that particular answer happened to be the correct one.

The genuine cleverness of quantum computing, developed by computer scientists and physicists over several decades starting in the 1980s and 1990s, lies in designing quantum algorithms, carefully structured sequences of quantum operations that use a phenomenon called interference, closely related to the wave like behavior of quantum systems, to amplify the probability of measuring the correct answer while simultaneously suppressing the probability of measuring incorrect answers, so that when the final measurement collapses the superposition, the outcome that emerges is overwhelmingly likely to be the useful one. This is a genuinely difficult and specialized kind of programming, and it is why quantum computers are not, and are not expected to become, faster replacements for conventional computers across every kind of task. For most everyday computing tasks, browsing the internet, writing a document, editing a photograph, conventional computers remain, and are expected to remain indefinitely, the more practical and more efficient tool. Quantum computers offer a genuine, dramatic advantage only for a specific, relatively narrow category of problems where a suitable quantum algorithm, exploiting superposition and entanglement in just the right way, happens to exist.

The most famous of these specialized quantum algorithms is Shor's algorithm, developed by the mathematician Peter Shor in 1994, which we touched on briefly in the previous chapter's discussion of quantum cryptography. Shor's algorithm provides a method for factoring large numbers into their prime components dramatically faster than any known algorithm running on a conventional computer, a capability that, if ever implemented on a sufficiently large and reliable quantum computer, would undermine the mathematical hardness assumption underlying RSA encryption, the same assumption that currently protects a substantial fraction of the world's encrypted digital communications. It is important to note, as security researchers regularly emphasize, that running Shor's algorithm at the scale needed to break real world encryption keys, which typically involve numbers thousands of digits long, requires a quantum computer with a number of stable, low error qubits far beyond what any quantum computer built to date has achieved, and most serious estimates from researchers in the field suggest this capability, sometimes referred to as a cryptographically relevant quantum computer, remains at minimum years away, and quite possibly considerably longer, depending on the pace of continued hardware development.

A second influential quantum algorithm, Grover's algorithm, developed by computer scientist Lov Grover in 1996, offers a more modest but still meaningful speedup for a broad category of search problems, essentially allowing a quantum computer to search through an unsorted list of possibilities using roughly the square root of the number of steps a conventional computer would require. This has narrower but still practically significant implications across fields ranging from database search to certain kinds of optimization problems encountered in logistics, finance, and materials science.

Building an actual, physical quantum computer capable of running these algorithms at a useful scale has proven to be one of the most demanding engineering challenges in contemporary physics, and it is worth understanding why, because the difficulty connects directly back to the concept of decoherence introduced in chapter four. Qubits, like the entangled photons and electron spins discussed throughout this book, are extraordinarily delicate. Any unwanted interaction with their surrounding environment, a stray vibration, a fluctuation in temperature, an errant electromagnetic field, tends to destroy the delicate superposition and entanglement that the quantum computation depends on, introducing errors into the calculation. Building a quantum computer, in practice, means building a system capable of maintaining dozens, hundreds, or eventually many thousands of qubits in a carefully controlled, isolated state, entangled with each other in precisely the configurations a given algorithm requires, for long enough to complete a useful calculation, all while remaining sufficiently isolated from the ordinary, noisy physical environment that surrounds every laboratory on Earth.

Several genuinely different physical approaches to building qubits have been pursued by different research groups and companies, each with distinct advantages and challenges. Superconducting qubits, pursued by companies including IBM and Google, use tiny circuits made from superconducting materials, cooled to temperatures colder than deep space, within a fraction of a degree of absolute zero, using specialized dilution refrigerators, to create quantum states that can be precisely controlled using microwave pulses. Trapped ion qubits, pursued by companies including IonQ, use individual charged atoms, held in place using electromagnetic fields and manipulated using precisely tuned lasers, an approach that tends to offer qubits with longer coherence times, meaning they can maintain their delicate quantum states for longer before decoherence sets in, though often at the cost of slower operation speeds compared to superconducting approaches. Photonic quantum computing, pursued by companies including PsiQuantum, uses photons themselves, the same particles of light discussed throughout this book's account of entanglement testing, as qubits, offering the advantage of naturally low interaction with the surrounding environment, since photons, as we discussed regarding the Micius satellite, interact only weakly with most materials, though this same weak interaction makes it correspondingly more difficult to make photons interact with each other in the controlled ways a computation requires.

IBM, one of the most established and publicly transparent participants in this competitive landscape, has pursued a publicly documented roadmap of steadily increasing superconducting qubit processors over the past several years, and Google's quantum computing team, working through its Google Quantum AI division, achieved a widely reported milestone in 2019 when its Sycamore processor completed a specific, carefully chosen computational task that the team calculated would have taken an enormous amount of time on the most powerful conventional supercomputers available at that time, a claim that was itself the subject of considerable technical debate among specialists regarding exactly how the comparison should be made and exactly how much of an advantage had genuinely been demonstrated, debate that reflects the genuinely difficult and still evolving state of the art in fairly benchmarking quantum against conventional computing performance. D-Wave, a Canadian company, has pursued a somewhat different technical approach called quantum annealing, specifically optimized for certain kinds of optimization problems rather than the general purpose, gate based quantum computing model pursued by IBM, Google, and IonQ, illustrating that even within the broader field of quantum computing, meaningfully different technical philosophies and target applications coexist.

It is worth being candid, in the spirit of honesty this book has tried to maintain throughout, about where the field of quantum computing genuinely stands today, as opposed to where enthusiastic press coverage sometimes suggests it stands. The current generation of quantum computers, generally described using the term noisy intermediate scale quantum, or NISQ, devices, a phrase coined by physicist John Preskill in 2018, contain enough qubits to be scientifically interesting and to demonstrate genuine quantum advantages on certain narrow, carefully chosen benchmark tasks, but not yet enough qubits, and not yet low enough error rates, to run the most commercially and scientifically anticipated algorithms, including Shor's algorithm at a cryptographically meaningful scale, or the detailed quantum chemistry simulations that many researchers believe represent one of the most promising near term practical applications of the technology, simulations that could, if fully realized, dramatically accelerate the discovery of new pharmaceutical compounds, new industrial catalysts, and new materials for batteries and solar cells, by directly simulating the genuinely quantum mechanical behavior of molecules and chemical reactions in a way that conventional computers, despite decades of clever approximation techniques, still struggle to do with full accuracy for anything beyond fairly small molecules.

Bridging the gap between today's noisy, error prone quantum devices and the large scale, fully error corrected quantum computers that many of the field's most anticipated applications require is widely regarded, among researchers actually working in the field, as the central engineering challenge of the coming decade. Quantum error correction, the collection of techniques designed to detect and correct the inevitable errors that decoherence and imperfect control introduce into a quantum computation, typically requires combining many physical qubits together to form a smaller number of more reliable logical qubits, meaning that the raw qubit counts companies frequently announce in press releases substantially overstate the actual computational power available once the overhead of error correction is properly accounted for. Serious estimates from researchers in the field suggest that running some of the most anticipated large scale quantum algorithms at a genuinely useful scale may require systems with somewhere in the range of one million or more physical qubits, once error correction overhead is included, a scale that remains, as of the time of this writing, a significant distance beyond the largest quantum computers currently operating anywhere in the world.

None of this is meant to suggest that quantum computing represents an overhyped technology destined to disappoint, a conclusion some more skeptical commentary occasionally reaches. It is meant, rather, to situate the genuine, well documented progress the field has made, progress built directly on the same entanglement this book has spent its opening chapters carefully establishing as real and reliable, within an honest, unhyped timeline, one that most serious researchers in the field describe not in terms of an imminent, world transforming breakthrough arriving next year, but in terms of a steady, multi decade engineering campaign, not unlike the multi decade campaign of loophole closing Bell test experiments recounted in earlier chapters of this book, gradually building toward a technology whose full potential likely remains a decade or more away from broad, transformative deployment, even as narrower, more specialized applications continue to reach practical maturity well before that broader horizon arrives.

What quantum computing shares, at its deepest conceptual level, with every other technology and phenomenon discussed in this book, from the coins in Farah and Idris's wooden boxes to the diamonds buried beneath a Dutch bicycle path to the satellite named after an ancient philosopher, is a direct, unbroken lineage back to the strange, whispered connection between particles that Einstein found so troubling in 1935, that Schrodinger named that same year, that Bell converted into testable mathematics in 1964, and that generations of careful, patient experimentalists spent the following five decades proving, beyond reasonable doubt, to be a genuine and permanent feature of the physical world. Every qubit entangled inside an IBM superconducting processor, every trapped ion manipulated by a precisely tuned laser inside an IonQ device, is drawing on exactly the same physical phenomenon that a Berkeley graduate student measured with hand built photon detectors in a converted basement more than five decades ago, now harnessed, deliberately and with steadily increasing engineering sophistication, to build machines capable of computation no classical device could ever match.

In the next chapter, we turn to a quieter, less publicly celebrated but arguably even more immediately practical application of entanglement, one already improving the precision of the instruments that keep global financial markets synchronized, that make GPS navigation possible, and that may, within the coming years, reshape how doctors detect disease at its earliest, most treatable stages.

CHAPTER THIRTEEN

The Body That Measures Time in Silence

There is a room at JILA, a joint research institute run by the University of Colorado Boulder and the National Institute of Standards and Technology, where a cloud of strontium atoms, cooled to a temperature within a whisper of absolute zero and trapped in a precisely structured lattice of laser light, ticks with a regularity so exact that if such a clock had been running since the beginning of the universe, roughly thirteen point eight billion years ago, it would still not have drifted by a single second. This is not a rhetorical flourish or a rounded up approximation for dramatic effect. It is, in substance, the kind of stability that modern optical atomic clocks, the most precise timekeeping devices ever built by human beings, now genuinely achieve, and entanglement, the same phenomenon this book has traced from a 1935 thought experiment through a Berkeley basement, a Parisian optics bench, a Dutch bicycle path, and a satellite orbiting five hundred kilometers overhead, is now being harnessed to push that already extraordinary precision even further.

To understand why entanglement helps here, and why researchers at JILA, at NIST, and at similar institutions around the world consider this one of the most promising near term practical applications of everything this book has described, it helps to understand what fundamentally limits the precision of an atomic clock in the first place, a limit that has nothing to do with engineering imperfections and everything to do with the same quantum uncertainty introduced back in chapter two.

An atomic clock works by measuring the frequency of light that causes electrons within an atom to jump between two specific, precisely defined energy levels, a frequency that is, for a given type of atom, a fixed and universal constant of nature, identical for every atom of that type anywhere in the universe, making it a far more reliable and reproducible standard for defining the length of a second than any mechanical or even quartz crystal based timekeeping device could ever be. The modern international definition of the second, in fact, is based on exactly this kind of measurement, defined since 1967 in terms of a specific number of oscillations of microwave radiation corresponding to a transition between two energy levels in a cesium atom, and newer, even more precise optical atomic clocks, using transitions in atoms such as strontium or ytterbium that oscillate at much higher, optical frequencies rather than microwave frequencies, are widely expected to eventually replace cesium as the international standard.

The precision of these measurements, however formidable, runs into a fundamental barrier imposed directly by quantum mechanics, a barrier researchers call the standard quantum limit. When physicists measure the frequency of a single atom's energy transition, quantum uncertainty, the same fundamental, irreducible randomness discussed in chapter two, introduces a degree of unavoidable noise into each individual measurement. The traditional way to reduce this noise, and thereby improve the clock's overall precision, is simply to average across measurements from a very large number of independent, unentangled atoms simultaneously, since averaging over many independent random measurements reduces the overall statistical noise in a well understood, predictable way, in exactly the same manner that averaging many independent, noisy measurements of any physical quantity tends to converge toward a more precise estimate of the true value. But this averaging strategy runs into diminishing returns: to improve precision by a further factor of two using this method, a clock needs four times as many atoms, and the mathematics of this relationship, well established since long before entanglement based approaches were seriously considered, imposes a hard ceiling, the standard quantum limit, on how much precision can be squeezed out of any given number of independent, unentangled atoms.

This is precisely where entanglement offers a way forward, and the underlying idea, once explained, has a pleasing symmetry with everything the preceding chapters of this book have established about how entangled particles share a single, unified description rather than behaving as fully independent objects. If, instead of measuring a large collection of independent, unentangled atoms, physicists instead entangle those atoms with one another first, correlating their quantum states so that they no longer behave as separate, individually noisy measuring devices but as a single, unified quantum system, the resulting collective measurement can, under the right conditions, achieve a precision that genuinely exceeds the standard quantum limit, a feat that is, importantly, mathematically impossible using any collection of unentangled atoms, no matter how numerous. Researchers refer to this technique, when applied in one of its most common forms, as spin squeezing, a term describing how the quantum uncertainty is deliberately redistributed, reduced in the specific property being measured for the clock's timekeeping purpose at the cost of increased uncertainty in some other, less relevant property, exploiting the same fundamental trade off structure that Heisenberg's uncertainty principle, discussed in chapter one, established nearly a century ago, but doing so deliberately and to specific advantage rather than treating it merely as an obstacle.

Researchers at JILA, working under Jun Ye, one of the most prominent figures in modern optical atomic clock research, along with colleagues including Adam Kaufman and James Thompson, published research in 2024 demonstrating a clock built from strontium atoms in which entanglement between groups of atoms allowed the device to divide a given stretch of time into smaller, more precisely measured units than would be possible using the same number of unentangled atoms, results described by Kaufman, in comments accompanying the published research, as approaching what he called the holy grail for optical atomic clock performance. A separate but related line of research, involving trapped strontium ions separated by a physical distance of roughly two meters and connected using a photonic link, an approach directly related to the entanglement distribution and entanglement swapping techniques discussed in earlier chapters covering the Delft loophole free experiments and the Micius satellite, demonstrated that entangling two physically separate atomic clocks reduces the number of measurements needed to achieve a given level of precision, halving the required measurement count in a manner consistent with a theoretical prediction called the Heisenberg limit, the absolute best precision quantum mechanics allows for a given number of measurements, a limit considerably more favorable than the older standard quantum limit that unentangled measurement strategies remain confined to.

Why does any of this matter beyond the satisfaction of pushing a scientific instrument to ever more impressive extremes of precision, a question worth asking honestly, since not every scientific achievement translates into practical benefit on any meaningful timescale. Atomic clocks, it turns out, are not simply of academic interest. They already sit at the center of an enormous amount of infrastructure that modern life depends on without most people ever thinking about it. The Global Positioning System, which underlies not only the navigation app on your phone but also the precise timing signals that synchronize electrical power grids, financial trading systems, and telecommunications networks around the world, depends fundamentally on a constellation of satellites, each carrying multiple atomic clocks, whose extraordinary precision allows a receiver on the ground to calculate its position by measuring tiny differences in the arrival time of radio signals traveling at the speed of light from multiple satellites simultaneously. Even a timing error as small as a few billionths of a second, translated through the speed of light, corresponds to a positioning error of several meters, which is why the continual improvement of atomic clock precision translates directly into continual improvement in the accuracy of navigation and positioning systems that a substantial fraction of the global economy now quietly depends on.

Beyond navigation, more precise atomic clocks open the door to entirely new categories of scientific measurement, an application area researchers refer to as quantum sensing, and it is worth walking through a few concrete examples, because they illustrate how directly this seemingly abstract improvement in timekeeping precision connects to tangible, real world capability. Because atomic clocks measure time using the frequency of light emitted by atoms, and because Einstein's general theory of relativity establishes that the passage of time itself runs at a very slightly different rate depending on the strength of the gravitational field an observer experiences, sufficiently precise atomic clocks can actually be used to measure tiny variations in gravity itself, a technique called relativistic geodesy. Two identical, sufficiently precise atomic clocks placed at slightly different elevations, even a difference of only a few centimeters, will tick at very slightly different rates because they experience very slightly different gravitational field strengths, a difference precise modern optical clocks are now sensitive enough to detect directly. Jun Ye and other researchers in the field have specifically pointed to this capability as a promising direction for future entanglement enhanced clocks, potentially enabling extraordinarily precise, portable measurements of local variations in Earth's gravitational field, with applications ranging from detecting underground geological structures, including mineral deposits, aquifers, and even underground tunnels or cavities, to monitoring subtle changes in ice sheet mass connected to climate research, to searching for the extremely subtle deviations from Einstein's predictions that some theoretical physicists believe might eventually reveal new fundamental physics beyond our current understanding.

Quantum enhanced sensing extends well beyond atomic clocks and timekeeping into other domains of precision measurement as well, and it is worth surveying a few of these briefly, since they illustrate how broadly the underlying entanglement enhanced measurement principle applies. Gravitational wave detectors, the enormous laser interferometer facilities that first directly detected ripples in spacetime itself in 2015, produced by the collision of distant black holes, an achievement that earned its own Nobel Prize in Physics in 2017, have themselves incorporated a specific form of quantum enhanced measurement called squeezed light, closely related in its underlying physics to the spin squeezing techniques described earlier in this chapter, to reduce quantum measurement noise and improve the detectors' sensitivity to the almost inconceivably faint spacetime distortions these instruments are built to detect. Quantum enhanced magnetometers, instruments for measuring extremely faint magnetic fields, are being explored for applications ranging from detecting mineral deposits and archaeological structures underground to, in a medical context that connects directly to human health in a way that feels appropriately grounding after several chapters focused on satellites and laboratory apparatus, potentially improving the sensitivity of magnetoencephalography, a technique for measuring the extraordinarily faint magnetic fields produced by electrical activity in the human brain, used in both neuroscience research and, in some clinical settings, in the diagnosis and surgical planning for conditions including epilepsy.

It is worth being appropriately careful and precise about the current maturity of these medical and geophysical applications, in keeping with this book's commitment throughout to distinguishing established, already demonstrated capability from promising but still developing research directions. As the NIST research overview on this topic states plainly, no entangled atomic clock has yet surpassed the precision of the very best conventional, unentangled atomic clocks in absolute terms, though researchers in the field widely expect this milestone, sometimes described as achieving a genuine, unambiguous quantum advantage in real world clock performance rather than merely in a narrow, favorable laboratory demonstration, to be reached within the coming years given the current pace of progress. Similarly, quantum enhanced magnetometry and gravimetry, while already demonstrating genuine, measurable improvements over classical measurement techniques in specific laboratory and field trial settings, remain, for the most part, in a research and early field trial stage rather than routine, widely deployed clinical or commercial practice.

This measured, honest pace of development should not diminish the genuine significance of what has already been achieved, and it connects, in a way I find genuinely moving, back to the very beginning of this book's story. Einstein, writing to Max Born in the 1940s, described entanglement as spooky, a word chosen specifically to convey his deep discomfort with a phenomenon he found almost embarrassingly at odds with how a well behaved universe ought to work. Eight decades later, that same spooky phenomenon is being deliberately, carefully engineered into the most precise timekeeping devices human beings have ever built, devices that keep the world's financial markets synchronized, that guide airplanes and ships and, increasingly, autonomous vehicles to their destinations, and that may, within the coming years, help detect the earliest, most treatable signs of neurological disease, or reveal underground water reserves in regions facing water scarcity, or measure gravitational variations precise enough to test whether Einstein's own theory of general relativity, formulated in the same intellectual tradition that first led him to doubt entanglement, continues to hold exactly true at scales and precisions no experiment has yet been able to probe.

There is a particular kind of poetry in this trajectory that I do not think is accidental, or at least not merely coincidental. The very phenomenon that troubled Einstein most deeply, because it seemed to violate the orderly, locally causal universe his own greatest theories had described, is now being used to build instruments precise enough to test those very theories more rigorously than ever before, instruments that may, in the coming decades, either confirm general relativity's predictions to an even more exacting degree of precision than currently achieved, or, perhaps more excitingly for the physicists involved, reveal the first genuine cracks in a theory that has otherwise withstood more than a century of increasingly demanding experimental scrutiny.

In the next and final substantive chapter of this book, we step back from the specific technologies, cryptography, computing, and sensing, that entanglement now makes possible, and ask a broader question, one that returns us to the philosophical terrain where this story began, in a Brussels hotel in 1927: what does it actually mean, for our understanding of reality itself, that the universe is built this way, that separation and connection turn out to be far less absolute, and far more negotiable, than either common sense or classical physics ever led us to expect.

CHAPTER FOURTEEN

What the Whispers Mean for Us

A friend of mine, upon hearing me describe the contents of this book over dinner some months ago, asked a question I have not been able to stop turning over since. If entanglement is real, she asked, and if two particles really can share a single description across any distance, does that mean, in some deep sense, that separation itself is something of an illusion. Is the universe, underneath all its apparent distances and boundaries, actually one continuous thing, wearing a costume of separateness that our senses and our physics have simply been slow to see through.

It is a beautiful question, and I want to treat it with the seriousness it deserves, which means resisting the temptation to give it either of the two easy answers that popular science writing about entanglement sometimes reaches for. The first easy answer, breathless and mystical, says yes, entanglement proves that everything is connected, that the ancient intuitions of various contemplative traditions about the fundamental unity of all things have been vindicated by modern physics, and that a proper understanding of quantum mechanics should reshape not only how we do science but how we live, love, and relate to one another and to the world. The second easy answer, brisk and dismissive, says no, entanglement is a narrow, technical phenomenon confined to carefully prepared microscopic systems in specialized laboratory conditions, with no bearing whatsoever on anything beyond the specific mathematics of quantum measurement, and anyone drawing broader lessons from it has simply misunderstood the physics.

I think both of these easy answers get something right and something importantly wrong, and the honest position, the one I have tried to hold throughout this book, sits somewhere more interesting and more difficult between them.

The dismissive answer is correct about an important, specific technical point that this book has emphasized repeatedly: entanglement does not allow information or influence to be transmitted faster than light, it does not enable any kind of practical telepathy or mystical connection between separated human beings, and the particular kind of correlation it involves has a precise, mathematically well defined meaning that does not straightforwardly generalize to claims about human consciousness, spiritual unity, or the interconnectedness of all living things, claims that some popular treatments of quantum mechanics, going back to books published in the 1970s and continuing in various forms to the present day, have drawn in ways that most working physicists regard as unsupported extrapolations well beyond what the actual physics establishes. If you have picked up this book hoping it would end by validating a specific metaphysical or spiritual worldview through an appeal to quantum mechanics, I want to be honest with you now, near the end, that I do not think the physics itself licenses that kind of leap, however emotionally appealing the idea of a fundamentally unified universe might be.

And yet the mystical answer, however overreaching in its specific claims, is pointing at something the dismissive answer risks losing entirely: entanglement really does force a revision of some of the deepest assumptions embedded in what we might call folk physics, the intuitive, largely unexamined picture of how objects and causation work that most human beings, including most physicists before the twentieth century, carried around without ever feeling the need to question it. That folk picture holds that objects have definite properties independent of observation, that influence between objects requires some kind of signal or contact traveling between them, and that sufficiently separated objects are, for all practical purposes, entirely independent of one another, connected only by whatever shared history brought them to their current, separate positions. Entanglement, established beyond reasonable doubt by the eight decades of theoretical and experimental work this book has traced, shows that this folk picture is not merely incomplete in some minor, technical way. It is, at the deepest level physics has yet been able to probe, simply false, or at minimum, radically incomplete as a description of how the universe is actually built.

What replaces it is not a simple, comforting story about universal oneness, but something considerably stranger and, I would argue, more genuinely interesting: a universe in which the relationships between particles can be, in a precise mathematical sense, more fundamental than the particles considered as separate individual objects. Recall the wave function discussion from chapter one. For two entangled particles, there is no way to write down a separate, independent description of particle one and a separate, independent description of particle two. There is only a single, unified description of the pair together, a description that resists being carved apart into independent pieces no matter how far the particles travel from one another. This is not a claim about mysticism or unity in any spiritual sense. It is a precise, technical claim about the mathematical structure that best describes how these particles actually behave, confirmed through exactly the kind of rigorous, skeptical, loophole closing experimental program recounted in the middle chapters of this book. But it is also, undeniably, a claim that unsettles the deep intuition that individual objects, considered separately, are the truly fundamental building blocks of reality, with relationships between them merely a secondary, derived feature.

Some physicists and philosophers of physics have taken this observation quite far, developing what is sometimes called relational quantum mechanics, an interpretive framework, associated particularly with the physicist Carlo Rovelli, which proposes that quantum properties should be understood not as intrinsic features that particles possess in some absolute sense, but as features that only become meaningful relative to a specific interaction with another system. Under this view, asking what an electron's spin actually is, independent of any relationship to a measuring device or another particle, may be a bit like asking what a person's height is independent of any unit of measurement or any comparison to another object, a question that sounds meaningful but may, on closer inspection, be subtly malformed. This is a genuinely serious, actively debated position within contemporary philosophy of physics, not a fringe speculation, though, as with the competing interpretations of quantum mechanics discussed back in chapter two, it remains one live option among several, rather than an established consensus.

I find myself drawn, without wanting to overstate my own certainty on a question genuinely contested by people who have devoted entire careers to it, toward a more modest but still, I think, genuinely significant conclusion: entanglement demonstrates that the universe does not respect a clean, absolute distinction between separate objects and the relationships connecting them, at least not at the fundamental level where quantum mechanics operates. Separateness and connection turn out to be more like two ends of a spectrum, or perhaps two different ways of describing the same underlying reality depending on how you choose to look at it, rather than two mutually exclusive categories with a sharp boundary between them. This is not the same claim as everything is one, a claim so broad and unspecific that it risks explaining nothing at all. It is a more precise, more interesting claim: that the specific kind of separateness classical physics assumed, in which distant objects are always, in principle, fully independent unless actively connected by some signal or force, does not hold at the deepest level physics has yet probed, and the replacement is not simple undifferentiated oneness but something considerably more textured, a universe of particles that can share a single mathematical description while remaining, in every operationally meaningful sense, individually measurable and individually locatable.

I want to close this chapter, and set up the shorter reflections that follow in the epilogue, by returning one final time to Einstein and Bohr, whose disagreement opened this book's central historical narrative back in chapter three. It has become something of a convention, in popular retellings of this story, to cast Einstein as the loser of this particular argument, the great genius who, for once, got something importantly wrong, while Bohr emerges as the prescient winner whose stranger, more counterintuitive picture of reality was eventually vindicated by decades of careful experimentation. I want to resist that framing, gently, in these closing pages, not because it is factually incorrect, since Einstein's specific hope for a local hidden variable theory restoring a fully classical picture of reality was indeed decisively ruled out by the experimental record this book has traced, but because I think it undersells what Einstein's persistent, rigorous skepticism actually achieved.

It was Einstein's refusal to accept quantum mechanics at face value, his insistence on constructing the EPR thought experiment specifically to expose what he believed was a fatal flaw, that gave entanglement its name, its precise mathematical characterization, and ultimately its experimental testability. Without Einstein's stubborn, world class skepticism, there is a genuine possibility that entanglement might have remained a curious mathematical footnote within quantum theory for decades longer than it did, discussed occasionally by specialists but never subjected to the rigorous, loophole closing experimental scrutiny that has made it, by 2022, worthy of the Nobel Prize in Physics. Being wrong about the ultimate conclusion, in other words, is not the same as having failed to contribute something essential to how that conclusion was reached, tested, and confirmed. Einstein lost the argument. He did not lose his role as one of the two indispensable architects, alongside Bohr, of the very question that entanglement research has spent the past nine decades so productively, and so rigorously, answering.

There is a further dimension to this story worth sitting with before we close, one that concerns not what entanglement means for the universe at large but what it means for how we think about knowledge itself, and about the particular kind of humility that good science requires. Throughout this book, we have watched physicists hold two things in tension that might, on the surface, seem contradictory: an absolute commitment to following the mathematics and the evidence wherever it leads, even when the destination is deeply uncomfortable, and an equally absolute commitment to skepticism, to refusing any conclusion, however strongly suggested by early results, until every reasonable alternative explanation has been exhausted. Clauser did not simply announce that he had proven quantum mechanics correct in 1972 and move on. He was explicit, in his own published work, about the loopholes his experiment left open. Aspect did the same in 1982. It took the collective, decades long labor of an entire international community of physicists, each generation building on and correcting the last, to finally arrive at the loophole free experiments of 2015, and even then, the physics community's acceptance of the result came not from any single dramatic announcement but from the careful, converging agreement of three independent experimental teams working with entirely different physical systems.

This is a model of how to hold a strange and uncomfortable truth that I think has value well beyond physics. It suggests that the proper response to encountering something that violates your deepest intuitions is neither uncritical acceptance nor reflexive dismissal, but patient, structured inquiry: find the specific point where your intuition and the evidence diverge, design a way to test that specific divergence as rigorously as possible, remain honest about the limitations of each test as you perform it, and keep refining the test until the ambiguity has been genuinely resolved, one way or the other. Bell did not simply declare that Einstein was wrong. He built a mathematical tool precise enough that nature itself could answer the question. That is a considerably harder and more honest thing to do than simply arguing more forcefully for one's preferred position, and it is, I think, the single most valuable habit of mind this entire story has to offer a reader who will likely never personally test a Bell inequality or trap a single photon in a laboratory.

There is one more thread worth gathering up before we move to the epilogue, and it concerns the sheer international, collaborative character of this entire scientific journey, a detail easy to lose amid the focus on individual physicists that a narrative history naturally tends to emphasize. Einstein was a German born physicist working in America. Bohr was Danish. Bell was Irish, working at a pan European laboratory in Switzerland. Clauser and Freedman were American. Aspect was French. Zeilinger was Austrian, and his most significant doctoral student, Jian Wei Pan, was Chinese, trained partly in Vienna before returning home to build the research program that eventually produced the Micius satellite. Hanson's decisive 2015 experiment was Dutch, performed in close intellectual dialogue with simultaneous experiments in Colorado and Vienna. This is not a story that belongs to any single nation or any single scientific tradition. It is a genuinely global, multi generational collaboration, conducted across political boundaries that shifted dramatically over the ninety years the story spans, including a world war, a cold war, and the recent reemergence of serious great power competition specifically focused on quantum technology itself. The physics of entanglement does not care about any of these boundaries, and the physicists who uncovered it, whatever their individual nationalities and the political circumstances of their individual eras, were engaged in a shared project that transcended all of them, each contributing a piece that only makes full sense in light of the pieces that came before and after.

In the epilogue that follows, I want to leave you with a final image, one that returns us to where this whole story began, and to the question that a friend of mine asked me over dinner, a question I still do not think has a fully settled answer, and that I suspect, if you have read this far, you may now be equipped to sit with more productively than you could have before you opened this book's cover.

EPILOGUE

The Question We Are Still Asking

Somewhere above your head right now, whatever time of day or night you happen to be reading these final pages, a satellite is likely passing overhead carrying, among its various instruments, at least one atomic clock keeping time with an accuracy that would have seemed like fantasy to the physicists who gathered in that Brussels hotel in 1927. Somewhere in a laboratory, possibly on a different continent entirely, a pair of photons is being coaxed into existence from a nonlinear crystal, sent along separate paths toward separate detectors, ready to violate a mathematical inequality that a quiet Irish physicist wrote on paper more than sixty years ago. Somewhere, a graduate student not so different from the young John Clauser, or the young Alain Aspect, is encountering this material for the first time, in a lecture hall or a textbook or perhaps, now, in a book very much like this one, and beginning to feel the particular, disorienting pull that this subject has exerted on nearly everyone who has taken it seriously since Einstein first sat down to write his objection to it in 1935.

I began this book with Farah and Idris, standing in an airport in Lisbon, a wooden box containing a single coin passing between them before a flight to Tokyo. I want to end, briefly, by returning to them, not because their story ever told the whole truth about entanglement, since I flagged its limitations honestly in the very first chapter, but because I think a good metaphor, even an imperfect one, deserves to be revisited once you have earned a fuller picture of exactly where it succeeds and exactly where it fails.

The coins in Farah and Idris's boxes, we now know with the benefit of everything the intervening chapters have established, do not actually resemble entangled particles in the most important respect. Real entangled particles do not carry a predetermined, fixed answer sealed away and merely awaiting discovery, the way a coin, once placed in its box, simply sits there being whatever it already is. Real entangled particles remain in a genuine, unresolved superposition until the moment of measurement, and the correlation between them, confirmed across five decades of increasingly rigorous experimentation from a Berkeley basement to a satellite in orbit, cannot be explained by any predetermined answer fixed in advance at their shared point of origin. This is the single most important thing this book has tried to establish, patiently, chapter by chapter, and I hope it has landed with the full force it deserves: the universe, at its most fundamental level, is not simply hiding information from us that a cleverer instrument might someday reveal. It is, in a manner that unsettled Einstein for the remainder of his career and that continues to unsettle nearly everyone who encounters it honestly for the first time, genuinely, irreducibly undetermined until the moment observation forces a resolution, and that resolution, for entangled particles, arrives correlated across any distance in a way that no story about shared, predetermined history can fully explain.

What, then, should you actually do with this knowledge, now that you have it. I do not think the honest answer is that your daily life should change in any dramatic, immediately visible way. You will not need to recalculate your morning commute or reconsider your career plans on the basis of Bell inequality violations. But I do think something quieter and, in its own way, more durable can change: a certain expansion of what you consider possible, a loosened grip on the assumption that the world must behave the way your unaided intuition, built for tracking predators and prey across an ancient landscape, insists it should. Entanglement is proof, about as rigorous and as thoroughly tested as proof gets in any domain of human knowledge, that reality does not owe your intuitions anything, and that the proper response to this fact is not despair or confusion but a kind of expanded curiosity, a willingness to keep asking what else might be true that common sense has not yet prepared you to expect.

I think often of John Bell, dying suddenly in 1990 at the age of sixty two, before the fully loophole free experiments that would vindicate his 1964 mathematics beyond any reasonable remaining doubt, and before the Nobel Prize that would eventually, thirty two years after his death, formally recognize the experimental program his work had made possible. I think of Stuart Freedman, gone in 2012, a decade before that same recognition arrived. I think of Einstein himself, dead since 1955, never learning that the phenomenon he named spooky and worked for decades to explain away would become, within a century of his objecting to it, one of the load bearing pillars of an entire emerging technological era, underpinning cryptography, computing, and measurement in ways he could not have anticipated even as he was, unknowingly, laying the theoretical groundwork for all of it through the very paper he wrote to refute it.

This is, I think, the deepest lesson this story has to offer, deeper even than the physics itself, remarkable as that physics is. The questions worth asking are not always the ones that promise a clean, comfortable answer, or even an answer that arrives within a single lifetime. Einstein spent his final decades defending a position that would eventually be shown, empirically and decisively, to be mistaken, and he is remembered, rightly, not as a cautionary tale about stubbornness but as one of the two indispensable architects, alongside Bohr, of one of science's great unresolved questions, a question whose eventual resolution required the patient, cumulative work of physicists across four continents and nearly a full century.

You do not need to become a physicist to carry a version of that patience with you. You need only remain willing, as Bell was, as Clauser was, as Aspect and Zeilinger and Pan and Hanson were, each in turn, to take a question seriously even when the people around you have already decided it is settled, or unfashionable, or not worth the years it might take to answer properly. Somewhere in the world, right now, a question exactly that shape is waiting for someone willing to ask it again, more carefully, with better tools than the last person had available. I hope this book has, in some small way, made you more willing to be that person, whatever form that question eventually takes in your own life.

The particles are still whispering to each other, across whatever distance separates them, in a language that took physics the better part of a century to learn to read. It is, I think, one of the more beautiful things we have ever learned to overhear.

A Glossary of Whispers

Bell inequality: A mathematical limit, derived by physicist John Stewart Bell in 1964, on the strength of correlation that can exist between two separated systems if their behavior is governed by local hidden variables. Quantum mechanics predicts, and experiments confirm, that entangled particles violate this limit.

Bell test: Any experiment designed to measure whether the correlations between entangled particles violate a Bell inequality, thereby testing whether nature obeys local realism or the stranger predictions of quantum mechanics.

CHSH inequality: A practical, experimentally testable version of Bell's original inequality, developed in 1969 by John Clauser, Michael Horne, Abner Shimony, and Richard Holt.

Copenhagen interpretation: The interpretation of quantum mechanics developed largely by Niels Bohr and his colleagues, holding that the wave function represents complete information about a system and that quantum indeterminacy is a genuine, irreducible feature of nature rather than a gap in our knowledge.

Decoherence: The process by which a quantum system's delicate superposition breaks down due to interaction with its surrounding environment, explaining why large objects such as cats, chairs, and people do not display observable quantum superposition in everyday life.

Detection loophole: A gap in early Bell test experiments arising from imperfect photon detectors that failed to register every particle, leaving open the possibility that the detected subset of particles was not representative of the whole.

Entanglement: A quantum mechanical phenomenon in which two or more particles share a single, unified description that cannot be separated into independent descriptions of each particle individually, such that measuring one particle instantaneously affects what can be known about its partner, regardless of distance.

EPR paper: The 1935 paper by Albert Einstein, Boris Podolsky, and Nathan Rosen arguing that quantum mechanics must be an incomplete description of physical reality, which introduced the thought experiment that gave entanglement its name.

Hidden variables: Hypothetical predetermined properties that some physicists, including Einstein, proposed might underlie the apparent randomness of quantum mechanics, restoring a fully deterministic picture of reality beneath the quantum description.

Local realism: The combined assumption that objects possess definite properties independent of measurement, and that no influence can travel between separated objects faster than the speed of light. Bell test experiments have ruled out local realism as a complete description of nature.

Locality loophole: A gap in early Bell test experiments arising from measurement settings that were not changed quickly enough, or randomly enough, to rule out the possibility of an undetected, slower than light signal coordinating the results between two measurement stations.

Many worlds interpretation: An interpretation of quantum mechanics, proposed by Hugh Everett in 1957, holding that every possible outcome of a quantum measurement is realized, each in its own separate, non communicating branch of reality.

Photon: A single, discrete particle of light, frequently used in entanglement experiments because photons interact only weakly with their environment and can travel long distances while preserving their delicate quantum state.

Qubit: The basic unit of information in a quantum computer, capable of existing in superposition of the zero and one states, unlike a classical bit, which must be either zero or one.

Quantum key distribution: A cryptographic technique that uses the quantum properties of photons to allow two parties to establish a shared secret key, with any eavesdropping attempt detectable due to the unavoidable disturbance quantum measurement causes.

Schrodinger's cat: A thought experiment proposed by Erwin Schrodinger in 1935, illustrating the apparent absurdity of applying quantum superposition to macroscopic objects, in which a sealed cat's fate is linked to a genuinely random quantum event.

Superposition: The quantum mechanical condition in which a system exists as a genuine mathematical blend of multiple possible states simultaneously, prior to measurement, rather than being in one definite but merely unknown state.

Uncertainty principle: Formulated by Werner Heisenberg in 1927, the principle that certain pairs of physical properties, such as position and momentum, cannot both be measured with unlimited precision simultaneously, a limit built into the structure of nature rather than a limitation of measuring instruments.

Wave function: The mathematical object in quantum mechanics that describes everything that can be known about a quantum system, including the probabilities of different measurement outcomes.

Notes on Sources and Further Reading

This book draws on the published scientific record surrounding quantum entanglement, including original research papers, official materials from the Nobel Prize committee, and reporting from established scientific publications. Readers wishing to go deeper into any of the topics covered in these chapters may find the following starting points useful.

On the foundational papers themselves: Albert Einstein, Boris Podolsky, and Nathan Rosen, Can Quantum Mechanical Description of Physical Reality Be Considered Complete, published in Physical Review in 1935, remains available through most university library systems and forms the starting point for the entire debate covered in this book. John Stewart Bell's own 1964 paper, On the Einstein Podolsky Rosen Paradox, originally published in Physics Physique Fizika, has been reprinted in a collection of Bell's own writing titled Speakable and Unspeakable in Quantum Mechanics, which remains one of the clearest firsthand accounts of his reasoning available to a careful general reader.

On the history and experimental development of Bell tests, the official scientific background documents published by the Nobel Prize committee at nobelprize.org, particularly the 2022 popular science background and advanced scientific background documents covering the award to Alain Aspect, John Clauser, and Anton Zeilinger, offer an authoritative and accessible summary written for a broad audience, including working physicists outside the immediate specialty.

Readers interested in the original experimental papers may wish to seek out Stuart Freedman and John Clauser's 1972 paper in Physical Review Letters, Experimental Test of Local Hidden Variable Theories, Alain Aspect, Jean Dalibard, and Gerard Roger's 1982 paper in the same journal, Experimental Test of Bell's Inequalities Using Time Varying Analyzers, Gregor Weihs and colleagues' 1998 paper, also in Physical Review Letters, Violation of Bell's Inequality Under Strict Einstein Locality Conditions, and the trio of 2015 loophole free experiments published in Nature by Ronald Hanson's group at Delft, and in Physical Review Letters by the groups led by Krister Shalm at the National Institute of Standards and Technology and by Anton Zeilinger's collaboration in Vienna.

On the Micius satellite program, Jian Wei Pan and colleagues' 2017 paper in Science, Satellite Based Entanglement Distribution Over 1200 Kilometers, provides the original technical account, while the review article Micius Quantum Experiments in Space, published in Reviews of Modern Physics in 2022, offers a comprehensive summary of the full program's achievements written for a physics literate but non specialist audience.

On quantum computing, the National Institute of Standards and Technology and the individual research pages maintained by IBM Quantum, Google Quantum AI, and IonQ provide current, regularly updated technical overviews of ongoing hardware development, appropriate for readers wanting the most current information beyond what any printed book can offer given the pace of change in this particular field.

On quantum sensing and atomic clocks, the National Institute of Standards and Technology maintains a public facing overview of atomic clock research, including its connections to entanglement, at its official website, offering regularly updated summaries of ongoing research from JILA and collaborating institutions.

Readers seeking a broader, more general history of quantum mechanics and its philosophical implications may find value in Manjit Kumar's Quantum: Einstein, Bohr, and the Great Debate About the Nature of Reality, which covers much of the same historical ground as the early chapters of this book in greater depth, and in Louisa Gilder's The Age of Entanglement: When Quantum Physics Was Reborn, which offers an unusually vivid, character driven account of the physicists whose personal and professional lives intersected across the decades this book has traced.

Finally, readers with access to an academic library or willing to seek out preprint archives such as arXiv.org can find the full text of most of the specific research papers referenced throughout this book, allowing direct engagement with the primary scientific literature underlying the narrative told here.

About the Author

Lior Bennathorne writes narrative nonfiction focused on making difficult scientific ideas genuinely accessible without diluting their substance. Trained originally outside of physics, Bennathorne has spent more than a decade translating dense, technical subject matter into prose intended for curious general readers who want to understand not just what scientists have discovered but why those discoveries matter and how they were made. Whispers Between Particles is Bennathorne's contribution to that ongoing effort, born from a conviction that the story of quantum entanglement, properly told, deserves a place alongside the great narrative histories of scientific discovery.

Fluid Mechanics, 2nd Edition

Topic Editors

Sihui Hong
Chaobin Dang
Shuangfeng Wang

MDPI

Basel • Beijing • Wuhan • Barcelona • Belgrade • Novi Sad • Cluj • Manchester

Topic Editors
Sihui Hong
School of Electrical Engineering
Zhejiang University
Hangzhou
China

Chaobin Dang
Graduate School of Engineering
University of Fukui
Fukui
Japan

Shuangfeng Wang
Key Lab of Heat Transfer and Energy Conservation of Education Ministry
South China University of Technology
Guangzhou
China

Editorial Office
MDPI AG
Grosspeteranlage 5
4052 Basel, Switzerland

This is a reprint of the Topic, published open access by the journals *Applied Sciences* (ISSN 2076-3417), *Energies* (ISSN 1996-1073), *Fluids* (ISSN 2311-5521), *Materials* (ISSN 1996-1944), *Mathematics* (ISSN 2227-7390) and *Micromachines* (ISSN 2072-666X), freely accessible at: https://www.mdpi.com/topics/8K79WPF645.

For citation purposes, cite each article independently as indicated on the article page online and as indicated below:

Lastname, A.A.; Lastname, B.B. Article Title. *Journal Name* **Year**, *Volume Number*, Page Range.

ISBN 978-3-7258-7725-6 (Hbk)
ISBN 978-3-7258-7726-3 (PDF)
https://doi.org/10.3390/books978-3-7258-7726-3

Contents

Σ mathematics

MDPI

Article

A Physical Insight into Computational Fluid Dynamics and Heat Transfer

Sergey I. Martynenko * and Aleksey Yu. Varaksin

Joint Institute for High Temperatures of the Russian Academy of Sciences, Moscow 125412, Russia
* Correspondence: martyn_s@mail.ru or martynenko@icp.ac.ru; Tel.: +7-(495)-485-80-90

Abstract: Mathematical equations that describe all physical processes are valid only under certain assumptions. One of them is the minimum scales used for the given description. In fact, this prohibits the use of derivatives in the mathematical models of the physical processes. This article represents a derivative-free approach for the mathematical modelling. The proposed approach for CFD and numerical heat transfer is based on the conservation and phenomenological laws, and physical constraints on the minimum problem-dependent spatial and temporal scales (for example, on the average free path of molecules and the average time of their collisions for gases). This leads to the derivative-free governing equations (the discontinuum approximation) that are very convenient for numerical simulation. The theoretical analysis of governing equations describing the fundamental conservation laws in the continuum and discontinuum approximations is given. The article demonstrates the derivative-free approach based on the correctly defined macroparameters (pressure, temperature, density, etc.) for the mathematical description of physical and chemical processes. This eliminates the finite-difference, finite-volume, finite-element or other approximations of the governing equations from the computational algorithms.

Keywords: numerical simulation; heat and mass transfer; CFD; Knudsen number; Boltzmann equation; finite volume method

MSC: 76-10; 76A02; 76M12

1. Introduction: Three Eras in CFD

Three eras can be distinguished in the history of fluid dynamics depending on the governing equations and methods of their solution: analytical era (from 17th century to first half of the 20th century), transitional era (from first half of the 20th century to present) and computational era (starts today) [1–3].

The main goal of scientists in the analytical era of fluid dynamics was the establishment and analysis of the governing equations, verification of their solutions using experimental data and application of the obtained results for solving engineering problems. Johann Bernoulli, Daniel Bernoulli, Leonhard Euler, Jean-Baptiste le Rond d'Alembert and their followers considered an abstract mathematical object (infinitesimal volume of fluid) for describing the fluid macroflows. In fact, a fictitious continuous medium (also called a continuum) continuously distributed in space and time instead of the real liquid or gas was used for solving macroproblems of the fluid dynamics by known analytical methods of 17th century. This incorrect approach makes it possible to use the differential and integral calculus not only in theoretical (mathematical) fluid dynamics, but also in all continuum mechanics. Since any fraction of the continuum is continuum, only the sufficient differentiable functions can be used to describe the continuum motion by the differential equations.

Modern science distinguishes three structural levels of matter, as follows:

(1) The microcosm is composed of molecules, atoms, elementary particles, i.e., space of extremely small microobjects from 10^{-18} up to 10^{-8} metres. The characteristic distance of studied physical phenomena are called scale, and various measures (from the Planck length up to the micrometre) are used as microscales. The molecular and atomic physics describe the microcosm phenomena (heat conduction, phase transitions, chemical reactions, etc.) in the scale range of 10^{-18}–10^{-8} metres;

(2) The macrocosm is a space of objects comparable to humans. Metre, second, kilogram and degree Kelvin are used as macroscales of length, time, mass and temperature;

(3) The megacosm involves planets, stars, galaxies, metagalaxies, i.e., huge cosmic scales and speeds where the distance is measured in light years and the time of existence of cosmic objects is millions and billions of years.

The differential and integral calculus formed the basis of mathematics of the XVII century. The application of this calculus to solve engineering problems required non-physical assumptions. The main assumption is a continuity hypothesis, i.e., a fictitious continuous medium (continuum) continuously distributed in space and in time is used instead of the real liquid or gas. The continuum approximation ignores molecular structure of substances, but the continuum has physical properties of a real liquid or gas on the macroscale lengths. In spite of this theoretical inaccuracy, the continuum approximation coupled with phenomenological laws makes it possible to solve many applied problems. Namely, the continuum allows us to constrain the volume of a liquid or gas to an interior point to obtain partial derivative equations that contradict the macrocosm laws.

The continuity hypothesis allows for using the partial differential equations to describe the properties and fluid flows. It was a new type of differential equations, and Euler described some of their types and also formulated methods for their integration. In 1755, Euler formulated the equations of an inviscid incompressible fluid flows (so called «Euler equations»)

$$\nabla\vec{V} = 0,$$
$$\rho\frac{d\vec{V}}{dt} = \nabla p,$$

where $\rho = const$ is the density of liquid or gas. These equations expresses the conservation of mass and momentum.

In 1822, Claude Louis Marie Henri Navier formulated the equations of a viscous incompressible fluid flows. Navier introduced an additional term accounting the viscosity of liquid or gas to the momentum equations. In 1829, Siméon Denis Poisson, using model concepts of molecular forces, formulated viscous compressible fluid flows. Later, a phenomenological derivation of these equations was given by Adhémar Jean Claude Barré de Saint-Venant and George Gabriel Stokes:

$$\frac{\partial\rho}{\partial t} + \nabla\cdot(\rho\vec{V}) = 0, \tag{1a}$$

$$\frac{\partial(\rho\vec{V})}{\partial t} + (\rho\vec{V}\cdot\nabla)\vec{V} = -\nabla p + \eta\,\triangle\vec{V} + \left(\zeta + \frac{\eta}{3}\right)\nabla div\,\vec{V}, \tag{1b}$$

where t is time, ∇ is the nabla operator, $\triangle$ is the vector Laplace operator, η is the dynamic viscosity coefficient, ζ is the second viscosity, ρ is the density, p is the pressure, and $\vec{V} = (v^1, \ldots, v^n)$ is the velocity vector. The unknown p and $\vec{V}$ are the functions of time t and coordinate $x \in \Omega$, where $\Omega \subset \mathbb{R}^n$, $n = 2,3$ is a domain where the fluid or gas is moving.

The differential and integral calculus defines the differential form of the governing equations in spite of rough physical assumptions. The nonlinear Navier–Stokes equations can be solved analytically in the simplest cases, so some semi-empirical theories such as the boundary layer approach have been proposed and developed to simplify mathematical description of the fluid flows [4,5].

The transitional era of fluid dynamics is based on the solver replacement. Approximate numerical methods for solving the (initial-)boundary value problems have replaced the differential and integral calculus. Computational physics is a branch of theoretical physics that uses numerical analysis and data structures to analyse and solve problems that involve various physical and chemical processes [6,7].

The current computational fluid dynamics (CFD) focuses on the more accurate numerical solution of governing partial differential equations (PDEs) [8]. For the given purpose, the (initial-)boundary value problems are replaced by their discrete analogues. The current target of computational mathematics is to obtain an sufficiently accurate approximation to the exact solution of the governing PDEs as the mesh size and time step approach to zero. This contradicts theoretical physics. The individual molecules have the standard physical microparameters: mass, momentum and energy. The standard physical macroparameters are defined by averaged microparameters of the individual molecules in a finite volume. For example, the density of a gas is the sum of the mass of the molecules divided by the finite volume which the gas occupies. The pressure and temperature of a gas are measure of the mean linear momentum and the mean kinetic energy of the molecules in the finite volume, respectively. It is clear that the averaging of the microparameters is correct only if this finite volume contains a sufficiently large number of molecules.

It is necessary to clarify the contradiction between the computational mathematics and the computational physics. Let the dimensions of the finite volumes be such that the gas or liquid contained in it can be considered as a continuous medium. The geometric scale is the representative size $\hbar$ of the least volume, i.e., smaller scale processes cannot be distinguished. Therefore, it not possible to use classic definition of the derivative

$$g'(a) = \lim_{h\to 0} \frac{g(a+h) - g(a)}{h}$$

in point a for description of the physical processes because $\hbar \gg h$. On the other hand, the temperature, pressure or density of a gas at the point a cannot be defined according to classical physical theories. The derivatives and the macroparameters (density, pressure, temperature, etc.) cannot be correctly defined simultaneously [9].

The computational era is based on a dichotomy between the correct definition of derivatives and macroparameters. Since modern mathematical modelling does not imply the analytical solution of governing differential equations, derivative-free models with correctly defined macroparameters should be employed for the numerical investigation of physical phenomena. In this case, the mathematical model actually coincides with the finite volume-based scheme [10–12].

The goal of this article is to demonstrate the influence of correctly defined macroparameters on a mathematical description of the physical and chemical processes for computer-based solvers. The results obtained will be useful for black-box software used for numerical simulation of various physical and chemical processes.

2. The Discontinuum Approximation of Conservation Laws

First, let us consider the phenomenological derivation of the continuity equation, which expresses the fundamental law of mass conservation in the continuum approximation. Density is the most important parameter for describing flows of a compressible liquid or gas. Next, we will discuss and analyse the disadvantages of the classical continuity equation.

To illustrate the phenomenological technique, an arbitrary finite volume V in a continuous medium can be used [13,14]. Since the resulting equations depend on the volume shape, we will choose a rectangular parallelepiped as the volume V for clarity:

$$V = \{(t,x,y,z) \mid 0 \leqslant t \leqslant h_t,\ 0 \leqslant x \leqslant h_x,\ 0 \leqslant y \leqslant h_y,\ 0 \leqslant z \leqslant h_z\}. \qquad (2)$$

Here h_x, h_y, h_z and h_t denote the spatial and temporal sizes of the volume V (Figure 1), respectively.

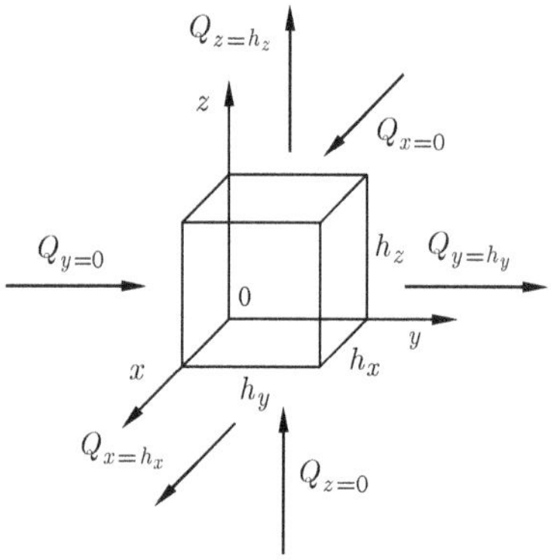

Figure 1. The finite volume$V$ in space.

The mass conservation is written as

$$M(t) = M(0) + Q_{x=0} + Q_{y=0} + Q_{z=0} - Q_{x=h_x} - Q_{y=h_y} - Q_{z=h_z} + S, \tag{3}$$

where $M(0)$ is mass of the liquid or gas in the volume V (2) in starting time $t = 0$, $M(t)$ is mass of the liquid or gas in the volume V (2) in time t $(0 \leqslant t \leqslant h_t)$, $Q_{x=0}$, $Q_{y=0}$ and $Q_{z=0}$ refer to the mass of the liquid or gas going to V (2) in time t $(0 \leqslant t \leqslant h_t)$ through the faces $F_{x=0}$, $F_{y=0}$ and $F_{z=0}$

$$F_{x=0} = \{(0, y, z) \,|\, 0 \leqslant y \leqslant h_y,\ 0 \leqslant z \leqslant h_z\}, \tag{4a}$$
$$F_{y=0} = \{(x, 0, z) \,|\, 0 \leqslant x \leqslant h_x,\ 0 \leqslant z \leqslant h_z\}, \tag{4b}$$
$$F_{z=0} = \{(x, y, 0) \,|\, 0 \leqslant x \leqslant h_x,\ 0 \leqslant y \leqslant h_y\}, \tag{4c}$$

$Q_{x=h_x}$, $Q_{y=h_y}$ and $Q_{z=h_z}$ refer to the mass of the liquid or gas coming from V for t $(0 \leq t \leq h_t)$ through the faces $F_{x=h_x}$, $F_{y=h_y}$ and $F_{z=h_z}$

$$F_{x=h_x} = \{(h_x, y, z) \,|\, 0 \leqslant y \leqslant h_y,\ 0 \leqslant z \leqslant h_z\}, \tag{4d}$$
$$F_{y=h_y} = \{(x, h_y, z) \,|\, 0 \leqslant x \leqslant h_x,\ 0 \leqslant z \leqslant h_z\}, \tag{4e}$$
$$F_{z=h_z} = \{(x, y, h_z) \,|\, 0 \leqslant x \leqslant h_x,\ 0 \leqslant y \leqslant h_y\}, \tag{4f}$$

S is a source term. Let us also assume that $S = 0$ and the shape of the volume V (2) remains unchanged.

The mass of the liquid or gas going to V (2) through the faces (4) for time t is

$$Q_{x=0} = \int_0^t m_{x=0}(\xi)\, d\xi, \qquad Q_{y=0} = \int_0^t m_{y=0}(\xi)\, d\xi, \qquad Q_{z=0} = \int_0^t m_{z=0}(\xi)\, d\xi, \tag{5a}$$

$$Q_{x=h_x} = \int_0^t m_{x=h_x}(\xi)\, d\xi, \quad Q_{y=h_y} = \int_0^t m_{y=h_y}(\xi)\, d\xi, \quad Q_{z=h_z} = \int_0^t m_{z=h_z}(\xi)\, d\xi, \tag{5b}$$

where $m_{x=0}(t)$, $m_{y=0}(t)$, $m_{z=0}(t)$, $m_{x=h_x}(t)$, $m_{y=h_y}(t)$ and $m_{z=h_z}(t)$ is the mass flow rates through the faces (4).

The mass balance (3) can be rewritten as

$$\frac{1}{h_t}\left(\frac{M(h_t)}{\mathrm{V}}-\frac{M(0)}{\mathrm{V}}\right)+\frac{1}{h_t}\int\limits_0^{h_t}\frac{m_{x=h_x}(t)-m_{x=0}(t)}{h_xh_yh_z}\,dt \tag{6}$$
$$+\frac{1}{h_t}\int\limits_0^{h_t}\frac{m_{y=h_y}(t)-m_{y=0}(t)}{h_xh_yh_z}\,dt+\frac{1}{h_t}\int\limits_0^{h_t}\frac{m_{z=h_z}(t)-m_{z=0}(t)}{h_xh_yh_z}\,dt=0,$$

where $\mathrm{V}=h_xh_yh_z$.

The classical definition of the continuous medium density (ρ) is ratio of the medium mass M in the finite volume V on $\mathrm{V}=h_xh_yh_z$, if $\max(h_t,h_x,h_y,h_z)\to 0$

$$\lim_{V\to 0}\frac{M(V)}{V}=\rho(t,x,y,z). \tag{7}$$

Limiting $\max(h_t,h_x,h_y,h_z)\to 0$ and taking into account that

$$\lim_{h\to 0}\frac{1}{h}\int\limits_0^h f(x)\,dx=f(0),$$
$$\lim_{h\to 0}\frac{f(h)-f(0)}{h}=f'(0),$$

Equation (6) becomes

$$\left.\frac{\partial\rho}{\partial t}\right|_{\substack{t=0\\x=0\\y=0\\z=0}}+\left.\frac{\partial(\rho u)}{\partial x}\right|_{\substack{t=0\\x=0\\y=0\\z=0}}+\left.\frac{\partial(\rho v)}{\partial y}\right|_{\substack{t=0\\x=0\\y=0\\z=0}}+\left.\frac{\partial(\rho w)}{\partial z}\right|_{\substack{t=0\\x=0\\y=0\\z=0}}=0.$$

Due to arbitrary choice of the finite volume V (2), this continuity equation will be valid for any point in domain where the liquid or gas flows. Therefore, the continuity equation takes the final form (1a) [15,16].

It is clear that the density cannot be defined correctly as (7). Let V and $M(V)$ be a finite volume and mass of liquid or gas in this volume, respectively. The shrinking of the finite volume to internal point results in uncertainty

$$\lim_{V\to 0}\frac{M(V)}{V}=\left[\frac{0}{0}\right],$$

because $M(V)\to 0$ as $V\to 0$. In addition, the pressure and temperature of a gas are the measure of the mean momentum and the mean kinetic energy of the molecules in the finite volume, respectively. Hence, the mass conservation equation must be used in form of (6), i.e., without shrinking of the finite volume to an internal point.

The basic idea is to generate a computational grid and to formulate the governing equations over the finite volumes of this grid. The eight sets of points

$$\alpha_l^{\mathrm{v}}=\frac{l-1}{\aleph_\alpha}=(l-1)h_\alpha,\qquad l=1,2,\ldots,\aleph_\alpha,\aleph_\alpha+1, \tag{8a}$$
$$\alpha_l^{\mathrm{f}}=\frac{2l-1}{2\aleph_\alpha}=(2l-1)\frac{h_\alpha}{2},\quad l=1,2,\ldots,\aleph_\alpha-1,\aleph_\alpha, \tag{8b}$$

define a Cartesian grid, where $h_\alpha=1/\aleph_\alpha$ and $\alpha=(t\ x\ y\ z)^T$. The points $(t_n^{\mathrm{v}},x_i^{\mathrm{v}},y_j^{\mathrm{v}},z_k^{\mathrm{v}})$ are the grid vertices and the finite volume is defined by

$$\mathcal{V}_{ijk}^{(n)}=\{(t,x,y,z)\mid t_{n-1}^{\mathrm{f}}<t<t_n^{\mathrm{f}},\ x_{i-1}^{\mathrm{f}}<x<x_i^{\mathrm{f}},\ y_{j-1}^{\mathrm{f}}<y<y_j^{\mathrm{f}},\ z_{k-1}^{\mathrm{f}}<z<z_k^{\mathrm{f}}\}, \tag{9}$$

where the points t_n^{f}, x_i^{f}, y_j^{f} and z_k^{f} define the finite volume faces. The mesh size and the minimum spatial scale $\hbar_s$ satisfy to

$$\min(h_x, h_y, h_z) \geq \hbar_s \gg l_{\mathrm{mol}}, \tag{10}$$

where l_{mol} is the mean free path of a molecule of the liquid or gas. The minimum size of the finite volume $\mathcal{V}_{ijk}^{(n)}$ (9) is chosen such that

$$\mathsf{Kn} = \frac{l_{\mathrm{mol}}}{\min(h_x, h_y, h_z)} < 10^{-3},$$

where Kn is the Knudsen number, i.e., the medium inside $\mathcal{V}_{ijk}^{(n)}$ is continuous.

The time step for the numerical solution of non-stationary equations should be limited by

$$h_t \geq \hbar_t \gg t_{mol} > 0, \tag{11}$$

where t_{mol} is the average time between collisions of two molecules and $\hbar_t$ is the minimum time scale.

Since all macroparameters are defined over the finite volumes (i.e., all functions are constant inside $\mathcal{V}_{ijk}^{(n)}$ (9)), the functions ρ, p, T can be assigned to any point of the volume (for example, to the vertex $(t_n^{\mathrm{v}}, x_i^{\mathrm{v}}, y_j^{\mathrm{v}}, z_k^{\mathrm{v}}) \in \mathcal{V}_{ijk}^{(n)}$). Due to limitations (10) and (11), the density

$$\frac{M(\mathcal{V}_{ijk}^{(n)})}{\mathrm{V}} = \rho_{ijk}^{(n)}$$

is defined correctly.

Note that the discontinuous conservative variables (ρ, p, T, etc.) are assigned to the vertices, but the continuous flux variables (ρu, ρv, ρw, etc.) are assigned to the faces of $\mathcal{V}_{ijk}^{(n)}$ (9). In general, it is necessary to express the flux variables in terms of the conservative ones using some interpolation.

The linear interpolation of the flux variables and the composite method for the integral evaluation leads to final form of the law of conservation of mass over the finite volume (9)

$$\begin{aligned} &\frac{\rho_{ijk}^{(n+1)} - \rho_{ijk}^{(n)}}{h_t} + \frac{1}{2}\left(\frac{\rho_{i+1jk}^{(n)} u_{i+1jk}^{(n)} - \rho_{i-1jk}^{(n)} u_{i-1jk}^{(n)}}{2h_x} + \frac{\rho_{i+1jk}^{(n+1)} u_{i+1jk}^{(n+1)} - \rho_{i-1jk}^{(n+1)} u_{i-1jk}^{(n+1)}}{2h_x}\right) \\ &\quad + \frac{1}{2}\left(\frac{\rho_{ij+1k}^{(n)} u_{ij+1k}^{(n)} - \rho_{ij-1k}^{(n)} u_{ij-1k}^{(n)}}{2h_y} + \frac{\rho_{ij+1k}^{(n+1)} u_{ij+1k}^{(n+1)} - \rho_{ij-1k}^{(n+1)} u_{ij-1k}^{(n+1)}}{2h_y}\right) \\ &\quad + \frac{1}{2}\left(\frac{\rho_{ijk+1}^{(n)} u_{ijk+1}^{(n)} - \rho_{ijk-1}^{(n)} u_{ijk-1}^{(n)}}{2h_z} + \frac{\rho_{ijk+1}^{(n+1)} u_{ijk+1}^{(n+1)} - \rho_{ijk-1}^{(n+1)} u_{ijk-1}^{(n+1)}}{2h_z}\right) = 0, \end{aligned} \tag{12}$$

where $h_t = t_{n+1}^{\mathrm{v}} - t_n^{\mathrm{v}}$, $h_x = x_i^{\mathrm{f}} - x_{i-1}^{\mathrm{f}}$, $h_y = y_j^{\mathrm{f}} - y_{j-1}^{\mathrm{f}}$, $h_z = z_k^{\mathrm{f}} - z_{k-1}^{\mathrm{f}}$ are the mesh sizes, and $\mathrm{V} = h_x h_y h_z$. Since all integrable functions are piecewise constant, all integrals are evaluated exactly. Really, it is similar to the Crank–Nicolson scheme with the piecewise constant functions (the conservative variables). From a physical point of view, the explicit scheme

$$\frac{\hat{\rho}_{ijk}^{(n+1)} - \hat{\rho}_{ijk}^{(n)}}{h_t} + \left(\frac{\partial(\rho u)}{\partial x}\right)_{ijk}^{(n)} + \left(\frac{\partial(\rho v)}{\partial y}\right)_{ijk}^{(n)} + \left(\frac{\partial(\rho w)}{\partial z}\right)_{ijk}^{(n)} = 0,$$

and the implicit scheme

$$\frac{\hat{\rho}_{ijk}^{(n+1)} - \hat{\rho}_{ijk}^{(n)}}{h_t} + \left(\frac{\partial(\rho u)}{\partial x}\right)_{ijk}^{(n+1)} + \left(\frac{\partial(\rho v)}{\partial y}\right)_{ijk}^{(n+1)} + \left(\frac{\partial(\rho w)}{\partial z}\right)_{ijk}^{(n+1)} = 0,$$

are not accurate, since the mass transfer $((\rho u)'_x + (\rho v)'_y + (\rho w)'_z)$ occurs only at one time step. The law of conservation of mass (6) on the unstaggered grid is given in [17].

The laws of conservation of momentum and energy over the finite volume $\mathcal{V}_{ijk}^{(n)}$ (9) can be formulated in the same way.

3. Remarks on Phenomenological Laws

Many phenomenological laws (Fourier's law, Fick's law, Newton's law and others) have a gradient-based form. Due to limitations (10) and (11), the integral form of the phenomenological laws should be used in the governing equations. For example, instead of differential Fourier's law

$$\vec{q} = -\lambda(T)\nabla T$$

the integral form

$$q = \langle\lambda\rangle_i \frac{T_{i+1}^{(n)} - T_i^{(n)}}{x_{i+1}^{\mathrm{v}} - x_i^{\mathrm{v}}}$$

should be used in the energy equations, where

$$\langle\lambda\rangle_i = 2\frac{\lambda(T_i^{(n)})\,\lambda(T_{i+1}^{(n)})}{\lambda(T_i^{(n)}) + \lambda(T_{i+1}^{(n)})}.$$

is the averaging coefficient of thermal conductivity $\lambda(T)$. The interface temperature T^* in the contact point $x^* = (x_i^{\mathrm{v}} + x_{i+1}^{\mathrm{v}})/2$ is defined by

$$T^* = \frac{\lambda(T_i^{(n)})T_i^{(n)} + \lambda(T_{i+1}^{(n)})T_{i+1}^{(n)}}{\lambda(T_i^{(n)}) + \lambda(T_{i+1}^{(n)})}.$$

4. Numerical Experiments

First of all, it is necessary to emphasise the difference between continuum and discontinuum approximations using the simplest 1D equations. Let an equation in the continuum mechanics

$$u' = f(x)$$

be an exact problem and a discrete equation

$$\frac{u(x_i^{\mathrm{f}}) - u(x_{i-1}^{\mathrm{f}})}{h} = f(x_i^{\mathrm{v}})$$

be an approximate (discrete) problem. Here, x_i^{v} and x_i^{f} are the vertices and the finite volume faces of the uniform grid (8), and h is a mesh size. The difference between the differential equation and their discrete analogue on $[x_{i-1}^{\mathrm{f}}, x_i^{\mathrm{f}}]$ becomes

$$\mathcal{E} = \max_{[x_{i-1}^{\mathrm{f}}, x_i^{\mathrm{f}}]} \left| u'(x_i^{\mathrm{v}}) - \frac{u(x_i^{\mathrm{f}}) - u(x_{i-1}^{\mathrm{f}})}{h} \right|.$$

In the continuum approximation, the differential equations are assumed to be exact problems, but their discrete analogues are assumed to be approximate problems. If $u(x)$

is a sufficiently differentiable function and $|f^{(n)}(x)| < Const$, one can apply a Taylor expansion to $u(x^{\mathrm{f}}_{i}) - u(x^{\mathrm{f}}_{i-1})$ with respect to h

$$\mathcal{E}(h) \approx \frac{h^2}{24} \max_{[x^{\mathrm{f}}_{i-1}, x^{\mathrm{f}}_{i}]} |f'(x^{\mathrm{v}}_{i})|.$$

The method is said to be convergent if $\mathcal{E}(h) \to 0$ as $h \to 0$, i.e., ultimately converging to the corresponding solution of the original differential equation at every vertex of the computational grid (8).

In the discontinuum approximation, the discrete equations are assumed to be exact problems, but their differential analogues as $h \to 0$ are assumed to be approximate problems. In this case, the error estimation becomes

$$\frac{\hbar_s^2}{24} \min_{[x^{\mathrm{f}}_{i-1}, x^{\mathrm{f}}_{i}]} |f'(x^{\mathrm{v}}_{i})| \leqslant \mathcal{E} \leqslant \frac{h^2}{24} \max_{[x^{\mathrm{f}}_{i-1}, x^{\mathrm{f}}_{i}]} |f'(x^{\mathrm{v}}_{i})|,$$

where $\hbar_s$ is the least space scale. Since $h \geq \hbar_s > 0$ and $f^{(n)}(x) \neq 0$, the solutions of discrete and differential problems are generally different. It means that the exact solution of the Navier–Stokes (1) is considered as an approximate solution in discontinuum mechanics.

We start our discussion with the discrete boundary value problem

$$\frac{u^h_{i-1} - 2u^h_i + u^h_{i+1}}{h^2} + 4\pi^2\omega^2 \sin(2\pi\omega x^{\mathrm{v}}_i) = 0, \qquad u^h_1 = u^h_{\aleph+1} = 0, \tag{13}$$

where $\omega = 1, 2, 3, \ldots$ is an integer parameter. The approximate boundary value problem

$$u''(x) + 4\pi^2\omega^2 \sin(2\pi\omega x) = 0, \qquad u(0) = u(1) = 0, \tag{14}$$

has the exact solution

$$u(x) = \sin(2\pi\omega x). \tag{15}$$

To illustrate this convergence, the criterion

$$\chi = \max_{i=1,2,\ldots,\aleph+1} |u(x^{\mathrm{v}}_i) - u^h_i| \tag{16}$$

is used to measure the difference between exact solutions of the approximate (14) and the discrete (13) boundary value problems. It is expected that $\chi \leq Ch^2$, where C is some h-independent constant and $h > \hbar$.

The finite volume method is the best discretisation approach for the geometric multigrid methods based on agglomeration of the finite volumes. The Robust Multigrid Technique (RMT) had been proposed and developed for parallel solving nonlinear (initial-)boundary value problems in a black-box manner [18,19]. RMT uses the multiple triple coarsening for problem-independent transfer operators. Figure 2 represents fine (l) and coarse grids ($l-1$ and $l-2$) obtained by the triple coarsening. The fine and coarse grid vertices are given by

$$x^{\mathrm{v}}_{\{i\}} = x^{\mathrm{v}}_i,\ i = 1, 2, \ldots, \aleph_l + 1,$$

where i and $\{i\}$ are the indices of the coarse and fine grid vertices

$$\{i\} = 3i - 2.$$

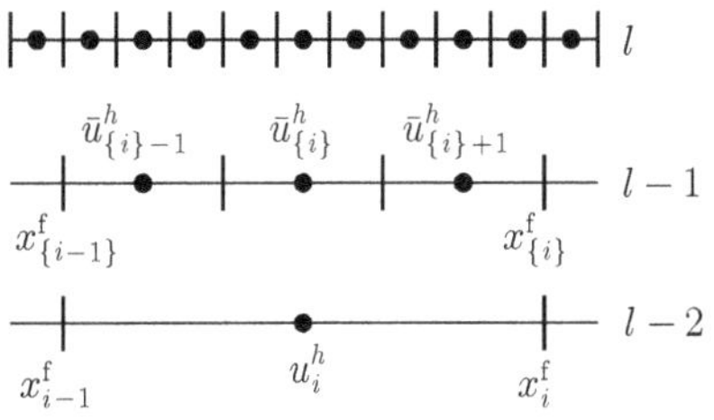

Figure 2. The finer and coarse grids of the Robust Multigrid Technique.

Since each function is constant inside the finite volume (Figure 2), the criterion

$$s = \max_{i=2,3,\ldots,\aleph_{l-2}} \left| u^h_i - \frac{\bar{u}^h_{\{i\}-1} + \bar{u}^h_{\{i\}} + \bar{u}^h_{\{i\}+1}}{3} \right| \tag{17}$$

can be used to measure the grid convergence. It is easy to see that

$$s \leq 10\chi.$$

Figure 3 demonstrates the function (15) and the solution of (13) as a piecewise constant function having only finitely many pieces. The grid convergence in solving the discrete boundary value problem (13) is demonstrated in Figure 4, which shows the reduction of criteria χ and s as a function of the number of grid vertices $\aleph + 1$ ($h = 1/\aleph$) for $\omega = 2$ and $\omega = 6$. The linear behaviour of S is obtained for the $\aleph$-range $10^2 < \aleph < 10^5$. It is expected that $\chi \approx S/10$ in this range, where the criterion s can be measured in computations.

A similar approach based on the mathematical formalism was proposed for Godunov-type schemes. The article shows that mathematical modelling of physical and chemical processes based on the physical formalism (physically meaningful macroparameters on the finite volumes) is more general than that based on the mathematical formalism.

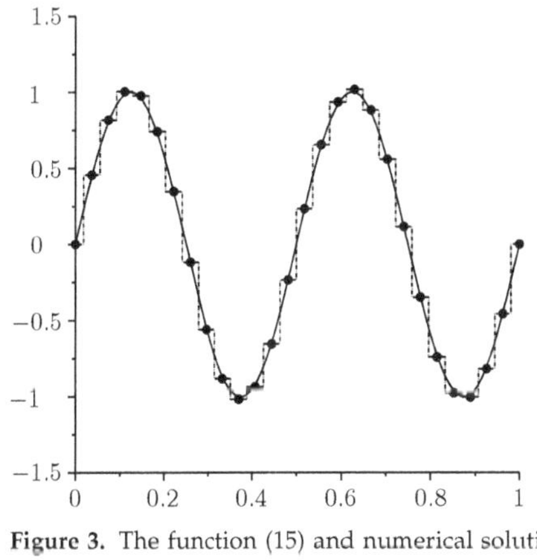

Figure 3. The function (15) and numerical solution of the discrete boundary value problem (13) ($\omega = 2, \aleph = 27, h = 1/27$).

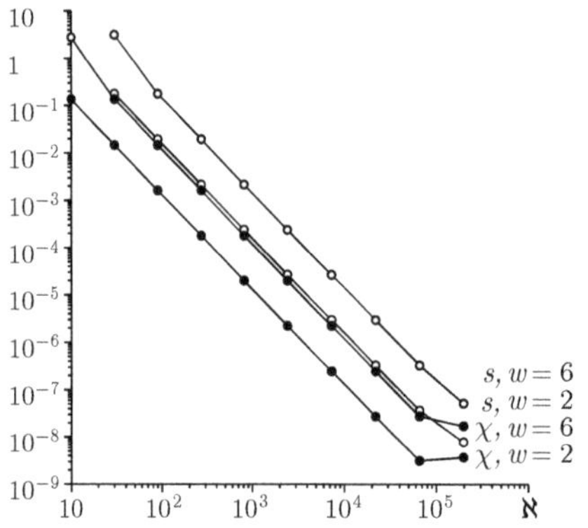

Figure 4. Grid convergence in solving the discrete boundary value problem (13).

The behaviour of grid convergence makes it possible to propose a robust criterion for grid refinement. Some of the grid generation issues that occur in the context of convection–diffusion equations can be illustrated by the linear 1D boundary value problem

$$\varepsilon\frac{u^h_{i-1} - 2u^h_i + u^h_{i+1}}{h^2} + \frac{u^h(x^{\mathrm{f}}_i) - u^h(x^{\mathrm{f}}_{i-1})}{h} = 0, \qquad u^h_1 = 0,\ u^h_{\aleph+1} = 1, \tag{18}$$

where $0 < \varepsilon \ll 1$ is a small parameter. The numerical solution of this problem has a sharp gradient (boundary layer) near the boundary $x = 0$, the width of the layer depends on the parameter ε.

In a weighted-order upwind scheme, the second (convection) term is approximated by weighted linear interpolation such that only upstream and central grid points are used in the discretisation

$$u^h(x^{\mathrm{f}}_i) = (1-\gamma)u^h_i + \gamma u^h_{i+1},$$

where $\gamma \in [0,1]$ is the weighted parameter. Equation (18) can be rewritten in matrix form

$$Au^h_{i-1} + Bu^h_i + Cu^h_{i+1} = 0, \qquad i = 2,3,\dots,\aleph,$$

where

$$A = \frac{\varepsilon}{h} + \gamma - 1, \quad B = -2\frac{\varepsilon}{h} - 2\gamma + 1, \quad C = \frac{\varepsilon}{h} + \gamma, \quad \gamma = \max\left(\frac{1}{2}, 1 - \frac{\varepsilon}{h}\right).$$

The approximate boundary value problem

$$\varepsilon u'' + u' = 0, \qquad u(0) = 0,\ u(1) = 1,$$

has the exact solution

$$u(x) = \frac{1 - \exp(-\frac{x}{\varepsilon})}{1 - \exp(-\frac{1}{\varepsilon})}. \tag{19}$$

The exact solution (19) of the differential problem is defined in the domain $[0,1]$. The classical numerical solution is defined at the grid vertices x^{v}_i. The piecewise constant solution is defined over the domain $[0,1]$. Therefore, it is possible to compare the numerical solutions obtained on different grids over the domain $[0,1]$. Figure 5 represents the convergence history for different mesh sizes $h = 1/\aleph$. The small value of the parameter s shows that the numerical solution of (18) on the coarse grids ($\aleph = 10$ and $\aleph = 30$) does not have the boundary layer near $x = 0$. The refinement of grid ($\aleph = 90$) shows an increase in the parameter s value. This means that the numerical solution of (18) has the boundary layer near $x = 0$, but the accuracy of computations inside the layer is low. The linear decrease of s ($\aleph > 270$) shows that the accuracy of numerical solution depends only on the mesh size $h = 1/\aleph$ (Figure 5). Figure 6 represents all solutions of the model convection–diffusion problem.

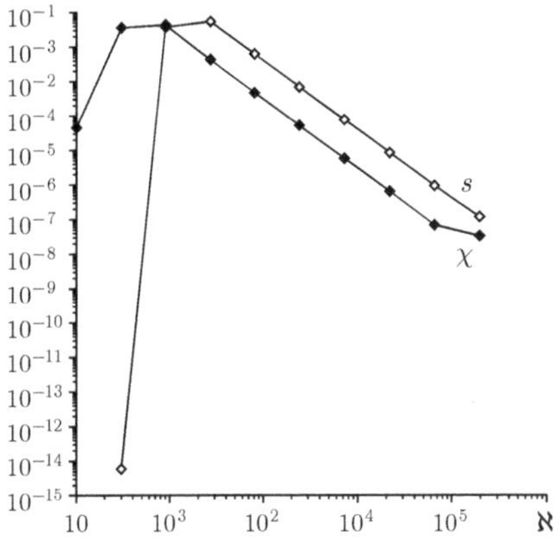

Figure 5. Grid convergence in solving the discrete boundary value problem (18) ($\varepsilon = 0.01$).

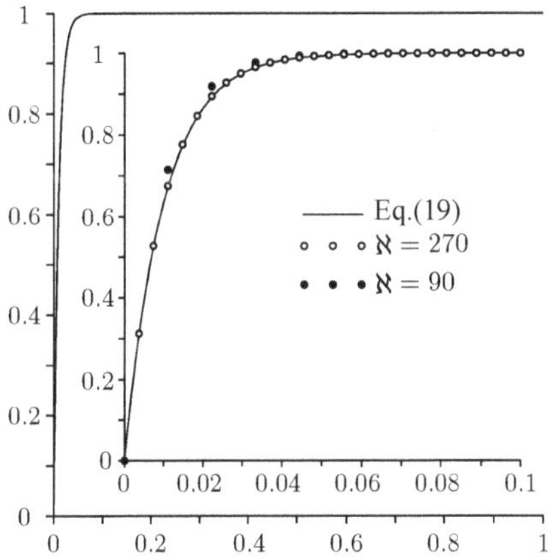

Figure 6. The numerical solution of (18) ($\varepsilon = 0.01$).

The criterion s (17) makes it possible to generate the adaptive grids in a black-box manner [20–23]. In addition, a robust stopping criterion based on the parameter s value can be proposed for multigrid methods [24].

5. From Boltzmann Equation to Equations of the Computational Macromechanics

The derivative-free description can be used for other phenomena [25,26]. Consider the kinetic Boltzmann equation for the expansion of the derivative-free approach. Let us define the distribution function $f(\vec{x}, \vec{v}, t)$ such that $f(\vec{x}, \vec{v}, t)\, d^3x\, d^3v$ is probability of finding a particle in phase space volume $d^3x\, d^3v$ centred on $\vec{x}$, $\vec{v}$ at time t. The Boltzmann equation becomes

$$f'_t + \vec{\nabla}_x \cdot (\vec{v} f) + \vec{\nabla}_v \cdot (\vec{g} f) = J,$$

where J represents discontinuous motion of particles through phase space because of collisions. Let X be a finite volume in physical space. The particle velocity space can be discretised in the same way. The volume-averaged density is defined by

$$\rho(\boldsymbol{x}_*, t) = \int\limits_V \frac{1}{X} \int\limits_X mf(\vec{x}, \vec{v}, t)\, d^3x\, d^3v,$$

where $\boldsymbol{x}_* \in X$ and m is the particle mass. In the simplest case of the same particle, local mass conservation leads to $J = 0$ and the Boltzmann equation multiplied by m becomes

$$\frac{\partial(mf)}{\partial t} + \sum_{i=1}^{3} \frac{\partial(mfv_i)}{\partial x_i} + \sum_{i=1}^{3} \frac{\partial(mfg_i)}{\partial x_i} = 0.$$

Integration over the finite volume in the phase space and over $t^{(n+1)} - t^{(n)}$ in time leads to the following volume-averaged form of the Boltzmann equation

$$\int\limits_V \frac{1}{X} \int\limits_X \frac{1}{t^{(n+1)} - t^{(n)}} \int\limits_{t^{(n)}}^{t^{(n+1)}} \frac{\partial(mf)}{\partial t}\, dt\, d^3x\, d^3v$$

$$+\sum_{i=1}^{3} \int\limits_V \frac{1}{X} \int\limits_X \frac{1}{t^{(n+1)} - t^{(n)}} \int\limits_{t^{(n)}}^{t^{(n+1)}} \left(\frac{\partial(mfv_i)}{\partial x_i} + \frac{\partial(mfg_i)}{\partial x_i} \right) dt\, d^3x\, d^3v = 0.$$

Using the divergence theorem, the second term can be written as

$$\int\limits_V \frac{1}{X} \int\limits_X \frac{m}{2} \left(\vec{\nabla}_x \cdot (\vec{v}f)^{(n)} + \vec{\nabla}_x \cdot (\vec{v}f)^{(n+1)} \right) d^3x\, d^3v$$

$$+\frac{1}{2X} \int\limits_X \int\limits_V m \left(\vec{\nabla}_v \cdot (\vec{g}f)^{(n)} + \vec{\nabla}_v \cdot (\vec{g}f)^{(n+1)} \right) d^3v\, d^3x$$

$$= \int\limits_V \frac{1}{2X} \int\limits_{S_x} \vec{n}_x \cdot \left((m\vec{v}f)^{(n)} + (m\vec{v}f)^{(n+1)} \right) dA + \frac{1}{2X} \int\limits_X \int\limits_{S_v} \vec{n}_v \cdot \left((m\vec{g}f)^{(n)} + (m\vec{g}f)^{(n+1)} \right) dB.$$

Finally, the volume-averaged conservation equation of mass

$$\frac{\rho(\boldsymbol{x}_*, t^{(n+1)}) - \rho(\boldsymbol{x}_*, t^{(n)})}{t^{(n+1)} - t^{(n)}}$$

$$+\frac{1}{2X} \int\limits_{S_x} \left(\rho(\boldsymbol{x}_*, t^{(n)}) \vec{V}_n(\boldsymbol{x}_*, t^{(n)}) + \rho(\boldsymbol{x}_*, t^{(n+1)}) \vec{V}_n(\boldsymbol{x}_*, t^{(n+1)}) \right) dA = 0$$

can be obtained from the Boltzmann equation averaged over the finite volumes in physical space.

A more attractive approach is direct modelling in a discretised space [27]. From the point of view of physical modelling, the derivative-free Boltzmann equation is more general than the classical one because the continuity assumption is not needed for flow variables.

6. Discussion

The classical mathematical description of physical and chemical processes in the continuum approximation is based on the mutually contradictory use of differential calculus and minimal spatial and temporal scales applied to ignore the molecular structure of matter. The resulting governing (partial) differential equations (PDEs) connect physically senseless functions assigned to points (t, x, y, z). The differentiability or continuity assumption is needed for all variables in the governing PDEs. The numerical solution of these PDEs implies the transformation of (initial-)boundary value problems into algebraic ones, i.e., to approximate them by algebraic equations that involve a finite number of unknowns. There are two the most popular ways to discretise PDEs. The finite element method is based on replacing the original function with a function that has some degree of smoothness over the global domain, but is piecewise polynomial on simple subdomains. The finite volume method is based on the approximation of continuous conservation laws [10].

Now it is necessary to completely abandon the use of differential calculus for the mathematical modelling of physical and chemical processes in the continuum approximation [28]. For these purposes, the global domain is divided into a number of finite volumes of such size that the liquid or gas in them can be considered continuous. In this case, all variables (density, pressure, temperature, etc.) have a physical sense in strict accordance with theoretical definitions, and therefore they are not assigned to points, but to finite volumes. As a result, all physical and chemical processes can be described only in a finite volume-averaging sense, i.e., all variables are piecewise constant functions (so-called discontinuum approximation: constant functions in the finite volume and discontinuous on its faces). The mathematical models of physical and chemical processes in each finite volume consist of fundamental conservation laws, discrete phenomenological laws and additional hypotheses. Really, the finite volume schemes are the mathematical models in the discontinuum approximation. The derivative-free description of physical and chemical processes is very preferable from a computing point of view. All variables must be integrable functions, they are not required to be differentiable. This is especially noticeable for the Godunov-type schemes [29,30]. From a physical point of view, the explicit and implicit schemes cannot be used for mathematical modelling.

The current target of computational mathematics is to recover the solution of original PDEs as the mesh size and time step approach zero. This leads to physically meaningless macroparameters (density, pressure, temperature and others defined over a finite volume of a continuous medium), since there are minimum spatial and temporal scales for each model. The minimum mesh size and time step cannot be less than the minimum spatial and temporal scales. This leads to derivative-free models. Then differential equations form an approximate model, the discrete equations form an exact model, and the finite difference, the finite volume and the finite element methods become unnecessary. A similar approach based on the mathematical formalism was proposed for Godunov-type schemes [29,30]. The article shows that mathematical modelling of physical and chemical processes based on the physical formalism (physically meaningful macroparameters on finite volumes) is more general than modelling based on the mathematical formalism.

The governing equations such as (12) in the above-mentioned discontinuous approximation depend on the shape of finite volume. To clarify, the formal form of the continuity equation

$$\frac{1}{t^{(n+1)}-t^{(n)}}\int\limits_{t^{(n)}}^{t^{(n+1)}}\frac{1}{V_{ijk}^{(n+1)}}\int\limits_{\mathcal{V}_{ijk}^{(n+1)}}\left(\frac{\partial\rho}{\partial t}+\nabla\cdot(\rho\vec{V})\right)dv\,dt=0$$

can be used in theoretical analysis, where integration is performed under the condition that all functions are constant in each finite volume $\mathcal{V}_{ijk}^{(n+1)}$.

7. Conclusions

The dilemma of correctly defined derivatives versus correctly defined macroparameters presents itself. In the correct derivatives approach, one tries to use differential and integral calculus in order to analyse the differential problems related to physically senseless variables. On the other hand, the idea of a physically meaningful solver based on the correctly defined macroparameters is to use conservation and phenomenological laws for a derivative-free description of the physical and chemical processes on each finite volume. The mesh size and time step are limited by the continuity hypothesis. The numerical solution of the governing discrete equations is the piecewise constant functions (density, pressure, temperature, etc.) corresponding to the classic physical theories. From a practical point of view, this approach results in impressive simplification of the mathematical modelling due to the absence of discretisation of the (initial-)boundary value problems. The best solver of these governing finite volume-based equations is the agglomeration multigrid.

Future work will be an analysis of the discretisation error of the governing equations in the discontinuous approximation, solution of 2D benchmark problems of CFD and comparison with the continuum approximation.

Author Contributions: Conceptualization, methodology, software: S.I.M.; formal analysis, review and editing: A.Y.V. All authors have read and agreed to the published version of the manuscript.

Funding: This work was supported by the Ministry of Science and Higher Education of the Russian Federation (State Assignment No. 075-00270-24-00). The research is funded by the Russian Foundation for Basic Research (project "Developement and application of high-performance parallel algorithms for supercomputer modeling of complex reaction flows", project code 21-51-46007) and Scientific and the Technological Research Council of Türkiye (TÜBİTAK), Grant №: ARDEB-220N170.

Data Availability Statement: The data are available upon request from the corresponding author.

Conflicts of Interest: The authors declare no conflicts of interest.

References

1. Sedov, L.I. *A Course in Continuum Mechanics*; Groningen: Wolters-Noordhoff, The Netherlands, 1971; Volume 1.
2. Moukalled, F.; Mangani, L.; Darwish, M. *The Finite Volume Method in Computational Fluid Dynamics. An Advanced Introduction with OpenFOAM and Matlab*; Springer: Cham, Switzerland, 2016. [CrossRef]
3. Jayanti, S. *Computational Fluid Dynamics for Engineers and Scientists*; Springer: Dordrecht, The Netherlands, 2018. [CrossRef]
4. Caltagirone, J.-P. An alternative to the concept of continuous medium. *Acta Mech.* **2021**, *232*, 4691–4703. [CrossRef]
5. Schlichting, H. *Boundary Layer Theory*, 7th ed.; McGraw-Hill: New York, NY, USA, 1979.
6. Runchal, A. *50 Years of CFD in Engineering Sciences. A Commemorative Volume in Memory of D. Brian Spalding*; Springer: Singapore, 2020. [CrossRef]
7. Bhattacharyya, S.; Abraham, J.P.; Cheng, L.; Gorman, J. Introductory Chapter: A Brief History of and Introduction to Computational Fluid Dynamics. In *Applications of Computational Fluid Dynamics Simulation and Modeling*; Bhattacharyya, S., Ed.; IntechOpen: Rijeka, Croatia, 2021. [CrossRef]
8. Suchde, P., Kuhnert, J., Tiwari S. On meshfree GFDM solvers for the incompressible Navier-Stokes equations. *Comput. Fluids* **2018**, *165*, 1–12. [CrossRef]
9. Gallagher, I. From Newton to Navier-Stokes, or how to connect fluid mechanics equations from microscopic to macroscopic scales. *Bull. Am. Math. Soc.* **2019**, *56*, 65–85. [CrossRef]
10. Kitamura, K. Introduction: Brief Review of Finite Volume Method (FVM) in Computational Fluid Dynamics. In *Advancement of Shock Capturing Computational Fluid Dynamics Methods*; Springer: Singapore, 2020. [CrossRef]
11. Kuzenov, V.V.; Ryzhkov, S.V. Estimation of the neutron generation in the combined magneto-inertial fusion scheme. *Phys. Scr.* **2021**, *96*, 125613. [CrossRef]
12. Kuzenov, V.V.; Ryzhkov, S.V. Numerical simulation of the interaction of a magneto-inertial fusion target with plasma and laser drivers. *High Temp.* **2022**, *60*, S7–S15 . [CrossRef]
13. Pyanylo, Y.D. Mathematical Modeling of Mass Transfer in Complex Engineering and Biomedical Systems. *J. Math. Sci.* **2023**, *273*, 163–180. [CrossRef]
14. Ngo, S.I.; Lim, Y.-I. Multiscale Eulerian CFD of Chemical Processes: A Review. *ChemEngineering* **2020**, *4*, 23. [CrossRef]
15. Deville, M.O. Exact Solutions of the Navier–Stokes Equations. In *An Introduction to the Mechanics of Incompressible Fluids*; Springer: Cham, Switzerland, 2022; pp. 51–89. [CrossRef]
16. Kuzenov, V.V.; Ryzhkov, S.V. Thermoplastic Parameter Estimation of a Neutron Source Based on the Action of Broadband Radiation on a Cylindrical Target. *Fusion Sci. Technol.* **2023**, *79*, 399–406. [CrossRef]
17. Martynenko, S.I. Modeling of heat and mass transfer in the discontinuum approximation. *Vestn. Udmurt. Univ. Mat. Mekhanika Komp'Yuternye Nauk.* **2024**, *34*, 137–164. (In Russian) [CrossRef]
18. Martynenko, S.I. *The Robust Multigrid Technique: For Black-Box Software*; De Gruyter: Berlin, Germany, 2017.
19. Martynenko, S.I. *Numerical Methods for Black-Box Software in Computational Continuum Mechanics*; De Gruyter: Berlin, Germany, 2023.
20. George, P.L. *Automatic Mesh Generation*; Wiley: New York, NY, USA, 1991.

21. Xu, K. Direct modeling for computational fluid dynamics. *Acta Mech. Sin.* **2015**, *31*, 303–318. [CrossRef]
22. Frey, P.; George, P.L. *Mesh Generation*; Wiley: New York, NY, USA, 2010.
23. Kuzenov, V.V.; Ryzhkov, S.V.; Varaksin, A.Y. The Adaptive Composite Block-Structured Grid Calculation of the Gas-Dynamic Characteristics of an Aircraft Moving in a Gas Environment. *Mathematics* **2022**, *10*, 2130. [CrossRef]
24. Hackbusch, W. *Multi-Grid Methods and Applications*; Springer: Berlin, Germany, 1985.
25. Wang, J. Continuum theory for dense gas-solid flow: A state-of-the-art review. *Chem. Eng. Sci.* **2020**, *215*, 115428. [CrossRef]
26. Liu, H.; Kang, Q.; Leonardi, C.; Schmieschek, S.; Narváez, A.; Jones, B.D.; Williams, J.R.; Valocchi, A.J.; Harting, J. Multiphase lattice Boltzmann simulations for porous media applications. *Comput. Geosci.* **2016**, *20*, 777–805. [CrossRef]
27. Xu, K. *A Unified Computational Fluid Dynamics Framework from Rarefied to Continuum Regimes*; Cambridge University Press: Cambridge, UK, 2021.
28. Heinz, S. A Mathematical Solution to the Computational Fluid Dynamics (CFD) Dilemma. *Mathematics* **2023**, *11*, 3199. [CrossRef]
29. Busto, S.; Dumbser, M.; Escalante, C.; Favrie, N.; Gavrilyuk, S. On High Order ADER Discontinuous Galerkin Schemes for First Order Hyperbolic Reformulations of Nonlinear Dispersive Systems. *J. Sci. Comput.* **2021**, *87*, 48. [CrossRef]
30. Busto, S.; Chiocchetti, S.; Dumbser, M.; Gaburro, E.; Peshkov, I. High order ADER schemes for continuum mechanics. *Front. Phys.* **2020**, *8*, 32. [CrossRef]

materials

MDPI

Article

Study on the Tribological Properties of Multilayer Concentric Hexagonal Laser Texturing on Rubber Surfaces of Screw Pumps

Xinfu Liu [1,*,†], Xinglong Niu [1,†], Chunhua Liu [2], Xiangzhi Shi [1], Yi Sun [1], Zhongxian Hao [3], Shouzhi Huang [3], Yuan Wang [4] and Hua Tao [4]

1 Key Lab of Industrial Fluid Energy Conservation and Pollution Control (Ministry of Education), Qingdao University of Technology, Qingdao 266520, China
2 College of Mechanical and Electronic Engineering, China University of Petroleum (East China), Qingdao 266580, China
3 Research Institute of Exploration & Development, PetroChina, Beijing 100083, China
4 Lusheng Petroleum Development Co., Ltd., China Petroleum & Chemical Corporation, Dongying 257077, China
* Correspondence: upcdoctor@126.com
† These authors contributed equally to this work.

Abstract: Given the friction and drag reduction effects observed in various biological hexagonal structures in nature, a new design was implemented on the rubber surface of the stator of a submersible screw pump. This design featured a multilayer concentric hexagonal groove structure. Furthermore, a composite multilayer hexagonal structure integrating grooves and pits was also developed and applied. This study investigated the influence of groove layer number, groove depth, pit depth, and multilayer hexagonal groove texture arrangement on the rubber surface flow characteristics. Additionally, the pressure field state, the degree of influence on the oil film-bearing capacity, and the biomimetic and hydrodynamic lubrication theories were tested using the finite element analysis method. Tribological experiments were conducted on nanosecond laser-processed rubber textures under simulated liquid lubrication conditions, reflecting actual shale oil well experiments. These experiments aimed to investigate the influence of multilayer hexagonal shape parameters on the tribological characteristics of the stator-rotor friction pair of a submersible screw pump. The results indicated that with a constant overall size, a multilayer hexagonal structure with ~0.1 mm groove depth enhanced the oil film-bearing capacity, providing significant friction and drag reduction. For composite textures, a deeper pit depth within the study area enhanced the oil film-bearing capacity. Furthermore, a gradient arrangement of groove textures featuring wider outer grooves and shallower depth exhibited superior performance in terms of bearing capacity.

Keywords: stator rubber; multilayer concentric arrangement; hexagonal texture; reduce friction and drag; fluid lubrication

1. Introduction

Submersible screw pumps are rodless artificial lifting devices that combine the advantages of both volumetric and centrifugal pumps [1–6]. They are widely employed in onshore oil recovery wells with high sand content and high gas-to-liquid ratios [7], as well as offshore oil platforms [8,9]. In the oil recovery process, the stator of the electric submersible screw pump is highly susceptible to wear due to stator-rotor friction, leading to screw pump failure. The rubber friction wear of the stator in electric submersible screw pumps is a significant factor in reducing the lifespan of these pumps. This issue is exacerbated under complex downhole conditions and the influence of lifting liquids. Researchers across various fields have conducted extensive studies on the friction wear of the rubber stator in screw pumps [10–17]. They aim to improve the tribological performance between the stator and rotor, enhancing the service life of the pump and achieving energy savings and consumption reduction.

In reality, there are no perfectly smooth, frictionless surfaces. However, the natural unevenness of biological surfaces often aids in lubrication and friction reduction. For instance, the scales on the skin of sharks significantly reduce resistance during swimming [18]. Inspired by nature, the development of bionic friction pairs prepared in regularly arranged geometric shapes has been proven to be an effective way to enhance the tribological performance of friction pairs [19–28].

For the study of the surface texture shape, the researchers made a lot of investigations using computational fluid dynamics (CFD) simulations and experiments [29–33]. Zhao et al. [34] discussed the friction reduction mechanisms and theoretical models under different lubrication conditions and reviewed the geometrical characteristics of surface texturing and recent progress in improving the tribological properties of material surfaces under practical working conditions. Zhang et al. proposed the preparation of a surface texture on the rubber surface of a stator to improve its tribological characteristics to reduce friction, prolong the service life, and improve the working efficiency [35]. Two main categories of surface texture patterns have been studied: pits and grooves [36]. Research indicates that groove-shaped textures are particularly effective in enhancing hydrodynamic pressure [37,38]. As research progressed, the study of surface textures has increasingly focused on complex structures within these textures [39–42].

In 1960, Theodore Meyman invented the first ruby laser. Laser processing technology has been extensively researched and applied in various industrial fields [43–46]. Furthermore, compared with other surface machining technologies, laser processing offers high precision, fast rate, low environmental pollution, and affordability. In fields of tool surfaces [47], gears [48], plain bearings [49], cylinder liners [50], pistons [50], and mechanical seals [51], laser-machined surface textiles are often employed with lubricating fluids to reduce friction [52–54]. The tribological properties of these textures are significantly influenced by different operating conditions. Moreover, the quality of the lase-processed texture is significantly affected by the laser processing power parameters. For instance, Andersson et al. found that suitable laser-textured surfaces can reduce friction on steel surfaces [55]. Wu et al. obtained the appropriate laser energy and scanning rate to make the processed surface structure clear and complete [56]. Mao et al. found that scanning power and scanning speed affect the processing depth of surface texture. Roushan [57] demonstrated that using an appropriate high-energy laser beam and scanning frequency to prepare linear textures on the surface of a cemented friction pair could reduce wear during movement [58,59](Figure 1). Therefore, the selection of appropriate laser processing parameters is crucial for enhancing the quality of rubber surface textures. Finally, this study aimed to explore the influence of multilayer concentric hexagonal groove structure shape parameters on the rubber surface pressure field and flow state. Additionally, it investigated the optimal laser power processing parameter through experimental research to address the wear issues of rubber surfaces in electric submersible screw pump stators.

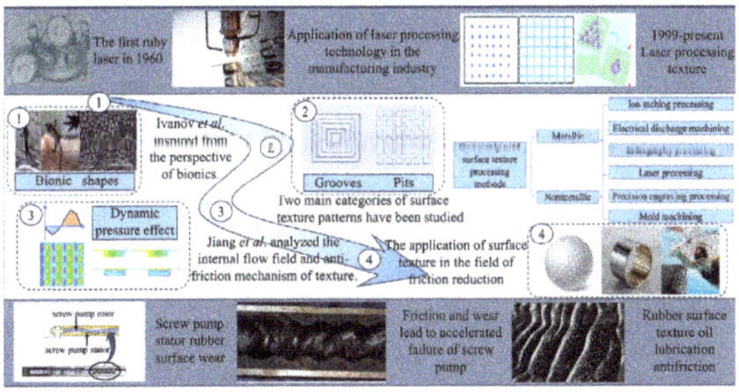

Figure 1. Research status of surface texture [40,60].

2. Modeling the Stator Internal Surface Texture of a Progressive Cavity Pump Using a Tribological Experimental Method

2.1. Simulation Model of a Multilayer Concentric Hexagonal Texture Structure

A hexagonal structure with multiple concentric grooves was designed and constructed using the bionic-derived single-layer hexagonal grooves as the base. Additionally, the hexagonal pit factor was introduced to construct a multilayer composite hexagonal braid with hexagonal grooves and pits. The number of hexagonal grooves, the groove depth, and the pit depth were controlled to study their influence on the tribological properties of the hexagonal braid. The multilayer hexagonal groove structure was deformed to experimentally explore the effect of two gradient arrangements of unequal grooves on the performance of the hexagonal structure, considering three layers of equal-sized grooves. The effects of concentric grooves of unequal width and depth on the lubrication properties of multilayer concentric hexagonal textures were investigated. Overall, the dynamic pressure lubrication effect and friction reduction performance of multilayer hexagonal groove structures were enhanced.

A fluid simulation model of a multilayer concentric hexagonal groove structure was established. Shale oil fluid medium with a density of 870 kg/m^3 and a viscosity of 0.08 mPa·s, was set to flow into the model from the x-axis in the positive direction at a speed of 1 m/s (Figure 2c). The inlet and outlet pressures were set to 101 kPa.

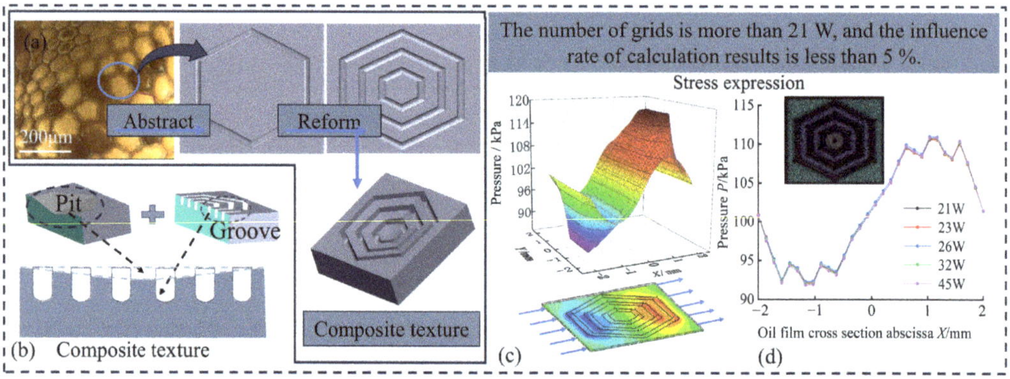

Figure 2. (**a–d**) Hexagonal texture simulation model.

A three-dimensional fluid model was established using the CFD finite element calculation and fluid analysis software (Workbench Fluent 2020). The average mesh quality was 6.21 (Figure 2d). When the mesh exceeded 200,000 elements, the influence rate of the increased mesh number on the calculation was <5%.

2.2. Mathematical Models of Fluid Dynamic Pressure Theory

The following assumptions were made:

(1) Volumetric forces, such as gravity and magnetism, were not considered.
(2) There was no relative sliding of the fluid surface at the solid interface.
(3) Pressure variation in the direction of film thickness was not considered, and the fluid at the surface of the friction pair was a Newtonian fluid of constant viscosity.
(4) Translational speed was considered instead of moving speed without accounting for the curvature of the oil film.
(5) The fluid on the surface of the friction pair experienced laminar flow.
(6) In comparison with viscous drag, inertial forces, including the force of fluid acceleration and the force of oil film curvature, were neglected.
(7) The viscosity along the direction of the lubricant film was constant.

The Reynolds equation for the fluid on the weaving surface was obtained under these assumptions:

$$\frac{\partial}{\partial x}\left(\frac{\rho h^3}{\eta}\frac{\partial p}{\partial x}\right)+\frac{\partial}{\partial y}\left(\frac{\rho h^3}{\eta}\frac{\partial p}{\partial y}\right)=6\left[\frac{\partial}{\partial x}(U\,]\rho h)+\frac{\partial}{\partial x}(V\rho h)+2\rho(w_{\mathrm{h}}-w_0)\right] \tag{1}$$

where η is the fluid viscosity (Pa·s), and p is the lubricating film pressure in N. Furthermore, U and V are the relative velocities of the friction pair along the x and y directions, respectively. w_{h} and w_0 are the velocities of the two surfaces of the friction pair along the direction of the lubricating film thickness.

Furthermore, assuming that the viscosity was constant in the gap and disregarding the effects of temperature and pressure on the viscosity, the Reynolds equation for a locally pressurized fluid film was derived (Equation (2)):

$$\frac{\partial}{\partial x}\left(\frac{\rho h^3}{\eta}\frac{\partial p}{\partial x}\right)+\frac{\partial}{\partial y}\left(\frac{\rho h^3}{\eta}\frac{\partial p}{\partial y}\right)=6U\frac{\partial}{\partial x}(\rho h) \tag{2}$$

When considering only the dynamic pressure effect, the Reynolds equation was simplified to Equation (3). This simplified Reynolds equation was adopted in the theoretical model of the weaving structure under a small gap, where an incompressible fluid dominated the laminar lubrication state (Equation (3)):

$$\frac{\partial}{\partial x}\left(h^3\frac{\partial p}{\partial x}\right)+\frac{\partial}{\partial y}\left(h^3\frac{\partial p}{\partial y}\right)=6U\eta\frac{\partial h}{\partial x} \tag{3}$$

The standard K-epsilon turbulence model [38] is as follows:

$$\frac{\partial}{\partial \mathrm{t}}(\rho k)+\frac{\partial}{\partial x_i}(\rho k u_i)=\frac{\partial}{\partial x_i}\left[\left(\mu+\frac{\mu_t}{\sigma_k}\right)\frac{\partial k}{\partial x_i}\right]+G_k+G_b-\rho_\varepsilon-Y_M+S_k \tag{4}$$

$$\frac{\partial}{\partial \mathrm{t}}(\rho\varepsilon)+\frac{\partial}{\partial x_j}(\rho\varepsilon u_i)=\frac{\partial}{\partial x_j}\left[\left(\mu+\frac{\mu_t}{\sigma_\varepsilon}\right)\frac{\partial \varepsilon}{\partial x_j}\right]+C_{1\varepsilon}\frac{\varepsilon}{k}(G_k+C_{3\varepsilon}G_b)-C_{2\varepsilon}\rho\frac{\varepsilon^2}{k}+S_\varepsilon \tag{5}$$

where the constants are set as follows: (i) 1.44 for the $C_{1\varepsilon}$; (ii) 1.92 for the $C_{2\varepsilon}$; (iii) 1.0 for the σ_k; and (iv) 1.3 for the σ_k.

Additionally, by integrating the pressure of the lubricant film, the bearing capacity of the lubricant film was obtained, and the expression for bearing capacity was derived as follows:

$$W=\iint p dx dy \tag{6}$$

The shear stress on the surface of the friction pairs was integrated, and the magnitude of the resistance was obtained.

$$F=\iint \tau \mathrm{d}x dy \tag{7}$$

Given the resistance and load-bearing force, the friction factor of the friction pair was obtained through their ratio, with the expression provided as follows:

$$\mu=\frac{F}{W} \tag{8}$$

2.3. *Experimental Method for Tribological Properties of Textured Rubber*

A specific type of high-acrylonitrile rubber was selected as the experimental material. The structure of the rubber surface was textured and simulated under oil extraction conditions (Figure 3a). The rubber samples were soaked in the actual extraction fluid from shale oil wells at 70 °C for 7 days. Subsequently, friction experiments were conducted under lubrication conditions with the actual extraction fluid. The stator rubber was textured using an IPG nanosecond laser, and the textured rubber material was observed using a three-dimensional (Figure 3c) confocal microscope (VK-X1050), Japan KEYENCE (Osaka, Japan).

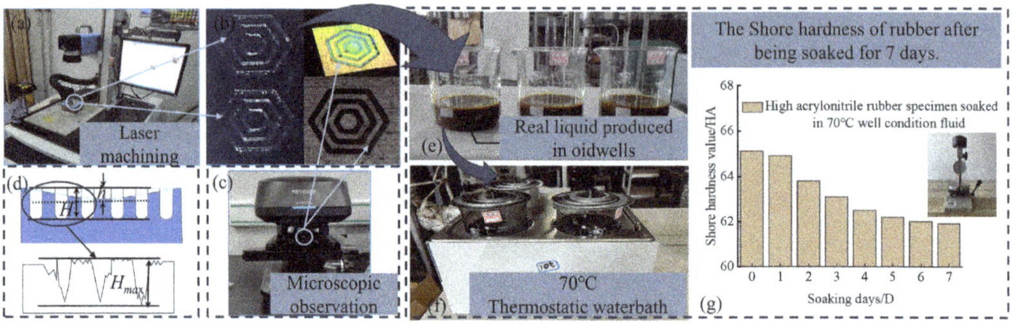

Figure 3. (**a–g**) Actual well condition and 70 °C immersion condition.

As shown in Figure 3d, H is the depth of the groove designed in the simulation, h is the depth of the pit designed in the simulation, and H_{max} is the maximum depth of the laser-machined grooves.

The nanosecond laser processing of rubber specimens was characterized by its simplicity and effective modeling, meeting the requirements for experimental processing. To simulate underground oil production conditions, the rubber specimen was soaked in the actual production liquid at 70 °C for 7 days (Figure 3e,f), ensuring the experimental results correlated with the actual application conditions. The variation of the average hardness values of the high-acrylonitrile rubber specimens utilized in the experiments is shown in Figure 3g. The hardness exhibited a decreasing trend, stabilizing after 7 days of immersion in the actual well-conditioned fluid at 70 °C, decreasing from 65.1 HA initially to 61.9 HA.

Tribological tests were conducted on screw pump stator rubber with a textured surface after 7 days of immersion using a UMT-3 reciprocating friction and wear tester, U.S. CETR products, with the textured screw pump stator rubber specimen placed in the lower part and a 25 mm diameter Gar 15 bearing steel ball in the upper part of the specimen. The two specimens underwent relative sliding at a fixed frequency of 3 Hz (reciprocating distance of 6 mm). Friction experiments were conducted at a constant speed, with a load of 10 N applied under lubrication conditions to simulate actual well conditions in shale oil wells for 1200 s (Figure 4).

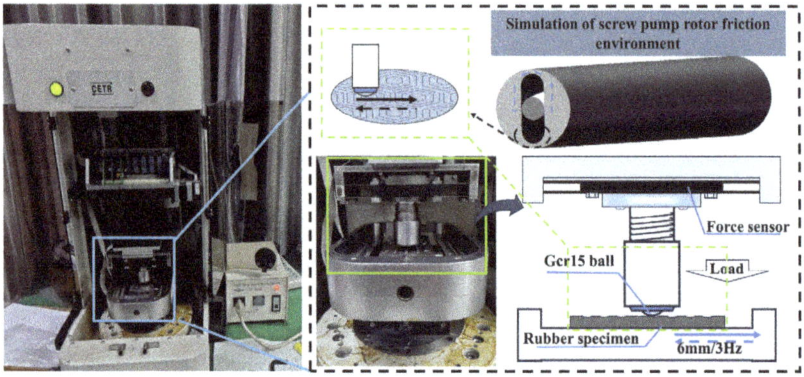

Figure 4. Experimental conditions of UMT friction testing machine.

3. Fluid Simulation Characterization of Woven Surfaces

3.1. Effect of Inflow Direction on the Properties of Multilayer Hexagonal Textures

Figure 5 depicts different inflow angles for the variation of pressure in the fluid domain of the weave, using a three-layer hexagonal groove weave as an example. The hexagonal shape has good symmetry, so it can be simplified to the effect of angle on the dynamic pressure effect of the weave in the range of 0° to 30°. Among them, 30° inflow direction has the best pressure performance, the maximum pressure on the upper wall is larger compared to other angles, and the oil film has a better carrying capacity.

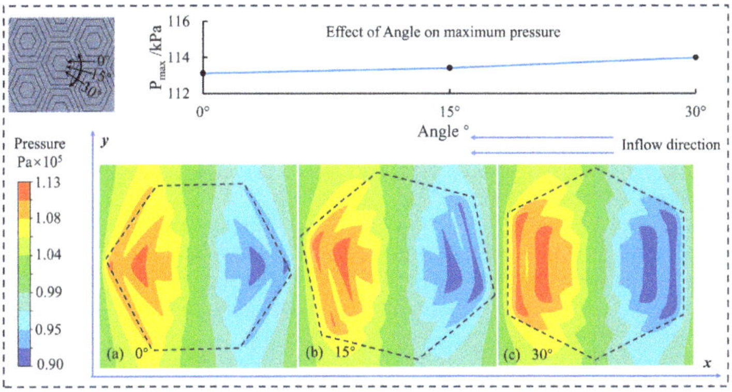

Figure 5. Influence of inflow direction ((**a**) 0°, (**b**) 15°, (**c**) 30°) on the dynamic pressure effect of multilayer hexagonal weaves.

Therefore, in the subsequent sections of this paper, the inflow direction of the fluid for the CFD simulation and the friction direction for the friction test are explored in the direction of 30°.

3.2. Effect of Groove Depth on the Properties of Multilayer Hexagonal Textures

The graphs in Figure 6 show the relationship between maximum pressure and groove depth on the upper wall of the multilayer concentric hexagonal texture, varying with different numbers of turns. The maximum pressure experienced a rapid increase from 0.01 mm to 0.09 mm deep, peaking at 0.1 mm dep. Subsequently, it decreased and plateaued at 0.3 mm deep.

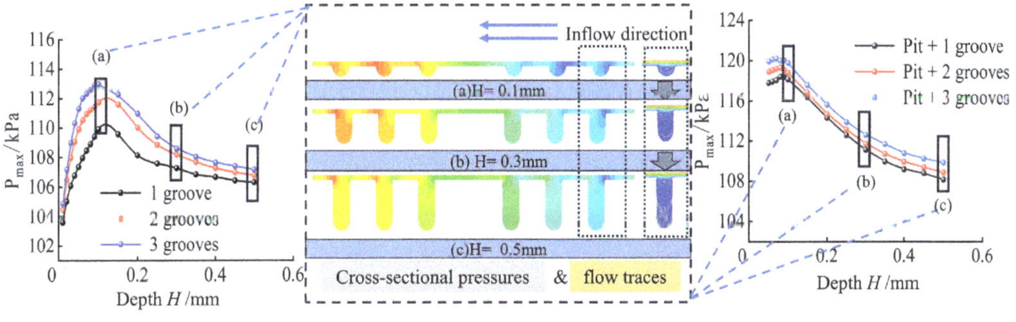

Figure 6. Pressure cloud map of trench section in different layers.

Meanwhile, Figure 6 shows the pressure situation of the hexagonal groove texture with a different number of layers in the cross-section of the groove, along with a change in the flow field within the groove, corresponding to the change of the groove depth. Pressure variations were presented by a sequential lightening of pressure colors from the outer to the inner laps. Darker colors signify areas of higher pressure, indicating a larger additional net-bearing capacity and superior oil film-bearing capacity. In the three-layer groove hexagonal structure, the outer two layers of grooves predominantly enhanced pressure. The change in the cross-section of the texture oil film included variations in fluid dynamic pressure during the flow process. Despite the axisymmetric texture positive and negative pressure areas being roughly equal in area, the wedge-shaped cross-section amplified the effect of fluid dynamic pressure. Consequently, the deviation from the standard pressure value was more pronounced in the high-pressure zone compared with the low-pressure zone, thereby enhancing the overall carrying capacity. This differential pressure contributed to an additional net load-bearing capacity across the entire oil film, consequently affecting the performance of the structure. A higher oil film-bearing capacity enhanced the tribological performance of the structure, leading to friction and drag reduction.

The increases in groove depth initially enhanced the pressurizing effect of the outermost groove, expanding its high-pressure area beyond that of the inner two cycles of the grooves. However, as the groove depth increased, the high-pressure area within the groove occupied a smaller proportion of the entire groove. Consequently, the pressurizing effect of the groove became weak.

The groove flow line exhibited profound changes in fluid flow state with increasing groove depth. For instance, between a groove depth of 0.01 mm and 0.1 mm, the groove formed a vortex, resulting in complete formation beyond a depth of 0.1 mm. The changes in oil film cross-section due to enhanced dynamic pressure mitigated energy loss induced by vortex formation, consequently increasing surface pressure on the oil film. Subsequently, an increase in groove depth led to further expansion of the vortex, accompanied by an increase in energy loss, ultimately resulting in a decrease in maximum pressure on the film surface. However, when the groove depth exceeded 0.3 mm, the influence of the bottom fluid on the upper fluid diminished and pressure at the groove tended to stabilize. At a groove depth of ~0.5 mm, a new vortex was formed, which rapidly increased in volume.

Conclusively, increasing the groove depth altered the fluid flow state in the groove, leading to vortex formation and energy loss in the oil film flow. The cross-section of the pressure curve indicated a decline in maximum oil film pressure at a groove depth of >0.1 mm, thereby reducing the oil film-bearing capacity and tribological performance of the texture.

3.3. *Effect of Pit Depth on the Properties of Composite Multilayer Hexagonal Texture*

The addition of pit elements, equivalent in size to the outermost grooves, into the concentric hexagonal grooves of the multilayers enhances the pressure-bearing capacity of the texture to a certain extent. The depth of these pits resulted in different performances of the pressure-bearing capacity of the texture. Figure 7 shows the variation in the oil film pressure with pit depth for a hexagonal texture with pits and multilayer grooves. The presence of pits increased the oil film pressure across the texture, and the maximum oil film pressure of the composite hexagonal grooved texture increased with increasing pit depth.

The effect of pit depth on the performance of the composite hexagonal texture was investigated under the conditions of constant groove depth (Figure 7). The maximum pressure of the fluid on the surface of the texture increased with increasing pit depth. An increase in pit depth from 0.01 mm to 0.05 mm gradually enlarged the high-pressure area formed by the grooves on the surface until the divided high-pressure region was connected. This facilitated the more complete development of dynamic pressure effects due to cross-sectional changes, consequently enhancing the load-carrying capacity of the oil film.

Meanwhile, the maximum pressure of the composite hexagonal texture of pits and multilayer grooves correlated with the change of the multilayer hexagonal texture concerning groove depth. Furthermore, with groove depths exceeding 0.1 mm, the maximum pressure of oil film tended to decrease with increasing groove depth. Thus, the rate of decrease diminished gradually.

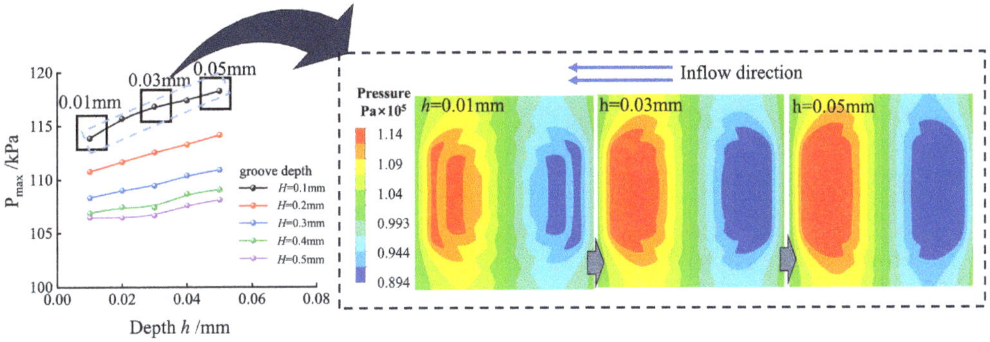

Figure 7. Variation of oil film pressure with pit depth.

3.4. Analysis of Multilayer Weaving Flow Field with Trench Gradient Arrangement

Figure 8 shows the maximum pressure of a multilayer hexagonal groove texture with different width gradients with increasing groove depth. The three-layer groove widths were represented as 1:2:3 and 3:2:1, respectively. The maximum pressure on the upper surface of the oil film for the multilayer hexagonal groove weave with a width gradient of 1:2:3 from outer to inner was lower compared with a width gradient of 3:2:1. Both sets of width gradient groove weaves exhibited similar pressure characteristics to the conventional multilayer hexagonal groove weave. As the depth of the grooves increased from 0.1 mm to 0.5 mm, the pressure gradually decreased and then plateaued.

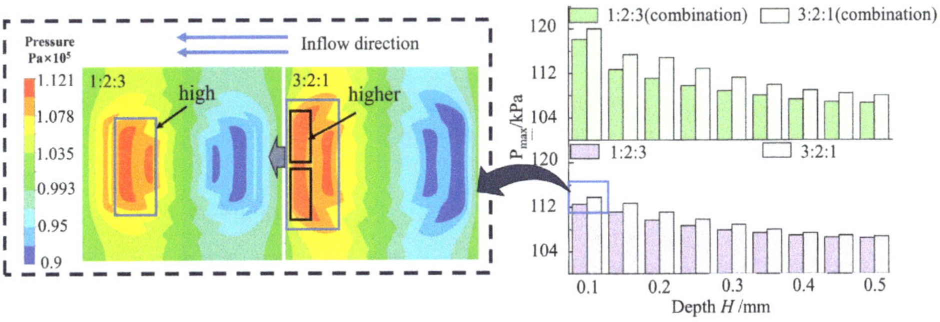

Figure 8. Variation of pressure of oil film with gradient texture of different widths and groove depth.

The pressure distribution shown in Figure 8 indicates the arrangement of the 1:2:3 and 3:2:1 groove width at a groove depth of 0.1 mm. In these configurations, high pressure tends to concentrate in the two wilder grooves. The wilder the outermost circles of the grooves, the higher the maximum pressure value. A wider groove width enhanced the dynamic effect on the groove cross-section, leading to more effective development and a greater likelihood of achieving higher maximum pressure. In the multilayer grooves analyzed above, the primary pressure-bearing role was assumed by the outer two circles of grooves, necessitating wider outer grooves in the groove arrangement and facilitating increased bearing capacity of the structure.

Figure 9 illustrates the variation of wall pressure on the oil film for two sets of multi-layered hexagonal groove weaves with varying groove depth gradient rows as the depth of the shallowest groove increased sequentially.

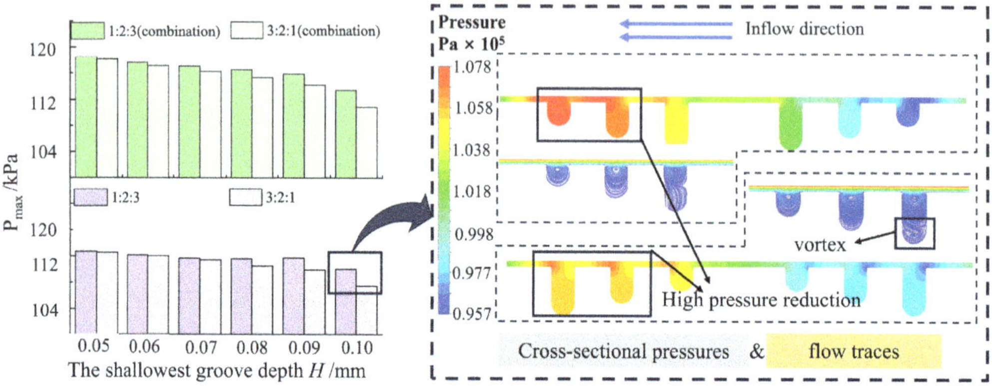

Figure 9. Variation of oil film pressure with grooves depth in different gradients.

A multilayer hexagonal groove texture with a 1:2:3 groove depth gradient from outer to inner exhibited higher maximum pressure on the upper wall of the oil film and superior load-bearing capacity of the oil film compared with a multilayer hexagonal groove texture of 3:2:1 groove depth gradient. Maximum pressure decreased with increasing depth.

The pressure maps at the cross-sections of the grooves elucidated the above phenomenon. The pressure map of the groove cross-section, combined with the description of the pressure map at the cross-section of the multilayer hexagonal texture by the depth of the grooves in Section 3.1 above, indicated that the outer two grooves primarily bear pressure in the multilayer hexagonal texture with three turns of grooves. In the 1:2:3 groove depth arrangement, the pressure was greater in the outermost and middle groove sections, with minimal energy loss in the deepest grooves in the absence of vortices. Conversely, in the 3:2:1 groove depth gradient arrangement, the outermost grooves were affected by the energy loss due to the formation and development of vortices. Thus, it was difficult for hydrodynamic effects to counteract the energy loss from the vortices, resulting in lower maximal pressure on the oil film.

A comparison of two hexagonal textures with two different groove widths and depth gradients revealed that multilayer groove textures with wider outermost groove widths and shallower depths, as well as groove gradient rows, exhibited exceptional performance.

4. Rubber Friction Experiment on Fluid Texturing in Real Well Conditions

The selection of appropriate laser processing parameters was crucial for enhancing the quality of surface textile processing. The surface morphology of the rubber texture was processed using a nanosecond laser with (a) ten times scanning (Figure 10a), (b) five times scanning (Figure 10b), (c) one times scanning (Figure 10c). Notably, at a constant laser power of 20 W, the surface texture of the rubber sample exhibited superior processing quality with <5 scans.

The study used the Keyence three-dimensional confocal microscope (VK-X1050). Japan KEYENCE, to observe and analyze the depth of the grooves of the multilayer hexagonal weave. The study took two auxiliary sections in each of the six directions of the weave and measured the perpendicular distance between the highest point and the lowest point of the groove at the maximum depth of the grooves here, $H_{max}(i)$. The average value was the maximum depth of the multilayer hexagonal groove weave, *Hmax*.

Multilayer hexagonal textures with depths of 0.01 mm, 0.05 mm, 0.1 mm, and 0.15 mm were selected for nanosecond laser processing and tribological experiments. Table 1 gives the depths of the rubber weave grooves after nanosecond laser processing as observed and analyzed with Keyence three-dimensional confocal microscope (VK-X1050), Japan KEYENCE.

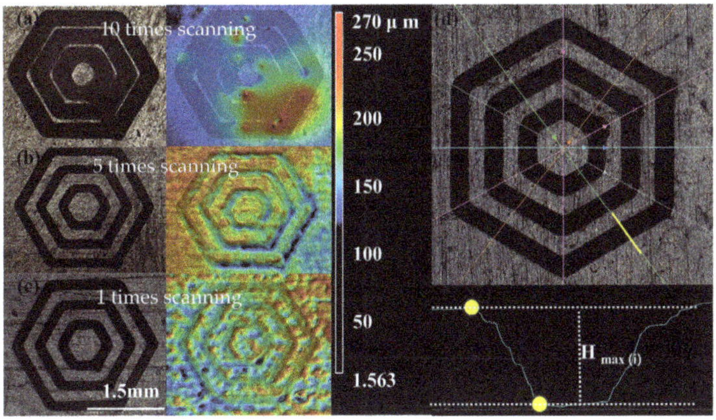

Figure 10. Influence of laser processing parameters on the morphology of rubber textures (**a–c**) and measurement of groove depth (**d**).

Table 1. Machining depth in nanoseconds.

Number	Design Depth (mm)	Number of Scans	Scanning Power (W)	Maximum Machining Depth (mm)	SD
1	0.01	2	20	0.037	0.00947
2	0.05	2	20	0.061	0.00723
3	0.1	5	20	0.107	0.00682
4	0.15	5	20	0.126	0.00578

Standard deviation was calculated as follows:

$$\sigma = \sqrt{\frac{\sum_{i=1}^{12} \left(H_{max}(i) - \overline{\mathrm{H}_{\mathrm{max}}}\right)^2}{12}} \quad (9)$$

The experiment was conducted under actual well conditions, employing liquid lubrication in an oilfield with shale oil. This setup simulated the working environment of a downhole screw pump for tribological experiments. Applying different loads to better simulate the friction characteristics of rubber after pressure deformation during downhole screw pump stator-rotor interference fit (Figure 11a). The friction coefficient of the rubber specimen exhibited a decreasing trend during the first 200 s of the friction experiment due to the high initial pressure. Subsequently, it gradually reached a plateau after 200 s. The friction factor value after the stabilization of the friction curve served as the standard for evaluating the tribological performance of different textures.

Figure 11b shows the tribological characteristic curves of four nanosecond laser-processed multilayer hexagonal texture types with different groove depths under a 10 N load condition. Among these, the smallest friction factor was observed for the multilayer hexagonal texture with a groove depth of 0.107 mm, followed by the multilayer hexagonal texture with a groove depth of 0.126 mm. Conversely, the highest friction factor value was observed in the multilayer hexagonal texture with a groove depth of 0.037 mm. This indicated that the friction and drag reduction effects of multilayer grooves were initially enhanced and subsequently weakened within ~0.1 mm groove depth. Thus, with the groove depth range of ~0.1 mm, the friction and drag reduction effects of multilayer grooved hexagonal textures exhibited a clear relationship; deeper groove depths of 0.1 mm resulted in an enhanced effect.

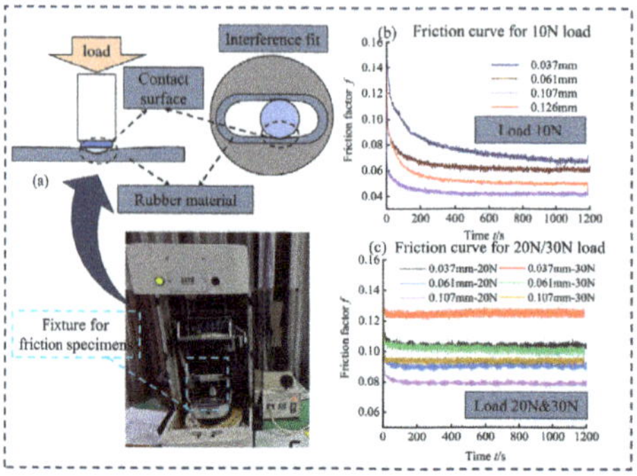

Figure 11. (**a–c**) Friction test of textured rubber specimen after soaking.

As a type of positive displacement pump, the fit between the stator and rotor of the screw pump is typically an interference fit. To achieve better sealing and thereby obtain higher volumetric efficiency, the UMT reciprocating tester, which applies load caused by deformation, can simulate very well the rubber deformation of the stator-rotor interference fit of the screw pump. Different applied loads result in different deformations of the rubber, which leads to different friction coefficients of the rubber. For example, we changed the applied load to 20 N and 30 and found that the higher the load, the higher the coefficient of friction. The reason for this phenomenon is that, as the rubber weave grooves receive pressure extrusion, the amount of lubricant stored inside the grooves decreases, which affects the dynamic pressure lubrication effect and causes the coefficient of friction to increase in the presence of increased loads.

Figure 12 gives the surface image of the laser-machined texturized rubber after an experiment with actual well-fluid lubricated friction under 30 N load conditions. Laser ablation marks on the groove edges are evident in Figure 12d. Tiny corrosion pits are more likely to appear in the laser heat radiation area inside and around the edges of the grooves in Figure 12b after oil soaking. The rubber used in the experiment is a special wear-resistant rubber for the stator bushing of the screw pump. The presence of the weaving structure also enhances the pressure-bearing capacity of the rubber surface. There are no obvious wear traces after the 1200 s friction experiment.

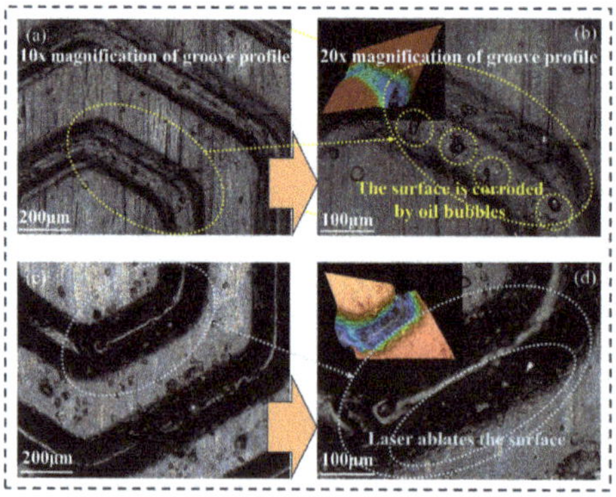

Figure 12. (**a–d**) Microscopic image of laser-machined rubber surface after oil immersion(10× & 20×).

5. Conclusions

(1) The stator rubber surface of the screw pump was designed with woven textures. Simulations of oil film fluid on the textured surface and tribological characteristics under actual well lubrication conditions indicated that a suitable multilayer hexagonal

groove texture enhanced the bearing capacity of the surface, exhibiting a more effective friction and drag reduction.

(2) In simulation studies and tribological experiments, it was observed that increasing the depth of hexagonal grooves to ~0.1 mm significantly enhanced the load-carrying capacity as well as friction and drag reduction. However, the increase in groove depth at $\geq$0.1 mm led to larger vortex volume inside the grooves, forming new vortexes and resulting in a greater energy loss and a decrease in load-carrying capacity.

(3) For composite multilayer hexagonal textures combining pits and grooves, the depth of the pits significantly affected the load-carrying capacity. The study revealed that deeper pits led to higher maximum pressures on the oil film, thereby increasing the load-carrying capacity.

(4) Among the four types of multilayer grooved textures with unequal groove arrangements, a quantitative gradient arrangement method was employed. Fluid simulation analysis indicated that the outermost groove was wider in width and shallower in depth compared with the inner groove, exhibiting superior load-carrying capacity and a more effective friction and drag reduction.

(5) Through the friction experiments of weaving structure oil lubrication under different loads, this study found that the larger the normal load suffered by the rubber specimen in the study range, the larger the coefficient of friction, which corresponded to the stator rotor of the screw pump being more likely to cause serious wear with more interference fit.

Author Contributions: Conceptualization, X.L. and X.N.; Data curation, X.L., X.N., C.L., X.S. and Y.S.; Formal analysis, X.L., X.N., Z.H. and S.H.; Funding acquisition, X.L. and C.L.; Investigation, X.L. and X.N.; Methodology, X.L., X.N. and C.L.; Project administration, X.L., X.N., Z.H. and S.H.; Resources, X.L., X.N., C.L., Z.H. and S.H.; Supervision, X.L., X.N. and C.L.; Validation, X.L., X.N., X.S. and Y.S.; Visualization, X.L., X.N., X.S. and Y.S.; Writing—original draft, X.L., X.N., X.S. and Y.S.; Writing—review & editing, X.L., X.N., Y.W. and H.T.; Investigation, X.S., Y.S., Y.W. and H.T. All authors have read and agreed to the published version of the manuscript.

Funding: This work was supported by the National Natural Science Foundation of China (Grant Nos. 52074161 and 52005281), the Taishan Scholar Project of Shandong Province (Grant No. tsqn202211177), the Shandong Provincial Plan for Introduction and Cultivation of Young Pioneers in Colleges and Universities (Grant No. 2021-Qing Chuang-30613019), the Natural Science Foundation of Shandong Province (Grant Nos. ZR2022ME173 and ZR2023QE011, and the Project of CNOOC Research Institute Ltd. (Grant No. CCL2023RCPS2129RQN).

Institutional Review Board Statement: Not applicable.

Informed Consent Statement: Not applicable.

Data Availability Statement: The raw data supporting the conclusions of this article will be made available by the authors on request.

Acknowledgments: The authors sincerely thank the anonymous reviewers for their insightful comments and suggestions.

Conflicts of Interest: Authors Y.W. and H.T. was employed by the Lusheng Petroleum Development Company Ltd., SINOPEC. The remaining authors declare that the research was conducted in the absence of any commercial or financial relationships that could be construed as a potential conflict of interest.

References

1. Alves, M.V.; Barbosa, J.R., Jr.; Prata, A.T.; Ribas, F.A., Jr. Fluid flow in a screw pump oil supply system for reciprocating compressors. *Int. J. Refrig.* **2011**, *34*, 74–83. [CrossRef]
2. Bi, H.B.; Wu, M.M.; Zhang, X.L. Design of parameters optimization system for screw pump well. *Int. J. Simul. Syst. Sci. Technol.* **2016**, *17*, 11–16.
3. Zhu, J.J.; Zhang, H.Q. A review of experiments and modeling of gas-liquid flow in electrical submersible pumps. *Energies* **2018**, *11*, 180. [CrossRef]

4. Hao, Z.X.; Zhu, S.J.; Pei, X.H.; Huang, P.; Tong, Z.; Wang, B.Y.; Li, D.Y. Submersible direct-drive progressing cavity pump rodless lifting technology. *Pet. Explor. Dev.* **2019**, *46*, 621–628. [CrossRef]
5. Shishlyannikov, D.; Zverev, V.; Ivanchenko, A.; Zvonarev, I. Increasing the Time between Failures of Electric Submersible Pumps for Oil Production with High Content of Mechanical Impurities. *Appl. Sci.* **2021**, *12*, 64. [CrossRef]
6. Dong, K.X.; Li, Q.E.; Zhang, Z.H.; Jiang, M.Z.; Xu, S.F. Submersible Screw Pump Fault Diagnosis Method Based on a Probabilistic Neural Network. *J. Appl. Sci. Eng.* **2022**, *25*, 1067–1075.
7. Ma, Z.F. The structure optimisation of hydraulic mechanical screw pump. *Int. J. Simul. Process Model.* **2019**, *14*, 28–35. [CrossRef]
8. Joel Romero, O.; Hupp, A. Subsea electrical submersible pump significance in petroleum offshore production. *J. Energy Resour. Technol.* **2014**, *136*, 012902. [CrossRef]
9. Han, C.J.; Ren, X.Y.; Zheng, J.P. Analysis of abrasion for stator bushings of equal thickness screw pump in high temperature and sand crude oil. *China Mech. Eng.* **2017**, *28*, 446–450.
10. Lv, X.R.; Wang, S.J.; Huo, X.Y. The effect of water content in crude oil on the tribological behavior of screw pump's stator rubber. *Chin. Sci. Bull.* **2014**, *59*, 2868–2874. [CrossRef]
11. Wang, H.M.; Wang, S.J.; Lv, X.R. The effects of temperature on the mechanical and tribological properties of progressing cavity pump NBR stator rubber. *Mechanics* **2016**, *22*, 308–312. [CrossRef]
12. Borisova, K.E.; Ivanova, T.N.; Latypov, R.G. Study of screw pump stator and rotor working capacity to increase the output. *Procedia Eng.* **2017**, *206*, 688–691. [CrossRef]
13. Liu, X.F.; Liu, C.H.; Li, Q.P.; Yao, H.Y.; Hao, Z.X.; Liu, Y.F. Two-phase flow chokeling results and control in high-pressure gas wells producing water. *Coal Geol. Explor.* **2024**, *52*, 48–55.
14. Wang, M.; Li, Q.; Huang, Z.; Qian, W.; Chen, X.; Li, Q.; Lai, T. Influence of surface texture parameters of screw pump rotor on tribological properties of its friction pairs. *Ind. Lubr. Tribol.* **2022**, *74*, 964–974. [CrossRef]
15. Li, Y.F.; Nie, D.L.; Cai, Z.Q. The performance analysis of screw pump stator elastomers: Polyamide 6/hydrogenated nitrile blends: Mechanical, oil resistance and tribological properties. *Polym. Test.* **2023**, *128*, 108226. [CrossRef]
16. Peng, S.J.; Huang, Z.Q.; Wang, M.L.; Xiong, C.; Chen, K. Study on the Influence of Surface Texture Parameters of Oil Production Screw Motor Rotor on the Tribological. *Processes* **2023**, *11*, 618.
17. Wang, M.L.; Huang, Z.Q.; Qian, W.; Li, G. Study of the effect of laser textured rotors on the starting performance of metal–rubber mating pairs under different lubricating media environments. *Proc. Inst. Mech. Eng. Part J J. Eng. Tribol.* **2024**, *238*, 450–463. [CrossRef]
18. Ivanov, T.; Bührig-Polaczek, A.; Vroomen, U. Casting of microstructured shark skin surfaces and possible applications on aluminum casting parts. *China Foundry* **2011**, *8*, 62–65.
19. Coblas, D.G.; Fatu, A.; Maoui, A.; Hajjam, M. Manufacturing textured surfaces: State of art and recent developments. *Proc. Inst. Mech. Eng. Part J J. Eng. Tribol.* **2015**, *229*, 3–29. [CrossRef]
20. Pawlus, P.; Reizer, R.; Wieczorowski, M. Functional importance of surface texture parameters. *Materials* **2021**, *14*, 5326. [CrossRef]
21. Huang, Q.; Shi, X.; Xue, Y.; Zhang, K.; Gao, Y.; Wu, C. Synergetic effects of biomimetic microtexture with multi-solid lubricants to improve tribological properties of AISI 4140 steel. *Tribol. Int.* **2022**, *167*, 107395. [CrossRef]
22. Qin, S.; Shi, X.; Xue, Y.; Zhang, K.; Huang, Q.; Wu, C.; Shu, J. Coupling effects of bionic textures with composite solid lubricants to improve tribological properties of TC4 alloy. *Tribol. Int.* **2022**, *173*, 107691. [CrossRef]
23. Costa, H.L.; Schille, J.; Rosenkranz, A. Tailored surface textures to increase friction—A review. *Friction* **2022**, *10*, 1285–1304. [CrossRef]
24. Nsilani Kouediatouka, A.; Ma, Q.; Liu, Q.; Mawignon, F.J.; Rafique, F.; Dong, G. Design methodology and application of surface texture: A review. *Coatings* **2022**, *12*, 1015. [CrossRef]
25. Song, J.Y.; Huang, H.; Wang, X.G.; Shi, W.B. Status and prospects of surface texturing: Design, manufacturing and applications. *Surf. Sci. Technol.* **2023**, *1*, 21. [CrossRef]
26. Wei, H.; Chen, G.; Chen, Z.; Yu, Z.; Huang, J. Progress on Bionic Textured Cutting Tools: A Review and Prospects. *J. Bionic Eng.* **2024**, *21*, 19–55. [CrossRef]
27. Luo, Y.Q.; Li, J.L.; Yao, B.H.; Luo, Q.; Zhu, Z.C.; Wu, W.B. Research progress and development trend of bionic harvesting technology. *Comput. Electron. Agric.* **2024**, *222*, 109013. [CrossRef]
28. Mei, X.Q.; Liu, C.J.; Li, Z.X. Research progress on functional, structural and material design of plant-inspired green bionic buildings. *Energy Build.* **2024**, *316*, 114357. [CrossRef]
29. Shi, X.; Ni, T. Effects of groove textures on fully lubricated sliding with cavitation. *Tribol. Int.* **2011**, *44*, 2022–2028. [CrossRef]
30. Arslan, A.; Masjuki, H.H.; Kalam, M.A.; Varman, M.; Mufti, R.A.; Mosarof, M.H.; Quazi, M.M. Surface texture manufacturing techniques and tribological effect of surface texturing on cutting tool performance: A review. *Crit. Rev. Solid State Mater. Sci.* **2016**, *41*, 447–481. [CrossRef]
31. Sugihara, T.; Enomoto, T. Performance of cutting tools with dimple textured surfaces: A comparative study of different texture patterns. *Precis. Eng.* **2017**, *49*, 52–60. [CrossRef]
32. Zhang, H.; Dong, G.; Hua, M.; Chin, K. Improvement of tribological behaviors by optimizing concave texture shape under reciprocating sliding motion. *J. Tribol.* **2017**, *139*, 011701. [CrossRef]
33. Antoszewski, B.; Kurp, P. Effect of surface texture on the sliding pair lubrication efficiency. *Lubricants* **2022**, *10*, 80. [CrossRef]
34. Zhao, L.X.; Zhang, B.L.; Liu, Y. Research progress in improving tribological properties based on surface texture technology. *Tribology* **2022**, *42*, 202–224. (In Chinese)

35. Zhang, B.L. Study on Surface Texturing Flow Field Characteristics and Anti-friction Performance of Screw Pumped NBR. Master's Thesis, Northeast Petroleum University, Daqing, China, 2021. (In Chinese).
36. Tan, N.; Xing, Z.G.; Wang, H.D. Research progress of geometric configuration and its function based on bionic principle. *J. Mater. Eng.* **2018**, *46*, 133–140.
37. Xing, Y.; Deng, J.; Wang, X.; Meng, R. Effect of laser surface textures combined with multi-solid lubricant coatings on the tribological properties of Al_2O_3/TiC ceramic. *Wear* **2015**, *342*, 1–12. [CrossRef]
38. Liu, X.F.; Liu, C.H.; Hao, Z.X.; Zheng, Y.; Zhang, K.; Wang, J.F.; Wei, S.B.; Hao, A.G.; Tao, J.L.; Cheng, H. Non-symmetric distributions of solids deposition for solid-water stratified flow in deviated tubing strings. *Pet. Sci.* **2023**, *20*, 3048–3061. [CrossRef]
39. Li, J.; Han, K.; Chen, W.; Tang, X.; Wang, Q. An improved TEHL analysis of textured roller bearings consider various texture parameters and slip. *Ind. Lubr. Tribol.* **2020**, *73*, 373–380. [CrossRef]
40. Jiang, Y.; Yan, Z.; Zhang, S.; Shen, Z.; Sun, H. Research on cavitation effect of microtextured array. *Sci. Rep.* **2022**, *12*, 13455. [CrossRef]
41. Fu, H.; He, Y.; Yang, J.; Fu, Y.; Yin, B.; Zhang, Y.; Zhou, Y. Enhancing adhesion strength of PVD AlCrN coating by novel volcano-shaped micro-textures: Experimental study and mechanism insight. *Surf. Coat. Technol.* **2022**, *445*, 128712. [CrossRef]
42. Feng, X.; Wang, R.; Wei, G.; Zheng, Y.; Hu, H.; Yang, L.; Zhou, H. Effect of a micro-textured surface with deposited MoS2-Ti film on long-term wear performance in vacuum. *Surf. Coat. Technol.* **2022**, *445*, 128722. [CrossRef]
43. Etsion, I. State of the art in laser surface texturing. *J. Tribol.* **2005**, *127*, 248–253. [CrossRef]
44. Hecht, J. Short history of laser development. *Opt. Eng.* **2010**, *49*, 091002. [CrossRef]
45. Ramesh, A.; Akram, W.; Mishra, S.P.; Cannon, A.H.; Polycarpou, A.A.; King, W.P. Friction characteristics of microtextured surfaces under mixed and hydrodynamic lubrication. *Tribol. Int.* **2013**, *57*, 170–176. [CrossRef]
46. Roushan, A. Influence of laser parameters on the machining performance of textured cutting tools. *Opt. Laser Technol.* **2023**, *165*, 109569. [CrossRef]
47. Obikawa, T.; Kamio, A.; Takaoka, H.; Osada, A. Micro-texture at the coated tool face for high performance cutting. *Int. J. Mach. Tools Manuf.* **2011**, *51*, 966–972. [CrossRef]
48. Mescola, A.; Paolicelli, G.; Ogilvie, S.P.; Guarino, R.; McHugh, J.G.; Rota, A.; Tripathi, M. Graphene confers ultralow friction on nanogear cogs. *Small* **2021**, *17*, 2104487. [CrossRef]
49. Qiu, Y.; Khonsari, M.M. Experimental investigation of tribological performance of laser textured stainless steel rings. *Tribol. Int.* **2011**, *44*, 635–644. [CrossRef]
50. Grabon, W.; Koszela, W.; Pawlus, P.; Ochwat, S. Improving tribological behaviour of piston ring–cylinder liner frictional pair by liner surface texturing. *Tribol. Int.* **2013**, *61*, 102–108. [CrossRef]
51. Yu, H.W.; Wang, X.L.; Zhou, F. Geometric shape effects of surface texture on the generation of hydrodynamic pressure between conformal contacting surfaces. *Tribol. Lett.* **2010**, *37*, 123–130. [CrossRef]
52. Shivakoti, I.; Kibria, G.; Cep, R.; Pradhan, B.B.; Sharma, A. Laser surface texturing for biomedical applications: A review. *Coatings* **2021**, *11*, 124. [CrossRef]
53. Etsion, I. Improving tribological performance of mechanical components by laser surface texturing. *Tribol. Lett.* **2004**, *17*, 733–737. [CrossRef]
54. Wahab, J.A.; Ghazali, M.J.; Yusoff, W.M.W.; Sajuri, Z. Enhancing material performance through laser surface texturing: A review. *Trans. IMF* **2016**, *94*, 193–198. [CrossRef]
55. Andersson, P.; Koskinen, J.; Varjus, S.; Gerbig, Y.; Haefke, H.; Georgiou, S.; Buss, W. Microlubrication effect by laser-textured steel surfaces. *Wear* **2007**, *262*, 369–379. [CrossRef]
56. Wu, X.F.; Zhan, J.M.; Mei, S.L. Optimization of micro-texturing process parameters of TiAlN coated cutting tools by femtosecond laser. *Materials* **2022**, *15*, 6519. [CrossRef]
57. Mao, B.; Siddaiah, A.; Liao, Y.L.; Menezes, P.L. Laser surface texturing and related techniques for enhancing tribological performance of engineering materials: A review. *J. Manuf. Process.* **2020**, *53*, 153–173. [CrossRef]
58. Kovalchenko, A.; Ajayi, O.; Erdemir, A.; Fenske, G.; Etsion, I. The effect of laser surface texturing on transitions in lubrication regimes during unidirectional sliding contact. *Tribol. Int.* **2005**, *38*, 219–225. [CrossRef]
59. Maressa, P.; Anodio, L.; Bernasconi, A.; Demir, A.G.; Previtali, B. Effect of surface texture on the adhesion performance of laser treated Ti6Al4V alloy. *J. Adhes.* **2015**, *91*, 518–537. [CrossRef]
60. Liu, X.; Shi, X.; Hao, Z.; Wei, S.; Sun, Y.; Niu, X.; Liu, C.; Li, M.; Li, Z. Effect of Complex Well Conditions on the Swelling and Tribological Properties of High-Acrylonitrile Stator Rubber in Screw Pumps. *Polymers* **2024**, *16*, 2036. [CrossRef]

fluids

MDPI

Article

Numerical Solution of the Newtonian Plane Couette Flow with Linear Dynamic Wall Slip

Muner M. Abou Hasan [1], Ethar A. A. Ahmed [2], Ahmed F. Ghaleb [3], Moustafa S. Abou-Dina [3] and Georgios C. Georgiou [4,*]

1 Faculty of Mathematics and Data Sciences, Emirates Aviation University, Dubai P.O. Box 53044, United Arab Emirates; muneere@live.com
2 School of Engineering and Applied Sciences, Nile University, Giza 12588, Egypt; ethar_ahmed54@yahoo.com
3 Department of Mathematics, Faculty of Science, Cairo University, Giza 12613, Egypt; afghaleb@gmail.com (A.F.G.); moustafa_aboudina@hotmail.com (M.S.A.-D.)
4 Department of Mathematics and Statistics, University of Cyprus, P.O. Box 20537, Nicosia 1678, Cyprus
* Correspondence: georgios@ucy.ac.cy

Abstract: An efficient numerical approach based on weighted-average finite differences is used to solve the Newtonian plane Couette flow with wall slip, obeying a dynamic slip law that generalizes the Navier slip law with the inclusion of a relaxation term. Slip is exhibited only along the fixed lower plate, and the motion is triggered by the motion of the upper plate. Three different cases are considered for the motion of the moving plate, i.e., constant speed, oscillating speed, and a single-period sinusoidal speed. The velocity and the volumetric flow rate are calculated in all cases and comparisons are made with the results of other methods and available results in the literature. The numerical outcomes confirm the damping with time and the lagging effects arising from the Navier and dynamic wall slip conditions and demonstrate the hysteretic behavior of the slip velocity in following the harmonic boundary motion.

Keywords: plane Couette flow; dynamic wall slip; Navier slip; weighted-average finite differences; hysteretic behavior

1. Introduction

In the past few decades, there has been a growing interest in the study of Newtonian and non-Newtonian viscometric flows in the presence of static or dynamic wall slip conditions for their importance in rheometry and in industrial applications [1–3]. Reviews of the slip conditions prevailing at the fluid–structure interface for various media of practical importance have been reported by Hatzikiriakos [2,3]. Malkin and Patlazan [4] also reviewed wall slip in complex fluids of different types, focusing on fluid–wall interaction and shear-induced fluid-to-solid transitions.

The simplest dynamic wall slip equation in the case of unidirectional one-dimensional Newtonian flow, such that the x-velocity component is $v = v(y,t)$ and the plane $y = 0$ represents a wall, reads as follows [1]:

$$v_s(0,t) + \lambda \frac{\partial v_s}{\partial t}(0,t) = \frac{\mu}{\beta}\left|\frac{\partial v}{\partial y}(0,t)\right|, \tag{1}$$

where v_s is the slip velocity, defined as the relative velocity of the fluid particles adjacent to the wall with respect to that of the wall, μ is the constant fluid viscosity, β is the slip coefficient, and λ is the slip relaxation parameter. When the latter parameter vanishes, Equation (1) is reduced to the Navier slip condition [5]:

$$v_s(0,t) = \frac{\mu}{\beta}\left|\frac{\partial v}{\partial y}(0,t)\right|. \tag{2}$$

If the wall is fixed, then $v_s = v$ and Equation (1) can be written as follows:

$$v(0,t) + \lambda_s \frac{\partial v}{\partial t}(0,t) = \frac{\mu}{\beta}\left|\frac{\partial v}{\partial y}(0,t)\right|,$$

Abbatiello et al. [6] presented the mathematical analysis of Navier–Stokes-like problems involving a boundary where dynamic slip applies. Ferrás et al. [7] presented analytical and semi-analytical solutions to some linear and nonlinear problems for Couette and Poiseuille flows for Newtonian and non-Newtonian media with slip boundary conditions at different walls. Thalakkottor and Mohseni [8] used molecular dynamic simulations to study slip at the fluid–solid boundary in an unsteady flow based on Stokes' second problem, when the wall undergoes an oscillatory motion. They showed the existence of dynamic wall slip and discussed the resulting hysteresis phenomena. Hysteresis was attributed to the unsteady inertial forces of the fluid. Kaoullas and Georgiou [9] derived analytical solutions to some Poiseuille and Couette problems including dynamic wall slip and discussed its effects on the flow development. More recent work on the flow of power law fluids in circular annuli and analytical approximate solutions was presented by Deterre et al. [10]. Ali et al. [11] solved the axial, annular Couette flow of a Newtonian viscous fluid of constant density, taking into account both Navier and dynamic slip boundary conditions, using the Laplace transform technique and inversion by Laguerre polynomials. Farragui et al. [12] used separation of variables to derive analytical solutions to the problem of cessation of annular Poiseuille and Couette flows of a Newtonian fluid with dynamic wall slip.

The analytical solution of flows with dynamic wall slip is possible only for linear problems, e.g., for Newtonian flows with a linear slip equation, such as Equation (1). However, the visualisation of the solution and the analysis of the flow still require numerical calculations, which may not be trivial (see, e.g., [13]). Moreover, the slip equation may be nonlinear [2,3]. Hence, it is necessary to use numerical methods to approximate the solution of this model. These approximation techniques require great effort. In the literature, there exist many methods used to numerically solve partial differential equations: spectral methods [14]; finite element [15]; finite difference, e.g., the weighted average finite difference method (WAFDM) [16,17]; and the collocation method [18,19].

The above literature review clearly shows the importance of using numerical schemes for the solution of Couette and Poiseuille flows of power-law fluids, side by side with the possibility of obtaining exact or approximate analytical solutions.

In the present work, we use an efficient WAFDM scheme to solve the Newtonian planar Couette flow when static or dynamic wall slip applies along the fixed plate and the other plate moves either at constant or oscillatory speed. The evolution of velocity and the volumetric flow rate are calculated for three cases: constant, oscillating, and single-period sinusoidal plate velocity. For the last two cases, the hysteretic behavior of the fluid in following the motion of the wall is put in evidence, as this is implies energy dissipation.

2. Governing Equations

Consider the time-dependent Couette flow of a viscous fluid between infinite parallel walls placed a distance d apart, as illustrated in Figure 1. The fluid is initially at rest and suddenly the upper wall starts moving with velocity $V_0 f(t)$ in the x direction. The x-momentum equation takes the form

$$\rho \frac{\partial v}{\partial t} = \mu \frac{\partial^2 v}{\partial y^2}, \quad t > 0, \quad 0 \leq y \leq d, \tag{3}$$

where ρ is the constant fluid density. Regarding the boundary conditions, the dynamic slip law applies at the fixed wall ($y = 0$) and no slip is assumed at the upper wall ($y = d$). Hence, the boundary conditions of the problem read

$$v(0,t) + \lambda_s \frac{\partial v}{\partial t}(0,t) = \frac{\mu}{\beta}\left|\frac{\partial v}{\partial y}(0,t)\right|, \qquad t > 0 \tag{4}$$

and

$$v(d,t) = V_0 f(t), \qquad t > 0. \tag{5}$$

The initial condition is

$$v(y,0) = 0, \qquad 0 \leq y \leq d. \tag{6}$$

We work with the dimensionless equations scaling lengths by d, time by $d^2\rho/\mu$, and the velocity by the characteristic velocity V_0 of the moving wall. For the sake of simplicity, we keep the same symbols for the dimensionless variables. Hence, the dimensionless form of the initial boundary value problem defined by Equations (3)–(6) is

$$\frac{\partial v}{\partial t} = \frac{\partial^2 v}{\partial y^2}, \quad t > 0, \quad 0 \leq y \leq 1, \tag{7}$$

$$v(0,t) + \Lambda_s \frac{\partial v}{\partial t}(0,t) = \frac{1}{B}\frac{\partial v}{\partial y}(0,t), \quad t > 0 \tag{8}$$

$$v(1,t) = f(t), \qquad t > 0 \tag{9}$$

$$v(y,0) = 0, \qquad 0 \leq y \leq 1 \tag{10}$$

where

$$B \equiv \frac{\beta d}{\mu} \tag{11}$$

is the dimensionless slip number, and

$$\Lambda \equiv \frac{\lambda \mu}{\rho} \tag{12}$$

is the dimensionless slip relaxation number. It should be noted that when $\Lambda_s = 0$, the slip equation is reduced to the static Navier slip equation. Moreover, when $B \to \infty$, the no-slip condition is recovered.

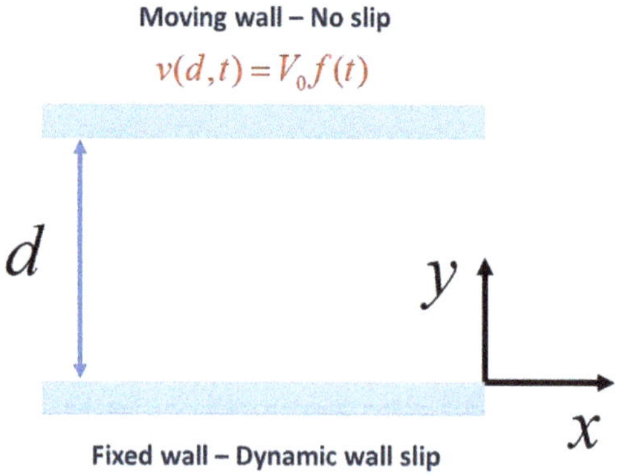

Figure 1. Geometry of plane Couette flow.

3. Weighted-Average Finite Difference Method

Here, we use a WAFDM scheme [20] to simulate and study the behavior of solutions for problems (7)–(10) of planar Couette flow with Navier and dynamic slip conditions prevailing at the fixed wall only. The motion is triggered by assigning a given motion at the other boundary in its own plane.

The WAFDM is a widely used numerical technique for solving differential equations by approximating the derivatives of a function at discrete points in space and time using finite differences. The discretized solution is eventually obtained by solving a system of algebraic equations [16]. The WAFDM is relatively simple to implement, computationally efficient, and can be applied to a wide range of problems [20]. Its formulation for the problem of Equations (7)–(10) is outlined below. The domain $[0,1] \times [0,T]$ in the (y,t)-plane is discretized by a uniform grid with steps $h = \Delta y$ and $k = \Delta t$ so that

$$h = \frac{1}{N}, \qquad k = \frac{T}{M}, \tag{13}$$

where N and M are the numbers of subintervals used for y and t, respectively. Hence, the coordinates of the grid points are

$$y_n = n\,\Delta y, \ \ n = 0,\ 1,\ 2,\ \dots,\ N, \qquad t_m = m\,\Delta t, \ \ m = 0,\ 1,\ 2,\ \dots,\ M.$$

The numerical values of the variable v and the function f at the general grid point (y_n, t_m) are denoted, respectively, by v_n^m and f_n^m. The following difference approximations are used for the time and spatial derivatives of the problem:

$$\left(\frac{\partial v}{\partial t}\right)_n^m = \frac{v_n^{m+1} - v_n^m}{k} + O(k), \tag{14}$$

$$\left(\frac{\partial v}{\partial y}\right)_n^m = \frac{v_{n+1}^m - v_n^m}{h} + O(h), \tag{15}$$

and

$$\left(\frac{\partial^2 v}{\partial y^2}\right)_n^m = \frac{v_{n+1}^m - 2v_n^m + v_{n-1}^m}{h^2} + O(h^2). \tag{16}$$

With the WAFDM, the spatial derivatives of v in the RHS of Equations (7) and (8) at time step $m+1$ are approximated as linear combinations of their corresponding values at time steps m and $m+1$ by means of a weight parameter $\theta, 0 \le \theta \le 1$. Substituting Equations (14)–(16) into the governing Equations (7)–(10) leads to a linear system of equations for the unknowns $v_n^m,\ n = 0,\ 1,\ 2, \dots,\ N, m = 0,\ 1,\ 2, \dots, M$:

$$\frac{v_n^{m+1} - v_n^m}{k} = \theta \frac{v_{n+1}^m - 2v_n^m + v_{n-1}^m}{h^2} + (1-\theta)\frac{v_{n+1}^{m+1} - 2v_n^{m+1} + v_{n-1}^{m+1}}{h^2}, \tag{17}$$

$$\theta v_0^m + (1-\theta)v_0^{m+1} + \Lambda_s \frac{v_0^{m+1} - v_0^m}{k} = \frac{\theta}{B}\frac{v_1^m - v_0^m}{h} + \frac{1-\theta}{B}\frac{v_1^{m+1} - v_0^{m+1}}{h}, m = 0,1,2,\dots,M \tag{18}$$

$$v_N^m = f(mk), \ \ m = 1,2,3,\dots,M \tag{19}$$

$$v_0^n = 0, \ \ n = 0,1,2,\dots,N. \tag{20}$$

Depending on the value of the weight factor θ, the method can be explicit (θ = 1; easy and simple for coding and can be used when the function being approximated is relatively smooth and well behaved) or implicit (θ = 0; more accurate, has a larger stability region, and can be used when the function being approximated is not smooth). When $\theta = 0.5$, the Crank–Nicolson implicit scheme is recovered [20]. More information on the stability and convergence of the numerical method can be found in Ref. [20].

After some manipulations and simplifications, the system of Equations (17)–(19) may be cast in matrix form as follows:

$$\mathbf{A}\mathbf{V}^{m+1} = \mathbf{B}\mathbf{V}^{m} + \mathbf{F}^{m}, \tag{21}$$

where $\mathbf{V}^{m+1}$ is the vector of unknown function values at time $m+1$,

$$\mathbf{A} = \begin{pmatrix} (1-\theta)(1+\frac{1}{Bh})+\frac{\Lambda_s}{k} & -\frac{1-\theta}{Bh} & 0 & 0 & \cdots & 0 & 0 \\ a & b & a & 0 & \cdots & 0 & 0 \\ 0 & a & b & a & \cdots & 0 & 0 \\ 0 & 0 & a & b & \cdots & 0 & 0 \\ \vdots & \vdots & \vdots & \vdots & \ddots & \vdots & \vdots \\ 0 & 0 & 0 & 0 & \cdots & b & a \\ 0 & 0 & 0 & 0 & \cdots & 0 & \alpha \end{pmatrix}_{N+1}, \tag{22}$$

$$\mathbf{B} = \begin{pmatrix} -\theta(1+\frac{1}{Bh})+\frac{\Lambda_s}{k} & \frac{\theta}{Bh} & 0 & 0 & \cdots & 0 & 0 \\ a' & b' & a' & 0 & \cdots & 0 & 0 \\ 0 & a' & b' & a' & \cdots & 0 & 0 \\ 0 & 0 & a' & b' & \cdots & 0 & 0 \\ \vdots & \vdots & \vdots & \vdots & \ddots & \vdots & \vdots \\ 0 & 0 & 0 & 0 & \cdots & b' & a' \\ 0 & 0 & 0 & 0 & \cdots & 0 & 0 \end{pmatrix}_{N+1}, \tag{23}$$

and

$$\mathbf{F}^{m} = (0,0,0,0,\ldots,f_n^m)^T_{N+1}, \tag{24}$$

where $f_n^m \equiv f(x_n, t_m)$,

$$a = -(1-\theta)\xi, \qquad b = 1+2(1-\theta)\xi, \qquad \xi = \frac{k}{h} \tag{25}$$

and

$$a' = \theta\xi, \qquad b' = 1-2\theta\xi. \tag{26}$$

The algorithmic steps of the proposed method are summarized in Algorithm 1. The system (21) is easily solved. The local truncation error of the scheme is of order $O(h+k)$. This scheme is conditionally stable when $\theta > 0.5$, and it is unconditionally stable when $\theta \leq 0.5$ [20].

Algorithm 1. Algorithmic steps of the proposed method

```
Initialize parameters and finite difference grid
Calculate initial condition V^0
Build A
% Time marching loop
for m = 1, ..., M
    % Node loop
    for n = 1, ..., N − 1
        Build BV^{m−1} + F^{m−1}
    end
    Solve Equation (21) for V^m
end
Visualize the results
```

4. Results and Discussion

In this section, we use the introduced numerical scheme (17)–(20) to simulate the approximation solution of (7)–(10). It will be shown that the introduced WAFDM provides good approximations for the solution of (7)–(10). It is applicable and efficient for solving the given system of governing equations with the accompanying boundary and initial conditions. Different test examples are carried out to approximate the solutions and compare between the solutions using different techniques. We compared the computational time of the WAFDM and the Legendre spectral collocation method and found that the WAFDM is more efficient.

The results assess the effect of the various parameters in the initial and boundary conditions to which the solution is subjected, demonstrate the efficiency of the introduced technique, and justify the accuracy in comparison with other methods. In our numerical experiments we choose different values for the weight factor and full agreement is reached with the theoretical stability condition.

4.1. Case 1: Plate Moving at a Constant Speed

In order to assess the efficiency of the proposed numerical scheme, we have compared our results (WAFDM) for the special case when $\theta = 0$ and $f(t) = 1$ with those obtained using the MATLAB2021b pdepe toolbox. This is one of the cases treated in [21], where one plate is kept fixed, while the other one is suddenly set to motion from rest with constant speed. In Figure 2, the predictions of pdepe and the WAFDM for $\Lambda_s = 0$ (Navier slip) and three values of the slip number B, i.e., $B = 0.1$ (strong), 1 (moderate), and 10 (weak slip), are compared. The agreement is excellent and this is perhaps more visible when comparing the slip velocities $v(0,t)$ in Figure 3. It should be noted that full agreement is also reached with the results presented in [21] based on a Fourier expansion and on the use of one-sided Fourier transform. The main feature of the solution is the reduction in the slip velocity at the boundary $y = 0$ as parameter B increases. The efficiency on the computational time of the WAFDM is demonstrated by comparing it with that of the Legendre spectral collocation method depending on the Legendre polynomial of degree M in Table 1.

Table 1. Comparison of the WAFDM with the spectral collocation method.

WAFDM		Spectral Method	
N	CPU	M	CPU
500	5.9 s	5	963.6 s
5000	18.3 s	12	7579.9 s
10,000	44.8 s	15	88,937.5 s
50,000	789.9 s	20	219,357.6 s

The volumetric flow rate, defined by

$$Q(t) = \int_0^1 v(y,t)dy, \tag{27}$$

is shown in Figure 4 for $\theta = 0$, $B = 0.1$ (strong slip) and 1 (moderate slip), and $\Lambda_s = 0$ (Navier slip), 1, and 10. The damping effect of the slip relaxation parameter in reaching a fully developed flow is clearly shown. As expected, the initial value of the volumetric flow rate is independent of Λ_s and decreases with the slip parameter B (since wall slip is reduced).

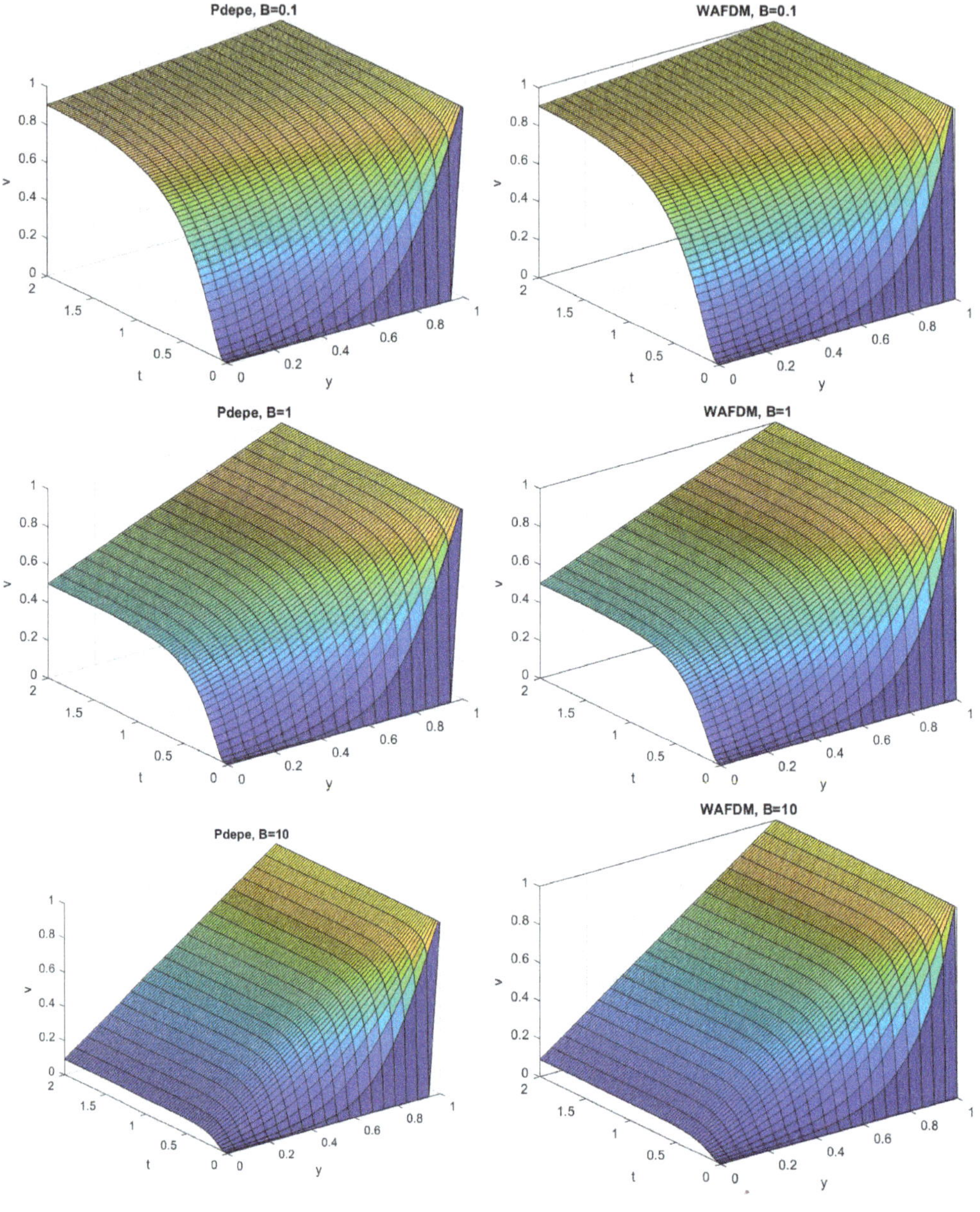

Figure 2. Evolution of the solution $v(y,t)$ in Case 1 (plate moving at constant speed) with $\theta = 0$ and Navier slip ($\Lambda_s = 0$) for $B = 0.1$ (strong slip), 1 (moderate slip), and 10 (weak slip). The results obtained with pdepe.m (**left column**) compare well with those of the WAFDM (**right column**).

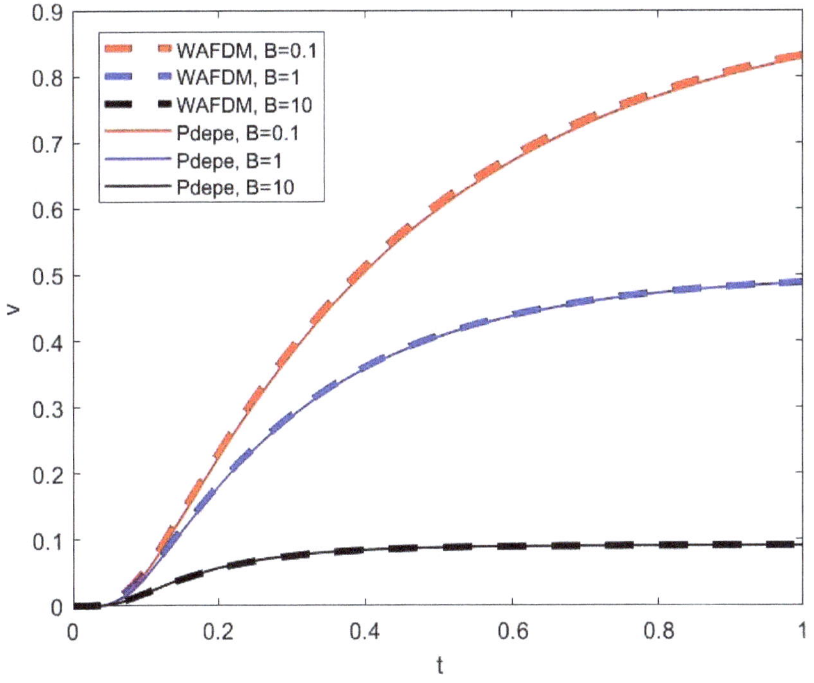

Figure 3. Evolution of the slip velocity $v(0,t)$ using the pdepe function of MATLAB (solid) and the WAFDM (dashed) in Case 1 (plate moving at constant speed and $\Lambda_s = 0$ (Navier slip)) with $\theta = 0$, $B = 0.1$ (strong slip), and 1 (moderate slip), and 10 (weak slip).

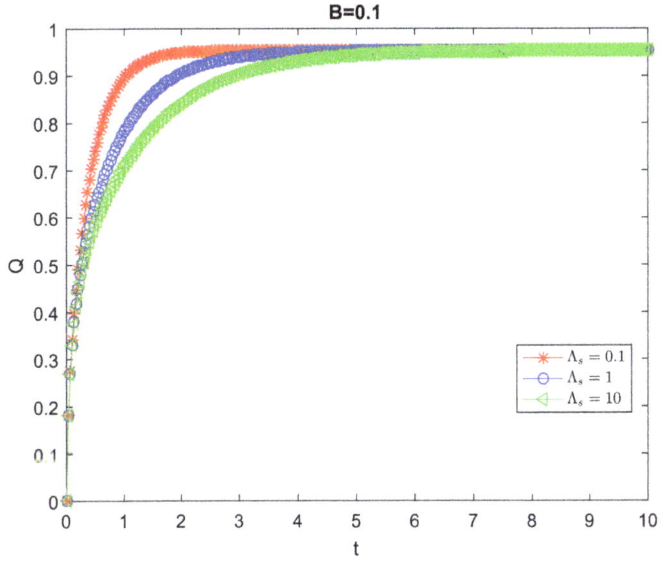

(a)

Figure 4. *Cont.*

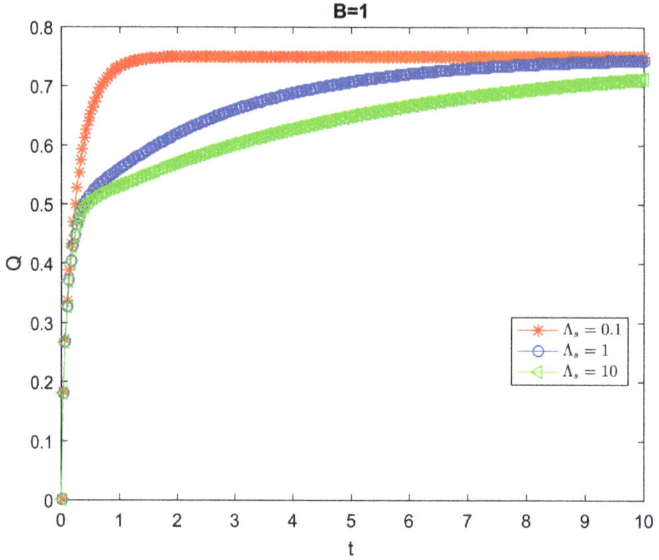

(**b**)

Figure 4. Evolution of the volumetric flow rate $Q(t)$ in Case 1 (plate moving at constant speed) with $\theta = 0$ and $\Lambda_s = 0$ (Navier slip), 1, and 10: (**a**) $B = 0.1$ (strong slip); (**b**) 1 (moderate slip).

4.2. Case 2: Oscillating Plate

Consider now the case where the moving plate is performing continuous harmonic oscillations with period equal to unity, i.e.,

$$f(t) = \sin 4\pi t. \tag{28}$$

A similar problem was addressed in [8] for unsteady flow to estimate the increment in slip at the boundary due to wall acceleration and uncover the hysteretic behavior of the slip velocity for this harmonic motion of the boundary. Here, it is required to evaluate the effect of the dynamic slip parameter Λ_s on the flow in general, and on the hysteretic behavior of the slip velocity due to oscillating wall motion.

Figure 5 illustrates the distribution of the solution in space and time with different values of Λ_s, as time is taken to run along two complete periods of the wall oscillations. The boundary condition at the moving wall is clearly satisfied. As Λ_s increases from 0.1 to 10, a damping of the amplitude of the oscillations with time takes place. The differences become more noticeable as one approaches the fixed wall at $y = 0$, where the velocity profile becomes more flattened.

Figure 6 illustrates the solutions at $y = 0$, i.e., the slip velocity, as functions of time for different values of Λ_s. The figure clearly shows time damping and lagging effects at the boundary $y = 0$. The amplitudes of oscillations in the slip velocity show a reduction of approximately 33% as the value of Λ_s increases from 0.1 to 10.

In order to put in evidence the crucial role played by the weight factor θ in the present numerical study, we have considered one case with a new value of this parameter to test the stability of the numerical scheme. As mentioned above, the scheme is stable provided that $\theta < 0.5$ [20]. The instability occuring when $\theta = 0.7 > 0.5$ is illustrated in the 3D plot in Figure 7.

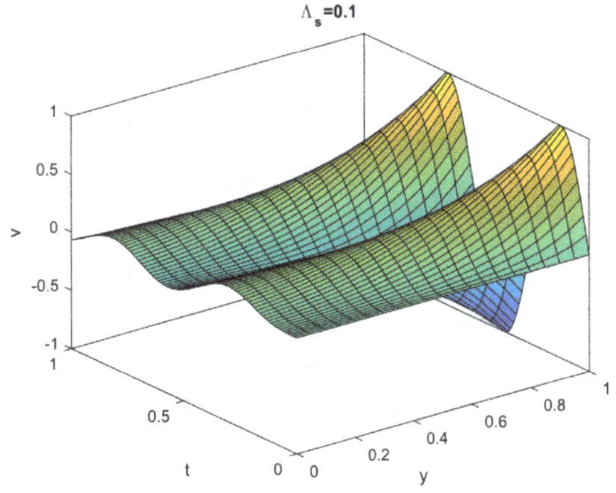

(**a**)

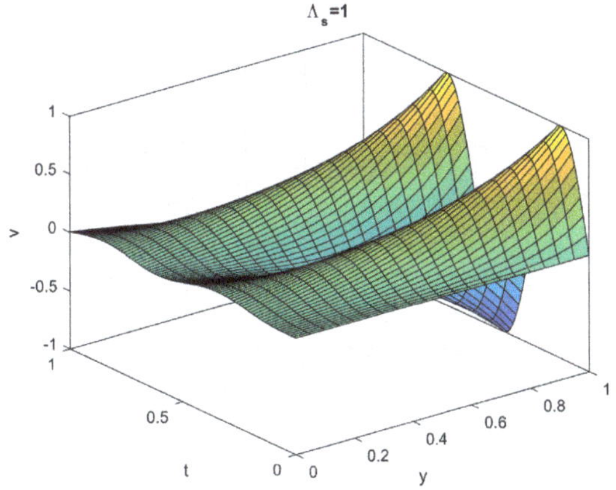

(**b**)

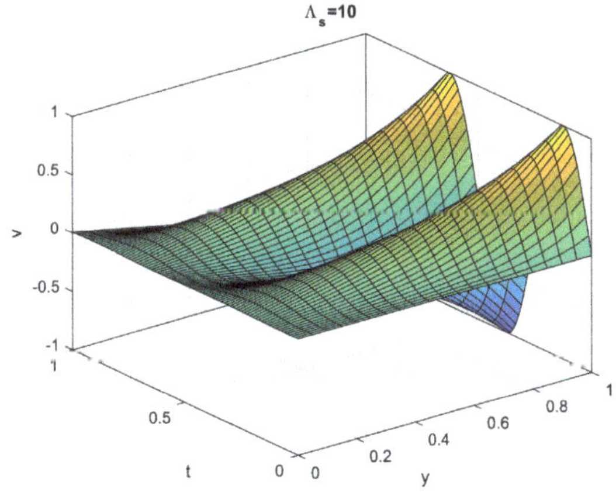

(**c**)

Figure 5. Evolution of the solution $v(y,t)$ in Case 2 (oscillating plate) with $\theta = 0$ and $B = 1$ (moderate slip): (**a**) $\Lambda_s = 0$ (Navier slip); (**b**) $\Lambda_s = 1$; (**c**) $\Lambda_s = 10$.

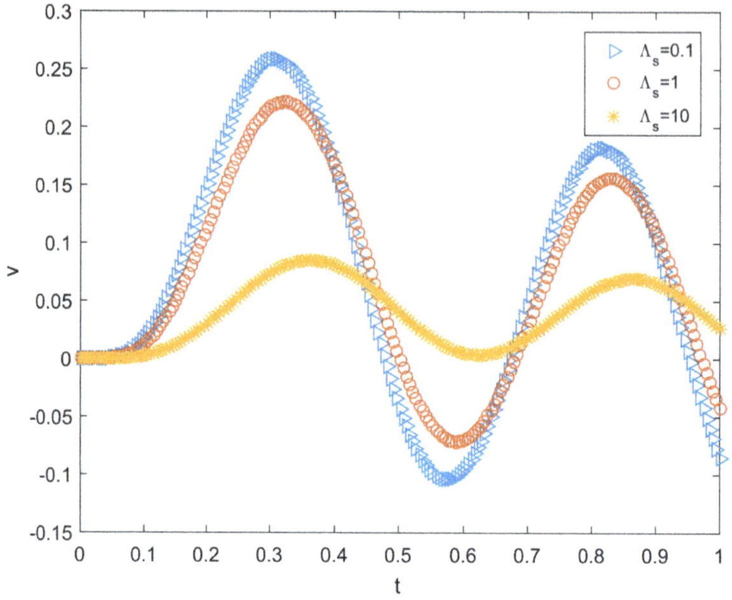

Figure 6. Evolution of the slip velocity $v(0, t)$ in Case 2 (oscillating plate) with $\theta = 0.5$ and $B = 1$ (moderate slip) for $\Lambda_s = 0.1$, 1, and 10. Time damping and lagging are observed.

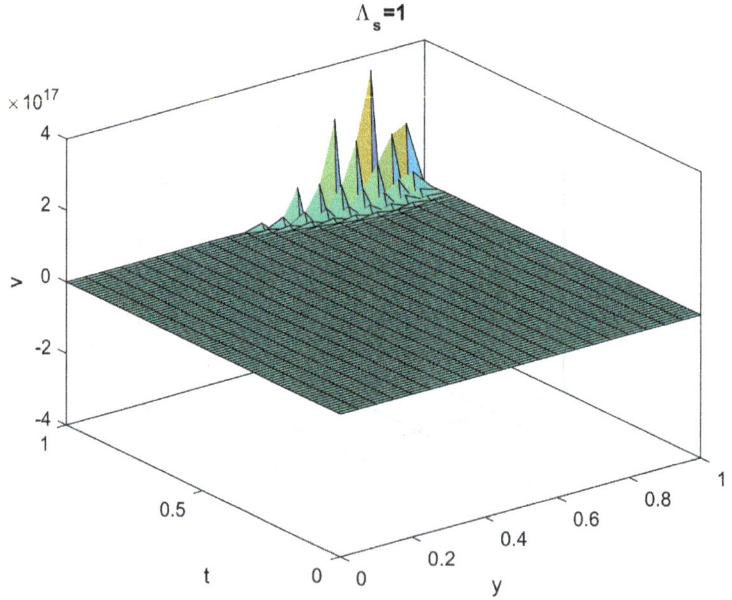

Figure 7. Instability of the solution $v(y, t)$ in Case 2 (oscillating plate) with $\theta = 0.7$, $B = 1$ (moderate slip), and $\Lambda_s = 1$.

The 3D plots in Figure 8 show how the solutions change in space and time when the slip parameter B assumes the values 0.1, 1, and 10. Figure 9 illustrates the dependence of the slip velocity on parameter B. In both plots, we have fixed the value $\Lambda_s = 0.5$. The effect of parameter B becomes more pronounced as the boundary $y = 0$ is approached. Here, again, one notes the presence of lag in following the boundary motion, and time damping of the peaks at the boundary $y = 0$ as the value of B increases.

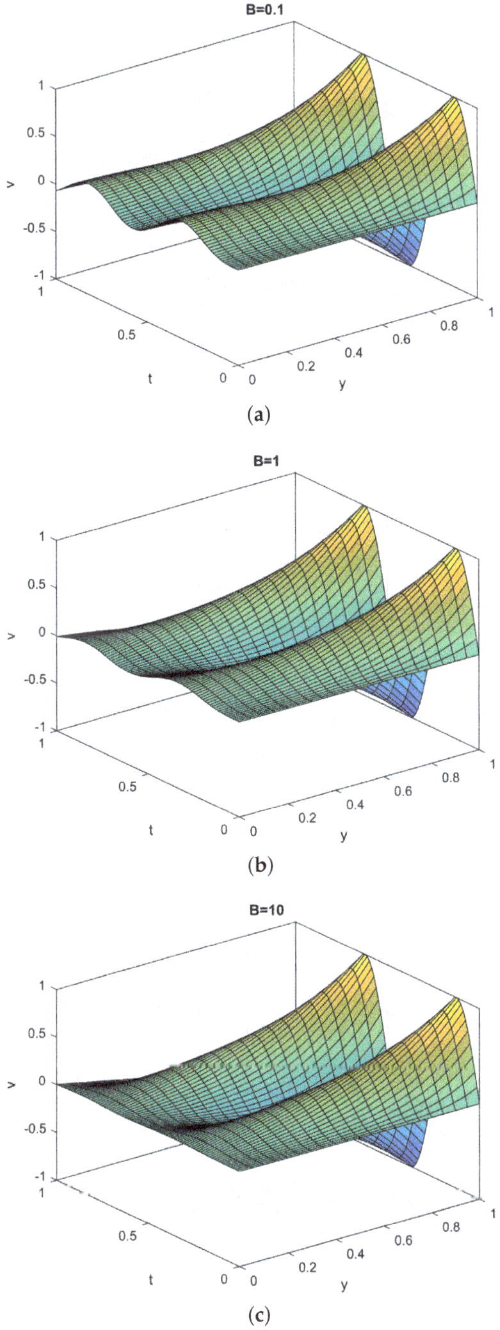

Figure 8. Evolution of the solution $v(y, t)$ in Case 2 (oscillating plate) with $\theta = 0$ and $\Lambda_s = 0.5$: (**a**) $B = 0.1$ (strong slip); (**b**) $B = 1$ (moderate slip); (**c**) $B = 10$ (weak slip).

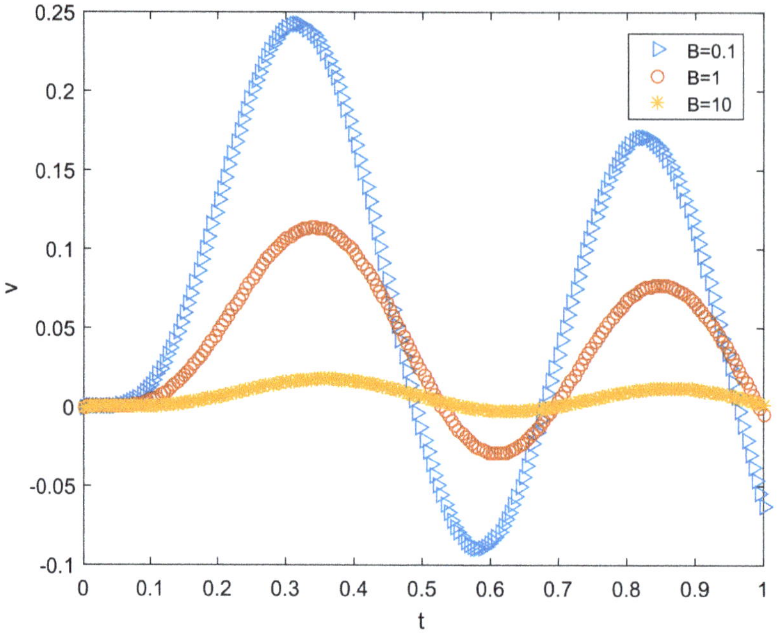

Figure 9. Evolution of the slip velocity $v(0,t)$ in Case 2 (oscillating plate) with $\theta = 0.5$ and $\Lambda_s = 0.5$ for $B = 0.1$ (strong slip), 1 (moderate slip), and 10 (weak slip).

In order to visualize the hysteretic behavior of the slip velocity and the lagging following the applied motion on one wall noticed in [8], plots have been provided in Figure 10 of the slip velocity against the boundary motion $\sin \omega t$ for three values of the dynamic slip parameter Λ_s. It is noticed that the hysteresis loops become narrower as the value of the dynamic slip parameter increases. In [8], the width of the hysteresis loop is related to the loss of energy transfer from the wall to the fluid.

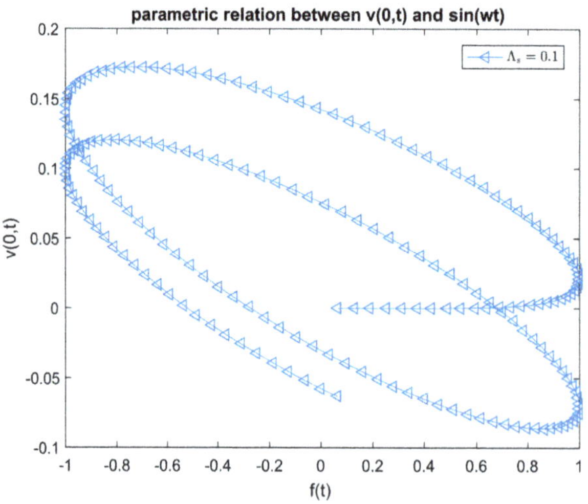

(a)

Figure 10. *Cont.*

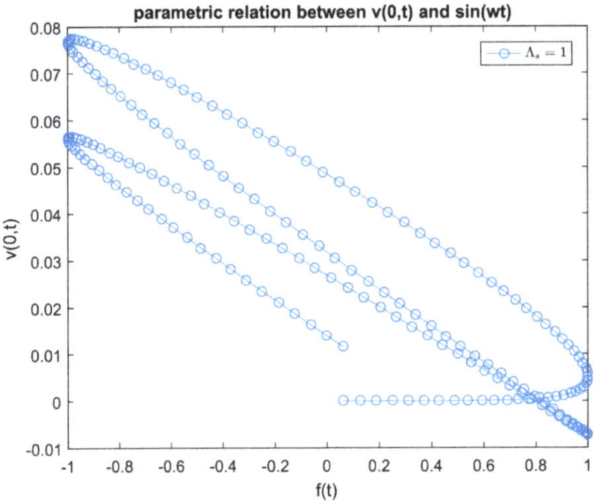

(**b**)

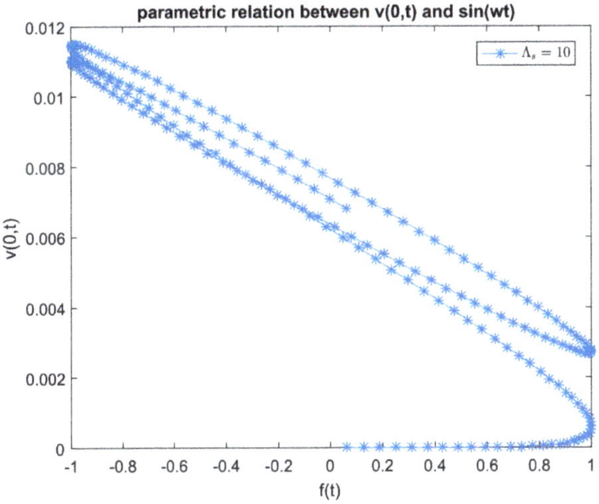

(**c**)

Figure 10. Hysteretic behavior of the slip velocity $v(0, t)$ in Case 2 (oscillating plate) with $\theta = 0$ and $B = 1$: (**a**) $\Lambda_s = 0.1$; (**b**) $\Lambda_s = 1$; (**c**) $\Lambda_s = 10$. Note that the scale of the vertical axis changes.

The volumetric flow rate is shown in Figure 11, where it is seen that this is oscillatory and is damped by the slip relaxation parameter. However, for sufficiently large values of Λ_s, it is seen that the amplitude of oscillations in $Q(t)$ will be less sensitive to any further increase in Λ_s. This phenomenon becomes more striking for larger values of the slip parameter B. Notice that after two complete oscillations, the volumetric flow rate has not vanished due to retardation.

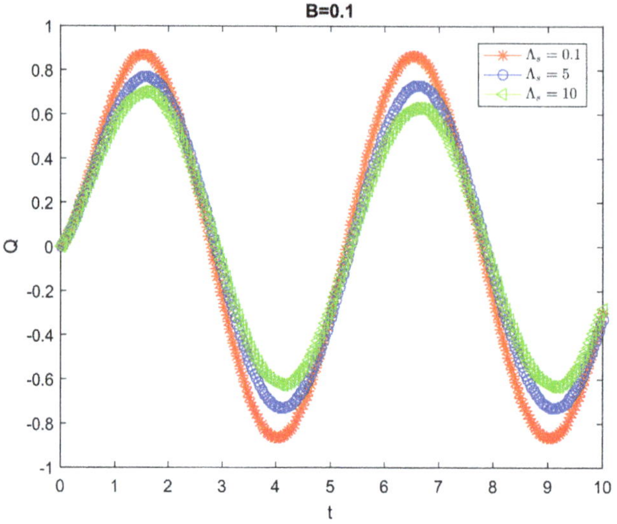

(**a**)

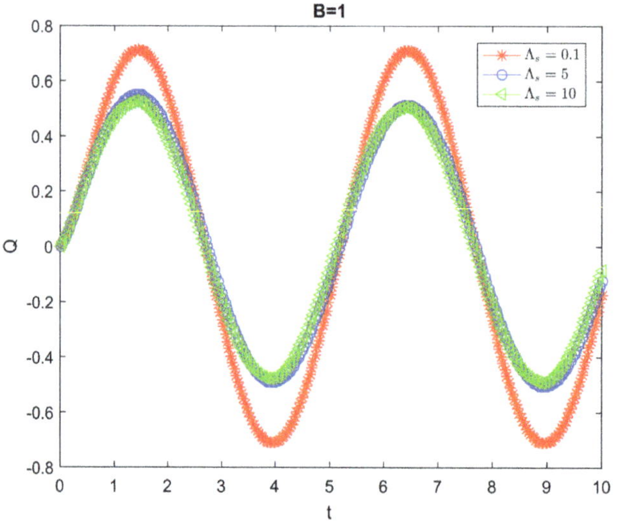

(**b**)

Figure 11. Evolution of the volumetric flow rate $Q(t)$ in Case 2 (oscillating plate) with $\theta = 0$ and $\Lambda_s = 0.1, 5$, and 10: (**a**) $B = 0.1$ (strong slip); (**b**) $B = 1$ (moderate slip).

4.3. Case 3: Single-Plate Oscillation

In this section, we assume that the upper plate oscillates only once and then comes to rest, i.e.,

$$f(t) = \begin{cases} \sin 4\pi t, & t \leq 0.5 \\ 0, & t > 0.5 \end{cases}. \tag{29}$$

Figure 12 illustrates the effect of the dynamic slip parameter Λ_s on the solution, where the value of the slip parameter was set to $B = 1$.

Figure 13 shows the effect of the relaxation parameter Λ_s on the slip velocity $v(0, t)$ for $B = 1$ (moderate slip). As for the case of continuous harmonic boundary motion, the effect of the dynamic slip parameter becomes stronger as the boundary $y = 0$ is approached as can be seen in Figure 12. The same remains valid for the effect of slip parameter B on the fluid motion as shown in Figures 14 and 15, where the value of the dynamic slip parameter was set $\Lambda_s = 0.5$.

Figure 16 represents the hysteretic behavior of the slip velocity in following the motion of the wall in Case 3 for $B = 1$ and three values of the dynamic slip parameter Λ_s. Here, again, it is noticed that the hysteresis loop becomes narrower as the value of Λ_s increases.

Finally, we show in Figure 17 the volumetric flow rate in Case 3 with $\theta = 0$ for two values of B and three values of Λ_s. The volumetric flow rate is almost independent of these two slip parameters.

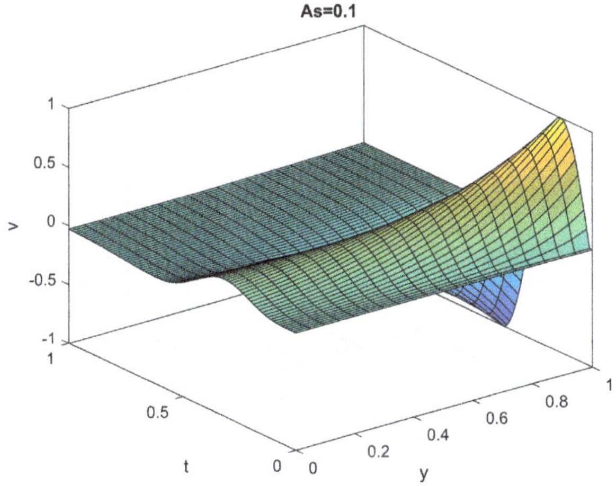

(a)

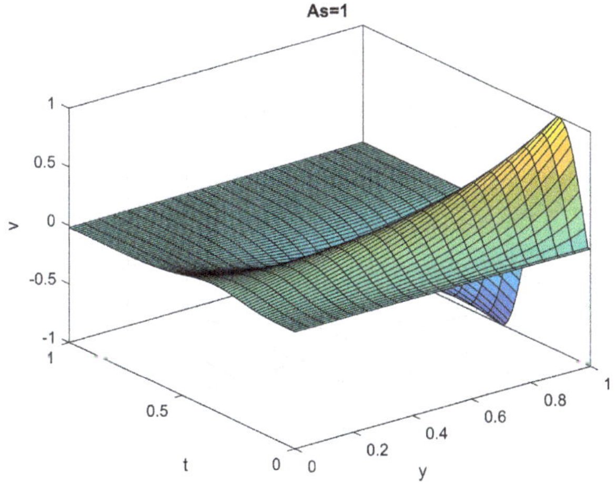

(b)

Figure 12. *Cont.*

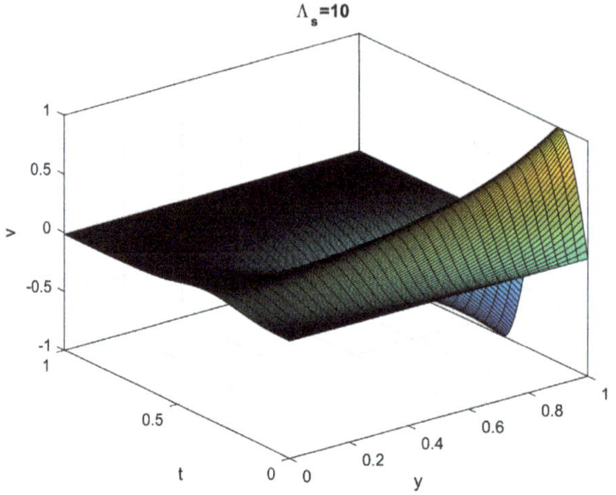

(c)

Figure 12. Evolution of the solution $v(y,t)$ in Case 3 (single-plate oscillation) with $\theta = 0$ and $B = 1$ (moderate slip): (**a**) $\Lambda_s = 0.1$; (**b**) $\Lambda_s = 1$; (**c**) $\Lambda_s = 10$.

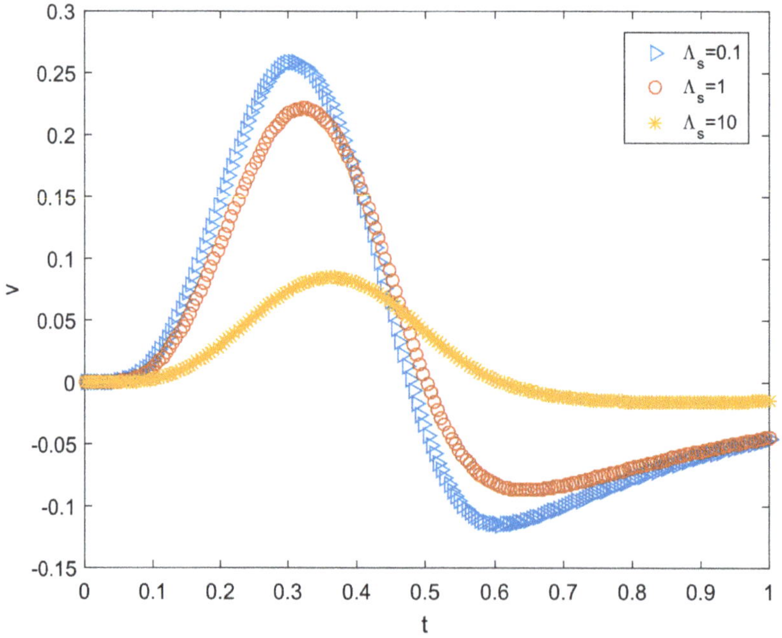

Figure 13. Evolution of the slip velocity $v(0,t)$ in Case 3 (single-plate oscillation) with $\theta = 0.5$ and $B = 1$ (moderate slip) for $\Lambda_s = 0.1$, 1, and 10.

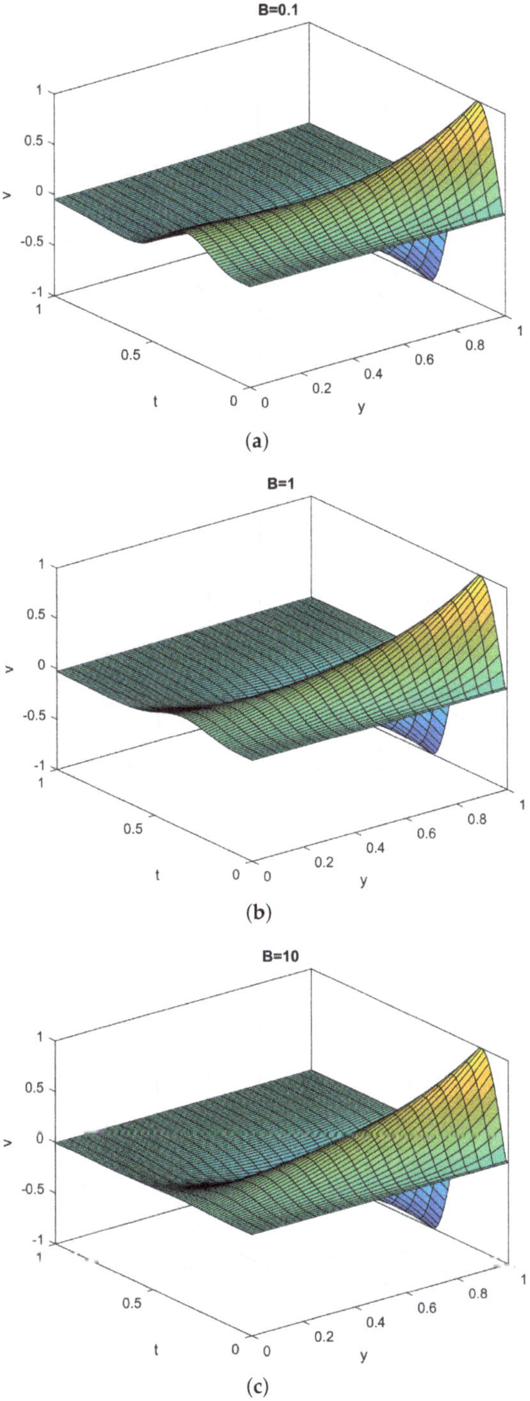

Figure 14. Evolution of the solution $v(y,t)$ in Case 3 (single-plate oscillation) with $\theta = 0$ and $\Lambda_s = 0.5$: (**a**) $B = 0.1$ (strong slip); (**b**) $B = 1$ (moderate slip); (**c**) $B = 10$ (weak slip).

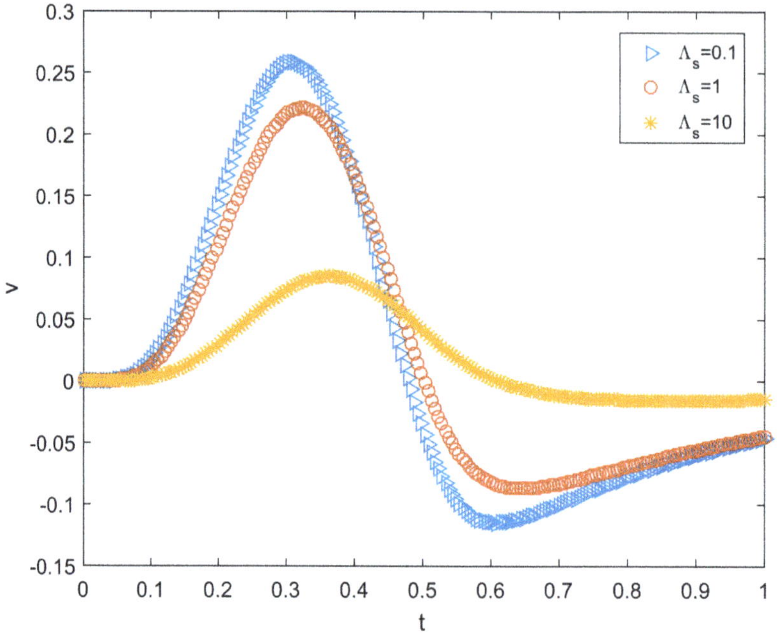

Figure 15. Evolution of the slip velocity $v(0,t)$ in Case 3 (single plate oscillation) with $\theta = 0.5$ and $\Lambda_s = 0.5$ for $B = 0.1$ (strong slip), 1 (moderate slip), and 10 (weak slip).

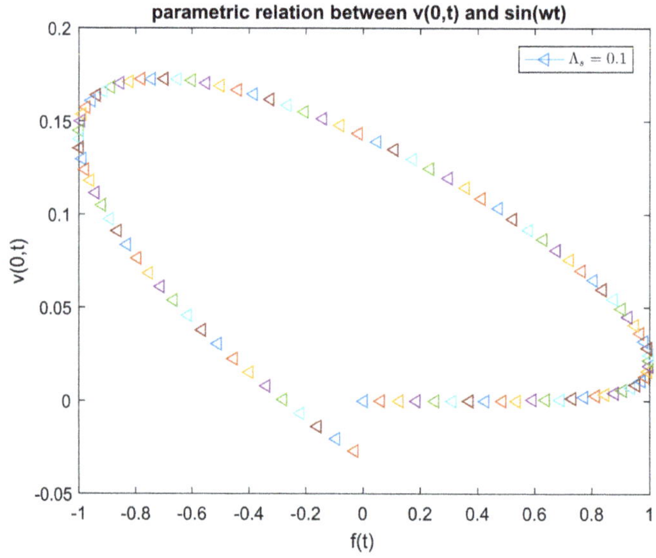

(**a**)

Figure 16. *Cont.*

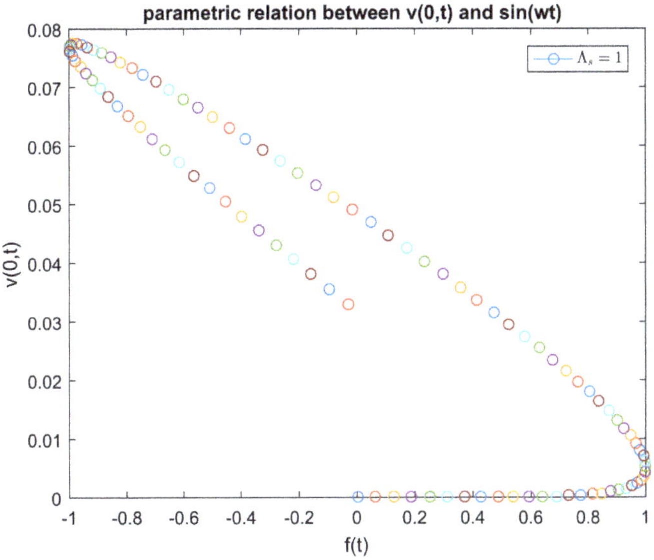

(**b**)

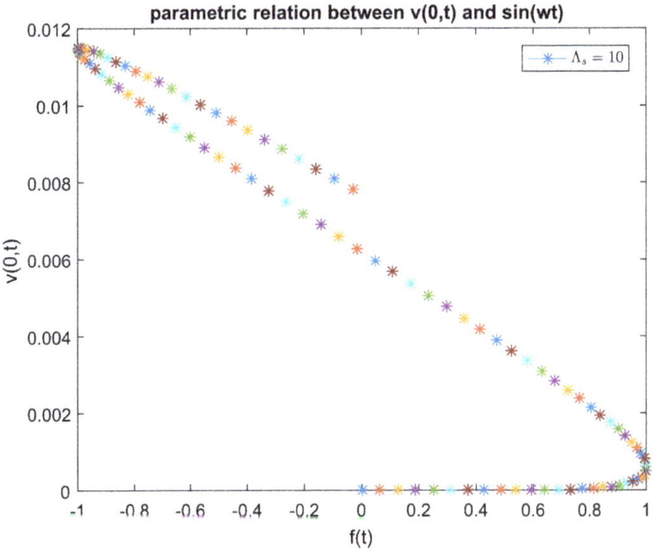

(**c**)

Figure 16. Hysteretic behavior of the slip velocity $v(0,t)$ in Case 2 (single-plate oscillation) with $\theta = 0$ and $B = 1$; (**a**) $\Lambda_s = 0.1$; (**b**) $\Lambda_s = 1$; (**c**) $\Lambda_s = 10$. Note that the scale of the vertical axis changes.

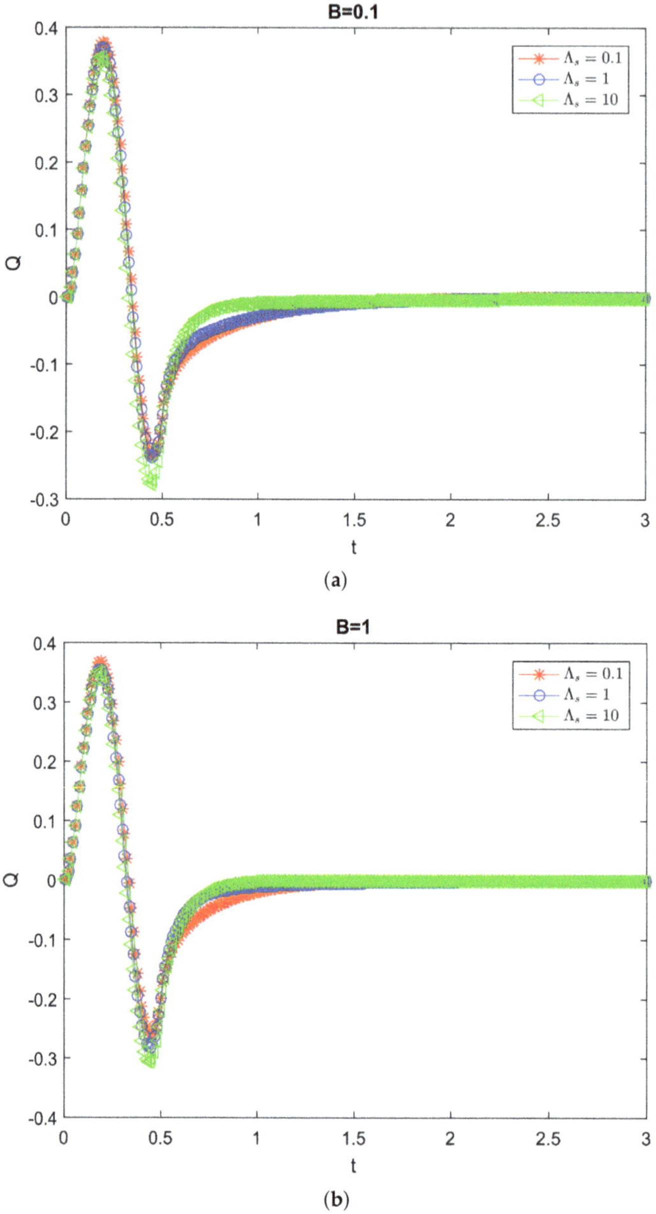

Figure 17. Evolution of the volumetric flow rate $Q(t)$ in Case 3 (single-plate oscillation) with $\theta = 0$ and $\Lambda_s = 0.1$, 1, and 10: (**a**) $B = 0.1$ (strong slip); (**b**) $B = 1$ (moderate slip, right).

5. Conclusions

A weighted-average finite difference scheme has been used to solve the time-dependent (start-up) plane Couette flow with dynamic slip along the fixed plate. Three different cases have been considered for the motion of the moving plate, i.e., constant-speed, sinusoidal, and single oscillation. The numerical solutions compare well with available analytical and numerical solutions. Both the slip and relaxation parameters appear to decelerate the evolution of the flow. For the two cases of accelerated (sinusoidal and one-period sinusoidal) boundary motion, the numerical results have clearly demonstrated a hysteretic behavior of the slip velocity that will be responsible for time lag and loss of energy transfer by the moving wall to the fluid.

Future work will be devoted to extending the numerical scheme for solving nonlinear extensions of the problem studied here. These include the non-Newtonian (e.g., power-law) flow under both Navier and dynamic slip at the fixed wall and the Newtonian flow with slip obeying a nonlinear dynamic slip law.

Author Contributions: Conceptualization, A.F.G., M.S.A.-D. and G.C.G.; Methodology, M.M.A.H., E.A.A.A., A.F.G. and M.S.A.-D.; Software, M.M.A.H. and E.A.A.A.; Validation, M.M.A.H., E.A.A.A. and G.C.G.; Investigation, E.A.A.A. and A.F.G.; Writing—original draft, M.M.A.H. and G.C.G.; Writing—review & editing, M.M.A.H., A.F.G. and G.C.G.; Visualization, M.M.A.H.; Supervision, A.F.G., M.S.A.-D. and G.C.G.; Project administration, A.F.G. and G.C.G. All authors have read and agreed to the published version of the manuscript.

Funding: This research received no external funding.

Data Availability Statement: All data generated or analyzed during this study are included in this article.

Conflicts of Interest: The authors declare that they have no conflicts of interest.

References

1. Pearson, J.R.A.; Petrie, C.J.S. On melt flow instability of extruded polymers. In *Polymer Systems: Deformation and Flow*; Wetton, R.E., Whorlow, R.E., Eds.; Macmillan: New York, NY, USA, 1968; pp. 163–187.
2. Hatzikiriakos, S.G. Wall slip of molten polymers. *Prog. Polym. Sci.* **2012**, *37*, 624–643. [CrossRef]
3. Hatzikiriakos, S.G. Slip mechanisms in complex fluid flows. *Soft Matter* **2015**, *11*, 7851–7856. [CrossRef] [PubMed]
4. Malkin, A.Y.; Patlazhan, S.A. Wall slip for complex liquids-Phenomenon and its causes. *Adv. Colloid Interface Sci.* **2018**, *257*, 42–57. [CrossRef] [PubMed]
5. Navier, C.L.M.H. Sur les lois du mouvement des fluides. *Mem. Acad. R. Des Sci.* **1827**, *6*, 389–440.
6. Abbatiello, A.; Bulíček, M.; Maringová, E. On the dynamic slip boundary condition for Navier-Stokes-like problems. *Math. Methods Appl. Sci.* **2021**, *31*, 2165–2212. [CrossRef]
7. Ferrás, L.L.; Nóbrega, J.M.; Pinho, F.T. Analytical solutions for Newtonian and inelastic non-Newtonian flows with wall slip. *J. Non-Newton. Fluid Mech.* **2012**, *175–176*, 76–88. [CrossRef]
8. Thalakkottor, J.J.; Mohseni, K. Analysis of boundary slip in a flow with an oscillating wall. *Phys. Rev. E* **2013**, *87*, 033018. [CrossRef]
9. Kaoullas, G.; Georgiou, G.C. Start-up and cessation Newtonian Poiseuille and Couette flows with dynamic wall slip. *Meccanica* **2015**, *50*, 1747–1760. [CrossRef]
10. Deterre, R.; Nicoleau, F.; Lin, Q.; Allanic, N.; Mousseau, P. The flow of power-law fluids in concentric annuli: A full analytical approximate solution. *J. Non-Newton. Fluid Mech.* **2020**, *285*, 104392. [CrossRef]
11. Ali, A.E.K.; Ghaleb, A.F.; Abou Dina, M.S.; Helal, M.A. Laplace transform solution of the time-dependent annular Couette flow with dynamic wall slip. *J. Braz. Soc. Mech. Sci. Eng.* **2023**, *45*, 586. [CrossRef]
12. Farragui, M.E.; Souhar, O.; Georgiou, G.C. Newtonian annular Poiseuille and Couette flows with dynamic wall slip. *Eur. J. Mech.-B/Fluids* **2024**, *103*, 136–144. [CrossRef]
13. Raju, S.; Gründing, D.; Maric, T.; Bothe, D.; Fricke, M. Computing hydrodynamic eigenmodes of channel flow with slip—A highly accurate algorithm. *Can. J. Chem. Eng.* **2022**, *100*, 3531–3547. [CrossRef]
14. Hesthaven, J.; Gottlieb, S.; Gottlieb, D. *Spectral Methods for Time-Dependent Problems*; Cambridge University Press: Cambridge, UK, 2007.
15. Reddy, J.N. *An Introduction to the Finite Element Method*; McGraw-Hill Education: New York, NY, USA, 2019.
16. Sweilam, N.H.; Ghaleb, A.F.; Abou-Dina, M.S.; Hasan, M.A.; Al-Mekhlafi, S.M.; Rawy, E.K. Numerical solution to a one-dimensional nonlinear problem of heat wave propagation in a rigid thermal conducting slab. *Indian J. Phys.* **2022**, *96*, 223–232. [CrossRef]
17. Ren, X.; Li, H.A. Reduced-dimension weighted explicit finite difference method based on the proper orthogonal decomposition technique for the space-fractional fiffusion qquation. *Axioms* **2024**, *13*, 461. [CrossRef]
18. Iserles, A. *A First Course in the Numerical Analysis of Differential Equations*; Cambridge University Press: Cambdrige, UK, 1996.
19. Singh, P.; Parumasur, N.; Singh, S. A review of collocation approximations to solutions of differential equations. *Mathematics* **2022**, *10*, 4438. [CrossRef]

20. Smith, G.D. *Numerical Solution of Partial Differential Equations: Finite Difference Methods*; Oxford University Press: Oxford, UK, 1985.
21. Abou-Dina, M.S.; Helal, M.A.; Ghaleb, A.F.; Kaoullas, G.; Georgiou, G.C. Newtonian plane Couette flow with dynamic wall slip. *Meccanica* **2020**, *55*, 1499–1507. [CrossRef]

applied sciences

MDPI

Article

Propagation Mechanism of Pressure Waves during Pulse Hydraulic Fracturing in Horizontal Wells

Yao Zhang, Jiye Zuo, Xinyu Fei and Shimin Dong *

School of Mechanical Engineering, Yanshan University, Qinhuangdao 066004, China;
zhangyao1207@stumail.ysu.edu.cn (Y.Z.); zjym@stumail.ysu.edu.cn (J.Z.); fxy8662@163.com (X.F.)
* Correspondence: ysudshm@ysu.edu.cn; Tel.: +86-13833575208

Abstract: Hydraulic fracturing, especially pulse hydraulic fracturing, is an important method for extracting oil and gas from low-permeability reservoirs, improving recovery rates significantly. Pulse hydraulic fracturing, which involves varying injection rates to create pressure waves, outperforms traditional constant-flow fracturing methods significantly. However, during pulse hydraulic fracturing operations, the flow properties of the fluid in the column change from moment to moment. Furthermore, current research on pulse hydraulic fracturing primarily focuses on vertical wells, while horizontal wells have become a common operational strategy. Therefore, a transient flow model of fluid within a horizontal well, considering variable-flow injection and unsteady friction conditions, is established in this paper. The model is solved using both the characteristic line method and the finite difference method. The hydrodynamic properties of the fracturing fluid were analyzed, and the propagation mechanisms of pressure waves within horizontal wells under various fluid injection schemes and well depths are analyzed to provide a reference for selecting appropriate fluid injection schemes in engineering practice. The study highlights the impact of fluid viscosity and injection flow amplitude on bottomhole pressure fluctuations, advancing the efficient development of low-permeability oilfields.

Keywords: horizontal wells; pulse hydraulic fracturing; nonconstant molar resistance; transient flow; fluctuation pressure; fluid mechanics

1. Introduction

With the continuous development of social economy, the demand for oil and gas resources has been increasing exponentially. However, the exploration of oil and gas resources has entered the middle and late stages, with increasingly high costs and difficulties. In particular, in recent years, low-permeability oilfields have accounted for more than half of the newly discovered reservoirs [1–3]. The reservoir environment of low-permeability reservoirs is complex, and traditional oil recovery methods have little effect; therefore, hydraulic fracturing technology is often used for reservoir transformation [4,5]. Hydraulic fracturing is a commonly used method for increasing production in low-permeability reservoirs. During the operation, high-pressure fracturing fluid is first injected into the wellbore, causing the rock in the reservoir to form cracks, making it easier for oil or natural gas to flow out of the reservoir [6–8].

In traditional hydraulic fracturing operations, multiple sets of fracturing pumps and pressure-resistant equipment are usually installed at the wellhead, allowing the fracturing fluid to be injected into the wellbore at a high pressure. However, due to the pressure-resistant equipment, the injection pressure at the wellhead cannot be too high, resulting in failure to achieve the expected fracturing effect. As a result, variable-flow hydraulic fracturing technology has emerged. Unlike traditional hydraulic fracturing technology, variable-flow hydraulic fracturing technology uses variable-flow circulation to inject fracturing fluid, generating fluctuating pressure at the bottom of the well, causing fatigue damage to the rock in the reservoir, and increasing the complexity of the cracks [9,10]. Thus, the wellhead injection pressure can be significantly reduced while ensuring the fracturing effect [11–14]. A comparison between constant-flow and variable-flow hydraulic fracturing is shown in Figure 1. Refs. [15–20] demonstrated in their 2019 study that the use of a variable-flow injection scheme could reduce the fracturing pressure by about 20%.

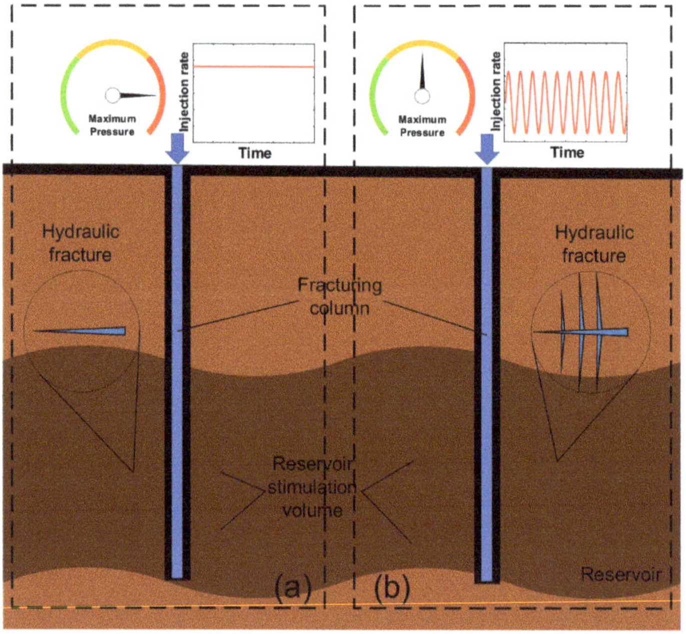

Figure 1. Comparison of constant-flow hydraulic fracturing and variable-flow hydraulic fracturing: (**a**) constant-flow hydraulic fracturing; (**b**) variable-flow hydraulic fracturing.

In variable-flow hydraulic fracturing operations, the length of the fracturing string typically extends several kilometers, and the key research questions revolve around the injection pressure of the fracturing fluid and the ability to generate fluctuating pressure at the well bottom. In traditional constant-flow hydraulic fracturing research, the pressure at the wellhead is essentially constant, and there are no significant fluctuations in the fluid flow velocity within the wellbore, which is considered steady-state flow [21]. The shear stress between the fracturing fluid and the pipe wall can be calculated using Darcy's formula, and the pressure transmission characteristics within the wellbore are relatively simple. Liu et al. [22] revised the parameters in the formula of the drag reduction ratio method in 2010 and used the drag reduction ratio method to calculate the friction along the fracturing string. In variable-flow hydraulic fracturing, the velocity of the fracturing fluid within the wellbore is constantly changing, and it is not possible to calculate the shear stress based on steady-state flow; instead, the impact of non-constant friction must be considered. Among the existing research, non-constant friction models mainly include the Zielke weighted function model and the Brunone model [23,24]. Henclik [25] proposed a time-domain numerical solution algorithm based on the characteristic line method in 2018, which has been instrumental in solving transient fluid flow within the pipe column. In 2020, Tong and Gao [26] established a pressure fluctuation model in a vertical wellbore under variable-flow injection, considering the effects of fluid gravity and non-steady friction.

However, the research objects in these papers were all conventional vertical wells [27], which are generally perpendicular to the reservoir and have a relatively short length of reservoir penetration. Current common operational schemes mainly employ horizontal wells. A typical horizontal well is composed of three parts: a vertical or slightly inclined section drilled downward, known as the vertical section; after reaching the reservoir, the well inclination angle gradually changes to 90°, parallel to the reservoir, known as the build-up section; and a horizontal section that extends for a certain length, known as the horizontal section. Since conventional vertical wells only have a vertical section, there is a lack of research on the build-up and horizontal sections in horizontal wells.

Fluctuating pressure also exists in long-distance oil pipelines, and the MacCormack algorithm is commonly used for simulating transient flow within pipelines [28,29]. Moreover, research on pipeline leakage monitoring based on pressure fluctuations within the pipeline has matured [30,31], providing a reference for the study of pressure wave propagation in the stable section of horizontal wells.

In the second chapter of this paper, a transient flow model of fracturing fluids within horizontal well pipelines will be established. The third chapter will introduce two methods used to solve the model and verify the accuracy of these methods. The fourth chapter will analyze the variation characteristics of bottom hole pressure in horizontal wells under different fluid injection schemes and well depths. The fifth chapter will present some conclusions drawn from this study. This research is helpful for engineers to choose appropriate fluid injection schemes.

2. Fluid Transient Dynamics Model

In this study, a transient flow model for fracturing fluid in horizontal wells has been established, taking into account the influence of non-constant friction on pressure transmission. Figure 2 illustrates a variable-flow horizontal well.

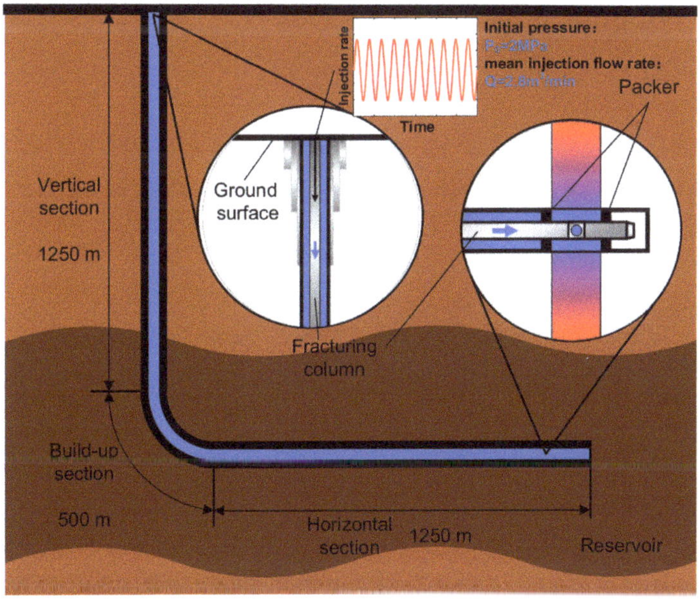

Figure 2. Diagram of a variable-flow horizontal well.

The model is based on the following assumptions:

(I) Axial flow only: it is assumed that the fracturing fluid exists only in the axial direction within the wellbore pipeline and does not have any radial flow.

(II) Adiabatic flow: the flow of fracturing fluid in the wellbore pipeline does not involve any heat exchange with the external environment, indicating an adiabatic process.

(III) Ignoring reservoir seepage effects: the influence of reservoir seepage on the fluid flow is neglected in this study.
(IV) Homogeneous and compressible fluid: The fluid within the pipeline is assumed to be homogeneous and compressible, with the viscosity remaining constant. The non-Newtonian characteristics of the fluid are neglected.
(V) Constant cross-sectional area and rigid wall: the cross-sectional area of the pipeline remains constant along its length, and the pipeline wall is considered to be rigid.
(VI) Ignoring the impact of azimuth: the azimuth remains constant, and only the influence of wellbore inclination on pressure transmission is considered.

2.1. Continuity Equation

The fluid element shown in Figure 3 is taken at any position within the operating column; according to the principle of mass conservation, the difference in fluid mass flowing into the element from cross-section One and flowing out of the element from cross-section Two within a unit time dt is equal to the change in fluid mass within the element during the same time interval. Based on this, a continuity equation can be established.

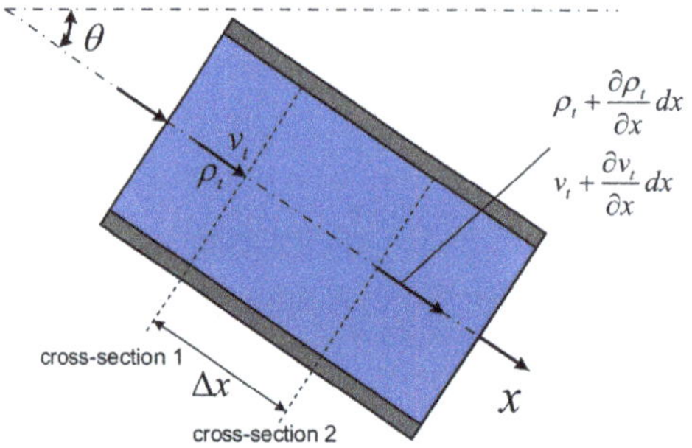

Figure 3. Schematic diagram of fluid element continuity analysis.

The fluid mass flowing into the microelement from cross-section One can be calculated using the following formula:

$$m_1 = \rho_t v_t A_t dt, \quad (1)$$

The fluid mass flowing into the microelement from cross-section Two can be calculated using the following formula:

$$m_2 = (\rho_t + \frac{\partial \rho_t}{\partial x} dx)(v_t + \frac{\partial v_t}{\partial x} dx) A_t dt, \quad (2)$$

where ρ_t is fluid density in element bodies, v_t is velocity of axial fluid motion in element bodies, A_t is cross-sectional area of working column, t is time, x is length of pipe.

The difference between the mass of fluid flowing into the element body from cross-section One and out of the element body from cross-section Two per unit time *dt* is

$$dm = m_1 - m_2 = \rho_t v_t A_t dt = -\frac{\partial(\rho_t v_t A_t dt)}{\partial x} dx, \quad (3)$$

The change in fluid mass within the element body per unit time *dt* is

$$dm = \frac{\partial(\rho_t v_t A_t dx)}{\partial t} dt, \quad (4)$$

This is easy to obtain according to the principle of conservation of mass:

$$\frac{\partial v_t}{\partial x} + \frac{v_t}{\rho_t}\frac{\partial \rho_t}{\partial x} + \frac{1}{\rho_t}\frac{\partial \rho_t}{\partial t} = 0, \quad (5)$$

which is obtained according to the differential operator method:

$$\frac{d\rho_t}{dt} = \frac{\partial \rho_t}{\partial t} + \frac{\partial \rho_t}{\partial x}\frac{dx}{dt} = \frac{\partial \rho_t}{\partial t} + v_t\frac{\partial \rho_t}{dx}, \tag{6}$$

$$\frac{dp_t}{dt} = \frac{\partial p_t}{\partial t} + \frac{\partial p_t}{\partial x}\frac{dx}{dt} = \frac{\partial p_t}{\partial t} + v_t\frac{\partial p_t}{dx}, \tag{7}$$

Substituting Equation (6) into Equation (5) gives

$$\frac{d\rho_t}{dt} + \rho\frac{\partial v_t}{\partial x} = 0, \tag{8}$$

The fluid in the operational pipe column satisfies the linear elasticity eigen structure relationship, which is obtained as

$$\frac{d\rho_t}{\rho_t} = \frac{dp_t}{K_t}, \tag{9}$$

where K_t is the volumetric modulus of elasticity of the fluid in the operating column.

The functions that represent ρ_t as p_t are

$$\rho_t = \rho_0 e^{\frac{1}{K_t}(p_t - p_0)}, \tag{10}$$

where ρ_0 is fluid density at standard pressure p_0.

Taking the derivative of Equation (9) with respect to time t, and combining Equations (7) and (8), the continuity equation for the compressible fluid within the working pipe can be obtained:

$$\rho_t c^2\frac{\partial v_t}{\partial x} + \frac{\partial p_t}{\partial t} + v_t\frac{\partial p_t}{\partial x} = 0, \tag{11}$$

where c is propagation velocity of pressure wave.

$$c = \sqrt{\frac{K_t}{\rho_t}}, \tag{12}$$

2.2. Momentum Equation

In Figure 4, we show a situation of taking a fluid element within the working pipe at an arbitrary location, wherein the inclination angle of the wellbore is θ. The forces acting on the fluid element include the pressure at both ends of cross-section One and cross-section Two, the self-weight of the fluid element, and the friction between the fluid element and the inner wall of the working pipe. When the force's direction is the same as the fluid movement direction within the element, the force magnitude is positive; otherwise, the force magnitude is negative. Based on Newton's second law, the motion equation for the compressible fluid within the working pipe could be established.

At cross-section One, the forces acting on the microelement are as follows:

$$F_1 = p_t A_t, \tag{13}$$

At cross-section Two, the forces acting on the microelement are as follows:

$$F_2 = (p_t + \frac{\partial p_t}{\partial x}dx)A_t, \tag{14}$$

The component of gravity force in the fluid flow direction for the microelement is

$$F_m = \rho_t g A_t dx \sin\theta_x, \tag{15}$$

The friction force acting on the microelement due to the inner wall of the working pipe is

$$F_f = \tau_t \pi D_t dx, \tag{16}$$

According to Newton's second law, the motion equation for the compressible fluid within the working pipe can be obtained as

$$\rho_t A_t dx(\frac{\partial v_t}{\partial t} + v_t \frac{\partial v_t}{\partial x}) = F_1 - F_2 + F_m - F_f, \tag{17}$$

After simplification, the motion equation for the compressible fluid within the working pipe becomes

$$\rho_t A_t dx(\frac{\partial v_t}{\partial t} + v_t \frac{\partial v_t}{\partial x}) = p_t A_t - [p_t A_t + \frac{\partial (p_t A_t)}{\partial x} dx] - \tau_t \pi D_t dx + \rho_t g A_t dx \sin\theta_x, \tag{18}$$

where p_t is the injection pressure within the microelement, τ_t is the shear stress between the fluid inside the microelement and the inner wall of the working pipe, D_t is the inner diameter of the working pipe, g is the gravitational acceleration, and θ_x is the angle between the axis of the pipe and the horizontal plane.

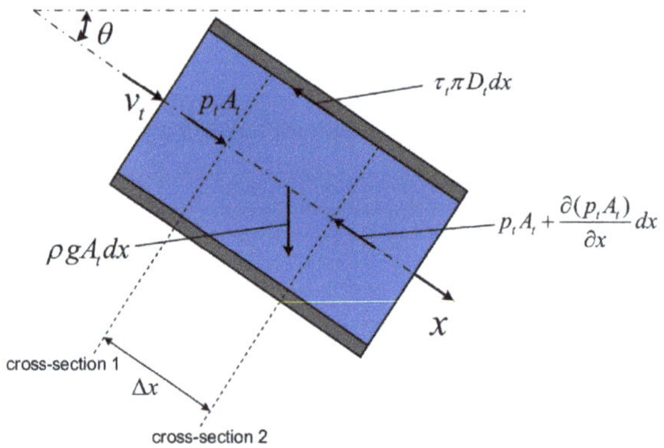

Figure 4. Schematic diagram of fluid element force analysis.

In the study of constant-flow hydraulic fracturing, the change rate of the fracturing fluid velocity inside the working pipe is not significant. Therefore, the shear stress between the fluid and the inner wall of the working pipe can be investigated under the assumption of steady-state flow. At this point, the shear stress between the fluid and the inner wall of the working pipe can be calculated using the Darcy formula.

The Darcy formula is as follows:

$$\tau_t = \frac{\rho_t v_t |v_t| f_d}{8}, \tag{19}$$

where f_d is the constant friction coefficient.

In constant-flow hydraulic fracturing, the rate of change in the actual fracturing fluid velocity is relatively small, and the error caused by simplifying the shear stress using the Darcy formula is acceptable. However, in variable-flow hydraulic fracturing, the injection rate of the fracturing fluid varies frequently, leading to rapid changes in the velocity. Continuing to use the Darcy formula to calculate the shear stress in this case would result in significant errors. Therefore, the influence of non-steady friction should be considered in variable-flow hydraulic fracturing operations.

At present, the main non-steady friction models include the Zielke weighted function model, the Brunone model, and modified models based on these two models. In Brunone's modified model, the friction coefficient consists of two parts: the constant friction coefficient and the non-steady friction coefficient. This formulation facilitates subsequent programming. The non-steady friction coefficient considers the influence of instantaneous local acceleration and instantaneous convection acceleration on friction loss. According to the Brunone model, the corrected friction coefficient can be expressed as

$$f_c = f_d + \frac{2k_t D_t}{v_t|v_t|}\left(\frac{\partial v_t}{\partial t} + c \cdot sign(v_t)\frac{\partial v_t}{\partial x}\right), \tag{20}$$

where f_c is the corrected friction coefficient, and k_t is the Brunone friction coefficient.

$$k_t = \frac{\sqrt{C^*}}{2}, \tag{21}$$

The flow state of the fluid is determined by the Reynolds number (Re):

$$\text{Re} = \frac{\rho_t v_t D_t}{\mu}, \tag{22}$$

where μ is the dynamic viscosity of the fracturing fluid.

When the fluid is in a laminar flow state,

$$C^* = 0.00476, \tag{23}$$

$$f_d = \frac{\text{Re}}{64}, \tag{24}$$

When the fluid is in a turbulent flow state [32],

$$C^* = \frac{7.41}{\text{Re}^{\log_{10}\left(\frac{14.3}{\text{Re}^{0.05}}\right)}}, \tag{25}$$

$$\frac{1}{\sqrt{f_d}} = -2\log\left[\frac{\delta_1/D_t}{3.7} - \frac{4.518}{\text{Re}}\log\left[\frac{6.9}{\text{Re}} + \left(\frac{\delta_1/D_t}{3.7}\right)^{1.11}\right]\right], \tag{26}$$

where C^* is an empirical parameter, and δ_1 is the pipe wall roughness.

By substituting the corrected friction coefficient of f_c into Equation (18), the motion equation for compressible fluid in the working pipe column considering non-uniform friction becomes

$$\frac{\partial v_t}{\partial t} + v_t\frac{\partial v_t}{\partial x} + \frac{1}{\rho_t}\frac{\partial p_t}{\partial x} - g\sin\theta_x + \frac{f_d v_t|v_t|}{2D_t} + k_t\left[\frac{\partial v_t}{\partial t} + c \cdot sign(v_t)\left|\frac{\partial v_t}{\partial x}\right|\right] = 0, \tag{27}$$

2.3. Boundary Condition

a. Entrance boundary condition

Variable-flow hydraulic fracturing necessitates precise control over the injection flow rate of fracturing fluid. Currently, electric hydraulic fracturing pumps and variable frequency control technology are capable of satisfying engineering requirements. The entrance boundary conditions of the working pipe column can be ascertained by the injection flow rate at the wellhead:

$$v|_{x=0} = \frac{4Q_t}{\pi D^2}, \tag{28}$$

where Q_t is the injection flow rate at the wellhead.

b. Exit boundary condition

In the present study, the influence of reservoir seepage at the bottom of the well is not considered. Therefore, the exit boundary condition can be expressed as

$$v|_{x=L_t} = 0, \tag{29}$$

where L_t is the total length of the working pipe column.

2.4. Initial Condition

Prior to the initiation of hydraulic fracturing, it is assumed that the pipe column is already filled with fracturing fluid, and there is no flow of the fluid within the pipe column.

$$\begin{cases} v_t|_{t=0} = 0 \\ p_t|_{t=0} = p_0 + \rho_t g z_x \end{cases}, \tag{30}$$

where p_0 is the initial entrance pressure, and z_x is the longitudinal depth of the pipe column.

When in the vertical section, z_x is numerically the same as x. However, when in the build-up and horizontal sections, z_x needs to be represented using a recursive algorithm.

$$z_x = z_{(x-1)} + \sin\theta_x, \tag{31}$$

3. Simulation Solution

In order to enhance the accuracy of the simulation results and verify the precision of the simulations, this chapter employs two numerical methods, the characteristic line method and the finite difference method, to solve the model.

First, we determine the space step: based on the length of the wellbore, we divide the space into grids with a spacing of 1m, such that Dx = 1.

Next, we determine the time step: to ensure simulation accuracy, we divide the time into grids based on the time it takes for a pressure wave to pass through one space grid, such that Dt = Dx/c.

3.1. Method of Characteristics

In the numerical simulation of transient fluid flow, the method of characteristics (MOCs) is frequently employed, as illustrated in Figure 5. In this representation, C_1 denotes the propagation of pressure waves from the upstream to the downstream direction, while C_2 indicates the opposite direction, from the downstream to the upstream. Further integration over space and time yields the following equation [33,34]:

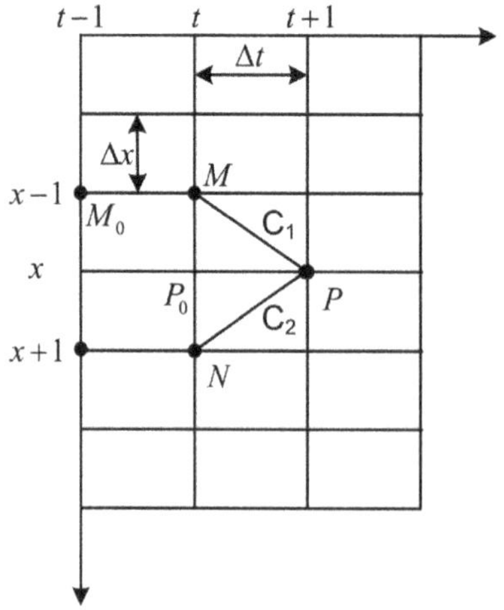

Figure 5. Spatial–temporal grid in the method of characteristics.

Case 1:

$$v_P + \frac{1}{\alpha\rho}p_P = v_M + \frac{1}{\alpha\rho}p_M + g\sin\theta_x\Delta t - \frac{v_M|v_M|f}{2D}\Delta t - k_t\Delta t\left(\frac{v_M - v_{M_0}}{\Delta t} - \alpha\frac{v_{P_0} - v_M}{\Delta x}\right), \tag{32}$$

Case 2:

$$v_P + \frac{1}{\alpha\rho}p_P = v_N + \frac{1}{\alpha\rho}p_N + g\sin\theta_x\Delta t - \frac{v_N|v_N|f}{2D}\Delta t - k_t\Delta t\left(\frac{v_N - v_{N_0}}{\Delta t} - \alpha\frac{v_N - v_{P_0}}{\Delta x}\right), \tag{33}$$

3.2. *Finite Difference Method*

The partial derivative with respect to time can be expressed in the following form:

$$\frac{\partial S}{\partial t} = \frac{S_j^i - \left[\alpha_1 S_j^{i-1} + \frac{1-\alpha_1}{2}\left(S_{j+1}^{i-1} + S_{j-1}^{i-1}\right)\right]}{\Delta t}, \tag{34}$$

In the equation, S represents the fluid pressure or velocity within the pipeline, S_j^i denotes the function value at time i and point j, and α_1 is the weighting coefficient. Through multiple simulation calculations, it is found that the best result is obtained when α_1 is set to 0.1 in the model established in this study, which exhibits good stability and convergence.

The partial derivative with respect to distance can be expressed in the following form:

$$\frac{\partial S}{\partial x} = \frac{S_{j+1}^{i-1} - S_{j-1}^{i-1}}{2\Delta x}, \tag{35}$$

By organizing the finite difference format and incorporating it into Equations (5), (11) and (27), along with the initial and boundary conditions of the pipe column, the flow state of the proppant within the column can be computed.

4. Results and Discussion

This section primarily focuses on validating the simulation accuracy of two numerical methods, the characteristic line method and the finite difference method, and analyzes the influence of various injection schemes on the pressure fluctuations at the bottom of the wellbore. The injection schemes considered include the following aspects: injection flow rate amplitude, injection flow rate waveform, injection flow rate mean value, and the dynamic viscosity of the fracturing fluid. In the simulation model employed within this study, the primary parameters selected for the oil well and construction are presented in Table 1.

Table 1. Main oil well and construction parameters.

Parameter	Value
$\rho/(\text{kg/m}^3)$	1050
$\mu/(\text{Pa·s})$	0.01
D/mm	89
L_v/m	1500
L_h/m	1500
β_v/rad	$\pi/2$
β_h/rad	0
P_0/Mpa	2
$Q/(\text{m}^3/\text{min})$	2.8

Where ρ is the fracturing fluid density, μ is the fracturing fluid dynamic viscosity, D is the inside diameter of the pipe column, L_v is the length of the pipe column's vertical section, L_h is the length of the pipe column's horizontal section, β_v is the average deviation angle of the pipe column's vertical section, β_h is the average deviation angle of the pipe column's horizontal section, P_0 is the initial pressure, and Q is the mean injection flow rate.

4.1. Accuracy Verification

For the sake of the computation, it is assumed that the fracture fluid flow rate at the wellbore is injected in a sinusoidal manner. The horizontal well consists of a 1500 m vertical section and a 1500 m horizontal section. At time zero, the fracture fluid has already filled the entire pipe column. The problem is solved using the characteristic line method and the finite difference method, and the results of the two solution approaches are compared. The comparison results are shown in Figure 6.

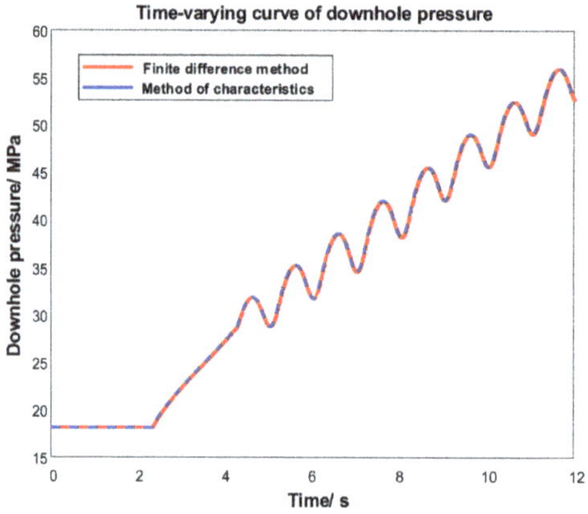

Figure 6. Comparison of results between finite difference method and characteristic line method.

The two methods yield nearly identical solutions for pressure transmission, with differences remaining within 1%. In terms of solution speed, the characteristic line method is faster than the finite difference method. However, due to the omission of the convection term in the solution process of the characteristic line method, its accuracy is inferior to the finite difference method when there is significant convection in the fracturing fluid within the pipe column. In the subsequent calculations, the finite difference method will continue to be used to study the problem.

As illustrated in Figure 7, when the sinusoidal waveform method is employed for fluctuating injections of the fracturing fluid, the wellhead pressure exhibits an immediate approximate sinusoidal variation. Due to the time required for pressure wave transmission, pressure fluctuations throughout the pipe column exhibit a certain degree of lag, with the lag time being directly proportional to the distance. As the distance increases, the amplitude of the pressure fluctuations decreases gradually due to the influence of friction losses along the pipe.

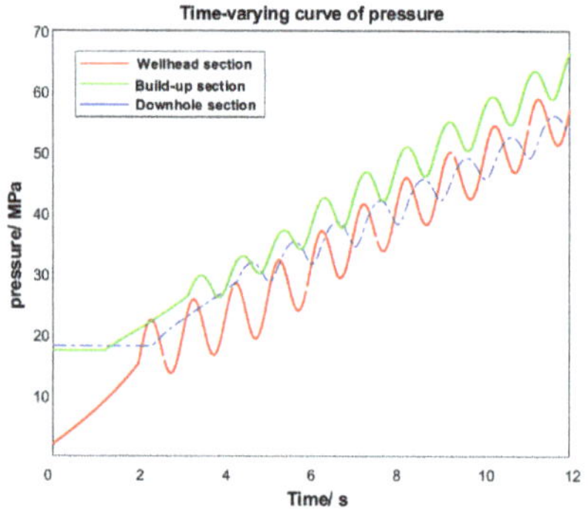

Figure 7. Pressure variation at different locations within the column.

4.2. Analysis of Fracturing Fluid Pressure and Velocity

In this section, it is assumed that the fracturing fluid flow rate at the wellhead varies in a sinusoidal pattern. Simulation calculations are performed using the finite difference method, resulting in the pressure variation with time and well depth presented in Figure 8, and the variation in velocity with time and well depth shown in Figure 9.

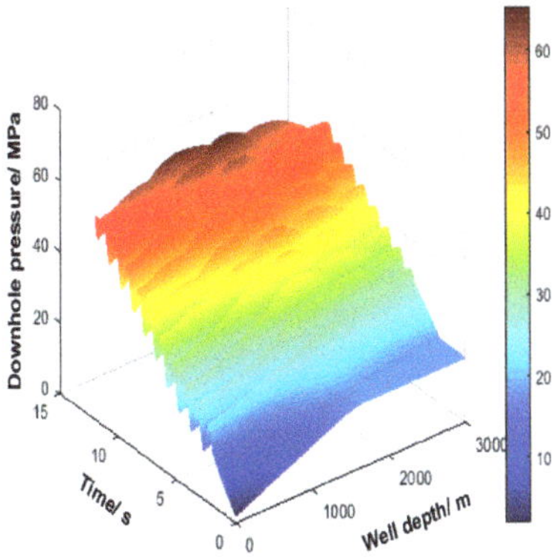

Figure 8. Pressure variation with time and well depth.

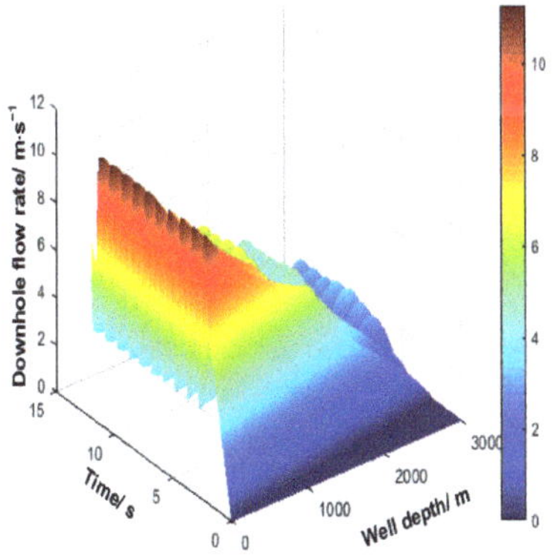

Figure 9. Velocity variation with time and well depth.

As illustrated in Figure 8, at t = 0, the pressure in the vertical section increases with increasing well depth, while the pressure in the horizontal section remains constant due to the constant wellbore depth. Ignoring the influence of reservoir seepage at the bottom of the wellbore, the pressure throughout the wellbore increases with time. Owing to the time required for pressure wave transmission, the pressure fluctuations at the bottom of the well exhibit different degrees of lag compared to the wellhead, with the lag time being proportional to the distance from the wellhead. During the transmission process, the fracturing fluid is influenced by friction with the inner wall of the pipe column, resulting in a gradual decrease in pressure along the transmission path.

As illustrated in Figure 9, the variation in fracturing fluid velocity at the wellbore mouth is identical to the change in injection flow rate. Ignoring the impact of leakage at the bottom of the pipe column, the velocity at the downhole is zero. Although there are fluctuations in the velocity with increasing well depth, the overall trend is downward.

4.3. Influence of Wellbore Section Length

4.3.1. Influence of Horizontal Section Length

In order to investigate the influence of horizontal section length on pressure fluctuations at the downhole, a constant liquid injection scheme and vertical section length of the well tubing are maintained, while the horizontal section lengths are varied to 1000 m, 1500 m, and 2000 m.

As evident from the simulation results in Figure 10, the downhole pressure and amplitude of pressure fluctuations gradually decrease with the increase in the horizontal section length of the well tubing. For every 500 m increase in the length of the horizontal section, the downhole pressure decreases by approximately 5 MPa, and the amplitude of pressure fluctuations reduces by 15% to 25%.

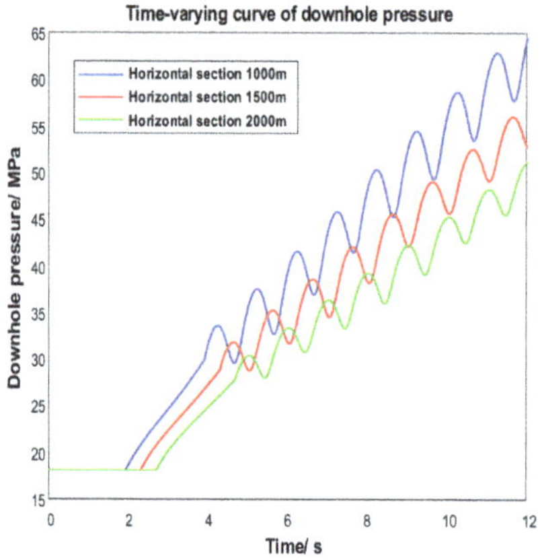

Figure 10. Impact of the horizontal section length on downhole pressure.

4.3.2. Influence of Vertical Section Length

In order to investigate the influence of vertical section length on pressure fluctuations at the downhole, a constant liquid injection scheme and horizontal section length of the well tubing are maintained, while the vertical section lengths are varied to 1000 m, 1500 m, and 2000 m.

As depicted from the simulation results in Figure 11, with the increase in the vertical section length of the well tubing, the initial value of downhole pressure gradually increases, the amplitude of pressure fluctuations at the downhole gradually decreases, and the rate of increase in downhole pressure slows down. With the passage of time, shorter vertical sections are more susceptible to generating higher downhole pressures. For every additional 500 m in the vertical segment length, the initial value of the downhole pressure increases by approximately 5.145 MPa, and the amplitude of pressure fluctuations decreases by 10–20%.

4.3.3. Analysis of Purely Vertical and Purely Horizontal Conditions

In this section, we assume the wellbore to be purely horizontal and purely vertical, directly comparing the pressure transmission patterns between the two. Assuming both wellbores are 3000 m long and other conditions remain unchanged, the simulation results are shown in Figure 12. As is evident from Figure 12, under the influence of gravity, pressure attenuates faster in a purely horizontal wellbore, with the average fluctuation amplitude of pressure reduced by about 13% compared to a purely vertical wellbore and the maximum fluctuation amplitude reduced by about 17%.

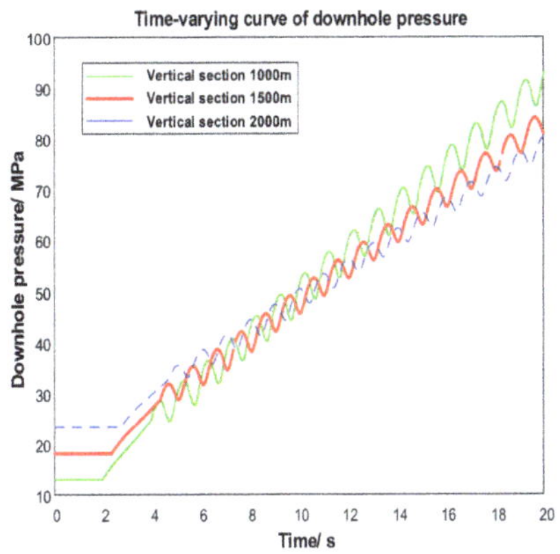

Figure 11. Impact of the vertical section length on downhole pressure.

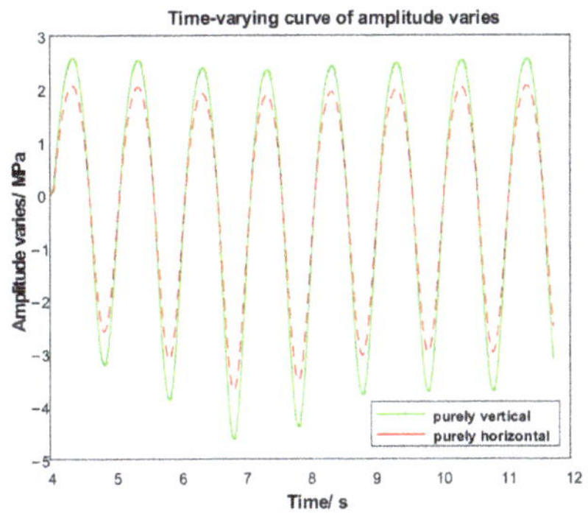

Figure 12. The amplitude of downhole pressure fluctuations in purely vertical and purely horizontal states.

4.4. Influence of Fluid-Injection Scheme

4.4.1. Influence of Injection Flow Rate Amplitude

When injecting fracturing fluid using a sinusoidal waveform, the mean injection flow rate of the fracturing fluid is kept constant, while the amplitude varies at 30%, 50%, and 80% of the mean value. The calculation results are presented in Figure 13. The simulation demonstrates that when the mean injection flow rate remains constant, the greater the amplitude variation in the injection flow rate, the more significant the change in downhole pressure, and the better the actual fracturing effect.

4.4.2. Influence of Injection Flow Rate Waveform

In variable-flow hydraulic fracturing operations, there are differences in the manner of injecting fracturing fluid. This sub-section will discuss three representative injection waveforms, namely rectangular wave injection, sinusoidal wave injection, and sawtooth wave injection. Figure 14 depicts the entrance flow rate waveforms of the three different injection schemes, while Figure 15 shows the simulation results of the three distinct injection waveforms.

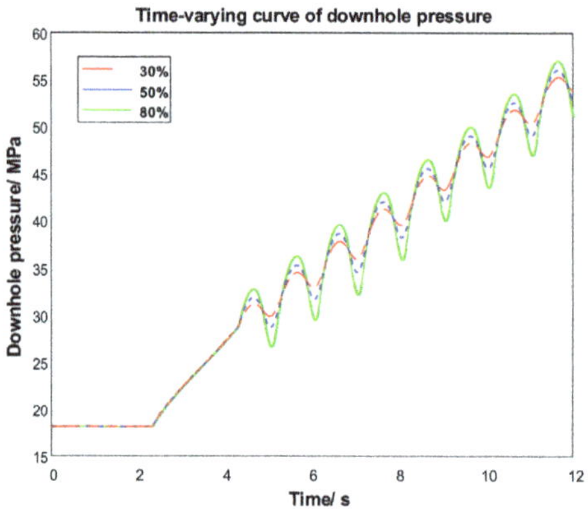

Figure 13. Influence of injection flow rate amplitude.

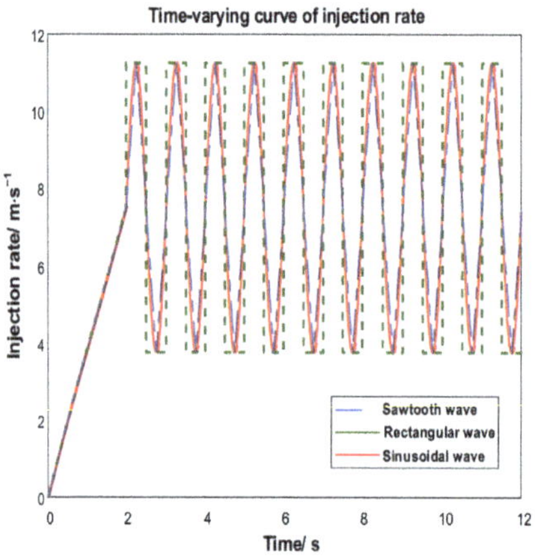

Figure 14. Flow waveform at the entrance.

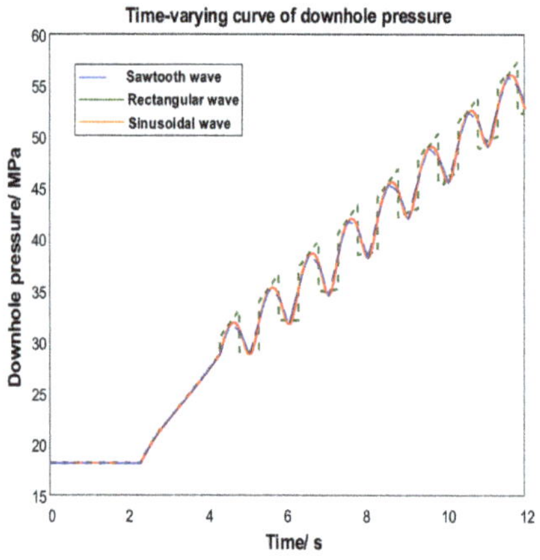

Figure 15. Influence of injection flow rate waveform.

It is noteworthy that when fracturing fluid is injected in the form of rectangular waves, the abrupt changes in injection lead to corresponding pressure transient responses at the downhole. The rectangular waves, compared to sinusoidal and sawtooth waves, are capable of generating greater pressure fluctuations at the downhole, resulting in more effective fracturing.

4.4.3. Influence of Injected Flow Rate Average Value

Under other constant parameters, the impact of varying injected flow rate mean values on the simulation results was examined. The mean values of injected flow rate were set at 2 m^3/min, 3 m^3/min, 4 m^3/min, and 6 m^3/min. The simulation results from Figure 16 reveal that the downhole pressure increases with the increase in the mean value of the injected flow rate. As the mean value of injected flow rate increases, the velocity of fracturing fluid in the pipe column also increases, resulting in increased pressure losses along the pipeline. Consequently, although the downhole pressure increases with the injected flow rate mean value, the fluctuation in downhole pressure decreases.

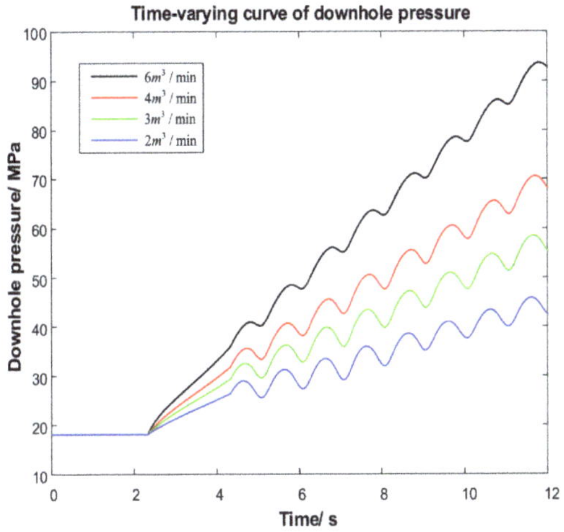

Figure 16. Influence of injected flow rate average value.

A further analysis was conducted on the average injection flow rate, with the average injection flow rate as the variable, simulating the amplitude of downhole pressure fluctuations at depths of 1000 m and 3000 m, respectively. The results indicate that a higher average flow rate is not always better, but is related to the well depth, and there exists an optimal solution. In Figure 17a,b, taking wells at depths of 1000 m and 3000 m as examples, it is evident that in the 1000 m deep well, with 10 m^3/min as the threshold, the fluctuation amplitude decreases as the average injection flow rate increases. Therefore, we can conclude that the optimal average injection flow rate for a 1000 m deep well is approximately 10 m^3/min; similarly, in the 3000 m deep well, this optimal solution occurs earlier, at 3 m^3/min.

4.4.4. Influence of Fracturing Fluid Dynamic Viscosity

In this sub-section, the dynamic viscosity of the fracturing fluid was varied and analyzed. The selected dynamic viscosity values were 0.01 Pa·s, 0.05 Pa·s, and 0.10 Pa·s. The simulation results from Figure 18 demonstrate that as the dynamic viscosity of the fracturing fluid increases, the amplitude of downhole pressure fluctuations decreases, indicating a decline in fracturing effectiveness.

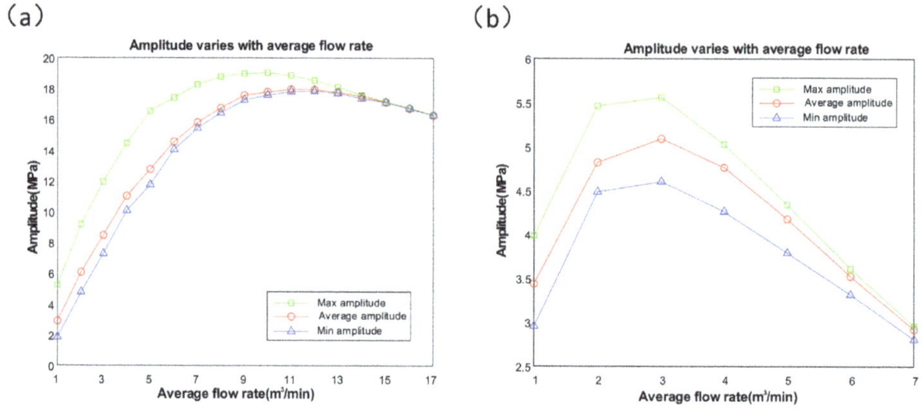

Figure 17. Influence of the mean flow rate on the amplitude of downhole pressure fluctuations: (**a**) 1000 m deep; (**b**) 3000 m deep.

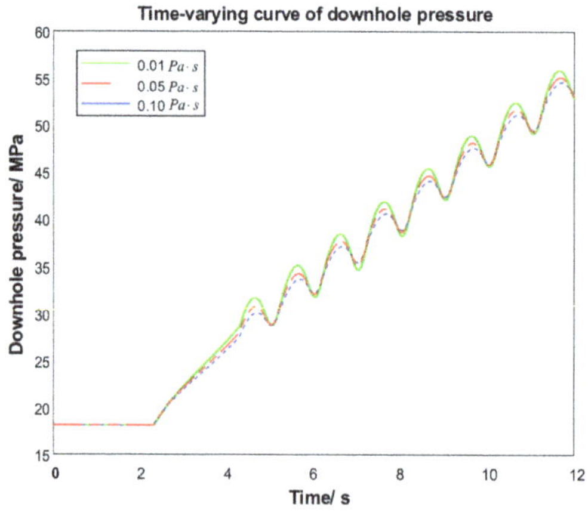

Figure 18. Influence of fracturing fluid dynamic viscosity.

4.4.5. Influence of Variation Frequency

In this section, the impact of injection frequency on the amplitude of downhole pressure fluctuations will be analyzed. With the injection frequency as the variable, simulations were conducted on the amplitude of downhole pressure fluctuations at depths of 1000 m and 3000 m. The simulation results are shown in Figure 19a,b. From the simulation results, it can be observed that when the well depth is smaller, the amplitude of downhole pressure fluctuations exhibits a fluctuating state, which is particularly evident in Figure 19a. The analysis suggests that the reason for this is the proximity of the injection frequency to the natural frequency of the wellbore. Therefore, it can be concluded that, similar to the average injection flow rate, there is an optimal solution for the injection frequency, and this optimal solution is related to the natural frequency of the wellbore.

In conclusion, the friction loss between the fracturing fluid and the inner wall of the pipe plays a dominant role in affecting the downhole pressure fluctuations in the variable-flow-rate hydraulic fracturing of horizontal wells. Therefore, exploring methods to reduce the friction loss between the fracturing fluid and the inner wall of the pipe should be one of the important research directions for future variable-flow-rate hydraulic fracturing.

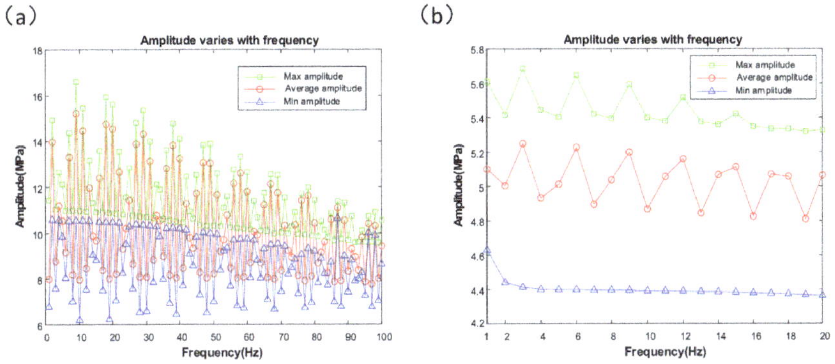

Figure 19. Influence of the frequency on the amplitude of downhole pressure fluctuations: (**a**) 1000 m deep; (**b**) 3000 m deep.

5. Conclusions

In this paper, a transient flow model for variable-flow-rate injection in horizontal wells was developed while considering the impact of unsteady friction. Based on this model, the effects of wellbore section lengths and liquid injection schemes on downhole pressure fluctuations were analyzed. The results of the analyses are as follows:

The vertical section length of the wellbore affects the magnitude of the downhole pressure and also has an impact on the amplitude of downhole pressure fluctuations. The horizontal section length, however, only affects the amplitude of downhole pressure fluctuations. Furthermore, the attenuation rate of pressure waves is faster in pure horizontal wells, with the average fluctuation amplitude of pressure reduced by about 13% compared to pure vertical wells.

The viscosity of the fracturing fluid and the amplitude of the injection flow rate have a significant impact on the amplitude of downhole pressure fluctuations. To improve the fracturing effectiveness, it is recommended to minimize the viscosity of the fracturing fluid and maximize the amplitude of the injection flow rate in engineering practice. Utilizing rectangular wave injection is more prone to induce fatigue failure in reservoir rocks.

The influence of the mean injection flow rate and injection frequency on downhole pressure fluctuations is not linear, but rather, there exists an optimal solution that is related to the wellbore parameters. This optimal solution can be calculated in advance before operation, thereby improving the operational efficiency.

Author Contributions: Writing—original draft preparation, Y.Z.; writing—review and editing, J.Z.; supervision, X.F.; funding acquisition, S.D. All authors have read and agreed to the published version of the manuscript.

Funding: This research was funded by the National Natural Science Foundation of China with the grant number 51974276.

Institutional Review Board Statement: Not applicable.

Informed Consent Statement: Not applicable.

Data Availability Statement: The datasets presented in this article are not readily available because the data are part of an ongoing study. Requests to access the datasets should be directed to zhangyao1207@stumail.ysu.edu.cn.

Conflicts of Interest: The authors declare no conflicts of interest.

References

1. Ding, S.; Xi, Y.; Liu, G.; Liu, S.; Yu, H. Adaptability of different injection methods of CO flooding in low permeability reservoirs to enhanced oil recovery and geological storage. *Pet. Geol. Recovery Effic.* **2023**, *30*, 104–111. [CrossRef]
2. Wang, H.; Dong, S.; Gan, Q.; Xin, H.; Zhu, G. Dynamic parameter simulation model of low-production pumping well and the ways to improve system efficiency. *Acta Pedol. Sin.* **2018**, *39*, 1299–1307.
3. Yu, J. Fracturing technology and development trend in low permeability oilfield. *China Pet. Chem. Stand. Qual.* **2023**, *43*, 175–177. [CrossRef]
4. Li, Q.; Xing, H.; Liu, J.; Liu, X. A review on hydraulic fracturing of unconventional reservoir. *Petroleum* **2015**, *1*, 8–15. [CrossRef]
5. Xu, J.; Zhai, C.; Qin, L. Mechanism and application of pulse hydraulic fracturing in improving drainage of coalbed methane. *J. Nat. Gas Sci. Eng.* **2017**, *40*, 79–90. [CrossRef]
6. Zhu, G. Study on Bottom Hole Dynamic Pressure, Reservoir Dynamic Stress and Fracture Propagation Law of Variable Flow Hydraulic Fracturing. Ph.D. Thesis, Yanshan University, Qinhuangdao, China, 2022. [CrossRef]
7. Zhu, G.; Dong, S. Motion characteristics of pump valve and water hammer phenomenon based on cavitation model. *J. China Univ. Pet.* **2020**, *44*, 124–132. [CrossRef]
8. Zhu, G.; Dong, S.; Pan, Z. Simulation model of dynamic stress in reservoir under unsteady excitation of hydraulic fracturing. *J. China Coal Soc.* **2021**, *46*, 149–156. [CrossRef]
9. Lu, P.; Li, G.; Huang, Z.; He, Z.; Li, X.; Zhang, H. Modeling and parameters analysis on a pulsating hydro-fracturing stress disturbance in a coal seam. *J. Nat. Gas Sci. Eng.* **2015**, *26*, 253–263. [CrossRef]
10. Sun, C.; Zheng, H.; Liu, W.D.; Lu, W. Numerical investigation of complex fracture network creation by cyclic pumping. *Eng. Fract. Mech.* **2020**, *233*, 107103. [CrossRef]
11. Hou, B.; Chen, M.; Li, Z.; Wang, Y.; Diao, C. Propagation area evaluation of hydraulic fracture networks in shale gas reservoirs. *Pet. Explor. Dev.* **2014**, *41*, 833–838. [CrossRef]
12. Hou, B.; Zhang, R.; Tan, P.; Song, Y.; Fu, W.; Chang, Z.; Kao, J.; Muhadasi, Y.; Chen, M. Characteristics of fracture propagation in compact limestone formation by hydraulic fracturing in central Sichuan, China. *J. Nat. Gas Sci. Eng.* **2018**, *57*, 122–134. [CrossRef]
13. Zhou, Z.; Zhang, G.; Dong, H.-R.; Liu, Z.-B.; Nie, Y.-X. Creating a network of hydraulic fractures by cyclic pumping. *J. Roc. Mech. Min. Sci.* **2017**, *97*, 52–63. [CrossRef]
14. Zhao, Z.; Liu, D.; Cui, Z.; Tang, T.; Han, W. Cyclic progressive pressure on the fracturing effect of shale. *J. Rock Mech. Geotech. Eng.* **2019**, *38*, 2779–2789. [CrossRef]
15. Kang, H.; Zhang, J.; Fan, X.; Huang, Z. Cyclic Injection to Enhance Hydraulic Fracturing Efficiency: Insights from Laboratory Experiments. *Geofluids* **2020**, *2020*, 1–10. [CrossRef]
16. Patel, S.M.; Sondergeld, C.H.; Rai, C.S. Laboratory studies of hydraulic fracturing by cyclic injection. *Int. J. Rock Mech. Min.* **2017**, *95*, 8–15. [CrossRef]
17. Zang, A.; Stephansson, O.; Stenberg, L.; Plenkers, K.; Specht, S.; Milkereit, C.; Schill, E.; Kwiatek, G.; Dresen, G.; Zimmermann, G.; et al. Hydraulic fracture monitoring in hard rock at 410 m depth with an advanced fluid-injection protocol and extensive sensor array. *Geophys. J. Int.* **2017**, *208*, 790–813. [CrossRef]
18. Zang, A.; Zimmermann, G.; Hofmann, H.; Stephansson, O.; Min, K.-B.; Kim, K.Y. How to Reduce Fluid-Injection-Induced Seismicity. *Rock Mech. Rock Eng.* **2019**, *52*, 475–493. [CrossRef]
19. Zhuang, L.; Jung, S.G.; Diaz, M.; Kim, K.Y.; Hofmann, H.; Min, K.-B.; Zang, A.; Stephansson, O.; Zimmermann, G.; Yoon, J.-S. Laboratory True Triaxial Hydraulic Fracturing of Granite Under Six Fluid Injection Schemes and Grain-Scale Fracture Observations. *Rock Mech. Rock Eng.* **2020**, *53*, 4329–4344. [CrossRef]
20. Zhuang, L.; Kim, K.Y.; Jung, S.G.; Diaz, M.; Min, K.-B.; Zang, A.; Stephansson, O.; Zimmermann, G.; Yoon, J.-S.; Hofmann, H. Cyclic hydraulic fracturing of pocheon granite cores and its impact on breakdown pressure, acoustic emission amplitudes and injectivity. *Int. J. Rock Mech. Min.* **2019**, *122*, 104065. [CrossRef]
21. Zou, J.; Chen, W.; Yuan, J.; Yang, D.; Yang, J. 3-D numerical simulation of hydraulic fracturing in a CBM reservoir. *J. Nat. Gas Sci. Eng.* **2017**, *37*, 386–396. [CrossRef]
22. Liu, H. Friction loss calculation and surface pressure prediction in oil well hydraulic fracturing. *J. Rock Mech.* **2010**, *29*, 2833–2839.
23. Brunone, B.; Ferrante, M.; Cacciamani, M. Decay of Pressure and Energy Dissipation in Laminar Transient Flow. *J. Fluids Eng.* **2004**, *126*, 928–934. [CrossRef]
24. Szymkiewicz, R. Developments in unsteady pipe flow friction modelling. *J. Hydraul. Res.* **2010**, *40*, 647–656. [CrossRef]
25. Henclik, S. Numerical modeling of water hammer with fluid–structure interaction in a pipeline with viscoelastic supports. *J. Fluid. Struct.* **2018**, *76*, 469–487. [CrossRef]
26. Tong, S.; Gao, D. Pressure behaviors of unstable flow in the wellbore during wave injection hydraulic fracturing. *Pet. Mach.* **2020**, *48*, 77–84. [CrossRef]
27. Zhu, G.; Dong, S.; Jia, H.-T.; Dong, Z.-L. Generation and propagation characteristics of fluctuation pressure in hydraulic fracturing with unstable fluid-injection. *J. Cent. South. Univ.* **2022**, *29*, 186–196. [CrossRef]
28. Mehmood, K. Unsteady Friction Fundamentals and Pressure Controlling Approaches of Transient Flow in Water Pipeline Systems. Ph.D. Thesis, Zhejiang University, Hangzhou, China, 2023.
29. Zhu, C. A study on the application of pressure wave in long-distance confined oil pipeline. *Sci. Technol. Innov.* **2023**, *15*, 164–166. [CrossRef]

30. Fang, G. Oil pipeline leakage monitoring technology. *Chem. Eng. Equip.* **2021**, *11*, 238–239. [CrossRef]
31. Wu, T.; Deng, Z.; Shen, L.; Xie, Z.; Chen, Y.; Liu, C.; Li, Y. Research Progress of Long-distance Oil Pipeline Leakage Monitoring Technology and Establishment of Evaluation System. *Oil Gas Storage Transp.* **2023**, *42*, 259–275.
32. Zigrang, D.J.; Sylvester, N.D. A Review of Explicit Friction Factor Equations. *J. Energy Resour. Technol.* **1985**, *107*, 280–283. [CrossRef]
33. Wiggert, D.C.; Hatfield, F.J.; Stuckenbruck, S. Analysis of Liquid and Structural Transients in Piping by the Method of Characteristics. *J. Fluids Eng.* **1987**, *109*, 161–165. [CrossRef]
34. Zhu, G.; Dong, S.; Wang, H. Reservoir stress analysis during simultaneous pulsating hydraulic fracturing based on the poroelastodynamics model. *J. Environ. Earth Sci.* **2024**, *83*, 421. [CrossRef]

applied sciences

MDPI

Article

Analyzing the Biomechanical Characteristics of Ski Jumping Take-Off Phase Based on CFD

Bojie Hou [1], Zhongqiu Ji [1,*], Yun Zhang [2] and Mingyan Yu [3]

1 College of Physical Education and Sports, Beijing Normal University, Beijing 100875, China; hbjrlrw@163.com
2 Winter Sports Management Center of the State General Administration of Sports, Beijing 100763, China; zhang10815@126.com
3 College of Competitive Sports, Tianjin University of Sport, Tianjin 300060, China
* Correspondence: jizhongqiu61@bnu.edu.cn

Abstract: This study aimed to analyze the aerodynamic characteristics of Chinese Nordic combined athletes during the ski jump take-off process, comparing them with elite athletes from the 2009 Nordic World Ski Championships using computational fluid dynamics (CFD) methods. Methods: Using 3D model analysis and continuous relative phase analysis, CFD methods were utilized to assess the mechanical characteristics of athletes during the take-off phase. Results: The analysis revealed that Chinese athletes displayed a lower dominance of the knee joint during the take-off phase, leading to increased air drag. Conclusion: Reduced knee joint dominance and an excessive ankle angle at the initiation of the ski jump take-off contribute to higher air drag. The lean angle of the body and the ankle angle post-take-off significantly affect the resultant lift and drag forces.

Keywords: fluid dynamics; human simulation modeling; coordination; 3D model; continuous relative phase

1. Introduction

Ski jumping, a critical component of Nordic combined, influences the starting order for the subsequent cross-country skiing event. This sport is divided into four distinct phases: the in-run stage, take-off, flight, and landing [1]. The final score in ski jumping is primarily determined by the athlete's flight distance. During the take-off phase, athletes must rapidly extend their legs and torso to achieve an optimal body posture within a short timeframe, setting the stage for the flight phase. The take-off phase typically lasts around 300 milliseconds [2], necessitating high coordination and muscular strength in the lower limbs. At take-off, athletes reach horizontal speeds of approximately 25 m/s, and rapid changes in joint angles at such high speeds create significant physical drag. Previous research has demonstrated that kinematic parameters during the take-off phase are crucial for determining the final flight distance [2,3]. Errors during the take-off phase cannot be corrected in the flight phase [4], highlighting the importance of take-off for overall performance. The effectiveness of the take-off phase directly influences the athlete's final score [5].

To analyze an athlete's coordination, quantitative methods can be used to measure the relationships among various joints during movement. This approach facilitates direct observation and comparison, offering valuable insights for athletic training [6]. Quantifying limb coordination provides a numerical representation of these relationships [7]. For example, continuous relative phase (CRP) analysis is employed to assess limb coordination. CRP analysis is widely used in sports to evaluate athletes' performance [8–10]. By applying CRP analysis to joint coordination, we gain a clear understanding of how different joints interact throughout a movement. This information is crucial for improving athletic techniques and training methods.

Keizo et al. found that different take-off techniques used by ski jumpers significantly affect the drag experienced during the take-off phase [11]. Although ski jumping is closely linked to aerodynamic characteristics, computational fluid dynamics (CFD) methods are primarily applied to the flight phase [1,12–15]. Research on the take-off phase is relatively limited.

This study investigates the kinematic and aerodynamic characteristics of Chinese athletes (CAs) in comparison to Nordic World athletes (NWAs) during the take-off phase using various analytical methods. We specifically analyzed the drag of the left and right lower legs (L/RLDs), left and right thighs (L/RTDs), drag of the torso (TOD), total drag (TD), and total lift (TL) at the start moment of the take-off phase (SMT) and the end moment of the take-off phase (EMT). We also calculated the lift-to-drag ratio (L/D) based on the total lift and total drag.

We employed Direct Linear Transformation (DLT) to process images and develop a three-dimensional motion model, which provided the kinematic data of the athletes. Additionally, Continuous Relative Phase (CRP) analysis was used to quantify intra-limb coordination during the take-off phase. Combined with Computational Fluid Dynamics (CFD) analysis, we performed a comprehensive examination of the aerodynamic features of athletes in various postures. By integrating these complementary methods, we created a robust research framework to elucidate the mechanical mechanisms of the take-off phase. This study offers new insights into ski-jumping biomechanics and provides scientific evidence to enhance training and performance.

This study investigates the impact of different joint motion characteristics on aerodynamic properties during the take-off phase using CFD and CRP methods. The findings provide a theoretical foundation for future sports training and research, enhance understanding of the aerodynamic effects of joint movements, and contribute valuable insights to the field of sports biomechanics.

The structure of this manuscript is as follows: First, we present the background and objectives of the study. Next, we detail the research methods and data-processing procedures. We then report the research results and, finally, summarize the main findings and discuss their implications in relation to previous research.

2. Method

This study collects kinematic data from Chinese ski jumpers during competitions and from Nordic elite athletes. Using CRP and CFD analyses, we investigate the biomechanical characteristics of athletes during the take-off phase. The detailed flowchart is shown in Figure 1.

2.1. Test Subjects and Venue

This study was conducted at the National Ski Jumping Training Base in Laiyuan, Baoding, China, to collect data. The subjects were five Chinese Nordic combined athletes, all first-class or higher-level athletes. Video recordings captured the athletes' take-off actions during ski-jumping training. During testing, each athlete wore the same ski-jumping suit, helmet, and skis. All the athletes performed two full ski jumps at their best, and the data from their optimal performances were selected for analysis.

This study adhered to the ethical principles outlined in the Declaration of Helsinki. All participants provided informed consent prior to testing, ensuring anonymity and confidentiality. Participation was entirely voluntary.

2.2. Data Collection and Processing

This study utilized a three-dimensional (3D) motion-capture system, which included two high-definition cameras, two camera stands, and a 3D calibration framework, operating at a sampling frequency of 50 Hz, to collect data from athletes on-site during competition. The coverage area of the two cameras was approximately 6 m long and 10 m wide.

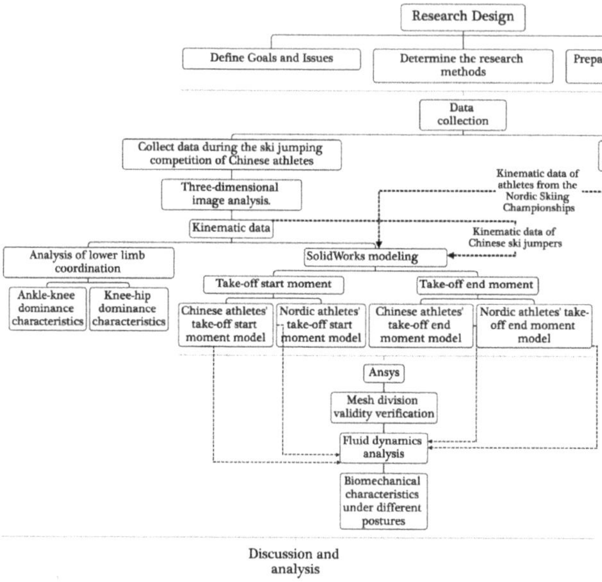

Figure 1. Experimental design flowchart.

Using a 3D video analysis system, a 3D coordinate system was established through 3D framework calibration, and the human joints were calibrated with automatic error correction to obtain a 3D human motion model. During the calibration process, the angle between the foot and tibia was defined as the ankle joint angle, the angle between the tibia and femur was defined as the knee joint angle, and the angle between the femur and trunk was defined as the hip joint angle. The kinematic data of the five athletes during the experiment was calculated using a 3D image analysis system (APAS, Ariel Dynamics, Inc., San Diego, CA, USA), including ankle, knee, and hip joint angles, as well as joint angular velocities.

2.3. Continuous Relative Phase (CRP)

After acquiring the kinematic data of the athletes at each moment during the take-off process, we processed the joint angles and joint angular velocities based on CRP analysis. Since the analysis focused on the dominance relationship between different joints (ankle–knee, knee–hip), we standardized the joint angles and angular velocities to account for the varying degrees of freedom among different joints. This standardization resulted in the determination of phase angles for the different joints of the athletes. Using these phase angles, the CRP values between different joints were calculated using the following specific equations [7,16,17]:

$$\overline{\theta} = 2\left[\frac{\theta - \min(\theta)}{\max(\theta) - \min(\theta)}\right] - 1 \quad (1)$$

$$\overline{\omega} = \left[\frac{\omega}{\max(|\omega|)}\right] \quad (2)$$

In Equation (1), θ represents the angle of a specific joint during continuous movement, while $\overline{\theta}$ represents the standardized value of the joint angle. In Equation (2), ω represents the angular velocity of a specific joint during continuous movement, and $\overline{\omega}$ represents the standardized value of the joint angular velocity during continuous movement.

$$\overline{\varphi}(i) = \tan^{-1}\left[\frac{\overline{\omega}(i)}{\overline{\theta}(i)}\right], = 1, 2, \cdots, n \quad (3)$$

Based on the standardized joint angular velocity and angle values obtained from Equations (1) and (2), Equation (3) calculates the phase angle of a joint at each moment in time. In Equation (3), $\overline{\omega}$ represents the phase angle of a specific joint at different moments during the movement.

$$\text{CRP} = \varphi_{\text{Distal}} - \varphi_{\text{Proximal}} \quad (4)$$

Equation (4) calculates the value of the CRP by taking the difference between the phase angles of the distal joint and the proximal joint. This allows us to determine the dominant joint at different moments during the movement. When the CRP value shows a decreasing trend, it indicates that the proximal joint is dominant. Conversely, when the CRP value shows an increasing trend, it suggests that the distal joint is dominant. If the CRP value remains relatively stable, it indicates that both joints are moving in synchrony, and neither joint is dominant during that specific period of time [16].

2.4. Fluid Dynamics Calculations

2.4.1. Mesh Density Independence Verification

To meet computational requirements, this study conducted an independence verification of different mesh densities, selecting four types ranging from 3 million to 12 million elements. The verification results, listed in Table 1 and Figure 2, show that variations in mesh density do not affect the accuracy of the final computational results. Based on the research by Liao Zhangwen et al. [18], which investigated the impact of different mesh generations on result accuracy, this study used a mesh density of approximately 8 million elements to ensure precision in the outcomes.

Table 1. Grid density independence verification results.

Parameters	Grid Density 1	Grid Density 2	Grid Density 3	Grid Density 4	Grid Density 5
Total number of grid	12,000,000	10,000,000	8,000,000	5,000,000	3,000,000
Lift–drag ratio	1.14	1.14	1.14	1.14	1.23

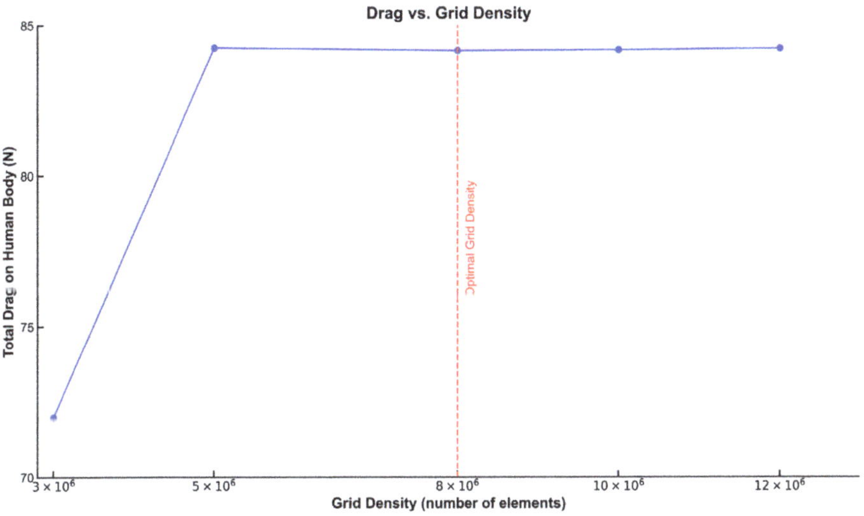

Figure 2. Independent grid test.

2.4.2. Establishment of Human Body Model and Fluid Domain

In this paper, we use direct linear transformation (DLT) to analyze the kinematic data during the take-off phase of five Chinese athletes. The "start moment" of the take-off phase is defined as the instant when athletes begin to rapidly extend their lower limbs from the in-run position into take-off. The "end moment" is defined as the stage when athletes have completed take-off and have left the glide path by approximately 5 m. The kinematic data of the athletes at the start moment of the take-off phase (SMT) and the end moment of the take-off phase (EMT) (Table 2) are compared with data obtained from Nordic World athletes during the 2009 Nordic World Ski Championships (NWA) by Miroslav Janura et al. [19]. These kinematic data are imported into SolidWorks 2023 SP1.0 modeling software to establish body poses corresponding to the respective kinematic data.

Table 2. Kinematic data of athletes at the start and end of the take-off phase.

	Elite Chinese/NW Athletes	Ankle Joint Angle (Mean ± SD)	Knee Joint Angle (Mean ± SD)	Hip Joint Angle (Mean ± SD)
SMT	CA	61.2 ± 13.8	90.9 ± 19.3	61.1 ± 20
	NWA	40.69 ± 3.67	95.45 ± 6.91	60.64 ± 7.36
EMT	CA	87.9 ± 8.37	151.19 ± 11.56	133.09 ± 14.7
	NWA	56.14 ± 5.56	180.93 ± 11.86	149.51 ± 10.95

The fluid domain of the human–board system measures 16 m in length, 2 m in width, and 6 m in height, as shown in Figure 3. The side of the fluid domain opposite the human body is designated as an air inlet with a wind speed of 29 m/s, based on the horizontal centroidal velocity during the athlete's take-off and the fluid velocity used by Huqi [20] to analyze the aerodynamic effects on asymmetrical ski-jumping postures. The wall facing the athlete's direction is set as an air inlet, the rear wall is designated as an air outlet, and the perimeter walls are treated as smooth boundaries. During grid generation, the mesh around the athlete is refined to enhance the accuracy of the aerodynamic analysis under these conditions.

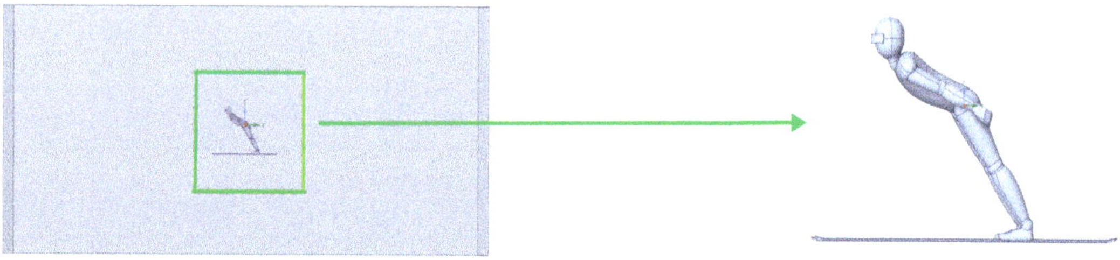

Figure 3. Schematic diagram of the computational region.

2.4.3. Governing Equations

This study examines the aerodynamic characteristics of athletes using the k-ε model, a widely adopted approach for fluid turbulence calculations, with its validity supported by previous verification studies [21]. This model is grounded in the Reynolds-Averaged Navier–Stokes (RANS) equations and incorporates two additional equations to close the turbulent stress calculation: one for turbulent kinetic energy (k) and another for the turbulent dissipation rate (ε). The governing equations are as follows:

$$\frac{\partial(\rho k)}{\partial t} + \frac{\partial(\rho k u_i)}{\partial x_i} = \frac{\partial}{\partial x_j}\left[\left(\mu + \frac{\mu_t}{\sigma_k}\right)\frac{\partial k}{\partial x_j}\right] + P_k - \rho\epsilon \quad (5)$$

Equation (5) is the k-equation (turbulent kinetic energy equation), which primarily describes the transport and generation of turbulent kinetic energy (k). Here, ρ represents the fluid density; u_i is the velocity component; x_i and x_j are spatial coordinates; μ is the molecular viscosity coefficient; μ_t is the turbulent viscosity coefficient; ∂k is the turbulent

Prandtl number for the k-equation (typically set to 1.0); P_k represents the production of turbulent kinetic energy; and ϵ denotes the dissipation rate of turbulent kinetic energy.

$$\frac{\partial(\rho\epsilon)}{\partial t}+\frac{\partial(\rho\epsilon u_i)}{\partial x_i}=\frac{\partial}{\partial x_j}[(\mu+\frac{\mu_t}{\sigma_\epsilon})\frac{\partial\epsilon}{\partial x_j}]+C_{1\epsilon}\frac{\epsilon}{k}P_k-C_{2\epsilon}\rho\frac{\epsilon^2}{k} \quad (6)$$

Equation (6) represents the ε-equation (turbulent kinetic energy dissipation rate equation), which describes the transport and generation of the turbulent kinetic energy dissipation rate ε. $\partial\epsilon$ is the turbulent Prandtl number for the ε-equation (typically set to 1.3), and $C_{1\epsilon}$ and $C_{2\epsilon}$ are model constants, commonly taken as 1.44 and 1.92, respectively.

In studying the aerodynamic characteristics of athletes using the k-ε model, the continuity equation ensures mass conservation throughout the analysis. The Navier–Stokes equation describes momentum conservation, while the k-ε model serves to close the turbulent stress term in the Navier–Stokes equation. In this study, these three equations jointly constitute a complete set of fluid dynamics equations, which are used to describe the behavior of fluids during flow.

$$\frac{\partial\rho}{\partial t}+\frac{\partial(\rho u_i)}{\partial x_i}=0 \quad (7)$$

Equation (7) represents the continuity equation, where ρ is the fluid density, u_i is the velocity component, t is time, and x_i is the spatial coordinate.

$$\frac{\partial(\rho u_i)}{\partial t}+\frac{\partial(\rho u_i u_j)}{\partial x_j}=-\frac{\partial p}{\partial x_i}+\frac{\partial\tau_{ij}}{\partial x_j}+\rho g_i \quad (8)$$

Equation (8) represents the Navier–Stokes equation, where p is the pressure, τ_{ij} is the stress tensor (including viscous stress and turbulent stress), and g_i is the component of gravitational acceleration.

2.4.4. Boundary Conditions and Computational Scenarios

In this study, the inlet velocity was set to 29 m/s, the outlet pressure was set to the normal atmospheric pressure of 101,325 Pa, and other boundaries were set as no-slip walls. The gas was assumed to be incompressible air, and the gravitational acceleration was set to g = 9.81 m/s^2. This study analyzes the drag and lift forces experienced by athletes based on their different body postures, and the equations are as follows:

$$F_L=0.5\times\rho\times V^2\times S\times C_l \quad (9)$$

$$F_D=0.5\times\rho\times A\times C_d\times V^2 \quad (10)$$

Equation (9) is the equation for calculating lift force, where ρ is the density of the fluid, V is the flow velocity, S is the frontal area of the object, and C_l is the lift coefficient. Equation (10) is the equation for calculating drag force, where ρ is the density of the fluid, A is the reference area of the object, C_d is the drag coefficient, and V is the velocity of the object. The ratio of lift-to-drag is defined as the lift-to-drag ratio generated by the athlete.

3. Results

3.1. Analysis of Joint Dominance within Limbs

3.1.1. Analysis of Ankle–Knee Joint Dominance

The results (Figure 4) reveal the dominance characteristics of the ankle and knee joints during the take-off phase of Chinese athletes as follows: Stable CRP values during the 0–20% and 90–100% phases suggest a synchronized movement pattern between the ankle and knee joints, indicating no clear dominance of either joint. During the 20–35%, 50–65%, and 80–90% phases, the CRP values exhibit a continuous increase, signifying that the changes in the ankle joint dominate during these intervals, meaning the ankle joint leads the knee joint in the take-off process. Conversely, during the 35–50% and 65–80% phases, the CRP values display a decreasing trend, indicating that the knee joint predominates during these stages, with the knee joint leading the ankle joint in movement.

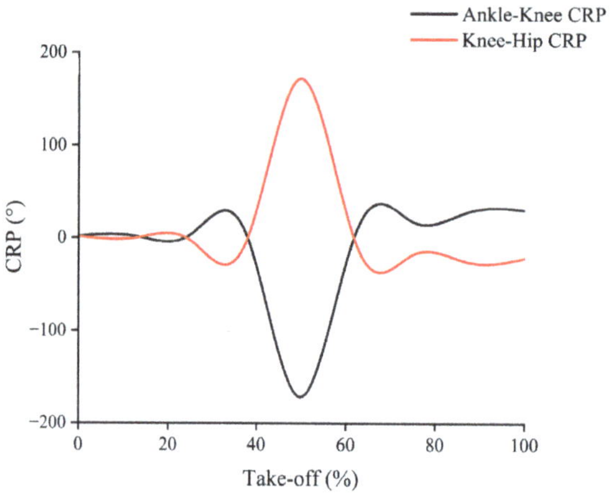

Figure 4. Coordination of limb joints in Chinese athletes during the take-off phase (CRP).

3.1.2. Analysis of Knee–Hip Joint Dominance

As shown in Figure 4, the dominance characteristics of the knee and hip joints during the take-off phase of Chinese athletes are evident. During the 0–20% and 90–100% phases, stable CRP values indicate a coordinated movement between the knee and hip joints, with no clear dominance of either joint. During the 20–35%, 50–65%, and 80–90% phases, the CRP values between the knee and hip joints exhibit a decreasing trend, indicating that the hip joint dominates during these intervals, leading the knee joint in the take-off process. Conversely, during the 35–50% and 65–80% phases, the CRP values show an increasing trend, suggesting that the knee joint becomes dominant, leading the hip joint in movement.

3.2. Drag Characteristics at Different Moments

3.2.1. Aerodynamic Characteristics at the Start Moment of the Take-Off Phase (SMT)

As depicted in Figures 5 and 6, the static pressure experienced by athletes at the SMT is predominantly concentrated on the inner thighs and torso. As the fluid flows over the athlete's body surface, it generates relatively low-pressure zones in the flow area, leading to reduced static pressure on the outer sides of the athletes' lower limbs during take-off.

The results indicate that the air drag experienced by various limbs of elite athletes varies based on the angle at the SMT. As shown in Table 3, during the initial stage of take-off, the drag on the left and right calves of Chinese athletes is significantly greater than that of NW elite athletes, measuring 13.26 N and 12.06 N, respectively. In contrast, the drag on the left and right calves of NW elite athletes is 2.08 N and 2.34 N, respectively. At the start moment of take-off, the drag on the left and right thighs of Chinese athletes is lower than that of NW elite athletes, measuring 0.02 N and 0.01 N, respectively, whereas the drag on the thighs of NW elite athletes is higher, with left and right drags of 2.88 N and 3.00 N, respectively. The discrepancy in drag borne by the trunk is even more pronounced: Chinese athletes experience a drag of 24.97 N on their trunk, compared to approximately 3.71 N for NW elite athletes. The total drag experienced by the athlete–ski system at the start moment of take-off for Chinese and NW elite athletes is 108.90 N and 55.40 N, respectively (Table 3).

Figure 5. Static pressure diagram of Chinese athletes at the start moment (SMT).

Figure 6. Static pressure diagram of NW elite athletes at the start moment (SMT).

Table 3. Drag characteristics at the start moment of the take-off phase (SMT).

	LLD	RLD	LTD	RTD	TOD	TD
	(N)	(N)	(N)	(N)	(N)	(N)
CA	13.26	12.06	0.02	0.01	24.97	108.90
NWA	2.08	2.34	2.88	3.00	3.71	55.40

3.2.2. Aerodynamic Characteristics at the End Moment of the Take-Off Phase (EMT)

The results reveal significant differences in the air drag experienced by Chinese athletes compared to NW elite athletes during the take-off phase. The discrepancies are most pronounced in the drag experienced by the lower legs, torso, and overall system, while the differences in drag between the left and right thighs of NW elite athletes are relatively minor. Specifically, at the end of take-off, the drag values for the left and right lower legs of Chinese athletes are 74.06 N and 37.02 N, respectively, whereas NW elite athletes experience considerably lower drags of 0.15 N and 0.79 N. Concurrently, the torso drag for Chinese athletes is also significantly higher, at 76.06 N, compared to 10.19 N for NW elite athletes (Table 4). Furthermore, Table 4 shows that the total drag, total lift, and lift-to-drag ratio of the athlete–board system vary between Chinese and NW elite athletes during take-off. For Chinese athletes, the total drag is 170.18 N, the total lift is 67.10 N, and the lift-to-drag ratio at the transition into the early flight phase is 0.39. In contrast, NW elite athletes experience a total drag of 84.15 N, a total lift of 95.18 N, and a lift-to-drag ratio of 1.14 during their transition into the early flight phase.

Table 4. Drag characteristics at the end moment of the take-off phase (EMT).

	LLD (N)	RLD (N)	LHD (N)	RHD (N)	TOD (N)	TD (N)	TL (N)	L/D
CA	74.06	37.02	8.31	9.54	76.06	170.18	67.10	0.39
NWA	0.15	0.79	6.48	9.47	10.19	84.15	95.18	1.14

As shown in Figures 7 and 8, there are notable differences in the static pressure experienced by Chinese and NW elite athletes during the EMT. NW elite athletes, during the completion of take-off, generally lean forward with adequate forward momentum and fully extend their lower limbs. This posture results in relatively lower static pressure across their body segments and the formation of low-pressure zones. In contrast, Chinese athletes do not lean forward as much during the EMT, resulting in comparatively higher static pressure on various body segments.

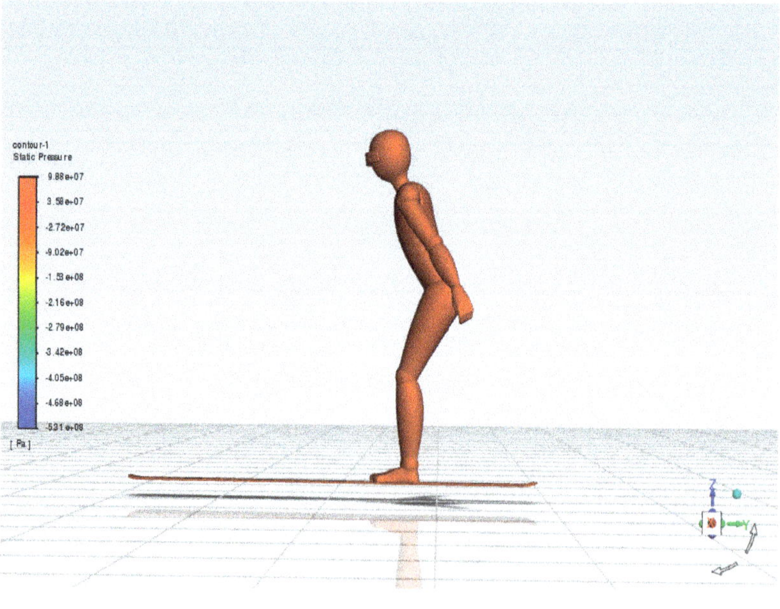

Figure 7. Static pressure diagram at the end moment of the take-off phase (EMT) for Chinese athletes.

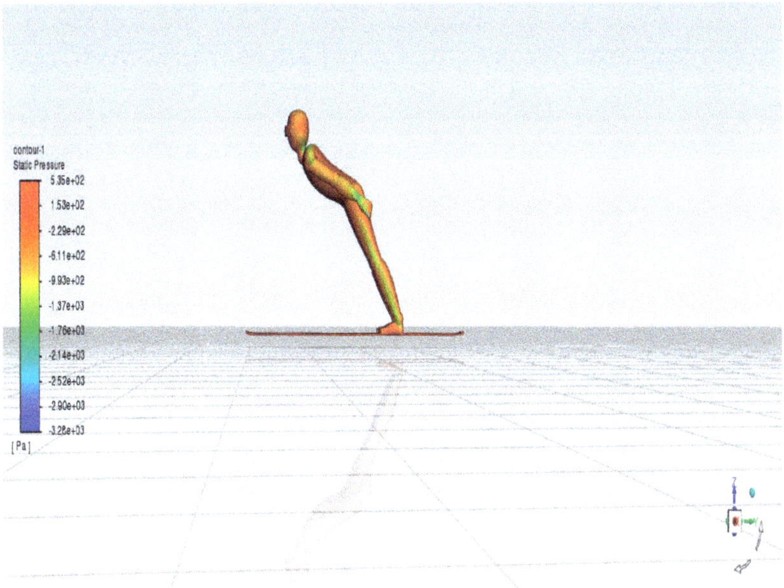

Figure 8. Static pressure diagram at the end moment of the take-off phase (EMT) for NW elite athletes.

4. Discussion

4.1. Dominant Characteristics of Limbs during the Take-Off Phase

This study utilizes the CRP method to examine the dominance characteristics of the ankle, knee, and hip joints during the take-off phase in athletes. By comparing these characteristics with those observed in NW elite athletes, the study seeks to offer a theoretical framework for improving athletes' take-off performance.

Based on the ankle–knee CRP curve, it can be observed that during the take-off phase, the dominant relationship between the ankle and knee joints of Chinese athletes shows that for 40% of the total take-off time, the ankle leads the knee, while for 30% of the time, the knee leads the ankle. This pattern suggests that kinematic changes in the ankle of Chinese elite athletes are relatively more pronounced compared to those in the knee throughout the take-off process, and the knee's dominance over the ankle is less prominent. As shown in Figure 9 [16], Julien Chardonnens analyzed the dominant characteristics of elite athletes from the Swiss national ski-jumping team during the take-off phase using the CRP method. Chardonnens' study assessed the coordination of the lower limbs by examining the angles between the limbs and the horizontal plane. Building on this research, we used the DLT method to develop a three-dimensional model and analyzed joint angle changes to explore lower-limb coordination during take-off. Chardonnens' findings revealed a continuous decrease in CRP values for Swiss elite athletes during the first 75% of take-off, indicating that the thigh dominates the lower leg, with the knee joint leading the ankle joint. This suggests that high-level athletes exhibit significant dominance of the thigh over the lower leg during take-off, attributed to the stability of the lower leg and delayed rotation of the lower leg [16]. Maintaining maximum ankle stability, enhancing lower-limb extension, and increasing knee joint angular velocity during take-off are crucial for optimal intra-limb dominance and performance improvement. Chardonnens also found that, in the latter half of take-off, the lower-leg rotation dominates because the initial rapid extension brings the knee joint angle closer to its optimal position, enhancing forward angular momentum for an optimal flight posture. However, Chinese elite athletes show more frequent changes in ankle and knee joint dominance throughout the take-off phase. In the initial phase (first 20%), their knee and ankle joints move synchronously. During the middle phase, dominance changes frequently, indicating potential issues with ankle angle stability and lower-limb extension. This may prevent the knee joint from reaching its optimal angle, impacting the lift-to-drag ratio in the subsequent flight. Previous studies suggest that elite athletes maintain lower-leg stability and minimal ankle angle changes during take-off, facilitating greater angular momentum and optimal lower-limb extension, resulting in a better flight posture and lift-to-drag ratio after take-off [22–24]. Research by Miroslav Janura et al. found that professional ski jumpers have a smaller ankle angle and maintain high stability throughout the take-off phase compared to Nordic combined athletes [19]. Data from Tables 2 and 3 indicate that an excessive ankle joint angle during take-off may cause athletes to lean back excessively, increasing air drag and affecting acceleration and the horizontal center of mass velocity. Further analysis of the data in Table 4 reveals that a large ankle joint angle increases air drag on the lower leg and torso, negatively impacting the aerodynamic characteristics of the athlete–ski system. An excessive ankle joint angle impedes achieving the optimal lift-to-drag ratio at the end of the take-off phase, which, in turn, affects aerodynamic efficiency during the flight phase that follows. This suboptimal lift-to-drag ratio leads to shorter flight times and reduced jump distances. Previous research indicates that while the knee joint often plays a dominant role during take-off, an excessively large ankle joint angle at the start, combined with significant changes in this angle throughout take-off, increases overall air drag. To enhance the quality of take-off and achieve the best lift-to-drag ratio during the flight phase, it is crucial to minimize the ankle joint angle and maintain stability throughout take-off. This approach will help reduce air drag and ensure optimal aerodynamic performance.

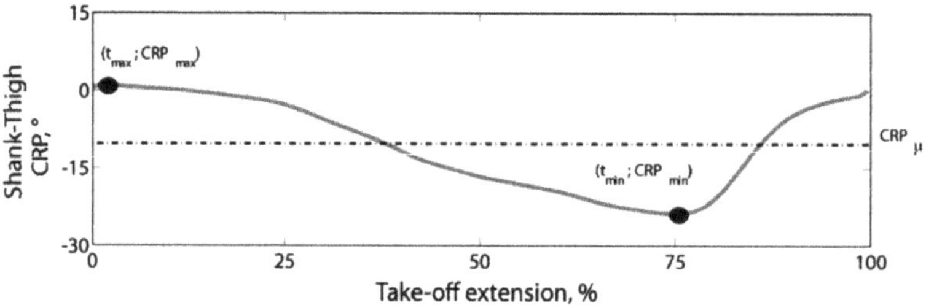

Figure 9. CRP characteristics of Swiss elite athletes' legs during takeoff.

The CRP curve of the knee and hip joints illustrates the dominance characteristics between these two joints during the take-off phase in Chinese elite athletes. This curve shows that the hip joint remains dominant over the knee joint for a longer duration throughout the take-off process (Figure 4). Zhang Dong et al. analyzed air drag on athletes' bodies in various take-off poses using wind-tunnel training. They concluded that a "knee-driven hip" approach minimizes air drag during the take-off process. Additionally, they suggested that the "synchronous hip–knee variable speed" approach results in the highest drag, likely due to the rapid opening of the hip joint while the lower limbs extend more slowly. This leads to uneven torso drag distribution and an early increase in the torso's force-bearing area, causing greater drag [25]. The study found that minimal variation in the CRP index values of the knee and hip joints during the initial 20% of the take-off phase indicates that Chinese elite athletes adopt a "synchronous hip–knee variable speed" approach. Throughout take-off, hip joint dominance accounts for 40% of the process, with the knee joint assisting. This movement pattern suggests that Chinese elite athletes might experience an excessively large hip joint angle during take-off, leading to an early and significant increase in torso air drag, thereby affecting their horizontal center-of-mass velocity. This finding highlights the need for optimizing the take-off technique to reduce aerodynamic drag and enhance performance. Janez Vodicar et al. argue that horizontal center-of-mass velocity is the most crucial factor affecting athletes during the take-off phase. Minimizing drag impact on horizontal center-of-mass velocity enables athletes to achieve greater flight distances during the flight phase [3]. Previous studies indicate that hip joint kinematics during take-off significantly impact drag experienced during the jump [26]. Maintaining a smaller hip joint angle during take-off is suggested to improve performance [27]. Although Chinese elite athletes also use a knee-led hip take-off posture, the duration of this phase is relatively short. Based on the data from Tables 2 and 3, we found that Chinese athletes have smaller knee joint angles during take-off, resulting in lower thigh drag compared to Nordic athletes. This indicates that maintaining the knee joint at approximately 90 degrees during the take-off and sliding phases can more effectively reduce air drag. However, this also suggests that athletes complete the push-off action more quickly within the same timeframe. Data from Tables 2 and 4 also show that at the end of take-off, Chinese athletes have smaller knee joint angles, correlating with a higher total drag at that moment. Therefore, during the SMT and sliding phases, athletes should reduce the knee joint angle to lower air drag and enhance knee joint dominance during take-off. This approach will help the knee joint achieve the optimal angle by the end of take-off, thereby improving performance. Furthermore, frequent shifts in dominance between the knee and hip joints during this process may generate additional drag, affecting horizontal center-of-mass velocity.

4.2. Aerodynamic Analysis during the Take-Off Phase

4.2.1. Aerodynamic Characteristics during the Start Moment of the Take-Off Phase

The initial phase of the ski jump take-off marks the transition from the approach phase to the take-off phase. Athletes' varying body postures at the SMT lead to differences in the air drag they encounter. Maximizing or even enhancing the horizontal center-of-mass velocity during the take-off process is crucial for achieving longer flight distances. Alterations in body posture during take-off result in varying degrees of air drag, a pivotal factor influencing the athlete's horizontal center-of-mass velocity. The approach phase serves as a critical stage for athletes to increase their horizontal center-of-mass velocity. This study aims to analyze the aerodynamic characteristics encountered at the SMT, thereby assessing the magnitude of air drag generated by different postures. As depicted in Table 3, the air drag endured by the athlete–ski system at the SMT for Chinese elite athletes is greater than that of NW elite athletes. Analysis of the kinematic data at the SMT reveals that while the angles of the knee and hip joints during the take-off process in Chinese elite athletes are comparable to those of NW elite athletes, the ankle joint angle exhibits a notable discrepancy. This particular kinematic characteristic results in an increased frontal area of the Chinese athletes' bodies at the instant of take-off, thereby subjecting them to greater air drag. In comparison to NW elite athletes, Chinese athletes experience relatively lower air drag on their bilateral thighs. This is primarily due to the comparatively smaller knee joint angles exhibited by Chinese athletes, which reduce the drag acting on their thighs during the SMT. Tang Weidi et al. analyzed the impact of different kinematic data on aerodynamic characteristics during the approach phase of ski jumping and found that ankle joint angles significantly affect the drag faced by athletes. They suggested that an ankle joint angle of 43–45° is more conducive to achieving optimal performance [28]. Based on this study, we found that while there are certain similarities in the kinematic data of the knee and hip joints between Chinese and NW elite athletes, the ankle kinematic data differs significantly. Moreover, the increased ankle angle leads to higher air drag experienced by athletes during the take-off process.

4.2.2. Aerodynamic Characteristics at the End Moment of the Take-Off Phase

The body posture of ski jumpers during the EMT significantly influences their ability to swiftly transition into the early flight phase and achieve an optimal lift-to-drag ratio. This, in turn, provides more sufficient flight time for a stable flight. Different body postures during the EMT have varying effects on the amount of drag encountered by the athlete and the lift generated. Xiong Li et al. found that hip angle and body–ski angle play a dominant role in aerodynamic performance, with the ski and torso contributing the most to aerodynamics and torque [13]. Ki-Don Lee et al. suggested that athletes achieve the optimal lift-to-drag ratio of approximately 1.5 when the angle between their body and the ski board is 15° during the flight phase [29]. As shown in Table 4, there are differences in the air drag experienced by various body parts between Chinese and Northwest (NW) elite athletes after take-off. The coordination characteristics and kinematic data during the athletes' take-off reveal that Chinese athletes have a relatively small proportion of time in which the knee joint dominates during take-off. This results in insufficient push-off and extension of the knee joint, preventing it from reaching the optimal angle. Additionally, the strong dominance of the ankle joint during take-off causes an excessively large ankle angle after take-off, resulting in inadequate forward leaning of the body. Anton and Arndt found that maintaining stability in the ankle joint during take-off allows the athlete's body to lean forward, gaining forward angular momentum [24]. Compared to NW elite athletes, Chinese athletes tend to have a larger ankle angle and a smaller knee angle, leading to inadequate forward leaning of the body and a larger windward area. Calculations of the air drag experienced by athletes at the EMT (Table 4) reveal that Chinese athletes experience higher air drag in various body segments than NW elite athletes, with significantly higher drag in the bilateral lower legs and torso. This is primarily due to the greater ankle angle during take-off, which results in less forward leaning and higher drag. Our analysis of the drag and lift forces during take-off, with respect to the lift-to-drag ratio generated by different postures at the EMT, shows that NW elite athletes, with smaller ankle angles and fully extended knee angles, achieve better forward leaning, less air drag, and higher lift force compared to Chinese athletes. Consequently, NW elite athletes achieve a higher lift-to-drag ratio (Table 4). Research by Stenseth suggests that disadvantages during the take-off phase are difficult to compensate for in the subsequent flight phase, implying that the body posture after take-off plays a crucial role in the stability and distance of the flight [4]. Mikko Virmavirta's study on the early flight phase of ski jumpers showed that achieving an optimal lift-to-drag ratio significantly impacts the flight distance [30]. Therefore, NW elite athletes' body posture at the EMT enables them to enter the stable flight phase faster, providing a longer flight time. In contrast, the kinematic characteristics of Chinese athletes result in higher air drag during take-off and subsequent flight adjustments, potentially leading to an unstable flight and a higher probability of injury.

5. Conclusions

During the take-off phase of ski jumping, adopting a knee-dominated approach while maintaining ankle stability enables athletes to execute the push-off motion more effectively, thereby minimizing air drag. However, an excessively large ankle angle at the SMT can significantly increase drag. Additionally, at the EMT, an excessively large ankle angle combined with insufficient forward leaning not only increases air drag, but also reduces lift generation. This prevents athletes from achieving the optimal lift-to-drag ratio during the flight phase, thereby affecting their performance and ultimately impacting their final score. This study focused solely on the aerodynamic characteristics at the SMT and EMT, without thoroughly investigating the aerodynamic effects of various body postures throughout the entire take-off process. To address these limitations, we employed CRP analysis to explore the dominant characteristics of the ankle, knee, and hip joints throughout the complete take-off process, integrating aerodynamic characteristics from both the SMT and EMT.

Author Contributions: Conceptualization, Z.J. and B.H.; methodology, B.H.; software, B.H.; validation, Z.J., B.H., Y.Z. and M.Y.; formal analysis, B.H.; investigation, Z.J., B.H., Y.Z. and M.Y.; resources, Z.J., Y.Z. and M.Y.; data curation, B.H.; writing—original draft preparation, B.H.; writing—review and editing, B.H., Z.J., Y.Z. and M.Y.; visualization, B.H. and Y.Z.; supervision, Z.J. and M.Y.; project administration, Z.J., Y.Z. and M.Y.; funding acquisition, Z.J., Y.Z. and M.Y. All authors have read and agreed to the published version of the manuscript.

Funding: This research received no external funding.

Institutional Review Board Statement: Although this study involved human participants, there were no ethical issues. The ethical review and approval were waived because the study did not present any ethical concerns. All participants signed informed consent forms and adhered to ethical standards.

Informed Consent Statement: Informed consent was obtained from all subjects involved in the study.

Data Availability Statement: The original contributions presented in the study are included in the article, further inquiries can be directed to the corresponding author.

Conflicts of Interest: The authors declare no conflicts of interest.

References

1. Hu, Q.; Liu, Y. The influence of athlete's posture on the aerodynamic characteristics of ski jumping during air flight. *J. Med. Biomech.* **2021**, *36*, 407–414.
2. Schwameder, H. Biomechanics research in ski jumping, 1991–2006. *Sports Biomech.* **2008**, *7*, 114–136. [CrossRef]
3. Vodicar, J.; Jost, B. The Factor Structure of Chosen Kinematic Characteristics of Take-Off in Ski Jumping. *J. Hum. Kinet.* **2010**, *23*, 37–45. [CrossRef]
4. Stenseth, O.M.R.; Barli, S.F.; Martin, R.K.; Engebretsen, L. Injuries in elite women's ski jumping: Surveillance through the 2017–2018 FIS World Cup season. *Br. J. Sports Med.* **2019**, *54*, 44–48. [CrossRef]
5. Ketterer, J.; Gollhofer, A.; Lauber, B. Biomechanical agreement between different imitation jumps and hill jumps in ski jumping. *Scand. J. Med. Sci. Sports* **2020**, *31*, 115–123. [CrossRef]
6. Irwin, G.; Kerwin, D.G. Inter-segmental coordination in progressions for the longswing on high bar. *Sports Biomech.* **2007**, *22*, 1303–1318. [CrossRef]
7. Burgess-Limerick, R.; Abernethy, B.; Neal, R.J. Relative phase quantifies interjoint coordination. *J. Biomech.* **1993**, *26*, 91–94. [CrossRef]
8. Floría, P.; Sánchez-Sixto, A.; Ferber, R.; Harrison, A.J. Effects of running experience on coordination and its variability in runners. *J. Sports Sci.* **2017**, *36*, 272–278. [CrossRef]
9. Seifert, L.; Leblanc, H.; Chollet, D.; Delignières, D. Inter-limb coordination in swimming: Effect of speed and skill level. *Hum. Mov. Sci.* **2010**, *29*, 103–113. [CrossRef]
10. Mazurek, C.M.; Pearsall, D.J.; Renaud, P.J.; Robbins, S.M. Differences in inter-segment coordination between high- and low-calibre ice hockey players during forward skating. *Sports Biomech.* **2020**, *22*, 1303–1318. [CrossRef] [PubMed]
11. Yamamoto, K.; Tsubokura, M.; Ikeda, J.; Onishi, K.; Baleriola, S. Effect of posture on the aerodynamic characteristics during take-off in ski jumping. *J. Biomech.* **2016**, *49*, 3688–3696. [CrossRef] [PubMed]
12. Jung, A.; Staat, M.; Müller, W. Flight style optimization in ski jumping on normal, large, and ski flying hills. *J. Biomech.* **2014**, *47*, 716–722. [CrossRef]
13. Li, X.; Wang, X.; Chen, L.; Zhao, T. Effects of body angles on the aerodynamic characteristics in the flight period of ski jumping: A simulation study. *Sports Biomech.* **2023**, 1–15. [CrossRef]
14. Hu, Q.; Tang, W.; Liu, Y. Computational Fluid Dynamics Simulation Study on Aerodynamic Characteristics under Unfavorable Conditions during Flight Phase in Ski Jumping. *Appl. Sci.* **2024**, *14*, 1390. [CrossRef]
15. Yu, J.; Liao, Z.; Ma, X.; Qi, S.; Liang, Z.; Wei, Z.; Zhang, S. Optimisation of stable flight posture of ski jumping based on computational fluid dynamics simulation technology. *Sports Biomech.* **2023**, 1–20. [CrossRef]
16. Chardonnens, J.; Favre, J.; Cuendet, F.; Gremion, G.; Aminian, K. Characterization of lower-limbs inter-segment coordination during the take-off extension in ski jumping. *Hum. Mov. Sci.* **2013**, *32*, 741–752. [CrossRef] [PubMed]
17. Tazji, M.K.; Ghale-Beig, A.V.; Sadeghi, H.; Koumantakis, G.A.; Chrysagis, N.; Abbasi, A. Effects of Running-induced Fatigue on the Trunk-pelvis-hip Coordination Variability During Treadmill Running at Different Speeds. *J. Musculoskelet. Neuronal Interact.* **2023**, *23*, 189–195.
18. Liao, Z.; Zhou, Y.; Yu, J.; Zhang, S.; Wei, S.; Zhang, C.; Jiang, F. Optimization and verification of aerial flight posture in ski jumping based on CFD simulation. *Tribology* **2023**, *43*, 1151–1164. [CrossRef]
19. Miroslav, J.; Lee, C.; Zdeněk, S.; Milan, E.; František, Z. Kinematic Analysis of the Take-Off and Start of the Early Flight Phase on a Large Hill (HS-134 m) during the 2009 Nordic World Ski Championships. *J. Hum. Kinet.* **2011**, *27*, 5–16. [CrossRef]
20. Hu, Q.; Liu, Y. The impact of posture asymmetry on aerodynamic characteristics in the flight phase of ski jumping. *China Sport Sci.* **2020**, *40*, 41–49. [CrossRef]
21. Chipongo, K.; Khiadani, M.; Lari, K.S. Comparison and verification of turbulence Reynolds-averaged Navier–Stokes closures to model spatially varied flows. *Sci. Rep.* **2020**, *10*, 19059. [CrossRef] [PubMed]
22. Virmavirta, M.; Isolehto, J.; Komi, P.; Schwameder, H.; Pigozzi, F.; Massazza, G. Take-off analysis of the Olympic ski jumping competition (HS-106 m). *J. Biomech.* **2009**, *42*, 1095–1101. [CrossRef]
23. Mikko, V.; Paavo, V.K. Takeoff analysis of a champion ski jumper. *J. Biomech.* **1994**, *27*, 695. [CrossRef]
24. Arndt, A.; Brüggemann, G.P.; Virmavirta, M.; Komi, P.V. Techniques Used by Olympic Ski Jumpers in the Transition from Takeoff to Early Flight. *J. Appl. Biomech.* **1995**, *11*, 224–237. [CrossRef]

25. Zhang, D.; Zou, X.; Liu, Y.; Xu, J.; Cao, C. The influence of body posture on aerodynamic resistance during the sliding and take-off phases of Nordic combined ski jumping. *China Sport Sci. Technol.* **2023**, *59*, 3–12. [CrossRef]
26. Fang, X.; Grüter, B.; Piprek, P.; Bessone, V.; Petrat, J.; Holzapfel, F. Ski Jumping Trajectory Reconstruction Using Wearable Sensors via Extended Rauch-Tung-Striebel Smoother with State Constraints. *Sensors* **2020**, *20*, 1995. [CrossRef] [PubMed]
27. Paavo, V.K.; Mikko, V. Determinants of Successful Ski-Jumping Performance. In *Biomechanics in Sport: Performance Enhancement and Injury Prevention*; Wiley: Hoboken, NJ, USA, 2000; pp. 349–362. [CrossRef]
28. Tang, W.; Suo, X.; Yang, C.; Cao, F.; Wu, X.; Liu, Y. Computational fluid dynamics simulation and optimization of the sliding phase in ski jumpin. *China Sport Sci.* **2022**, *42*, 62–70.
29. Lee, K.-D.; Park, M.-J.; Kim, K.-Y. Optimization of ski jumper's posture considering lift-to-drag ratio and stability. *J. Biomech.* **2012**, *45*, 2125–2132. [CrossRef]
30. Virmavirta, M.; Isolehto, J.; Komi, P.; Brüggemann, G.-P.; Müller, E.; Schwameder, H. Characteristics of the Early Flight Phase in the Olympic Ski Jumping Competition. *J. Appl. Biomech.* **2005**, *38*, 2157–2163. [CrossRef]

applied sciences

MDPI

Article

Strength and Vibration Analysis of Axial Flow Compressor Blades Based on the CFD-CSD Coupling Method

Haiwei Lin [1], Hong Bao [1], Huaihuai Zhang [2,3], Feifei Zhao [1,*] and Junli Wang [2,3,*]

[1] College of Mechano-Electronic Engineering, Xidian University, Xi'an 710071, China; linhaiwei@qhgas.com (H.L.); hbao@xidian.edu.cn (H.B.)
[2] College of Mechanical Engineering, Shaanxi University of Technology, Hanzhong 723000, China; 15991207493@163.com
[3] Shaanxi Key Laboratory of Industrial Automation, Shaanxi University of Technology, Hanzhong 723000, China
* Correspondence: zhaofeifei@xidian.edu.cn (F.Z.); wjl503@126.com (J.W.)

Abstract: During the operational process of an axial-flow compressor, the blade structure is simultaneously subjected to both aerodynamic loads and centrifugal loads, posing significant challenges to the safe and reliable operation of the blades. Considering both centrifugal loads and aerodynamic loads comprehensively, a bidirectional CFD-CSD coupling analysis method for blade structure was established. The Navier–Stokes governing equations were utilized to solve the internal flow field of the axial-flow compressor. The conservative interpolation method was utilized to couple and solve the blade's static equilibrium equation, and the deformation, stress distribution, and prestress modal behavior of compressor blades were mainly analyzed. The research results indicate that the maximum deformation of the blades occurred at the lead edge tip, while stress predominantly concentrated approximately 33% upward from the blade root, exhibiting a radial distribution that gradually decreased. As the rotational speed increased, the maximum deformation of the blades continuously increased. Furthermore, at a constant rotational speed, the maximum deformation of the blade exhibited a trend of first increasing and then decreasing with the increase in mass flow. In contrast, the maximum stress showed a trend of first increasing, then decreasing, and finally increasing again as the rotational speed continuously increased. Centrifugal loads are the primary factor influencing blade stress and natural frequency. During operation, the blades exhibited two resonance points, approximately occurring at 62% and 98% of the design rotational speed.

Keywords: axial flow compressor; bidirectional CFD-CSD coupling method; multiple field loads; structural characteristics

1. Introduction

With the advancement of aviation engine technology, axial flow compressors, as crucial components of aviation engines, face increasingly stringent demands on overall performance. Consequently, in recent years, there has been an unparalleled development in turbomachinery, with design trends gradually shifting towards higher pressure ratios, improved efficiency, and broader operating ranges. However, the working environment for axial flow compressor blades is highly hostile, subjecting them to prolonged exposure to high temperatures, velocities, and pressures. During operation, compressor blades encounter various factors such as centrifugal loads, aerodynamic loads, thermal loads, and vibrations [1]. If blade vibration or fracture occurs, it will result in significant economic losses for the entire unit. Studies indicate that over 65% of compressor failures are blade-related [2]. According to data from a complete operational cycle of aviation engines and external maintenance test runs, compressor rotational speeds vary based on the aircraft's flight status. Under different rotational speed conditions, the centrifugal and aerodynamic loads experienced by the blades exhibit notable variations. Therefore, conducting strength verification of compressor blades, assessing the influence of loads on blade stress and strain, and identifying potential resonance points during operation are of paramount practical and guiding significance. Such measures can prevent blade-resonance-induced damage, thereby enhancing equipment economy and safety.

In recent years, scholars both domestically and internationally have conducted extensive research on the structural performance of compressors. Ramamurti et al. [3] utilized finite element analysis to simulate the stress on a simplified impeller model, revealing that an increase in impeller rotational speed results in elevated stress levels, with centrifugal loads exhibiting a more pronounced effect. The interface between the blades and the hub emerges as the location of maximum stress. Day et al. [4] conducted a finite element analysis study on compressor blades. The research findings indicate that despite the large number of stages in the compressor, the methods utilized in the study are applicable across the entire compressor range, and vibration stress curves were plotted. Additionally, vulnerable areas of the impeller were detailed, guiding for assessing the lifespan of impeller blades. In recent years, research on the structural performance and vibration of axial flow compressor blades based on the CFD-CSD coupling method has primarily focused on coupling strategies and the numerical solution accuracy of aerodynamic forces, stresses, and strains [5,6]. Tao et al. [7] conducted numerical simulation studies on the aeroelasticity of compressor blades using fluid–structure coupling methods. They analyzed the dynamic stress and strain of the blades caused by aerodynamic impacts and assessed whether compressor blade flutter would occur. Tao [8] conducted simulation studies on the structure of turbochargers' compressors, analyzing the deformation and stress distribution of blades under fluid dynamics, thereby offering theoretical foundations for structural optimization. Wang et al. [9] and others studied the compressor impeller of a certain type of micro-turbine engine, analyzing the formation mechanism of thermal loads and their impact on impeller stress and strain. Li [10] utilized the DS module of the ANSYS Workbench platform and the APDL language as the interface for transferring surface load data of blades, conducting a numerical simulation study on the internal flow of a transonic fan in a certain type of turbofan engine with fluid–structure coupling. The research findings demonstrated that the fluid–structure coupling-based numerical simulation was closer to experimental results, validating the effectiveness of this method. Zhang et al. [11] conducted a study on solving internal flow fields and analyzing blade vibration of high-pressure five-stage moving blades in a certain axial-flow compressor under multiple operating conditions. The distribution of blade dynamic stress under hazardous working conditions was determined, and fatigue failure analysis of the blades was carried out. The research revealed that under critical conditions, the distribution of blade vibration stress was closely related to the resonant modal vibration, and it was pointed out that the first-order resonance of blades at critical rotational speeds was the primary cause of blade failure. Li et al. [12] conducted a study on the modal analysis of compressor blades under prestress based on fluid–structure coupling. The research revealed that the centrifugal inertial force of the compressor blades themselves and the aerodynamic disturbances they experience are the primary mechanical factors causing blade vibration and fatigue failure.

When studying the structural characteristics of compressor blades, most scholars only consider aerodynamic loads, and less consider the centrifugal loads. During the structural analysis of an axial compressor, the centrifugal load significantly impacts both the bending stress at the blade root and the radial deformation of the blade. This research method of overlooking the centrifugal load may result in an inability to accurately simulate the stress and structural deformation of the compressor under real-world operating conditions, ultimately impacting its operational efficiency and the safety of the axial compressor. This paper employed a bidirectional CFD-CSD coupling method to investigate the deformation and stress distribution of compressor blades under aerodynamic loads, centrifugal loads, and the coupling of these two loads. Based on these prestresses, a modal analysis of the blades was conducted to obtain the vibration frequencies under different loading conditions, and the Campbell diagram of the blades was plotted to judge whether the blade would resonate or not. The research findings can provide a theoretical basis for the optimized design of blade structures.

2. Computational Methods

This paper employed a bidirectional CFD-CSD coupled method to solve the deformation and stress distribution of blades. The Navier–Stokes equations were utilized to compute the flow field within the transonic axial flow compressor blades. Once the flow field calculation converged, the pressure field distribution across the compressor blades was obtained. The conservative interpolation method was applied to map the pressure field onto the structural mesh nodes of the blade. Subsequently, the structural static equilibrium equations of the blade were solved to obtain the structural deformation. If the structural deformation did not converge, the deformation on the structural nodes was transferred back to the flow field mesh nodes using the conservative interpolation technique. Subsequently, the Motion Mesh method was employed to adjust the mesh positions, ensuring that the aspect ratio and skewness of the mesh were maintained at approximately 1 and 0.8, respectively. The adjusted mesh was then used to recalculate the flow field. After the flow field converged, the pressure field was again transferred to the structural nodes through the coupling interface, and the structural deformation was recalculated. This iterative process was repeated until the structural deformation of the compressor blade converged. The computational flowchart is shown in Figure 1.

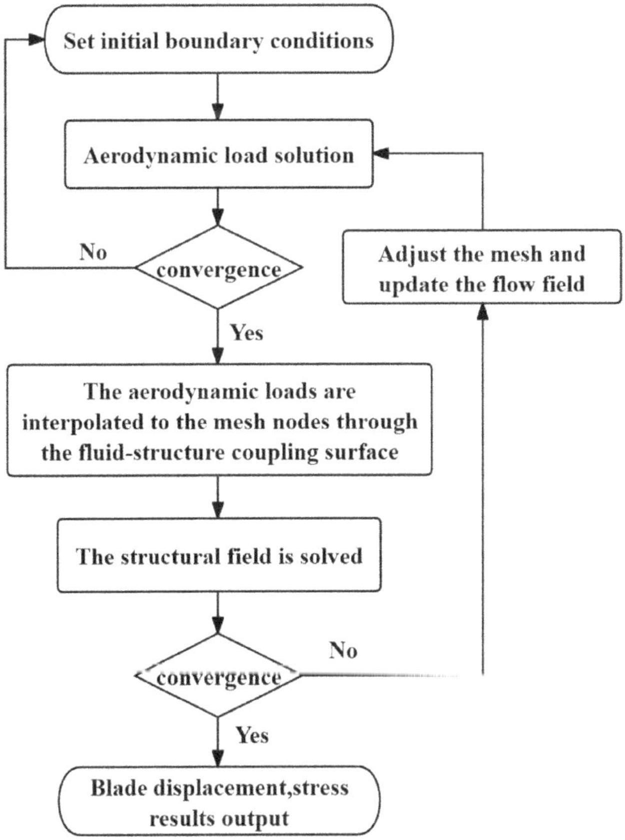

Figure 1. Process diagram of the bidirectional CFD-CSD coupled method.

2.1. Flow Field Solution Method

The Reynolds-averaged Navier–Stokes equations were employed for the flow field solution [13]. Spatial discretization primarily uses the first-order upwind difference scheme. The convection terms were treated with a high-resolution (HR) scheme, while the temporal discretization adopted the lower–upper symmetric Gauss–Seidel (LU-SGS) implicit time integration method.

$$\frac{\partial W}{\partial t}+\frac{\partial f}{\partial x}+\frac{\partial g}{\partial y}+\frac{\partial q}{\partial z}=\frac{\partial R}{\partial x}+\frac{\partial S}{\partial y}+\frac{\partial T}{\partial z} \quad (1)$$

In the Equation (1), W is the vector of conservative variables; $f,\ g,\ q$ is the vector of convective fluxes; and R, S, T are the vectors of viscous fluxes.

During numerical simulations of axial-flow compressors, turbulence models such as the k-ε model and the k-ω model are commonly used. To ensure computational accuracy, the k-ω shear stress transport (SST) model and the k-ε renormalization group (RNG) model were specifically used to simulate the performance of the compressor rotor blades. The simulation results, as presented in Figure 2, indicate that the SST model provided closer agreement with experimental data. Consequently, the SST model was selected for subsequent numerical simulations of the axial-flow compressor.

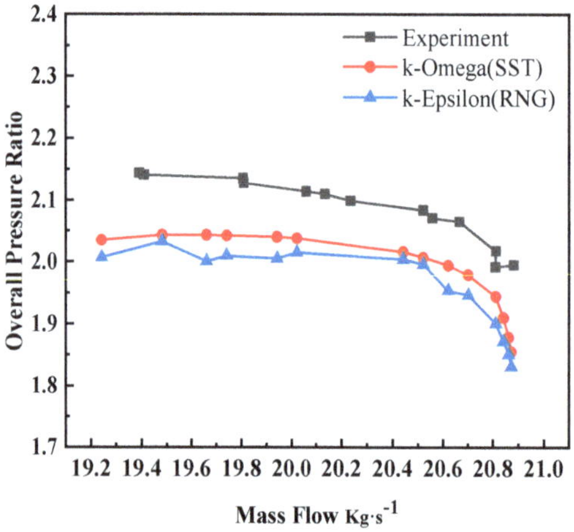

Figure 2. Computation results with different turbulence models.

2.2. Structural Solution Method

In the solid field, the primary concern is the deformation of solids caused by the forces exerted by the fluid field. The conservation equations for the solid part are derived from Newton's second law:

$$\rho_s \ddot{d}_s = \nabla \cdot \sigma_s + f_s \quad (2)$$

In the Equation (2), ρ_s represents the solid density, σ_s is the Cauchy stress tensor, f_s is the volume force vector, and $\ddot{d}_s$ is the local acceleration vector in the solid domain.

2.3. Data Transfer Method

In the coupled computation process, the method of conservative profile-preserving interpolation is utilized for data exchange across the coupling interface [14]. The data exchange process can be described as follows:

$$u_f = H u_s \quad (3)$$

$$\delta u_f \cdot f_a = \delta u_s^T \cdot f_s \tag{4}$$

where in (3) and (4), u_f represents the displacement of the aerodynamic points; u_s represents the displacement of the structural points; H is the interpolation matrix; f_a denotes the aerodynamic loads; f_s represents the structural equivalent loads; and δu_f, δu_s are the virtual displacements of the aerodynamic and structural points, respectively.

If H is independent of the displacement, then we have

$$\delta u_f = H \cdot \delta u_s \tag{5}$$

Finally, the relationship between the structural equivalent loads and the aerodynamic loads is derived, which can be expressed as

$$f_s = H^T \cdot f_a \tag{6}$$

3. Meshing and Computational Method Validation

3.1. Research Objectives

This research focused on Rotor37 [15], an axial flow compressor rotor designed and tested at the NASA Lewis Research Center. This rotor is commonly utilized in test cases for transonic axial flow compressor rotors, and it has been subject to extensive research and testing by numerous domestic and international research centers, yielding a wealth of experimental data. The basic design parameters are summarized in Table 1. Figure 3 shows the test positions and computational domain of the NASA Rotor 37 transonic rotor flow field [16].

Table 1. Basic design parameters of the NASA Rotor37.

Parameter	
Number of blades	36
Tip clearance/mm	0.356
Pressure ratio	2.106
Adiabatic efficiency	0.877
Inlet blade tip relative to Mach	1.48
Inlet blade root relative Mach	1.13
Design rotational speed/(rpm)	17,188.7
Design flow rate/(Kg/s)	20.19

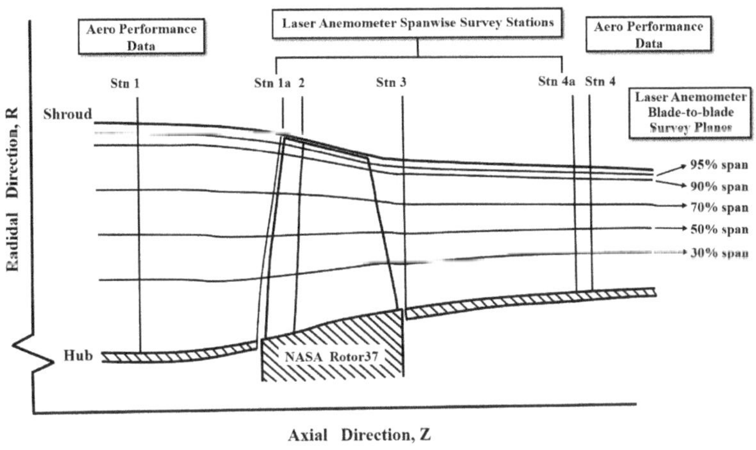

Figure 3. Test positions and computational domain of the NASA Rotor37 transonic rotor flow field.

3.2. *Boundary Conditions for the Fluid Domain*

In the numerical simulation, the inlet conditions were set with a total temperature of 288.15 K and a total pressure of 0.1 MPa. The outlet condition was specified with back pressure, and the distribution of back pressure along the spanwise direction was determined using radial equilibrium equations. The last condition point before the numerical stall was identified as the near-stall condition by gradually increasing the outlet back pressure towards the stall point [17]. The blade passage on both sides was defined as periodic boundary conditions, while the blades, casing, and hub surfaces were set with adiabatic and no-slip boundary conditions. The blade walls were set as the fluid–structure interaction interface, where pressure was transmitted to the structural solver and mesh deformation from the structural solver was received. The blade flow field model and boundary conditions are shown in Figure 4.

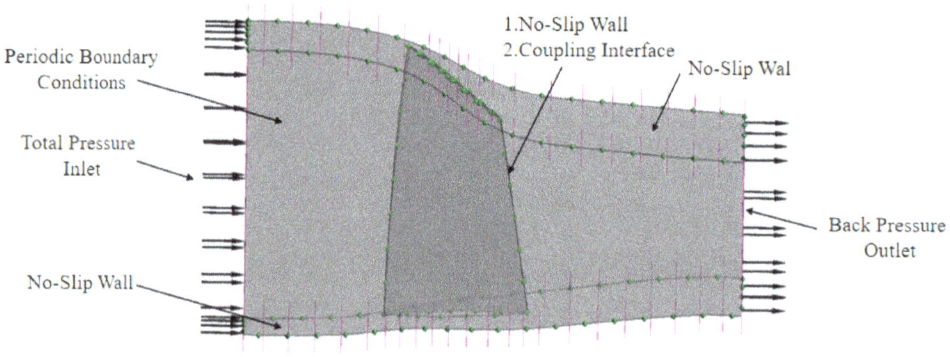

Figure 4. The blade flow field model and boundary condition schematic diagram.

3.3. *Boundary Conditions for the Structural Domain*

The boundary conditions for the structural solver were set as follows: A fixed constraint was applied at the interface where the blade connected to the hub. A rotational velocity was imposed on the entire blade. A fluid–structure interaction interface was added to the suction surface, pressure surface, and blade tip of the blade. This configuration enabled the blade surface to capture the aerodynamic forces transmitted from the fluid field during coupled calculations and simultaneously provided deformation feedback to the fluid field. The blade structural field model and boundary conditions are shown in Figure 5.

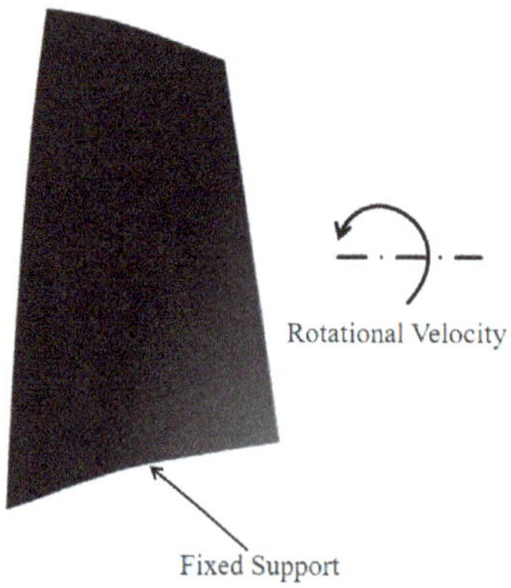

Figure 5. The blade structural field model and boundary condition schematic diagram.

3.4. Meshing

The single-channel spatial discrete mesh of the axial flow compressor is generated based on CFX/TurboGrid [18], a hexahedral structured mesh generator specially designed for rotating machinery. In this paper, sub-area meshing was used, with an O4H-type mesh topology for the interior of the channel, an H-type mesh topology for the inlet and outlet, and an O-type mesh topology for the blades. The leaf top gap adopted butterfly mesh topology, the inner layer was an H-type mesh structure, and the outer layer surrounded the O-mesh. The number of single-channel mesh nodes was 741,244, and the number of meshes was 706,419. The mesh schematic is shown in Figure 6.

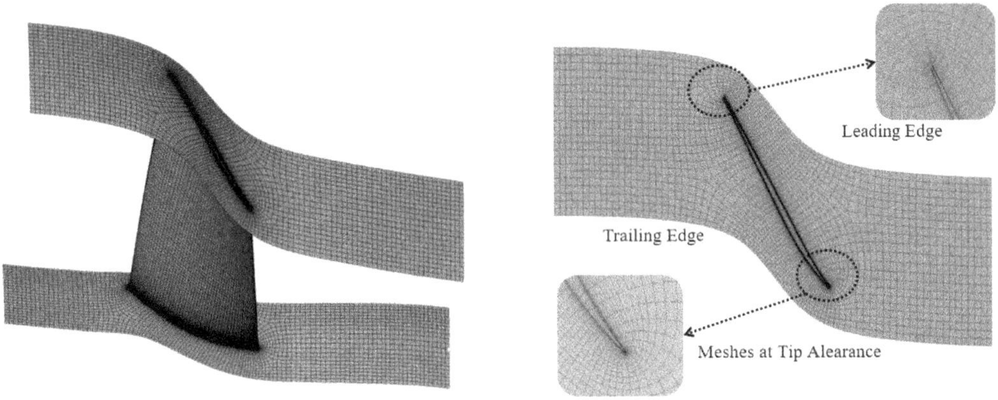

Figure 6. Schematic of blade single passage meshing.

To appraise the influence of the number of meshes on the numerical calculation results, eight sets of meshes with different numbers of nodes, 380,000, 470,000, 580,000, 740,000, 1,030,000, 1,360,000, 1,660,000, and 2,110,000, were obtained by adjusting the mesh sizes of the surface area of the blades and the top clearance of the blades, respectively. Considering that this paper focused on the near-stall condition of axial-flow compressors, different outlet back pressures (0.102 MPa, 0.108 MPa, 0.114 MPa, 0.120 MPa, and 0.125 MPa) were considered to ensure the accuracy of mesh irrelevance, and the total pressure ratio was used as a parameter to assess the impact of mesh density on performance. Figure 7 shows the mesh around the blade for different numbers of nodes at 0.120 MPa.

As shown in Figure 8, under varying outlet back pressure conditions, when the mesh node count reached 740,000, the maximum deviation in total pressure ratio compared to other mesh quantities was 0.5%. Therefore, to balance computational efficiency and accuracy, this study ultimately selected a mesh node count of 740,000.

3.5. Method Validation

Figure 9 presents a comparative analysis between the numerical simulation results and experimental data [19] under the designed rotational speed. It is evident that the variation trend of the numerical results closely aligns with that of the experimental findings. Upon comparison with the experimental data, it is noted that the numerical computation exhibited a discrepancy of 3.68% for the near-design point total pressure ratio and 3.67% for the isentropic efficiency. At the designed rotational speed, the flow rate at the rotor choke point was measured at 20.93 kg/s, while the numerical calculation yielded 20.90 kg/s, resulting in a marginal difference of 0.14%. Similarly, at the rotor near-stall point, the flow rate was recorded at 19.36 kg/s in contrast to the numerical prediction of 19.24 kg/s, indicating a slight deviation of 0.62%. Overall, the numerical results tended to underestimate the experimental values, which may be attributed to the limitations in experimental measurements, where the highest measurement position can only reach 95%

of the blade height, while numerical simulations can capture flow losses across the entire blade height.

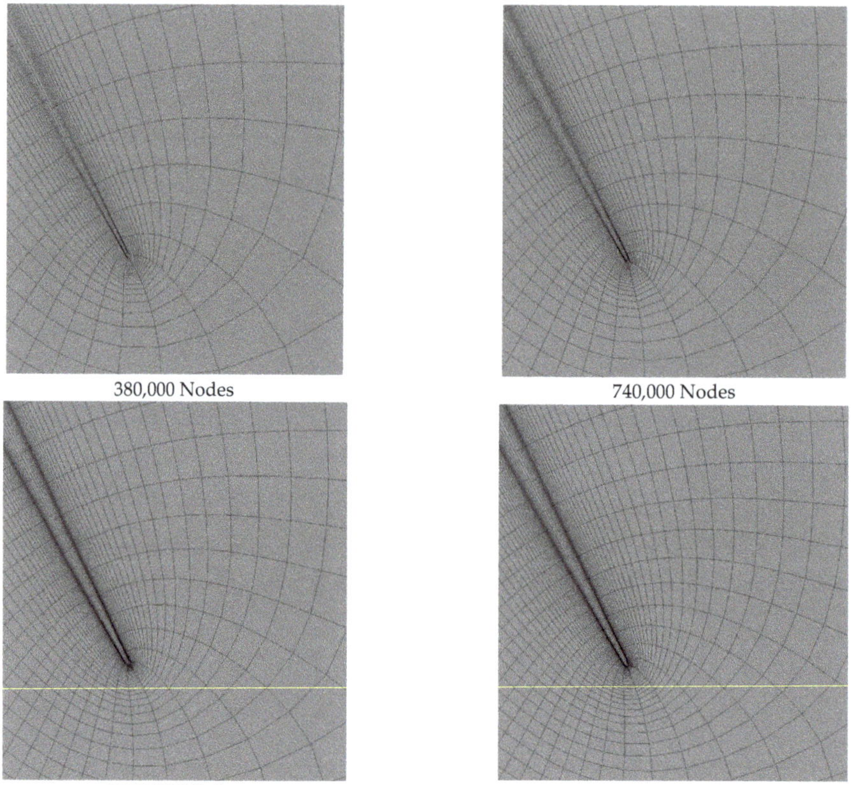

380,000 Nodes 740,000 Nodes

1,360,000 Nodes 2,110,000 Nodes

Figure 7. Mesh diagrams near the blade for different numbers of nodes.

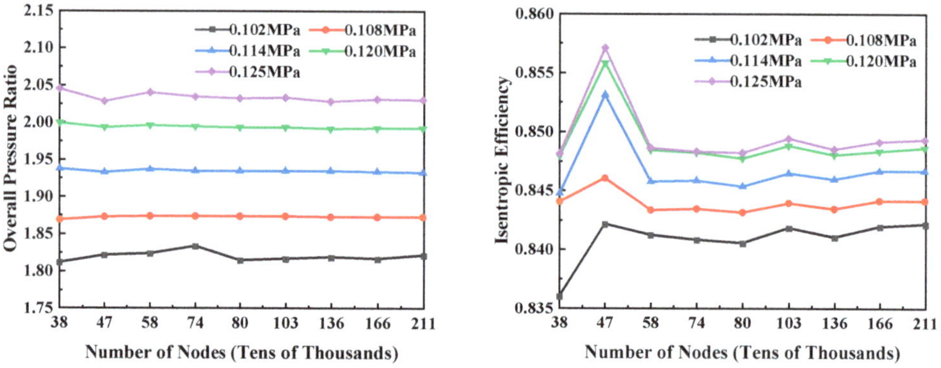

Overall Pressure Ratio. Isentropic Efficiency.

Figure 8. Mesh independence study.

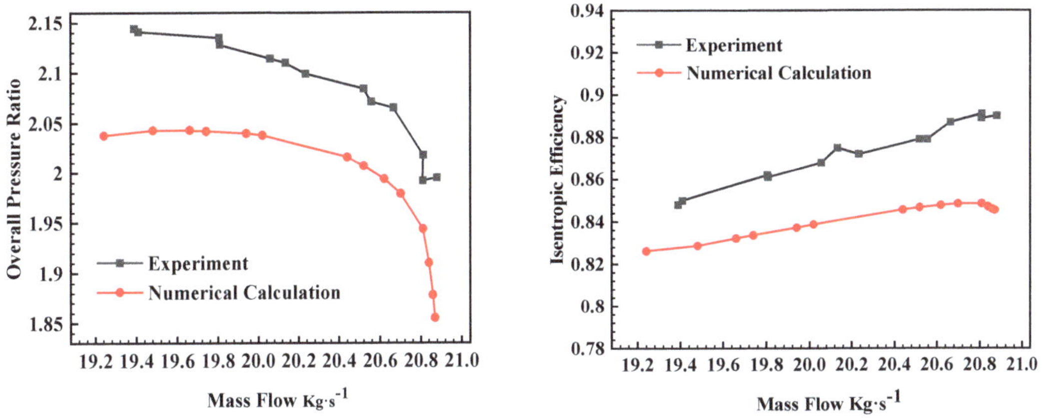

Overall Pressure Ratio. Isentropic Efficiency.

Figure 9. Comparison of calculated and experimental compressor characteristics.

The analysis indicates that the predicted characteristics and trends of the numerical simulation closely aligned with the experimental results. Moreover, the discrepancies in the flow choke point and stall point between numerical predictions and experimental values were minimal and fell within acceptable margins. Therefore, it can be concluded that the numerical computational method employed in this study is reliable.

4. Results and Discussion

4.1. Influence of Different Loads on Blade Stress and Strain

During the operation of compressor blades, they are subjected to various load factors. To gain a clearer understanding of how different loads affect the deformation and stress of the blades, this paper analyzed the blade under the following load conditions: aerodynamic load, centrifugal load, and a combination of aerodynamic and centrifugal load. Figure 10 , Figures 11 and 12 depict the blade's total deformation and equivalent stress contour maps under various loads. From Figure 10, it can be observed that the maximum equivalent stress on the blade under aerodynamic loading was relatively low. The maximum deformation occurred at the lead edge tip due to the generation of strong shock waves at the leading edge during high-speed rotation. The lead edge dimension was designed to be thinner to reduce shock wave resistance at the leading edge, resulting in reduced stiffness at the leading edge and significant deformation during rotation.

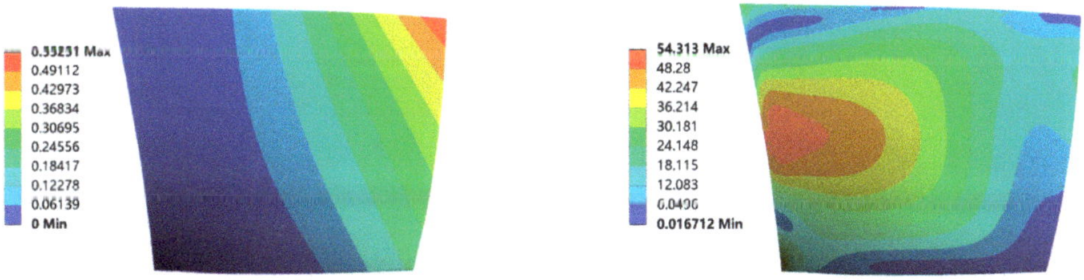

(**a**) Deformation. (**b**) Equivalent Stress.

Figure 10. Deformation and equivalent stress of blades under aerodynamic load.

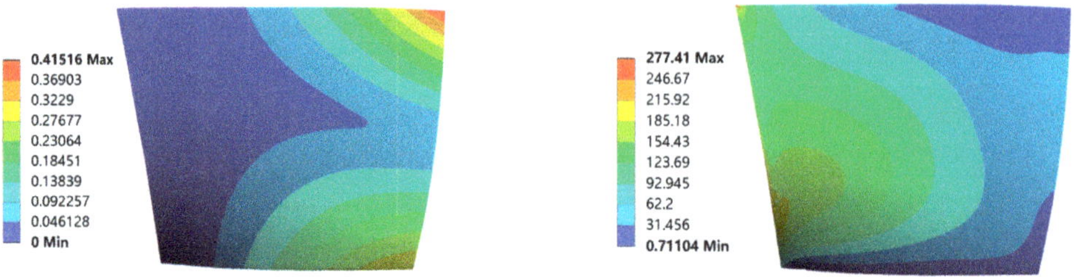

(**a**) Deformation. (**b**) Equivalent Stress.

Figure 11. Deformation and equivalent stress of blades under centrifugal load.

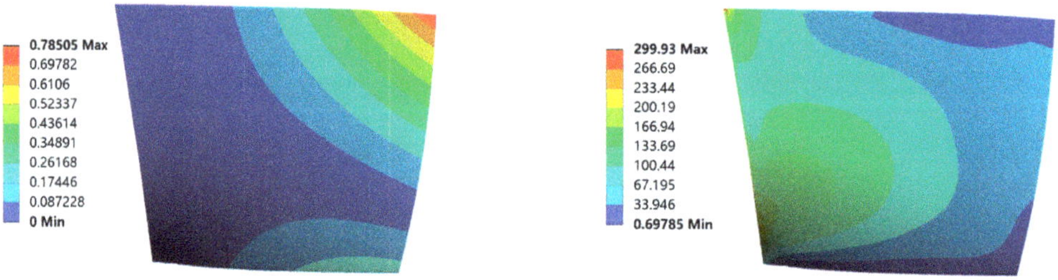

(**a**) Deformation. (**b**) Equivalent Stress.

Figure 12. Deformation and equivalent stress of blades under centrifugal and aerodynamic loads.

Figure 11 reveals that the maximum equivalent stress under centrifugal loading alone was 277.41 MPa, whereas, under aerodynamic loading alone, the maximum equivalent stress on the blade was 54.31 MPa, representing only 19.57% of the stress exerted under centrifugal loading alone. This indicates that centrifugal load predominantly influenced blade stress. Figures 11 and 12 show that the variation trend of equivalent stress remained consistent when the blade was subjected to centrifugal loading alone or in combination with aerodynamic loading, spreading radially along the blade height direction and concentrating mainly within the upper 33% of the blade length. This is attributed to pressure differences between the suction and pressure surfaces of the blade during high-speed rotation, causing bending of the blade, with the root end face experiencing higher stress as a fixed end, making it prone to fracture and stress fatigue.

The three scenarios show that the stress distribution on the blade did not simply follow a linear superposition of forces. A comparison of Figures 10–12 shows that the deformation of the leading edge under aerodynamic loading alone or centrifugal loading alone was less than that under the combined action of aerodynamic and centrifugal loading. Conversely, the deformation of the trailing edge under centrifugal loading alone was greater than that under the combined action of aerodynamic and centrifugal loading. This is because at the blade's leading edge, the forces from aerodynamic and centrifugal loading acted in the same direction, leading to increased deformation. Conversely, at the trailing edge of the blade, the troops from aerodynamic and centrifugal loading acted in opposite directions, partially canceling each other out and resulting in reduced deformation of the blade.

In summary, the fluid–structure coupling of the blade represents the optimal approach for studying the stress and strain of impeller blades, providing the most accurate representation of their distribution. It reflects the actual stress and strain distribution of the blades. For compressor blades to operate safely, they must adhere to strength safety criteria, whereby the maximum stress should not exceed the permissible stress [α]. Compressor blade working conditions involve high temperatures and pressures, so their strength must meet specified requirements; otherwise, they are susceptible to damage. The strength verification formula is

$$\sigma_{\max} \leq \frac{\sigma_s}{k} = [\sigma] \tag{7}$$

In Equation (7), $\sigma_{\max}$ represents the maximum stress that the blade can withstand (MPa), σ_s denotes the material's yield strength (MPa), and k is the safety factor. For specialized dynamic machinery such as impellers, due to their unique operational conditions under static loading, a safety factor of k = 1.2 was selected in this study. As per the preceding context, it is apparent that the material composition of the blades was TC4, distinguished by its exceptional ductility, with a yield strength of 825 MPa. With a safety factor of k = 1.2 applied, the allowable stress for the material was established at 687.5 MPa. At the rated rotational speed of 17,188.7 rpm, the maximum equivalent stress experienced by the blades was recorded at 299.93 MPa. Notably, this stress magnitude remained below the allowable stress threshold of 687.5 MPa for the material, thereby ensuring compliance with the strength criteria for the blades at 17,188.7 rpm.

For a clearer understanding of the influence patterns of different loads on the maximum deformation and maximum stress of the blades, this study contrasted the effects at 90%, 80%, and 70% of the design rotational speed. From Figure 13a, it is observed that under various design rotational speeds, the deformation solely induced by aerodynamic loads slightly exceeded that caused solely by centrifugal loads. Moreover, the deformation resulting from the combined action of centrifugal and aerodynamic loads surpassed that induced by either force acting alone. Figure 13b reveals that the maximum equivalent stress induced solely by aerodynamic loads was significantly lower than that induced solely by centrifugal loads or by the combined action of centrifugal and aerodynamic loads.

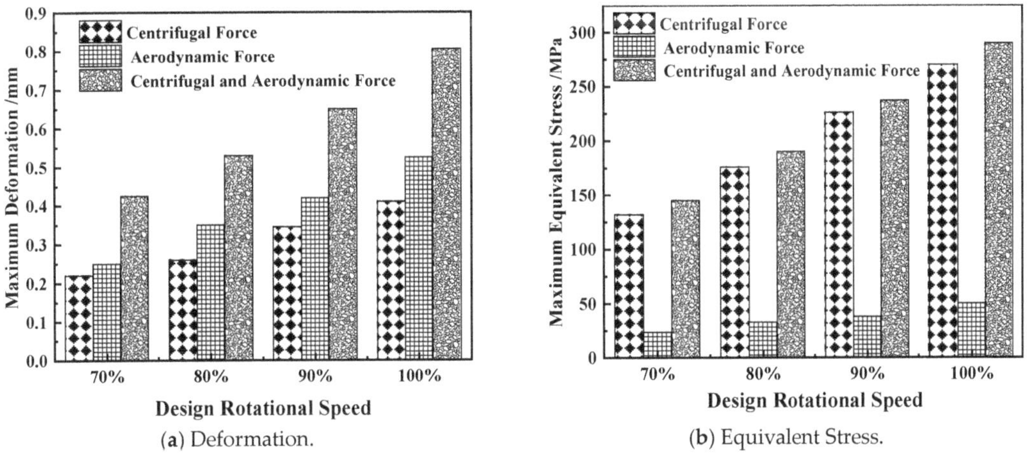

(a) Deformation.

(b) Equivalent Stress.

Figure 13. Deformation and stress of blades at different rotational speeds under load.

Figure 14 presents the variation of blades' maximum deformation and maximum stress with mass flow at different rotational speeds. Figure 14a shows that the blades' maximum deformation under different rotational speeds exhibited a trend of initially increasing and subsequently decreasing with the increasing mass flow, with a peak value present throughout the entire operational range. It is evident that with the continuous increase in rotational speed, the maximum deformation of the blade progressively increased. Notably, at each rotational speed, the maximum deformation under near-stall conditions exceeded that at the near-choke point, with the deformation trend of the blades intensifying as the rotational speed increased. Variations in rotational speed induced variations in blade deformation characteristics. From Figure 14b, it can be observed that the variation trend of maximum blade stress under different rotational speeds was relatively smooth, with minimal impact from variations in mass flow. As rotational speed increased, the variation in maximum blade stress did not exhibit a singular increasing trend; rather, it increased from 70% to 80% speed, decreased from 80% to 90% speed, and then increased again from 90% to 100% speed. Overall, the maximum stress exhibited a variation process of initial increase, subsequent decrease, and subsequent increase with the increase in rotational speed. Hence, selecting an appropriate rotational speed can effectively mitigate the blades' maximum deformation and stress.

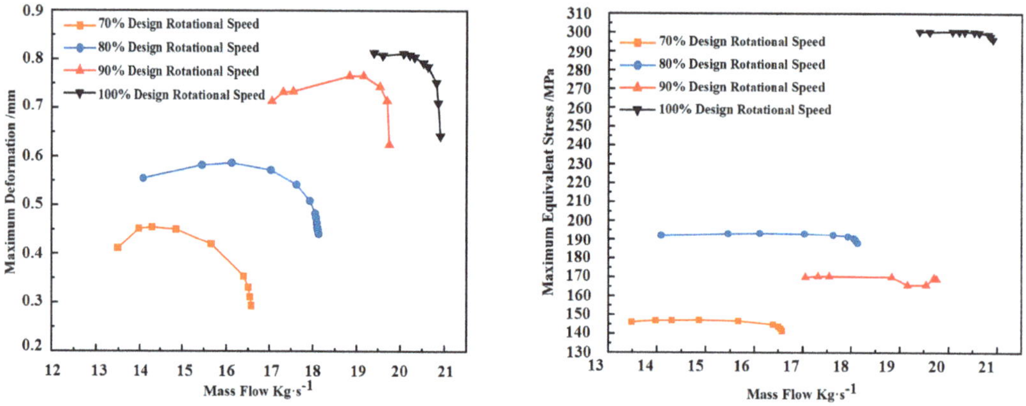

(**a**) Deformation. (**b**) Equivalent Stress.

Figure 14. Variation of maximum deformation and maximum equivalent stress of blades at different rotational speeds with mass flow.

4.2. Analysis of Blade Vibration Characteristics under Different Pre-Stresses

To explore the impact of prestress induced by load generation on the natural frequencies of blades and to prevent resonance during blade operation, this section performs a modal analysis of the blades based on the prestresses generated by the three load scenarios discussed in the previous section. This analysis provides the vibration frequencies of the blades at different loads. Table 2 reveals that with increasing rotational speed, the natural frequencies of the blades under different pre-stresses all increased. When the blades were subjected solely to aerodynamic loads, their natural frequencies were significantly lower than those under centrifugal loads alone or combined loads. This is because the blades undergo certain deformations under aerodynamic loads, but the stresses generated by aerodynamic loads are limited, consistent with the analysis of load effects on blade stress conducted earlier in this paper. By comparing the natural frequencies of the blades under the three types of loads, it is evident that the natural frequencies under combined loads exceeded those under individual loads. However, the natural frequencies under centrifugal loads were very close to those under combined loads, indicating that centrifugal loads are the primary factor influencing the natural frequencies of the blades.

When the natural frequency of a blade matches or is an integer multiple of the frequency of the excitation force, it can lead to resonance [11]. The Campbell diagram [20] can be used to determine whether the excitation force frequencies generated by rotating components in a compressor during normal operation coincide with the natural frequencies of some components, effectively preventing resonance. The Campbell diagram is created using modal frequencies at different speeds, lines for different rotational speeds, and excitation force data for the compressor [20]. As shown in Figure 15, the black horizontal lines represent the modal frequency lines for the first six modes of the blade, the red vertical lines represent different rotational speed lines of the impeller, and the blue diagonal lines represent the excitation lines. Since the amplitude of pressure fluctuations increases with lower excitation frequency orders (r), leading to higher vibration stress and increased risk of blade failure [21], this paper considered excitation frequencies corresponding to two, four, six, and eight times the rotational speed frequencies. From Figure 15, it can be observed that resonance points occurred around 62% and 97% of the rotational speed, with a particular resonance point near 97% indicating that resonance is very likely to occur at 100% rotational speed, potentially leading to blade damage. Therefore, it is necessary to either modify the blade structure to change its natural frequency or adjust the compressor's operational speed to avoid the resonance points as much as possible.

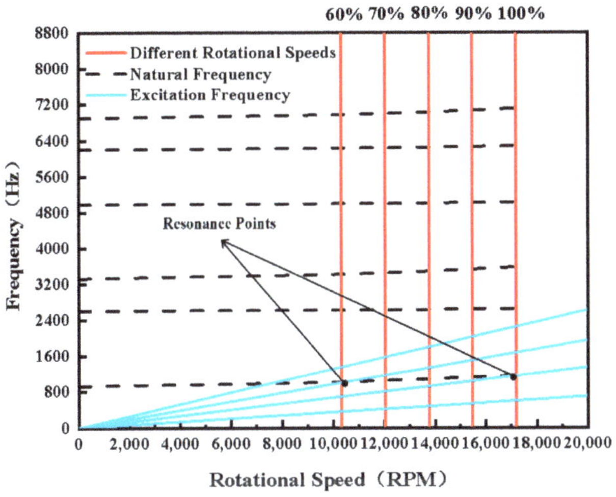

Figure 15. The Campbell diagram of an axial flow compressor blade.

Table 2. Natural frequencies of blades for the first six modes at various rotational speeds.

Order	Mode	Load	Frequency (Hz) Design Rotational Speed of 70%	Design Rotational Speed of 80%	Design Rotational Speed of 90%	Design Rotational Speed of 100%
1	First-order bending	Aerodynamic load	927.48	927.98	928.58	929.01
		Centrifugal load	1050.9	1086	1124.2	1165.4
		Centrifugal and Aerodynamic load	1051.6	1086.8	1125.3	1166.5
2	First-order torsion	Aerodynamic load	2603.5	2602.3	2602.9	2600
		Centrifugal load	2628.7	2635.3	2642.7	2651.3
		Centrifugal and Aerodynamic load	2624.6	2630	2638.4	2644.5
3	Second-order bending	Aerodynamic load	3324.8	3325.5	3327.3	3327.1
		Centrifugal load	3447.6	3484.6	3525.8	3571.5
		Centrifugal and Aerodynamic load	3447.7	3484.7	3527.2	3572.1

Table 2. *Cont.*

Order	Mode	Load	Frequency (Hz) Design Rotational Speed of 70%	Design Rotational Speed of 80%	Design Rotational Speed of 90%	Design Rotational Speed of 100%
4	Second-order torsion	Aerodynamic load	4969.7	4968.6	4970.5	4966.2
		Centrifugal load	5000.5	5008.6	5017.6	5028.4
		Centrifugal and Aerodynamic load	4995.2	5001.8	5012.7	5019.4
5	Third-order bending	Aerodynamic load	6195.6	6194.2	6194.3	6191.7
		Centrifugal load	6236.6	6248.6	6262.2	6277.6
		Centrifugal and Aerodynamic load	6235.7	6246.8	6261.2	6275.1
6	Third-order torsion	Aerodynamic load	6893.8	6894.0	6897.2	6894.6
		Centrifugal load	6999.0	7031.3	7068.0	7109.8
		Centrifugal and Aerodynamic load	6996.4	7028.0	7067.4	7105.9

5. Conclusions

This study established a bidirectional CFD-CSD coupling analysis method for blade structures, comprehensively considering the effects of aerodynamic and centrifugal loads on the structural characteristics of transonic axial flow compressor blades. It analyzed stress and deformation distributions under different conditions. Based on this, modal analysis of the blades under various prestresses was performed, and Campbell diagrams were plotted. The main conclusions are as follows:

(1) The deformation of the blades gradually increased along the blade height, with maximum deformation occurring at the lead edge tip. Regions of high stress were primarily concentrated at the upper 33% of the blade root, then radiated radially outward and gradually decreased. Both aerodynamic and centrifugal loads affected the magnitude and distribution of blade deformation and stress, but the influence of centrifugal loads on blade stress was significantly greater than that of aerodynamic loads, making centrifugal loads the primary consideration in stress calculations.
(2) With increasing rotational speed, the maximum deformation of the blades showed a trend of initially increasing and subsequently decreasing, with maximum deformation under near-stall conditions exceeding that at near-choke points. However, the maximum stress of the blades exhibited a trend of initially increasing, subsequently decreasing, and then increasing, suggesting that selecting an appropriate rotational speed can reduce both the maximum deformation and maximum stress of the blades.
(3) The natural frequencies of the blades increased with increasing rotational speed. Both centrifugal and aerodynamic loads had some degree of influence on the natural frequencies of the blades, but centrifugal loads were the primary factor affecting blade natural frequencies.
(4) There were resonance points for compressor blades around 62% and 98% of the rotational speed. Care should be taken during compressor operation to prevent resonance, which could lead to blade damage.

Author Contributions: Conceptualization, H.L., H.B. and H.Z.; methodology, H.L., H.B. and H.Z.; software, H.L.; validation, H.Z.; formal analysis, H.L. and H.Z.; investigation, H.L.; resources, H.B.; data curation, H.L. and H.Z.; writing—original draft preparation, H.L. and H.Z.; writing—review and editing, H.L., H.B. and H.Z.; visualization, H.L. and H.Z.; supervision, H.B., F.Z. and J.W.; project administration, H.B., F.Z. and J.W.; funding acquisition, H.B., F.Z. and J.W. All authors have read and agreed to the published version of the manuscript.

Funding: This research was funded by the National Natural Science Foundation of China (52305583); the Shaanxi Provincial Key R&D Project (2023-YBGY-385); the Shaanxi Provincial Natural Science Basic Research Project (2023-JC-YB-018); and the Fundamental Research Funds for the Central Universities (XJSJ23124).

Institutional Review Board Statement: Not applicable.

Informed Consent Statement: Not applicable.

Data Availability Statement: The raw data supporting the conclusions of this article will be made available by the authors upon request.

Acknowledgments: The author would like to express additional gratitude to the College of Mechanical Engineering at Shaanxi University of Technology for the 2024 Graduate Innovation Fund Project (SLGJX202402).

Conflicts of Interest: The authors declare no conflicts of interest.

References

1. Chen, G. *Structural Design and Analysis of Aerospace Engines*, 3rd ed.; Beijing University of Aeronautics and Astronautics Press: Beijing, China, 2014.
2. Jin, X.-M.; Gao, D.-P.; Cai, X.-X. Analysis of integrated centrifugal impeller blade vibration reliability. *J. Aerosp. Power* **2004**, *19*, 610–613.
3. Ramamurti, V.; Subramani, D.A.; Sridhara, K. Free vibration analysis of a turbocharger centrifugal compressor impeller. *Mech. Mach. Theory* **1995**, *30*, 619–628. [CrossRef]
4. Day, W.-D.; Fiebiger, S.W.; Patel, H.N. Parametric evaluation of compressor blade blending. *Proc. ASME Turbo Expo* **2012**, *7*, 1099–1110.
5. Li, L.; Xu, P.; Xu, W.-X. Multi-field coupling vibration patterns of the multiphase sink vortex and distortion recognition method. *Mech. Syst. Signal Process.* **2024**, *29*, 111624. [CrossRef]
6. Tan, Y.-F.; Ni, Y.-S.; Xu, W.-X. Key technologies and development trends of the soft abrasive flow finishing method. *J. Zhejiang Univ. Sci. A* **2023**, *24*, 1043–1064. [CrossRef]
7. Tao, H.-L.; Zhu, Y.-L.; Guo, B.-T. Numerical simulation of aeroelastic response in compressor based on fluid-structure coupling. *J. Aerosp. Power* **2012**, *27*, 1054–1060.
8. Tao, P. Analysis on compressor blade structure based on fluid-solid coupling. *Des. Manuf. Diesel Engine* **2014**, *20*, 8–11.
9. Wang, D.-B.; Han, L.-L.; Wang, G.-H. The flow simulation in the compressor with the effect of fluid-solid interaction. *J. Therm. Sci. Technol.* **2020**, *19*, 64–71.
10. Li, X. The Flow Simulation in the Compressor with the Effect of Fluid-Solid Interaction. Master's Thesis, Harbin Institute of Technology, Harbin, China, 2010.
11. Zhang, J.-H.; Fu, X.; Kou, H.-J. Numerical study on vibration characteristics of axial flow compressor blade under fluid-structure interaction. *J. Therm. Sci. Technol.* **2018**, *38*, 59–66.
12. Li, C.-W.; Wang, C.; Li, Q. Modal analysis of aero-engine blade based on thermal-structure coupling pre-stressing force. *J. Air Force Eng. Univ. (Nat. Sci. Ed.)* **2016**, *17*, 1–4.
13. Li, S. Preliminary Study on Axial Flow Fan Design Based on FLUENT Software. Master's Thesis, Northwestern Polytechnical University, Xi'an, China, 2004.
14. Guo, T.-Q.; Lu, Z.-L.; Zhou, D. *Coupled CFD/CSD Calculation of Aircraft Aeroelasticity*, 1st ed.; Science Press: Beijing, China, 2019.
15. Reid, L.; Moore, R.D. Design and Overall Performance of Four Highly Loaded, High-Speed Inlet Stages for An Advanced High-Pressure-Ratio Core Compressor. NASA TP-1337. 1978. Available online: https://ntrs.nasa.gov/citations/19780025165 (accessed on 20 July 2024).
16. Denton, J.D. Lessons from rotor 37. *J. Therm. Sci.* **1997**, *6*, 1–13. [CrossRef]
17. Wang, W.; Chu, W.-L.; Zhang, H.-G. The effects on stability, performance, and tip leakage flow of recirculating casing treatment in a subsonic axial flow compressor. In Proceedings of the ASME Turbo Expo, Seoul, Republic of Korea, 13–17 June 2016.
18. Wang, H. Study of Unsteady Tip Flow and Rotating Instability in Axial Compressors. Ph.D. Thesis, Shanghai Jiao Tong University, Shanghai, China, 2016.
19. Boretti, A. Experimental and computational analysis of a transonic compressor rotor. *Mar. Geol.* **2010**, *122*, 243–262.

20. Liu, Y.; Yang, C.; Ma, C.-C. Forced Responses on a Radial Turbine with Nozzle Guide Vanes. *J. Therm. Sci.* **2014**, *23*, 138–144. [CrossRef]
21. Zhang, H.; Wang, X.-Y.; Zhang, Y.-F. Vibration Failure Characteristic Analysis of Compressor Impeller Under Plateau Conditions. *Chin. Intern. Combust. Engine Eng.* **2024**, *45*, 47–56.

materials

MDPI

Article

Effects of Pressure, Surfactant Concentration, and Heat Flux on Pool Boiling Using Expanding Microchanneled Surface for Two-Phase Immersion Cooling

Yifei Hu [1], Dengwei Fu [1], Chaobin Dang [2] and Sihui Hong [1,3,*]

1 School of Physics and Astronomy, Sun Yat-sen University, Zhuhai 519082, China; huyf63@mail2.sysu.edu.cn (Y.H.); fudw3@mail2.sysu.edu.cn (D.F.)
2 Graduate School of Engineering, University of Fukui, 3-9-1 Bunkyo, Fukui-shi 910-8507, Fukui, Japan; dangcb@u-fukui.ac.jp
3 College of Electric Engineering, Zhejiang University, Hangzhou 310027, China
* Correspondence: hongsh23@mail.sysu.edu.cn

Abstract: Deionized water is replacing fluorinated liquids as the preferred choice for two-phase immersion cooling in data centers. Yet, insufficient bubble removal capability at low saturated pressure is a key challenge hindering the widespread application. To solve this issue, this study employs non-ionic surfactant (Tween 20) and asymmetric structures (expanding microchannel) to enhance the boiling performances of deionized water under sub-atmospheric pressure. The research examines the effects of pressure (8.8~38.5 kPa), surfactant concentration (0.1~0.5 mL/L), and heat flux density (10~180 W/cm^2) on the boiling heat transfer characteristics and analyzes the mechanism of unusual temperature oscillations induced by surfactants. It was found that the trade-off between the sub-atmospheric pressure, surface tension coefficient, and reduced static contact angle results in pronounced intermittent boiling on the heated surface. Even with the addition of surfactants, the improvement in heat transfer requires demanding conditions. Boiling enhancement throughout all heat flux conditions was achieved when the surfactant concentration was higher than 0.2 mL/L for the expanding microchanneled surface. The heat transfer coefficient reached 6.89 W·cm^{-2}·K^{-1} under 8.8 kPa, which was 45% higher than without the surfactant. Under the same heat flux and sub-atmospheric pressure, as the concentration increased from 0.1 to 0.5 mL/L, the amplitudes of temperature fluctuation of the plane surface and expanding microchanneled surface decreased from 10 K to 2 K and 18 K to 1 K, respectively. The onset of nucleate boiling and wall superheat of the expanding microchanneled surface gradually decreased with the increase in surfactant concentration, where the onset of nucleate boiling decreased by 10.54 K. When the heat flux is 160 W/cm^2, the wall superheat is reduced by 12.8 K.

Keywords: two-phase immersion cooling; sub-atmospheric pressure; non-ionic surfactants; temperature oscillation; bubble behaviors; boiling heat transfer

1. Introduction

1.1. Challenges of Two-Phase Immersion Cooling in Data Centers

Data centers consume enormous amounts of energy, making them high-energy-consuming industries. Energy-efficient green data centers typically require a Power Usage Effectiveness (PUE) of less than 1.3, while the PUE of air-cooled or air-conditioned systems is often above 1.7 [1]. Further optimization and improvements bring limited marginal benefits, failing to meet energy consumption requirements. Immersion liquid cooling systems are a thermal management technology in which electronic devices or other heat-generating components are fully immersed in a specially designed liquid coolant with a PUE lower than 1.1, minimizing system overheating, temperature fluctuations, fan failures, noise, dust, and air quality issues [2]. This enhances the reliability and lifecycle of data center systems, making immersion liquid cooling the most promising technology for data center cooling systems [3]. Immersion liquid cooling systems can be divided into single-phase liquid cooling and two-phase liquid cooling, depending on whether the cooling liquid undergoes phase change. Single-phase liquid cooling uses high-boiling-point (>100 °C) cooling liquids such as mineral oil or perfluorinated polyether, with lower cooling efficiency but reliable performance, currently at a high level of commercialization. Two-phase immersion cooling exponentially improves the heat transfer efficiency of the cooling liquid through the boiling and condensation processes. However, due to reliability issues with phase-change systems, actual commercial applications have not yet been realized. However, there are some initial applications, such as Microsoft's use of 3M liquid coolant in 2021, where the boiling liquid carries away heat generated by computer servers in Microsoft's data centers [4].

As mentioned before, compared to traditional cooling techniques, two-phase immersion cooling [5–8] has garnered increasing attention due to its high heat transfer efficiency. However, the development of this technology currently faces two main technical bottlenecks. First is the insufficient bubble removal rate. At low heat flux, the bubble detachment frequency that solely relies on buoyancy is low, resulting in poor boiling heat transfer performances. At high heat flux (>100 W/cm^2), the vapor generation rate sharply increases, rapidly forming a vapor film covering the heating surface [9,10]. Steam leaves in the form of continuous vapor columns that overlap each other, hindering the wetting of the liquid to the heating components. Mutual interference between the vapor and liquid phases leads to the "dry-out" of micro-liquid layers, resulting in a sharp deterioration of boiling heat transfer [11,12]. The second issue is the mismatch of fluid properties. Electronic fluorinated liquids have good insulation properties but poor heat transfer performance [13], with an evaporation latent heat only 5% to 10% of that of deionized (DI) water. Moreover, some suitable electronic fluorinated liquids have been prohibited worldwide as they are not friendly to the environment. It is exigent and important to explore available alternatives for thermal management utilizing liquid-cooling techniques. DI water, replacing fluorinated liquids as the primary working fluid, is preferred due to its high latent heat, environmental friendliness, and economic benefits. The use of water cooling has gradually become the primary choice for many data centers [14]. Moreover, existing technologies allow for the insulation of electronic devices from DI water [13]. However, the issue of high boiling temperature (>100 °C) using DI water in electronic cooling needs to be addressed. Low saturated pressure conditions greatly reduce the onset temperature of the phase change [15], which enables a reasonable operating temperature for two-phase immersion cooling. However, at low pressure (<101 kPa), the interfacial Gibbs free energy is significantly higher, and the diameter of detached bubbles increases by more than 10 times compared to atmospheric pressure conditions. This exacerbates the difficulty of bubble removal, leading to excessively low boiling heat transfer performance. Meanwhile, intermittent boiling occurs due to increased surface tension and decreased vapor density [16], leading to significant surface temperature fluctuations [17]. Currently, it is still challenging to apply this technology to two-phase immersion liquid-cooling systems. This study considers the use of additives (surfactants) in combination with machined surfaces (expanded channel structures) to enhance heat transfer under low-pressure conditions.

1.2. State of the Art on Pool Boiling Enhancement Using Surfactant

Pool boiling refers to the process of heating a liquid within a confined volume to its boiling point, resulting in the formation of bubbles on the heated surface and heat being transferred. As a critical process in immersion, phase-change cooling achieves efficient heat transfer through phenomena such as nucleation, bubble growth, and detachment on the heated surface. Therefore, studying how to enhance the pool boiling effect under different conditions is of great significance for improving the overall efficiency of immersion phase-change cooling. In studies of various pool boiling enhancement techniques, the use of surfactants as additives in water under atmospheric pressure has been proven to be highly effective [18–21]. The enhancement mechanism primarily involves two aspects [22–26]:

firstly, reducing surface tension to facilitate bubble detachment, and secondly, the interaction of hydrophilic and hydrophobic groups at the bubble interface to prevent bubble coalescence. The heat transfer enhancement is closely related to the concentration and type of surfactant. The maximum enhancement is generally achieved when the concentration is close to the critical micelle concentration (CMC) [25], and the types mainly differ among cationic, anionic, and non-ionic surfactants. It is reported that the anionic surfactant enhanced the boiling heat transfer efficiency of the nanofluid, with its heat transfer coefficient surpassing that of the cationic surfactant, which in turn exceeded that of the nonionic surfactant [27].

However, there are some inconsistencies in the literature regarding the role of surfactants in enhancement, making the effect less predictable. Yang [28], Lee [29], and Proper et al. [30] pointed out that the addition of surfactants reduces equilibrium surface tension, thereby enhancing boiling heat transfer, but also reduces the static contact angle, inhibiting part of the heat transfer process. Additionally, Yang et al. [31] and Jia et al. [32] observed that under high heat flux conditions, during the boiling of surfactant solutions, a phenomenon of bubble crowding occurs, where bubble clusters expand and contract. Periodic temperature fluctuations and heat transfer deterioration could occur.

Moreover, under low saturated pressure, enhancing boiling heat transfer with surfactants faces significant challenges. Bubble behavior is closely related to heat transfer performance, but as pressure decreases, surface tension increases and vapor density decreases, resulting in larger bubble detachment diameters and a corresponding reduction in bubble detachment frequency [16,33]. Under such conditions, intermittent boiling phenomena may occur. Giraud et al. [34] pointed out that boiling phenomena under sub-atmospheric pressure (0.8 kPa~15 kPa) are influenced by environmental non-uniformity, defining an "intermittent boiling regime", with wall temperature fluctuations reaching up to 20 K. Yamada et al. [35] reported that under sub-atmospheric pressure, when the heat flux is 6 W/cm^2, deionized water exhibits continuous boiling when the saturation pressure exceeds 14 kPa. However, as the pressure decreases from 11 kPa to 8.8 kPa, temperature fluctuations intensify, indicating a transition towards intermittent boiling.

In conclusion, under low-pressure conditions, the difficulty of using surfactants as a means of heat transfer enhancement significantly increases, particularly in maintaining stable boiling and preventing heat transfer deterioration, which may lead to unstable temperature oscillations and intermittent boiling. Previous studies on the impact of surfactants on pool boiling heat transfer have primarily focused on atmospheric pressure conditions, with relatively few investigations into the effects of surfactant concentration on pool boiling performance under low-pressure conditions.

Table 1 summarizes the comparisons of various studies under different conditions, listing the experimental conditions at different sub-atmospheric pressures and surfactant concentrations. selection of the four most common surfactants: SDS (CMC = 2300 ppm), Tween 20 (CMC = 600 ppm), TritonX-114 (CMC = 120 ppm), CTAB (CMC = 1500 ppm), all of the CMC were measured at room temperature. As shown in Figure 1, after organizing the existing literature using the four quadrant method, the solid points and semi-solid points, respectively, represent the results of changes in surfactant solution concentration and pressure. It can be observed that the literature lacks research on surfactant concentration under low-pressure conditions. As a matter of fact, under saturated conditions, pressure variations can alter the boiling temperature, subsequently affecting solute solubility and CMC [36]. Additionally, the coupled effects of pressure and surfactants can lead to changes in bubble growth and detachment parameters, such as the gas–liquid density ratio, surface tension, latent heat of vaporization, and the hydrophilic–hydrophobic balance of surfactant molecules. The intensity of bubble disturbances is directly related to heat transfer efficiency.

Table 1. Overview of research on pool boiling under various pressure ranges and with surfactants.

Reference	Pressure Variation (kPa) [a]
Luling Yuan (2019) [37]	12.3, 31.2, 101
Sandra Michaie (2017) [38]	4.2, 5.6, 7.4, 9.6, 12.4, 15.8, 20, 25.1, 31.2, 38.6, 47.4, 57.9, 70.2, 84.6, 101.4
Tomasz Halon (2017) [39]	0.75, 1, 2, 4,
Sang M. Kwark (2010) [40]	20, 47, 101, 200
Xiang Wang (2024) [41]	40, 60, 80, 100, 120
Ayşenur Ateş (2023) [42]	28.3,103.7,
Guanghan Huang (2022) [43]	25, 65
Aniruddha Pal (2008) [44]	9.7, 15, 21
Smreeti Dahariya (2019) [45]	103.4, 206.8, 310.2, 413.7
Hanzhi Chen (2017) [46]	150, 200, 300, 400
Staniszewski (1959) [47]	101, 193, 276,
Akiyama (1969) [48]	101, 203, 507, 807
Reference	**Concentration Range (Ratio to CMC) [b]**
Bing Li (2024) [26]	0.1, 1, 5, 10 (SDS)
Sasan Etedali (2019) [49]	0.480 (SDS) 0.732 (CTAB)
Chi Young Lee (2014) [29]	0.437 (SDS)
R.I. Elghanam (2011) [18]	0.087, 0.217, 0.435, 0.652 (SDS)
Zi-Cheng Hu (2022) [50]	0.087, 0.174, 0.348, 0.435, 0.783, 1.3 (SDS)
Jialun Yin (2020) [51]	0.135, 0.405, 0.946 (SDS)
Zhen Yang (2022) [31]	0.093, 0.93, 1.85 (Tween 20)
Mario R. Mata (2022) [52]	(0.02~31.3) (Tween 20)
H. Jeremy Cho (2013) [53]	0.61 (TritonX-114)
Dong-Sheng Guo (2020) [54]	1 (TritonX-114)
Tao Wen (2022) [25]	0.267, 0.533, 1.066, 2.133 (CTAB)

[a]: the works in the column are all for pure liquid (Key-116 was used in [41], methane was used in [46], others: DI water). [b]: the works in the column are all at 101 kPa.

According to the above survey, the heat transfer performance after adding surfactants under low-pressure conditions is of great significance for applications in the field of immersion phase-change cooling.

1.3. Purpose of the Present Work

As mentioned in Section 1.1, pool boiling enhancement at sub-atmospheric pressure is the primary target in two-phase immersion cooling systems. The role of surfactants in improving boiling performance under low pressure remains unclear. Given the electrical conductivity concerns in practical applications, this study examines the boiling behavior of a non-ionic surfactant (Tween 20) solution under low pressure (8.8–38.5 kPa). Temperature fluctuations from intermittent boiling are analyzed in relation to heat flux, concentration, and pressure. Bubble dynamics—departure diameter, growth time, and waiting time—are evaluated through visualization experiments, and the impact of channel structure on boiling improvement is also discussed.

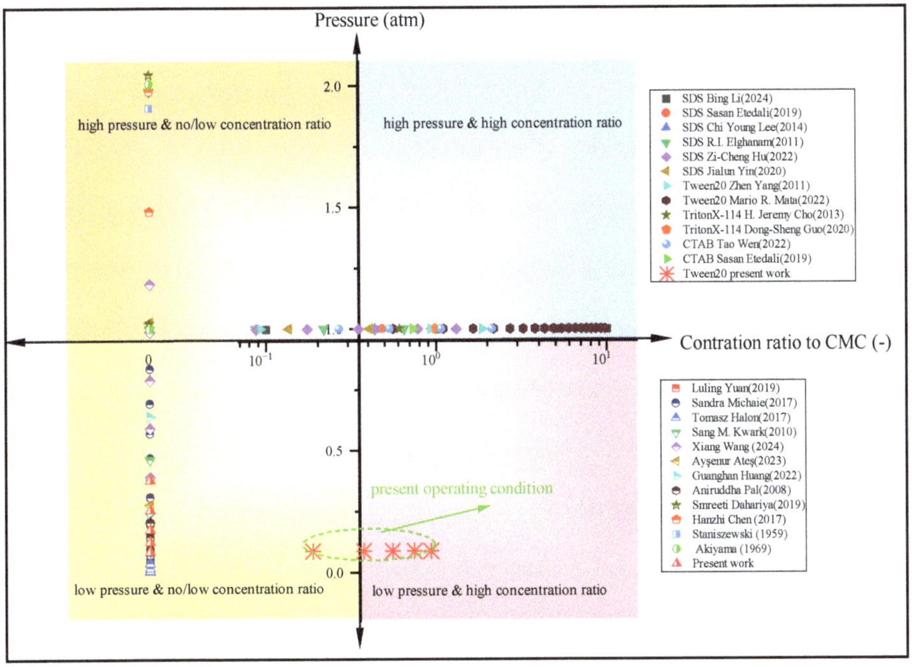

Figure 1. Experimental data under different working conditions.

2. Experimental Facilities

2.1. Experimental Set-Up

The saturation pressure and temperature of the system must be maintained stable. In this study, a stainless-steel liquid storage tank with a volume of 1 L is used as the container for the working fluid. The outer wall of the storage tank is wrapped with heating bands to control the saturation temperature of the fluid inside the tank, thereby controlling the saturation pressure. The temperature of the fluid inside the tank is measured using T-type thermocouples attached to the wall surface. The pressure inside the tank is measured using a pressure sensor (range 0~2000 kPa). The vacuum pump is connected to the liquid storage tank and the boiling pool via a two-way valve to complete preparations such as vapor removal and fluid flushing. During the experiment, the boiling tank is kept connected to the storage tank, the heating surface is continuously heated by inserting heating rods and generating a phase change, while the condensation coil condenses the rising vapor back into the boiling tank to control the stability of the system's saturation pressure.

As shown in Figure 2a, the boiling pool is constructed from acrylic material in a square space. It includes a condensation coil to maintain stable pressure, a copper block surface for testing heat transfer effectiveness, a hollow cylindrical base for inserting heating rods, an auxiliary heater for maintaining temperature stability, and APS and armored T-type thermocouples for real-time monitoring of pressure and temperature.

2.2. Surface Treatment

Before each experiment, detailed surface treatment is performed, including the following steps: (1) alcohol pre-degreasing treatment was carried out, followed by thorough cleaning with DI water to ensure surface cleanliness, (2) copper cleaner (MG440) was used to remove free iron, metallic residues, oxides, and other corrosion products from the surface of the metal, followed by another DI water rinse to ensure surface cleanliness, (3) copper sealer (CS100) was applied to protect the surface, and finally, drying in a vacuum oven to avoid contamination.

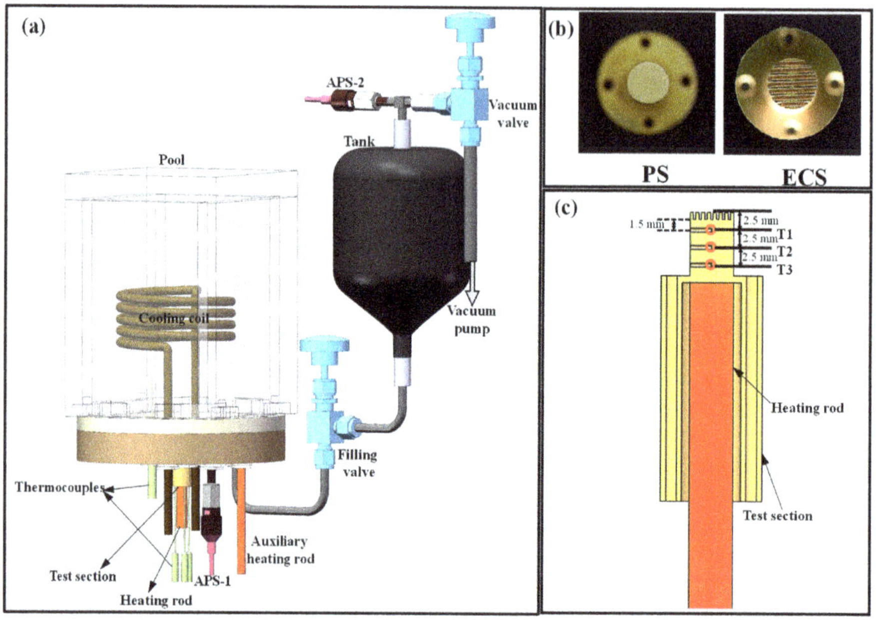

Figure 2. (**a**) The schematic map of experimental setup, (**b**) the image of the PS and ECS, and (**c**) the schematic map of the position of the thermocouples.

2.3. Imaging and Measurement

Images of the bubble dynamics behaviors on all test surfaces were captured by a high-speed camera (Phantom LAB110, Wayne, NJ, USA) paired with an autofocus lens (Nikon AF NiKKOR 24-85 mm 1:2.8-4 D, Tokyo, Japan) and a zoom lens (LEICA Z16 APO, Wetzlar, Germany). The main parameters of the high-speed camera were set as follows: frame rate of 1600 fps, resolution of 1280 × 800. Each video was captured for 10 s, totaling 16,000 frames. To obtain sufficiently bright and clear images, a flicker-free LED cold light source (VISICO LED-150T, Yaoyu, China) was used to illuminate the test surface. The image data captured by the high-speed camera were transmitted via Ethernet and saved to a computer. Bubble observation and measurement were conducted using high-speed camera software (Phantom Camera Control, 3.1.772.0, Wayne, NJ, USA).

2.4. Heated Surfaces and Solution Preparation

As shown in Figure 2b, both heating surfaces have a diameter of 6 mm and a surface roughness of less than 3.2 μm. In addition, considering the net additional pressure introduced by the microchannels and the flow resistance of the gas–liquid two-phase microchannels, the height of the microchannels was set to 1 mm. The ECS features seven grooves cut into the surface, with a groove width of 0.3 mm at the center and 0.5 mm at the tail. The groove width uniformly widens from the center to the tail. During the experiment, the sides of the copper block were wrapped with 10 mm thick insulation to minimize heat loss to the surroundings, approximating one-dimensional upward heat transfer. As shown in Figure 2c, the heating rod was inserted from the bottom, and three thermocouples were inserted from the top to measure the temperature, evenly spaced in the vertical direction with 2.5 mm intervals and a uniform insertion depth of 3 mm.

The solute is Tween 20 and the solvent is deionized water. A glass pipette was used to draw up 0.1–0.5 mL of solute into 1 L of water, which was stirred for more than 30 min using a magnetic stirring bar to ensure that the solute was uniformly distributed. The experiment should be completed within 4 h after the solution has been prepared to prevent deviations in concentration brought about by the volatilization of water. After completing each set of experiments, the boiling pool and storage tank were cleaned twice with deionized water to prevent solute residue.

3. Data Processing and Uncertainty Analysis

To confirm the repeatability of the experiment, two pool boiling experiments were conducted on the same test surface under identical conditions. The HTC curves showed a high degree of conformity, with a maximum difference of only 17%.

As shown in Figure 2c, there are three temperature measurement points arranged on the copper block, all positioned along the central axis, labeled as T_1, T_2, and T_3 from top to bottom. Figure 3 shows the temperature distribution during the boiling of deionized water at different heat flux densities. It can be observed that the temperatures of the three thermocouples inside the copper block exhibit nearly linear distribution, with R^2 values all exceeding 0.99845.

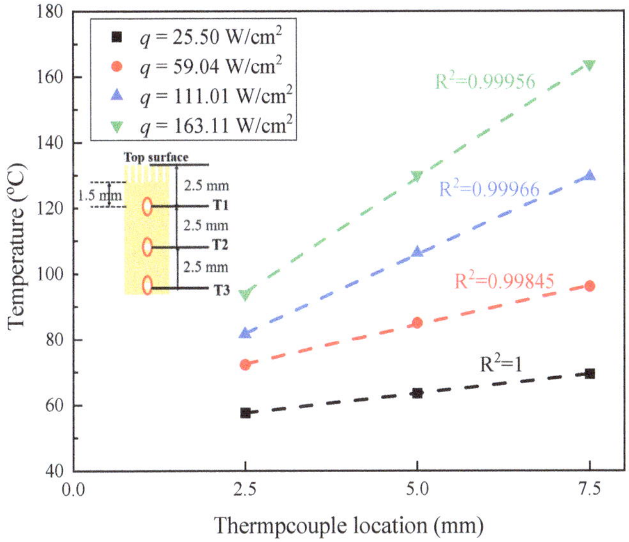

Figure 3. Measured temperature distribution at different heat fluxes.

According to Fourier's law, the heat flux q is calculated using Equation (1), where λ_{cu} is the thermal conductivity of copper. $\frac{dT}{dx}$ represents the slopes of the lines obtained by linear fitting of the temperature points from the three measuring points.

$$q = -\lambda_{cu} \cdot \frac{dT}{dx} \tag{1}$$

The wall temperature T_w can be determined using Equation (2), where z_w is the distance from the temperature measurement point T_1 to the heating surface.

$$T_w = T_1 - \frac{q\, z_w}{\lambda_{cu}} \tag{2}$$

Finally, the HTC (heat transfer coefficient) is calculated using Equation (3), where T_s is the saturation temperature of deionized water under system pressure.

$$HTC = \frac{q}{T_w - T_S} \tag{3}$$

The sensors and instruments used in this study were calibrated before testing. The main sources of experimental error are temperature measurement errors from the thermocouples and dimensional errors in machining the copper block. The temperature measurement accuracy of the T-type thermocouples is ±0.1 K. The temperature data represent the average values taken over one minute at steady state for each heat flux condition, with a typical input heat flux step size of 10 W. However, to more accurately determine the transition from intermittent boiling to continuous boiling for PS, a step size of 5 W was used at low heat flux in the surfactant solution. The machining accuracy of the distance between adjacent temperature measurement points on the copper block is ±0.1 mm. The accuracy of the pipette is 0.01 of its range.

After validating the reliability of the experimental setup, for five different concentrations, each concentration was tested three times on two surfaces, eliminating potential interferences such as improper surface treatment and concentration changes due to long-term exposure to surfactants, ensuring the stability and accuracy of the experimental data. Similarly, for five different pressure levels, each pressure was tested twice on two surfaces to ensure that the trend of experimental data with pressure changes remained consistent.

The bubble waiting time was determined through frame-by-frame video analysis, starting from the moment the last bubble leaves the heated surface, and ending when a new bubble appeared after a waiting period. In the 10 s video (16,000 frames), states with a complete waiting process were randomly selected for statistical analysis, and the average result was calculated from these statistics. The statistical minimum bubble wait time was 23 ms and the observed error was 1 frame (0.625 ms).

The specific calculation methods for determining the maximum uncertainty of heat flux and HTC are provided in Equations (4)–(6). As shown in Table 2, whicn is present maximum uncertainties of measured parameters

$$\frac{U(q)}{q} = \sqrt{\left(\frac{\partial(\ln q)}{\partial(\lambda_{cu})}U(k\lambda_{cu})\right)^2 + \left(\frac{\partial(\ln q)}{\partial\left(\frac{dT}{dx}\right)}U\left(\frac{dT}{dx}\right)\right)^2} \tag{4}$$

$$U(T_w) = \sqrt{\left(\frac{\partial T_W}{\partial(z_w)}U(z_w)\right)^2 + \left(\frac{\partial T_W}{\partial(T_1)}U(T_1)\right)^2 + \left(\frac{\partial T_W}{\partial\left(\frac{dT}{dx}\right)}U\left(\frac{dT}{dx}\right)\right)^2} \tag{5}$$

$$\frac{\mathrm{U(HTC)}}{\mathrm{HTC}} = \sqrt{\left(\frac{\partial(\ln \mathrm{HTC})}{\partial(q)}U(q)\right)^2 + \left(\frac{\partial(\ln \mathrm{HTC})}{\partial(T_{\mathrm{w}}-T_{\mathrm{s}})}U(T_{\mathrm{w}} - T_{\mathrm{s}})\right)^2} \tag{6}$$

Table 2. Maximum uncertainties of measured parameters.

Parameters	Maximum Relative Uncertainty
Heat flux (W/cm^2)	16.9%
HTC ($W/cm^2 \cdot K^{-1}$)	20.6%
Concentration of surfactant	1.0%
Static contact angle (°)	1.4%
Waiting time of bubble (ms)	2.7%

4. Results and Discussion

4.1. Effect of Saturated Pressure

Figure 4 compares the temperature oscillations along with the saturated pressure on (a) the PS and (b) the ECS under q = 40 W/cm^2 without surfactant. Increased saturated pressure led to a decreased surface tension coefficient and a reduced Gibbs surface energy. Therefore, the increase in saturated pressure enabled more steady temperature variations on both the PS and ECS. Only slight temperature oscillations were observed at P_s = 8.8 kPa, and barely no oscillation occurred after that on either the PS or ECS.

The increased saturated pressure leads to rapidly detached and smaller bubbles. The wall dissipates heat steadily through uniform agitation, resulting in stable temperature variation versus time and no more t_w is required. To illustrate this, Figure 5 compares the time for a single bubble to grow and detach on the (a) PS and (b) the ECS under different pressures, and q = 40 W/cm^2. The time step for each frame in the images is 6.25 ms (10 frames), until complete detachment is achieved. According to McGillis' correlation [17], i.e., $t_d = 0.266P_s^{-0.565}$, where the bubble departure time, t_d, was in seconds, and the saturated pressure, P_s, was in kPa, the bubble departure time was inversely correlated with the saturated pressure.

As shown in Figure 5c, as the saturated pressure increased, the time for a bubble to grow from nucleation to detachment gradually decreased. The results basically fit with the tendency change reported in the literature [16,38,55], compared with Li's correlation [26], for DI water, i.e., $fD_d{}^{0.5} = 3.34Ra^{0.15}$, where Ra is the wall roughness, D_d is the bubble departure diameter $D_d = P^{-1}\sqrt{\frac{\sigma}{(\rho_L-\rho_v)g}}$, and since the waiting time of the bubbles is very short, the bubble departure frequency f is considered to be the reciprocal of the bubble departure time. The experimental data fit well with the calculated results. For saturated pressure ranging from 8.8 to 38.5 kPa, the bubble departure time was reduced by 51.4% for PS and 38.0% for the ECS, aligning with the decreasing trends proposed by Li's correlation.

At the same pressure of 8.8 kPa, the growth time of the bubble on the ECS was less than that on PS, and the detachment diameter was smaller on the ECS compared to PS. However, when the pressure increases to 38.5 kPa, the advantage of the ECS diminishes. This is because the evaporation momentum force as an additional force applied on the bubbles along the detachment direction was inversely proportional to the vapor density. As the saturated pressure decreases, the vapor density decreases, allowing the evaporation momentum force to play a more significant role.

4.2. Effect of Surfactant Concentration

Figure 6 displays the variations of (a) the surface tension coefficient and (b) the static contact angle versus the surfactant concentration. The surface tension coefficient was measured by the Wilhelmy plate method at atmospheric pressure. Compared to other researchers [31], the surface tension coefficient generally decreases with increasing concentration. By analyzing the curves of the adjacent data points, it is found that the trend of decreasing surface tension coefficient with increasing concentration becomes gradually smoother, fitted with the reported critical micelle concentration (CMC) of Tween 20 that ranges from 0.48 to 0.6 mL/L [56]. Moreover, the measured static contact angle by using a static contact angle measuring instrument (DSA-X ROLL, Guangzhou, China) varied from 78.46° at c = 0 mL/L to 57.04° at c = 0.5 mL/L, reflecting the improvement in the hydrophilicity.

Figure 7 compares the detach diameters of the bubble on the two surfaces. Despite limitations imposed by the camera's range of observation, a trend can still be observed: as the concentration increased, the surface tension coefficient decreased, and bubbles were prone to detach, leading to a reduced bubble detachment diameter. The decline trend was even more pronounced on the ECS, especially when comparing the case of c = 0 mL/L to c = 0.5 mL/L.

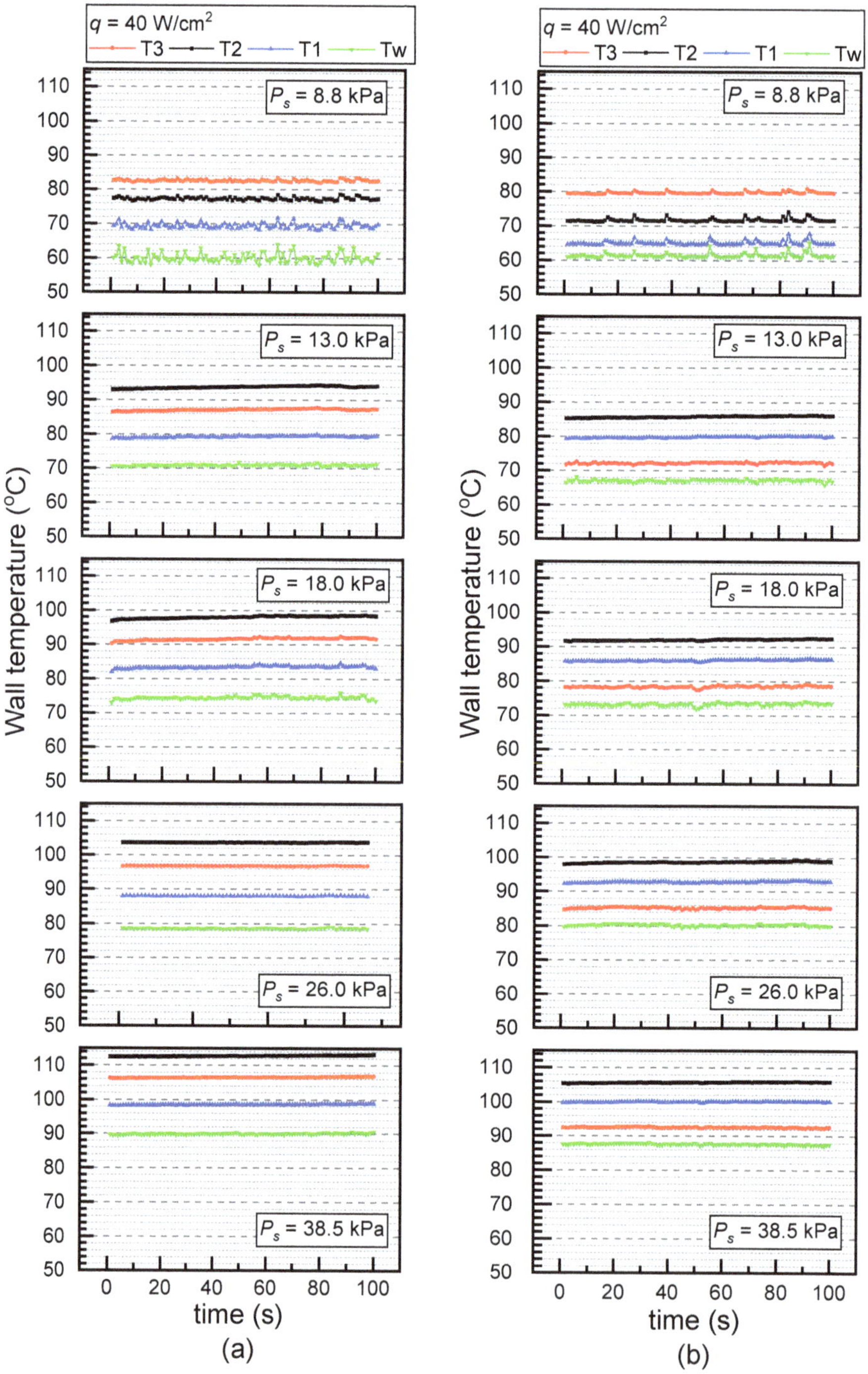

Figure 4. Temperature oscillations at different saturated pressures under q = 40 W/cm^2 and c = 0 mL/L (**a**) PS and (**b**) ECS.

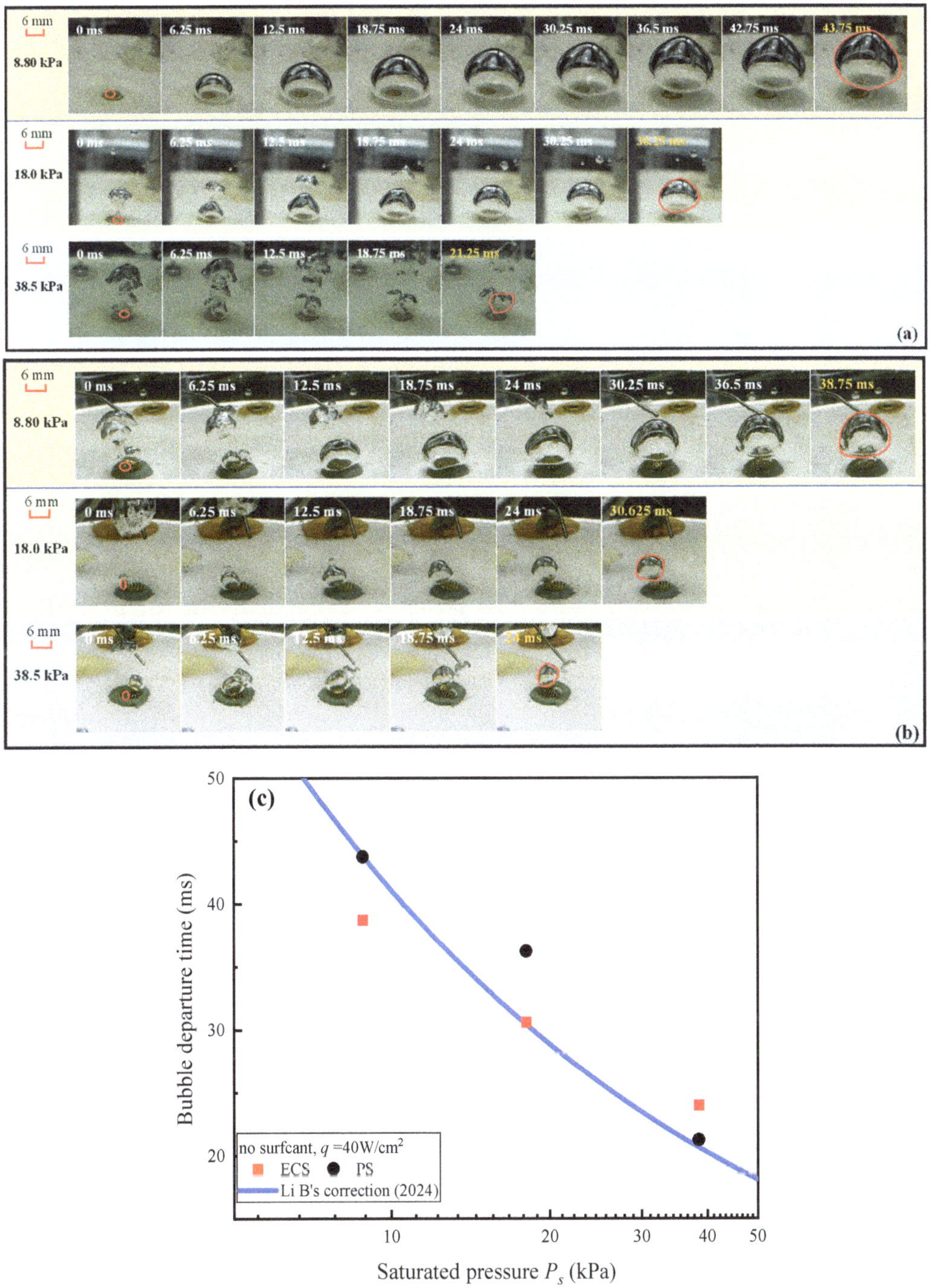

Figure 5. Visual results of bubble growth and detachment on (**a**) PS and (**b**) ECS under $q = 40$ W/cm^2 and $c = 0$ mL/L, and (**c**) variations of bubble detachment time versus with saturated pressures.

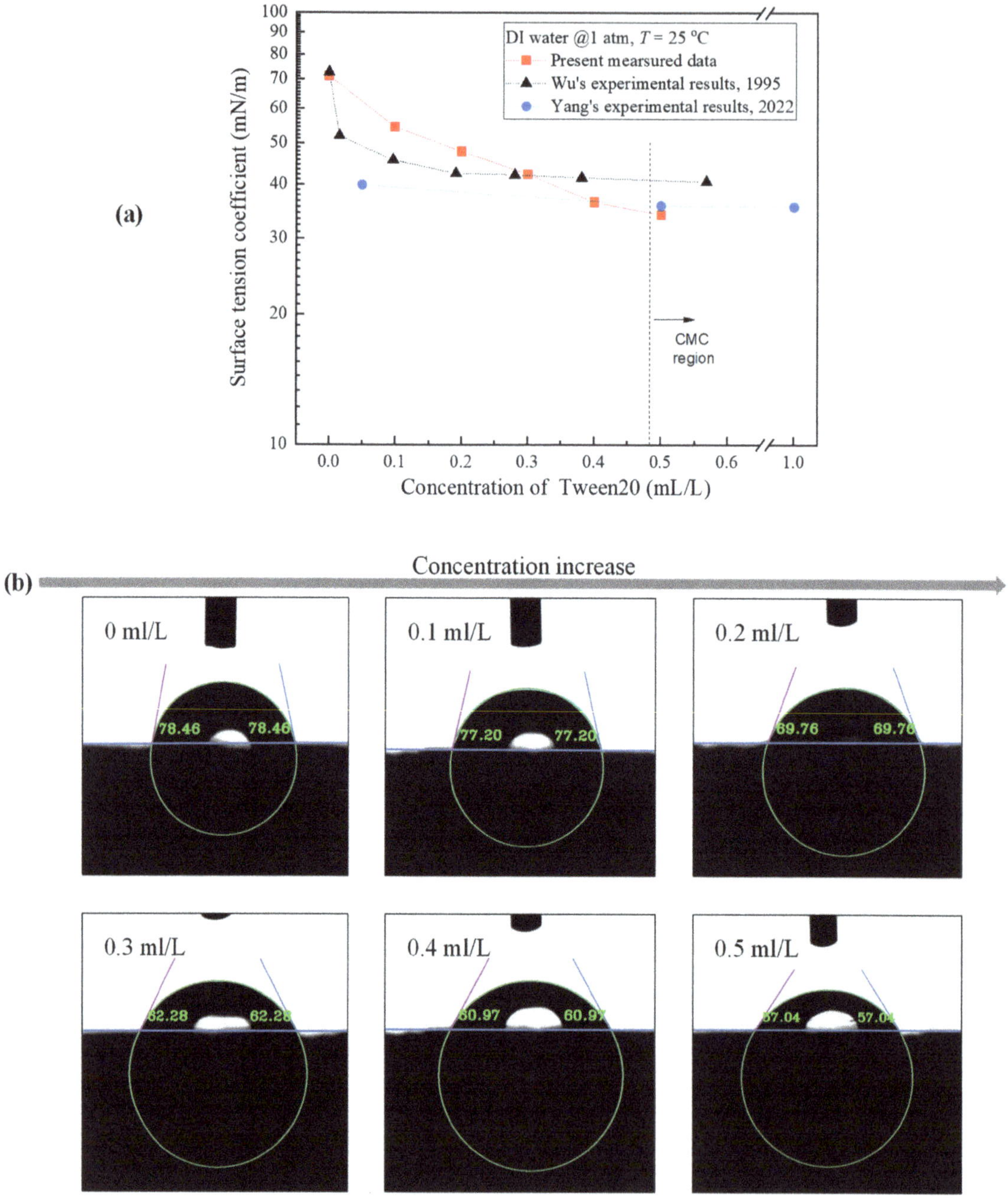

Figure 6. Variations of (**a**) surface tension coefficient and (**b**) static contact angle with Tween 20 concentrations.

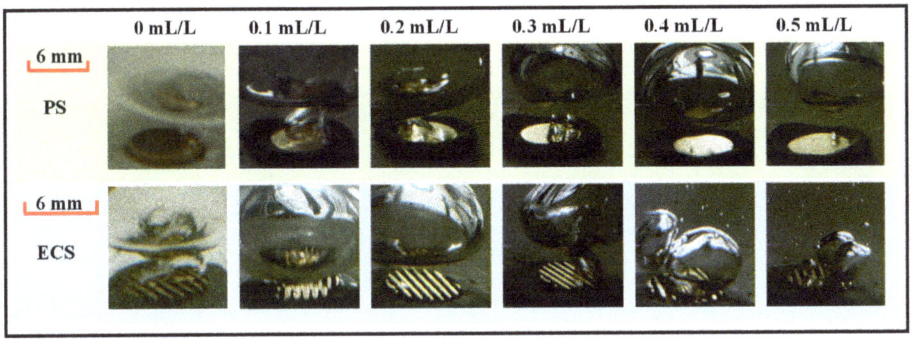

Figure 7. The effect of concentrations on detachment diameters of bubble (P_s = 8.8 kPa, q = 40 W/cm^2).

Figure 8 compared the temperature oscillations along with the surfactant concentration on the PS (a) and the ECS (b) under P_s = 8.8 kPa and q = 40 W/cm^2. At a concentration of 0.1 mL/L, the temperature fluctuations of the PS and ECS reached 10 K and 18 K, respectively. However, when the concentration increased to 0.5 mL/L, the temperature fluctuation of the PS decreased to 2 K, while that of the ECS was less than 1 K. It is shown that the temperature oscillation on either the PS or ECS was significantly weakened along with the increase in surfactant concentration. This is because the Gibbs surface energy for nucleation and bubbling decreased with the increase in surfactant concentration. The temperature oscillations on the ECS and PS were almost eliminated at c = 0.3 mL/L and c = 0.4 mL/L, respectively.

The waiting time for bubbling (t_w) was extended under low saturated pressure so as to accumulate sufficient superheat to overcome the barrier to nucleation. Hence, the t_w under various heat fluxes was captured as evidence to illustrate the variation of the temperature oscillation. The waiting time is defined as the time interval from the moment the last bubble leaves the surface to the end of the frame before the appearance of the next bubble. During this time interval, the heat transfer on the surface is only completed by natural convection. Figure 9 compares the t_w at different concentrations of (a) the PS and (b) the ECS under P_s = 8.8 kPa and q = 40 W/cm^2. As the concentration changed from 0.1 to 0.5 mL/L, the time intervals were gradually shortened. The t_w approached 0 s on the ECS at c = 0.3 to 0.5 mL/L, which made it difficult to determine the image of the waiting process in the recorded video as the bubbling and departure process persisted. Comparing the two surfaces at 0.1 mL/L, the t_w of the PS is smaller than that of the ECS, but as the concentration increases, the t_w of the ECS decreases more significantly. At 0.5 mL/L, there is a constant bubble production on the ECS, while the t_w on the PS was 43 ms. Currently, the ECS has the advantage of more nucleation points due to the enlarged area for heat diffusion, which effectively shortens the t_w.

Comparing the case of adding a surfactant to the one with no surfactants, the results are completely different from the above situations. The added Tween 20 not only changes the surface tension coefficient but also affects the static contact angles. Thereby, the surface became more hydrophilic and unfavorable for nucleation after adding the surfactant, leading to greater temperature oscillations at c = 0.1 mL/L than that at c = 0 mL/L. Correspondingly, the t_w extended at c = 0.1 mL/L.

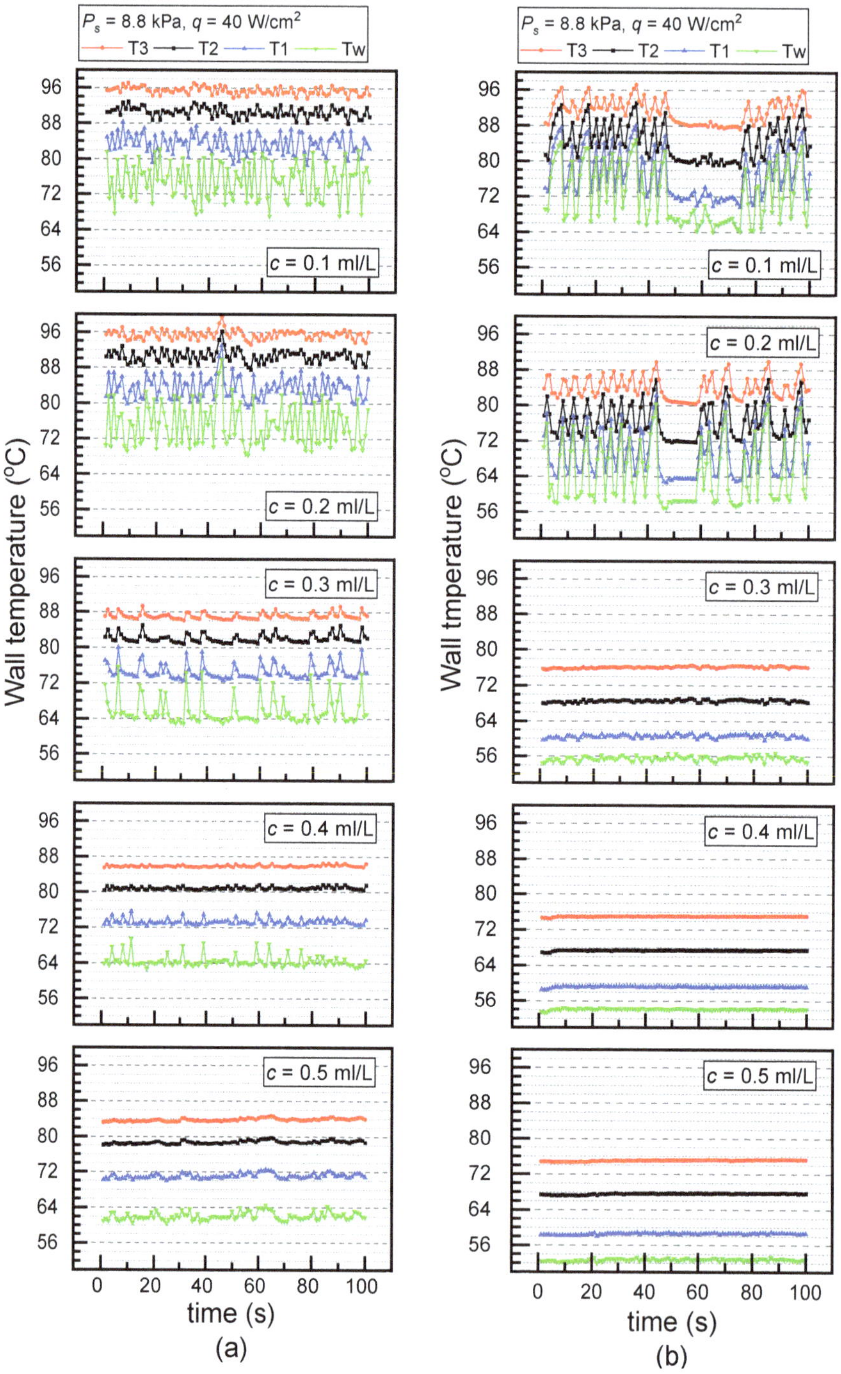

Figure 8. Temperature oscillations at different concentrations under P_s = 8.8 kPa and q = 40 W/cm^2 (**a**) PS and (**b**) ECS.

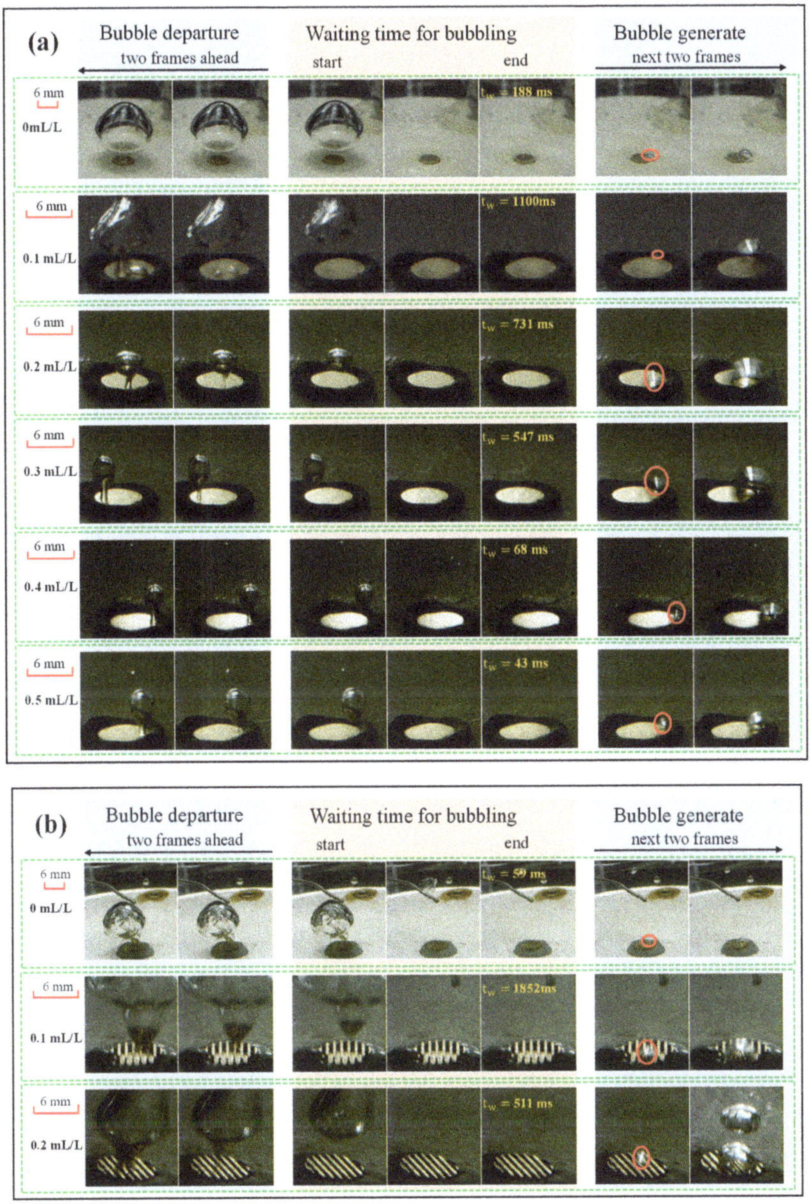

Figure 9. Waiting time for bubbling at different concentration under P_s = 8.8 kPa and q = 40 W/cm^2 (**a**) PS and (**b**) ECS.

4.3. Effect of Heat Flux

Heat flux provides extra energy for the nucleation and growth of bubbles. The increase in superheat leads to a higher nucleation rate of bubbles. Figure 10 compares the temperature oscillations along with the heat flux on the PS (a) and the ECS (b) under P_s = 8.8 kPa and c = 0.1 mL/L. As illustrated in Figure 10, for both the PS and ECS, the temperature oscillation becomes weakened with the increase in heat flux, illustrating that the intermittent bubbling phenomenon is more pronounced under low heat flux conditions.

It is found that the influence of heat flux on temperature oscillation on the PS is more significant than the ECS. For instance, the oscillation amplitude of T_w was about 10 K at q = 30 W/cm^2 and 2 K at q = 70 W/cm^2 on the PS, whereas the oscillation amplitude of T_w was about 14 K at q = 30 W/cm^2 and10 K at q = 70 W/cm^2 on the ECS. This may be because the liquid velocity and forced convection yielded on the channel surface are stronger than on the plane surface. Thus, the temperature oscillation on the ECS is less sensitive to the increase in the heat flux.

Figure 11 presents the visualization results of *the* t_w under P_s = 8.8 kPa and c = 0.1 mL/L. As shown in Figure 11a, the t_w on the PS declined from 1710 ms to 23 ms as the heat flux increased from 30 W/cm^2 to 70 W/cm^2. The sharp drop in the waiting time corresponds closely with the elimination of temperature oscillation in Figure 10a. Increased heat flux, increased wall superheat, and decreased bubble waiting time are consistent with the negative correlation between waiting time and wall superheat presented by Basu et al. [57]. It should be noted that the waiting time on the ECS at the same heat flux of 30 W/cm^2 was about 3.4 times longer than that on the PS, as shown in Figure 11b, which is consistent with the aggravated temperature oscillation exhibited in Figure 10b. This aligns with the view that the presence of microchannels leads to enhanced disturbance. Moreover, the waiting time on the ECS was 609 ms at 50 W/cm^2 and 1331 ms at 60 W/cm^2, which is accordant with the amplitude enlargement in temperature oscillation under these two cases.

Furthermore, as shown in Figure 10b, at q = 40 W/cm^2, a period of 20 s of stable temperature appeared on the ECS, maintaining the wall temperature of approximately 68 °C at time 50~70 s. Due to smaller bubbles continuously persisting and generated in one channel during the stable period, compared with multiple nucleation cores that were distributed on adjacent channels, they are observed during the temperature oscillation process. This contributed to the decreased length of the triple-phase contact line and the decreased detach diameter of the bubble. In this case, the disturbance of liquid, as well as the temperature oscillation, was relieved.

4.4. Boiling Curves and HTC

Figure 12a plots the boiling curves of DI water with various surfactant concentrations on the PS under P_S = 8.8 kPa, where the black square points refer to the condition without surfactant. It was found that the boiling curve of DI water with surfactant presented as an "S" shape. The superheat ΔT increased first and then decreased gradually. The results under sub-atmospheric pressure were different from the atmospheric conditions, as described in [24,31,58]. Compared to the case without surfactant, the heat transfer enhancement brought about by the addition of surfactant only occurred in high heat flux or high concentration regions (>0.3 mL/L). Under low heat flux conditions, severe temperature overshoot occurred when the surfactant concentration varied from 0.1 mL/L to 0.3 mL/L. For instance, at q = 10 W/cm^2, the temperature overshoot at c = 0.1 mL/L was about 32.5 °C, which was 17 °C higher than that at c = 0 mL/L. This corresponds to the intermittent bubbling behaviors on the heating surface. The temperature overshoot became relieved when the surface concentration increased to 0.4 mL/L. Moreover, at c = 0.1 mL/L, the boiling heat transfer improvement did not occur until q was larger than 50 W/cm^2, whereas the boiling enhancement worked only when q was larger than 20 W/cm^2 at c = 0.5 mL/L. In addition, considering the dual role of surfactants in reducing static contact angle and surface tension coefficient, the onset of nucleate boiling (ONB) was delayed compared to the case without surfactants, but the order of delay was not exactly consistent with the order of concentration reduction. Moreover, CHF appeared at about 40 W/cm^2 on the PS when no surfactant was added, but no CHF was recorded after adding Tween 20, indicating the positive role of surfactant in raising the boiling CHF under low saturated pressures. As shown in Figure 13a, the continuous increase in heat flux leads to the maximum HTC reaching 2.86 W·cm^{-2}·K^{-1} when P_S = 8.8 kPa, c = 0.4 mL/L, and q = 80 W/cm^2, compared to the case where c = 0 mL/L. This represents a 138% increment.

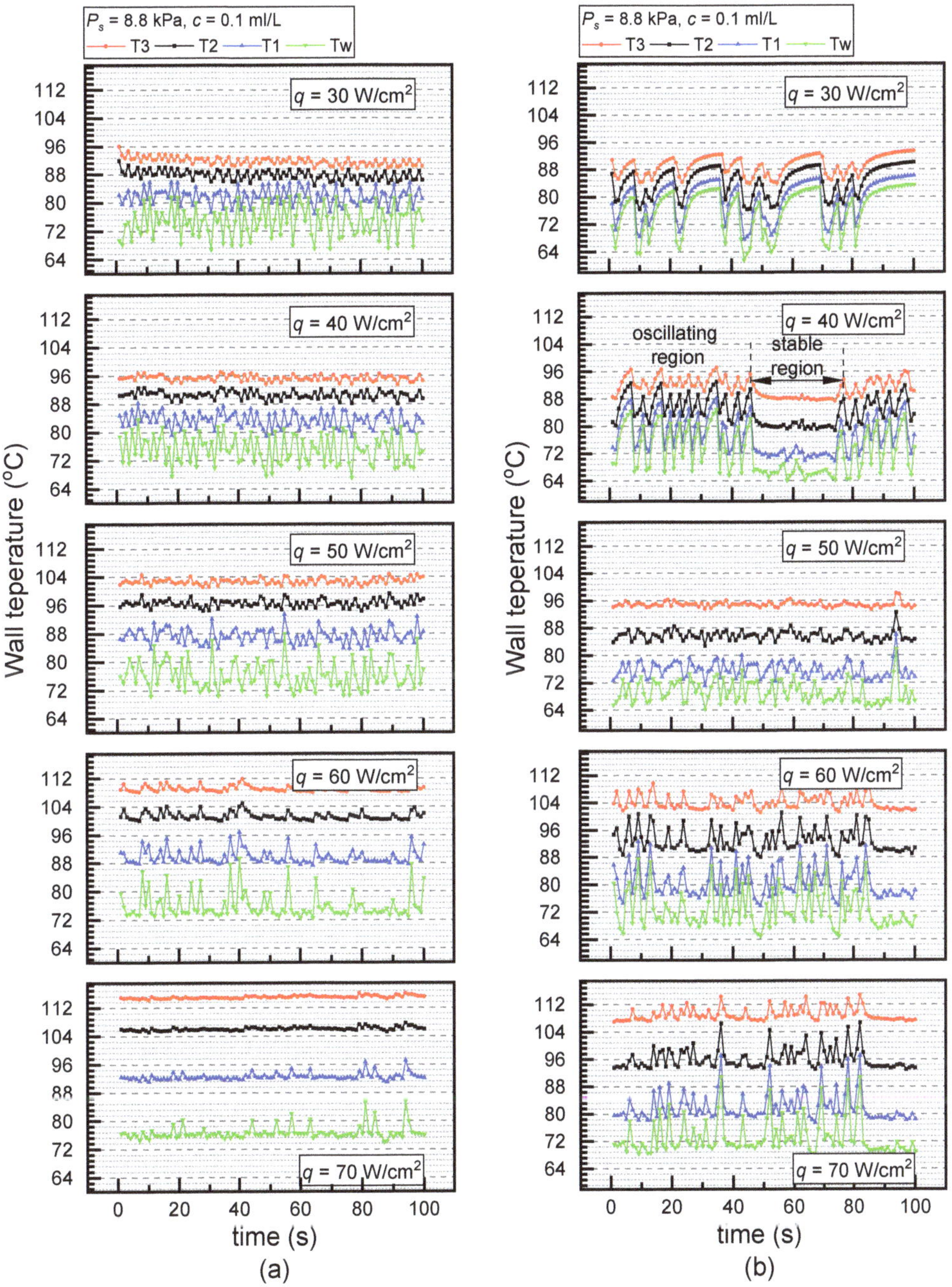

Figure 10. Temperature oscillations at different heat fluxes under P_s = 8.8 kPa and c = 0.1 mL/L (**a**) PS and (**b**) ECS.

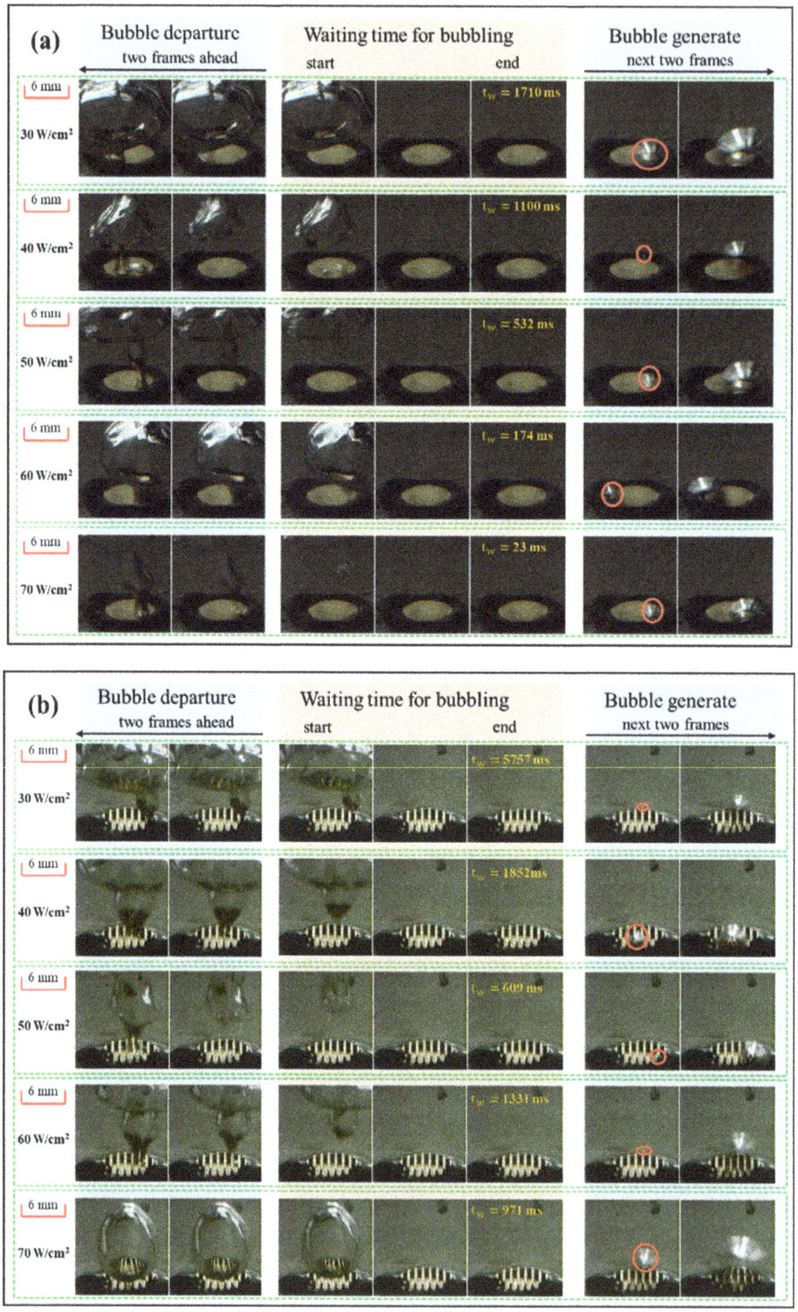

Figure 11. Waiting time for bubbling at different heat fluxes under P_s = 8.8 kPa and c = 0.1 mL/L on (**a**) PS and (**b**) ECS.

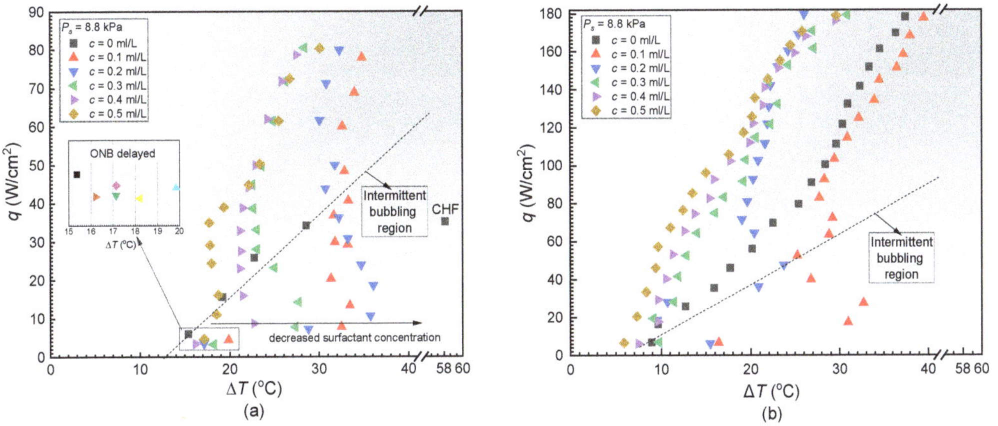

Figure 12. Pool boiling curves under the same saturated pressure (**a**) PS and (**b**) ECS.

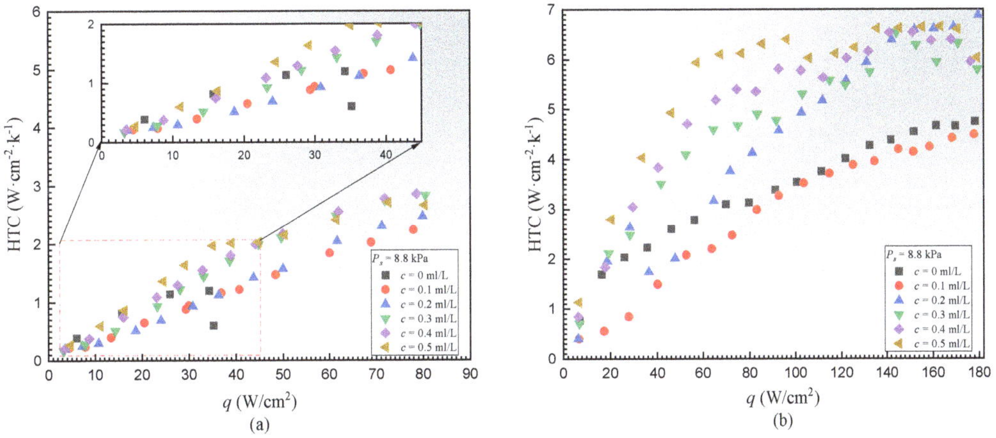

Figure 13. HTC curve under the same saturated pressure (**a**) PS and (**b**) ECS.

Similarly, Figure 12b displays the boiling curves on the ECS under the same conditions. At c = 0.1 mL/L, due to the excessively long t_w, the wall superheat remained consistently higher across the entire range of heat flux compared to the situation without the addition of the concentration, mirroring the behavior observed for the PS. However, as the concentration increases, the heat transfer capability gradually improves, eventually surpassing the case without the addition of the surfactant. For instance, at q = 160 W/cm^2, the wall superheat measured at c = 0.5 mL/L was 24.5 °C, which was 10 °C lower than the case when no surfactant was added. As the concentration increased from 0.1 to 0.5 mL/L, the onset of nucleate boiling and wall superheat gradually decreased, and the ONB decreased by 10.54 K. When the heat flux was 160 W/cm^2 the wall superheat was reduced by 12.8 K. The key difference between the boiling behaviors on the ECS and PS was that the temperature overshoot phenomenon for the ECS only occurred at low concentrations of 0.1 and 0.2 mL/L. Additionally, as shown in Figure 13b, the maximum HTC can reach 6.89 W·cm^{-2}·K^{-1}, which was increased by 45% compared to the case without surfactant Tween 20.

In summary, the change in heat transfer performance primarily depends on the intensity of the bubble disturbance per unit of time. Under low saturated pressure, compared with pure water, the declined surface tension coefficient still leads to smaller bubbles and increased departure frequency, but it also prolongs the t_w, causing noticeable temperature overshoot at low heat flux. It should be noted that although there is temperature overshoot at low heat flux when adding surfactant, no CHF is ever measured on the PS or ECS with surfactant added. As the occurrence of CHF requires a persistent vapor-film coverage, the temperature oscillation induced by the "bubble–liquid–bubble" cycle possibly strengthens the disturbance and delays the occurrence of CHF.

4.5. Mechanism of Intermittent Bubbling and Temperature Oscillation

Intermittent bubbling and temperature oscillation are observed under low saturated pressure conditions, and the phenomenon gets severe when surfactant is added, which is barely reported in the previous literature. The intermittent boiling and temperature oscillation occurs for two reasons: (1) the increased activation energy of nucleation ΔG_c; and (2) the intensified liquid disturbance during bubble departure.

According to classic nucleation theory [59], as shown in Figure 14a, the resistance to the formation of critical nuclei in superheated liquid (surface free energy ΔG_{surf}) is compensated by 2/3 of the decrease in the volume free energy ΔG_{bulk}, while the remaining 1/3 of the resistance is compensated by the energy fluctuation in the superheated liquid. As long as the structural fluctuation of the superheated liquid is $\geq r^*$, and the energy fluctuation is $\geq \Delta G_c$, the interface between the crystal nucleus and the liquid phase can be formed, and the stable crystal nucleus continues to grow and form a bubble. For heterogeneous nucleation, the required activation energy of nucleation $\Delta G_{c,Het}$ is closely related to the characteristics of the nucleation surface [59], i.e.,

$$\Delta G_{c,Het} = \frac{16\pi\sigma^3 T_s^2}{3(h_{lv}\Delta T)^2} f(\theta) = \frac{16\pi\sigma^3 T_s^2}{3(h_{lv}\Delta T)^2} \cdot \frac{2+\cos\theta-\cos^2\theta}{4} \quad (7)$$

where σ is the surface tension coefficient of the liquid, T_s is the saturated temperature, ΔT is the superheating degree of liquid, h_{lv} is the latent heat of liquid, and θ is the static contact angle. According to Equation (7), in the case of the same superheating ΔT, the decrease in P_s yields an increase in $\Delta G_{c,\,Het}$. Namely, under sub-atmospheric pressure, a larger ΔT is always required for liquid to form nucleation and bubble. As for the surfactant, the adding of surfactant reduces the surface tension coefficient σ on one hand, reducing the barrier to nucleation but decreases the static contact angle θ on the other hand, raising the required activation energy of nucleation. These two aspects are trade-offs in nucleation.

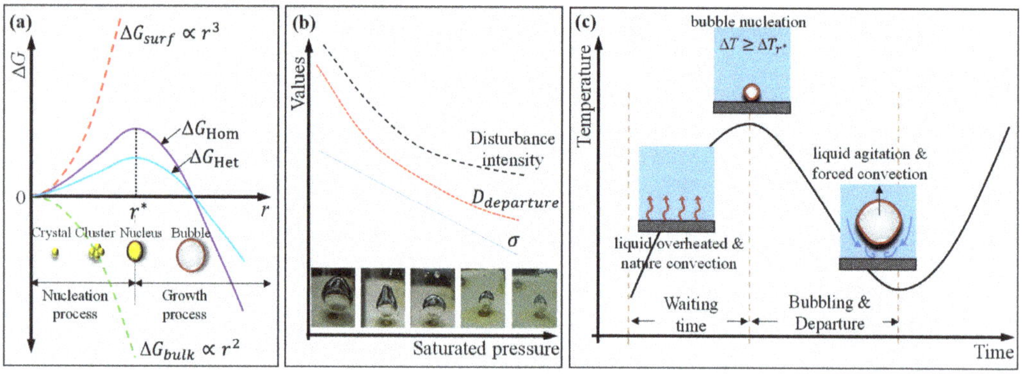

Figure 14. The mechanism of intermittent bubbling and temperature oscillation: (**a**) classical nucleation theory, (**b**) effect of saturated pressure on bubble behaviors, and (**c**) stages of temperature oscillation.

Besides the nucleation, P_s and surfactant concentration also have a significant effect on the departure behaviors of bubbles. As the surface tension is the main force that prevents bubbles from detachment [60], the addition of surfactant benefits the bubble departure. However, the bubble departure diameter is greatly increased under low saturated pressure, reaching 10 times that under atmosphere [61,62]. Considering the large detachment diameter of bubbles at low saturated pressure, as shown in Figure 14b, the disturbance of liquid became vigorous to the heating surface and caused a noticeable temperature drop due to the forced convection.

Figure 14c depicts the development stages of temperature oscillation versus time in one period, where the corresponding behaviors of vapor and liquid phases are also given. Before nucleation, the liquid is overheated to accumulate the energy to activate the nucleus. The wall temperature increases fast in this stage. If the superheating degree of liquid reaches the threshold for nucleation, boiling begins on the surface and the wall temperature decreases. During the bubbling and departure stage, liquid agitation and forced convection also contribute to reducing the wall temperature. The wall temperature continues to decrease and reaches the lowest when the boiling is stopped because of the insufficient liquid superheat. The liquid should be overheated again to be boiled; therefore, intermittent bubbling occurs, and temperature oscillation is observed.

5. Conclusions

In this paper, the effects of saturation pressure, surfactant concentration, and heat flux on nucleation, bubble growth and departure, temperature oscillations, and pool boiling enhancement are explored using experimental data. Theoretical analyses were also carried out from the point of view of the energy required for nucleation. The mechanism is proved with visualization evidence and temperature oscillation characteristics. Based on the analysis, the following conclusions are drawn:

(1) Increasing the saturated pressure can effectively promote the transition from intermittent boiling to continuous boiling. In this study, the saturated pressure was systematically reduced from 13 kPa to 8.8 kPa. The decrease in pressure resulted in changes to the increased activation energy of nucleation ΔG_c, leading to a reduction in system stability and making intermittent boiling more pronounced.

(2) At low saturated pressures, the boiling curve of deionized water with added surfactant exhibits an "S" shape. The surfactant reduces both surface tension and wettability, resulting in a more pronounced temperature overshoot at low heat flux compared to without surfactant. However, as the concentration increases, the effects of surface tension become dominant, leading to smaller bubbles that effectively alleviate temperature overshoot. Under 8.8 kPa, boiling enhancement was achieved across all heat flux conditions until a concentration of 0.3 mL/L was reached for the ECS.

(3) The increase in heat flux allows for more heat transfer to the surface for nucleation, which can shorten the waiting time for bubbling under sub-atmospheric pressure. The maximum HTC of the PS with added surfactant reached 2.86 $W \cdot cm^{-2} \cdot K^{-1}$, representing a 138% increase compared to deionized water. The maximum HTC of the ECS with added surfactant reached 6.89 $W \cdot cm^{-2} \cdot K^{-1}$, which was increased by 45%.

(4) When the pressure decreased from 38.5 kPa to 8.8 kPa, the bubble detachment time significantly increased due to a sharp increase in bubble size, extending by 105% for the PS. At this point, with the addition of a low surfactant concentration, more pronounced intermittent boiling and temperature oscillations occurred. When the concentration decreased from 0.5 mL to 0.1 mL, the bubble waiting time increased by more than 25 times.

In summary, the expanding microchanneled surface effectively intensifies the bubble removal capability and enhances pool boiling heat transfer performance, which is suitable for low-pressure two-phase immersion cooling of DI water. A higher surfactant concentration is recommended to mitigate temperature oscillations in sub-atmospheric pressure applications.

Author Contributions: Validation, S.H.; Formal analysis, D.F.; Investigation, Y.H.; Writing—original draft, Y.H.; Supervision, S.H.; formal analysis, visualization and Resource, C.D. All authors have read and agreed to the published version of the manuscript.

Funding: This work was financially supported by the Zhuhai Industry University Research Cooperation and Basic and Applied Basic Research Project (ZH22017003210001PWC), Guangdong Basic and Applied Basic Research Foundation (2023A1515010942), and the National Key Research and Development Program of China (2022YFC2204401).

Institutional Review Board Statement: Not applicable.

Informed Consent Statement: Not applicable.

Data Availability Statement: The original contributions presented in the study are included in the article, further inquiries can be directed to the corresponding author.

Conflicts of Interest: The authors declare no conflict of interest.

Nomenclature

C	surfactant concentration, mL/L
G	energy
h_{lv}	latent heat of liquid, kJ/kg
P_s	saturated pressure, kPa
q	heat flux, W/cm^2
r	radius, mm
T	temperature, K
ΔT	superheat, K
t_d	bubble departure time, ms
t_w	waiting time for bubbling, ms
Ra	wall roughness, um
Greek alphabet	
θ	static contact angle, °
σ	surface tension coefficient, mN/m
Abbreviations	
CTAB	cetyltrimethylammonium bromide
ECS	expanding microchanneled surface
ONB	onset of nucleate boiling
ppm	part per million
PS	plane surface
SDS	solution sodium dodecyl sulfate
Tween 20	polysorbate 20
TritonX-114	Tert-OCLylphenoxy poly ethanol

References

1. Liu, Y.; Wei, X.; Xiao, J.; Liu, Z.; Xu, Y.; Tian, Y. Energy consumption and emission mitigation prediction based on data center traffic and PUE for global data centers. *Glob. Energy Interconnect.* **2020**, *3*, 272–282. [CrossRef]
2. Wang, H.; Yuan, X.; Zhang, K.; Lang, X.; Chen, H.; Yu, H.; Li, S. Performance evaluation and optimization of data center servers using single-phase immersion cooling. *Int. J. Heat Mass Transf.* **2024**, *221*, 125057. [CrossRef]
3. Li, X.; Xu, Z.; Liu, S.; Zhang, X.; Sun, H. Server performance optimization for single-phase immersion cooling data center. *Appl. Therm. Eng.* **2023**, *224*, 120080. [CrossRef]
4. Roach, J. To Cool Datacenter Servers, Microsoft Turns to Boiling Liquid. 2024. Available online: https://news.microsoft.com/source/features/innovation/datacenter-liquid-cooling (accessed on 26 April 2021).

5. Liu, J.; Chen, H.; Huang, S.; Jiao, Y.; Chen, M. Recent progress and prospects in liquid cooling thermal management system for lithium-ion batteries. *Batteries* **2023**, *9*, 400. [CrossRef]
6. Larrañaga-Ezeiza, M.; Navarro, G.V.; Garmendia, I.G.; Arroiabe, P.F.; Martinez-Aguirre, M.; Arostegui, J.B. Parametric optimisation of a direct liquid cooling—Based prototype for electric vehicles focused on pouch-type battery cells. *World Electr. Veh. J.* **2022**, *13*, 149. [CrossRef]
7. Pulugundla, G.; Dubey, P.; Wu, Z.; Wang, Q.; Srouji, A.K. Thermal management of lithium ion cells at high discharge rate using submerged-cell cooling. In Proceedings of the 2020 IEEE Transportation Electrification Conference & Expo (ITEC), Chicago, IL, USA, 23–26 June 2020; pp. 1–5.
8. Dubey, P.; Pulugundla, G.; Srouji, A.K. Direct comparison of immersion and cold-plate based cooling for automotive li-ion battery modules. *Energies* **2021**, *14*, 1259. [CrossRef]
9. Gong, S.; Cheng, P. Lattice Boltzmann simulation of periodic bubble nucleation, growth and departure from a heated surface in pool boiling. *Int. J. Heat Mass Transf.* **2013**, *64*, 122–132. [CrossRef]
10. Katto, Y.; Yokoya, S. Behavior of a vapor mass in saturated nucleate and transitional pool boiling. *Trans. Jpn. Soc. Mech. Eng.* **1975**, *41*, 294–305. [CrossRef]
11. Cooper, M. The microlayer and bubble growth in nucleate pool boiling. *Int. J. Heat Mass Transf.* **1969**, *12*, 915–933. [CrossRef]
12. Coulibaly, A.; Lin, X.; Bi, J.; Christopher, D.M. Bubble coalescence at constant wall temperatures during subcooled nucleate pool boiling. *Exp. Therm. Fluid Sci.* **2013**, *44*, 209–218. [CrossRef]
13. Birbarah, P.; Gebrael, T.; Foulkes, T.; Stillwell, A.; Moore, A.; Pilawa-Podgurski, R.; Miljkovic, N. Water immersion cooling of high power density electronics. *Int. J. Heat Mass Transf.* **2019**, *147*, 118918. [CrossRef]
14. Gao, T.; Shao, S.; Cui, Y.; Espiritu, B.; Ingalz, C.; Tang, H.; Heydari, A. A study of direct liquid cooling for high-density chips and accelerators. In Proceedings of the 2017 16th IEEE Intersociety Conference On Thermal and Thermomechanical Phenomena in Electronic Systems (Itherm), Orlando, FL, USA, 30 May–2 June 2017; pp. 565–573.
15. Carey, V.P. *Liquid-Vapor Phase-Change Phenomena: An Introduction to the Thermophysics of Vaporization and Condensation Processes in Heat Transfer Equipment, Third ed.*; CRC Press: Boca Raton, FL, USA, 2020.
16. Gao, W.; Qi, J.; Yang, X.; Zhang, J.; Wu, D. Experimental investigation on bubble departure diameter in pool boiling under sub-atmospheric pressure. *Int. J. Heat Mass Transf.* **2019**, *134*, 933–947. [CrossRef]
17. McGillis, W.R.; Carey, V.P.; Fitch, J.S.; Hamburgen, W.R. Pool boiling enhancement techniques for water at low pressure. In Proceedings of the 1991 Proceedings, Seventh IEEE Semiconductor Thermal Measurement and Management Symposium, Phoenix, AZ, USA, 12–14 February 1991; pp. 64–72.
18. Elghanam, R.; Fawal, M.; Aziz, R.A.; Skr, M.; Khalifa, A.H. Experimental study of nucleate boiling heat transfer enhancement by using surfactant. *Ain Shams Eng. J.* **2011**, *2*, 195–209. [CrossRef]
19. Raza, Q.; Kumar, N.; Raj, R. Wettability-independent critical heat flux during boiling crisis in foaming solutions. *Int. J. Heat Mass Transf.* **2018**, *126*, 567–579. [CrossRef]
20. Du, J.; Wang, Y.; Yang, W.; Wang, J.; Cao, Z.; Sundén, B. Effect of nanoparticle concentration and surfactants on nanofluid pool boiling. *Int. J. Heat Mass Transf.* **2024**, *221*, 125080. [CrossRef]
21. Ammerman, C.N.; You, S.M. Determination of the boiling enhancement mechanism caused by surfactant addition to water. *J. Heat Transf.* **1996**, *118*, 429–435. [CrossRef]
22. Frost, W.; Kippenhan, C.J. Bubble growth and heat-transfer mechanisms in the forced convection boiling of water containing a surface active agent. *Int. J. Heat Mass Transf.* **1967**, *10*, 931–949. [CrossRef]
23. Kumar, N.; Raza, Q.; Raj, R. Surfactant aided bubble departure during pool boiling. *Int. J. Therm. Sci.* **2018**, *131*, 105–113. [CrossRef]
24. Wu, W.-T.; Yang, Y.-M.; Maa, J.-R. Enhancement of Nucleate Boiling Heat Transfer and Depression of Surface Tension by Surfactant Additives. *J. Heat Transf.* **1995**, *117*, 526–529. [CrossRef]
25. Wen, T.; Luo, J.; Jiao, K.; Lu, L. Pool boiling heat transfer enhancement of aqueous solution with quaternary ammonium cationic surfactants on copper surface. *Int. J. Heat Mass Transf.* **2022**, *190*, 122761. [CrossRef]
26. Li, B.; Zheng, J.; Dang, C.; Kang, Y.; Jia, H. Experimental investigation on bubble dynamic behaviors in pool boiling of surfactant solutions. *Int. J. Therm. Sci.* **2024**, *198*, 108858. [CrossRef]
27. Tian, Z.; Etedali, S.; Afrand, M.; Abdollahi, A.; Goodarzi, M. Experimental study of the effect of various surfactants on surface sediment and pool boiling heat transfer coefficient of silica/DI water nano-fluid. *Powder Technol.* **2019**, *356*, 391–402. [CrossRef]
28. Yang, Y.; Maa, J. On the criteria of nucleate pool boiling enhancement by surfactant addition to water. *Chem. Eng. Res. Des.* **2001**, *79*, 409–416. [CrossRef]
29. Lee, C.Y.; Zhang, B.J.; Kim, K.J. Influence of heated surfaces and fluids on pool boiling heat transfer. *Exp. Therm. Fluid Sci.* **2014**, *59*, 15–23. [CrossRef]
30. Proper, J.; Kulacki, F. Pool boiling of a dilute emulsion: Surfactant effect on heat transfer. *Int. Commun. Heat Mass Transf.* **2021**, *128*, 105586. [CrossRef]
31. Yang, Z.; Yao, Y.; Wu, H. Effects of surfactants on subcooled pool boiling characteristics: An experimental study. *Int. J. Heat Mass Transf.* **2022**, *199*, 123419. [CrossRef]
32. Jia, H.; Xu, L.; Xiao, X.; Zhong, K. Study on boiling heat transfer of surfactant solution on grooved surface. *Int. J. Heat Mass Transf.* **2021**, *181*, 121876. [CrossRef]

33. Sun, C.; Guo, D.; Wang, Z.; Sun, F. Investigation on active thermal control method with pool boiling heat transfer at low pressure. *J. Therm. Sci.* **2018**, *27*, 277–284. [CrossRef]
34. Giraud, F.; Rullière, R.; Toublanc, C.; Clausse, M.; Bonjour, J. Experimental evidence of a new regime for boiling of water at subatmospheric pressure. *Exp. Therm. Fluid Sci.* **2015**, *60*, 45–53. [CrossRef]
35. Yamada, M.; Shen, B.; Imamura, T.; Hidaka, S.; Kohno, M.; Takahashi, K.; Takata, Y. Enhancement of boiling heat transfer under sub-atmospheric pressures using biphilic surfaces. *Int. J. Heat Mass Transf.* **2017**, *115*, 753–762. [CrossRef]
36. Najim, A.; More, V.; Thorat, A.; Patil, S.; Savale, S. Enhancement of pool boiling heat transfer using innovative non-ionic surfactant on a wire heater. *Exp. Therm. Fluid Sci.* **2017**, *82*, 375–380. [CrossRef]
37. Yuan, L.; Hong, F.; Cheng, P. Pool boiling enhancement through a guidance structure mounted above heating surface. *Int. J. Heat Mass Transf.* **2019**, *139*, 751–763. [CrossRef]
38. Michaie, S.; Rullière, R.; Bonjour, J. Experimental study of bubble dynamics of isolated bubbles in water pool boiling at subatmospheric pressures. *Exp. Therm. Fluid Sci.* **2017**, *87*, 117–128. [CrossRef]
39. Halon, T.; Zajaczkowski, B.; Michaie, S.; Rulliere, R.; Bonjour, J. Experimental study of low pressure pool boiling of water from narrow tunnel surfaces. *Int. J. Therm. Sci.* **2017**, *121*, 348–357. [CrossRef]
40. Kwark, S.M.; Amaya, M.; Kumar, R.; Moreno, G.; You, S.M. Effects of pressure, orientation, and heater size on pool boiling of water with nanocoated heaters. *Int. J. Heat Mass Transf.* **2010**, *53*, 5199–5208. [CrossRef]
41. Wang, X.; Yang, C.; Xu, C.; Zhu, H.; Chen, X.; Liang, K.; Xu, J. Correlation of pool boiling heat transfer and enhancement by microstructure surfaces at sub-atmospheric pressures. *Int. J. Heat Mass Transf.* **2024**, *226*, 125465. [CrossRef]
42. Ateş, A.; Benam, B.P.; Mohammadilooey, M.; Çelik, S.; Serdyukov, V.; Surtaev, A.; Sadaghiani, A.K.; Koşar, A. Pool boiling heat transfer on superhydrophobic, superhydrophilic, and superbiphilic surfaces at atmospheric and sub-atmospheric pressures. *Int. J. Heat Mass Transf.* **2023**, *201*, 123582. [CrossRef]
43. Huang, G.; Tang, K.; Yu, S.; Tang, Y.; Zhang, S. Enhanced pool boiling heat transfer by metallic nanoporous surfaces under low pressure. *Int. J. Heat Mass Transf.* **2021**, *184*, 122382. [CrossRef]
44. Pal, A.; Joshi, Y. Boiling of water at subatmospheric conditions with enhanced structures: Effect of liquid fill volume. *J. Electron. Packag.* **2008**, *130*, 011010. [CrossRef]
45. Dahariya, S.; Betz, A.R. High pressure pool boiling: Mechanisms for heat transfer enhancement and comparison to existing models. *Int. J. Heat Mass Transf.* **2019**, *141*, 696–706. [CrossRef]
46. Chen, H.; Chen, G.; Zou, X.; Yao, Y.; Gong, M. Experimental investigations on bubble departure diameter and frequency of methane saturated nucleate pool boiling at four different pressures. *Int. J. Heat Mass Transf.* **2017**, *112*, 662–675. [CrossRef]
47. Staniszewski, B.E. *Nucleate Boiling Bubble Growth and Departure*; Massachusetts Institute of Technology, Division of Industrial Cooperation: Cambridge, MA, USA, 1959.
48. Akiyama, M.; Tachibana, F.; Ogawa, N. Effect of pressure on bubble growth in pool boiling. *Bull. JSME* **1969**, *12*, 1121–1128. [CrossRef]
49. Etedali, S.; Afrand, M.; Abdollahi, A. Effect of different surfactants on the pool boiling heat transfer of SiO2/deionized water nanofluid on a copper surface. *Int. J. Therm. Sci.* **2019**, *145*, 105977. [CrossRef]
50. Hu, Z.-C.; Liu, X.-Y.; Wang, Q. Calculation and analysis of excess adsorption and surface tension of aqueous surfactant solutions under pool boiling conditions. *Int. J. Heat Mass Transf.* **2021**, *185*, 122416. [CrossRef]
51. Yin, J.; Xiao, X.; Feng, L.; Zhong, K.; Jia, H. Experimental investigation of pool boiling characteristics of surfactant solutions on bi-conductive surfaces. *Int. J. Heat Mass Transf.* **2020**, *157*, 119914. [CrossRef]
52. Mata, M.R.; Ortiz, B.; Luhar, D.; Evereux, V.; Cho, H.J. How dynamic adsorption controls surfactant-enhanced boiling. *Sci. Rep.* **2022**, *12*, 1–10. [CrossRef]
53. Cho, H.J.; Sresht, V.; Blankschtein, D.; Wang, E.N. Understanding enhanced boiling with triton x surfactants. In Proceedings of the ASME 2013 Heat Transfer Summer Conference collocated with the ASME 2013 7th International Conference on Energy Sustainability and the ASME 2013 11th International Conference on Fuel Cell Science, Engineering and Technology, Minneapolis, MN, USA, 14–19 July 2013.
54. Guo, D.-S.; Li, X.-B.; Zhang, H.-N.; Li, F.-C.; Su, W.-T.; Zhu, H. Bubble behaviors during subcooled pool boiling in water and nonionic surfactant aqueous solution. *Int. J. Heat Mass Transf.* **2020**, *159*, 120087. [CrossRef]
55. Surtaev, A.; Serdyukov, V.; Malakhov, I. Effect of subatmospheric pressures on heat transfer, vapor bubbles and dry spots evolution during water boiling. *Exp. Therm. Fluid Sci.* **2019**, *112*, 109974. [CrossRef]
56. Shiau, B.-J.; Sabatini, D.A.; Harwell, J.H. Properties of food grade (edible) surfactants affecting subsurface remediation of chlorinated solvents. *Environ. Sci. Technol.* **1995**, *29*, 2929–2935. [CrossRef]
57. Basu, N.; Warrier, G.R.; Dhir, V.K. Wall heat flux partitioning during subcooled flow boiling: Part 1—Model development. *J. Heat Transf.* **2005**, *127*, 131–140. [CrossRef]
58. Ravikumar, S.V.; Haldar, K.; Jha, J.M.; Chakraborty, S.; Sarkar, I.; Pal, S.K.; Chakraborty, S. Heat transfer enhancement using air-atomized spray cooling with water–$Al2O_3$ nanofluid. *Int. J. Therm. Sci.* **2015**, *96*, 85–93. [CrossRef]
59. Kwak, H.-Y.; Kim, Y.W. Homogeneous nucleation and macroscopic growth of gas bubble in organic solutions. *Int. J. Heat Mass Transf.* **1998**, *41*, 757–767. [CrossRef]
60. Karri, S.B.R. Dynamics of bubble departure in micro-gravity. *Chem. Eng. Commun.* **1988**, *70*, 127–135. [CrossRef]

61. Halon, T.; Zajaczkowski, B.; Michaie, S.; Rulliere, R.; Bonjour, J. Enhanced tunneled surfaces for water pool boiling heat transfer under low pressure. *Int. J. Heat Mass Transf.* **2018**, *116*, 93–103. [CrossRef]
62. Wang, X.; Liang, K.; Xu, J.; Wang, J.; Chen, X. Experimental study on bubble dynamics and heat transfer of pool boiling at sub-atmospheric pressures. *Int. Commun. Heat Mass Transf.* **2023**, *148*, 107065. [CrossRef]

materials

MDPI

Article

Features of Structure and Flow Field in Homemade Co-Current Cavitation Water Jet Nozzle

Chenhao Guo [1,*], Xing Dong [2], Haorong Song [2] and Yun Jiang [3]

1 School of Safety Engineering, Heilongjiang University of Science and Technology, Harbin 150022, China
2 School of Mechanical Engineering, Heilongjiang University of Science and Technology, Harbin 150022, China; dongxingwrh@163.com (X.D.); 13754623001@163.com (H.S.)
3 School of Mining Engineering, Heilongjiang University of Science and Technology, Harbin 150022, China; jiangyun011010@163.com
* Correspondence: 18145974766@163.com; Tel.: +86-18945918376

Abstract: The cavitation water jet cleaning and coating removal technique represents an innovative sustainable method for cleaning and removing coatings, with the nozzle serving as a crucial component of this technology. Developing an artificially submerged nozzle with a reliable structure and excellent cavitation performance is essential for enhancing cavitation water jets' cleaning and coating removal efficacy in an atmosphere environment (non-submerged state). This study is based on the shear flow cavitation mechanism of an angular nozzle, the resonance principle of an organ pipe, and the jet pump principle. A dual-nozzle co-current cavitation water jet nozzle structure was designed and manufactured. The impact of the nozzle's inlet pressure on the vapor volume percentage, as well as the axial and radial velocities inside the flow field, were examined utilizing ANSYS Fluent software with the CFD method. The dynamic change rule of the cavitation cloud is derived by analyzing the picture of the cavitation cloud in the nozzle's outflow field utilizing pseudo-color imaging techniques. The results show that the maximum vapor volume percentage is more significant than 95% for different inlet pressures in the internal nozzle. The changes that occur in the cavitation cloud exhibit notable regularity, including the four stages of cavitation, which are inception, development, shedding, and collapse. A change period is 1.5 ms, which proves that the homemade co-current cavitation water jet nozzle can achieve good cavitation effects.

Keywords: co-current cavitation water jet nozzle; numerical simulation; pseudo-color image; cavitation cloud

1. Introduction

Cavitation is a prevalent occurrence in fluid dynamics and was first identified as a significant impediment to the effective functioning of hydraulic machinery. It adversely impacts domains such as engineering and medicine [1–3]. As the research progressed, the researchers not only found the cause of cavitation but also ingeniously utilized the high-intensity energy generated by the collapse of cavities, developing it into an inexpensive, clean, and environmentally friendly manufacturing technology, cavitation water jet technology [4,5]. Cavitation water jet technology uses high-pressure water as a carrier. When high-pressure water flows through a specific nozzle configuration, the surrounding absolute pressure falls below the saturated vapor pressure of water, forming numerous cavitation bubbles inside the water. When cavitation bubbles strike a target surface by a jet, it has a dual effect, namely the pressure pulsation from the collapse of the cavitation bubble and the erosive damage caused by the high-velocity jet. It has higher impact force, destructive power, and lower energy consumption than ordinary water jets. Consequently, it has been extensively utilized in areas such as effective cleaning and coating removal, shot peening, and material cutting [6–10].

The critical component of the water jet that generates cavitation is the nozzle, and the nozzle's structural design directly affects the jet's cavitation effect and cleaning and coating removal performance. According to the different principles of cavitation priming, the conventional submerged cavitation nozzle is divided into shear-type, bypassing-type, and oscillating-type nozzles [11]. Researchers can promote cavitation's initial formation and development by selecting a reasonable nozzle structure and adjusting the jet's target distance. Energy is released when the cavitation bubbles collapse at the jet's impact target, thus achieving the effect of cavitation damage [12]. Soyama H [13] conducted cavitation erosion tests using eight cavitation nozzles with different structures. The erosion rate of the target per unit of time was used as the evaluation index for examining the impact of various nozzle geometries upon the intensity of the cavitation jet. The test results showed that the submerged cavitation nozzle, throat diameter d, exit diameter D, and length l affect cavitation erosion intensity. When $d{:}D{:}l = 1{:}8{:}8$, the nozzle's cavitation intensity is maximal, proving that the nozzle's geometry significantly affects the cavitating jet's intensity. Panda et al. [14] designed a new submerged cavitation nozzle with an outlet diffusion to improve the jet's cavitation effect without increasing energy consumption. Comparative numerical simulations indicate that the new type of submerged cavitation nozzle with an outlet diffusion section generates three times the cavitation intensity of a regular cylindrical nozzle and 1.5 times the intensity of a venturi nozzle, proving that this structure can significantly enhance the jet cavitation effect. Dong et al. [15] employed CFD fluid dynamics software to numerically simulate an ordinary submerged cavitation nozzle, investigating the effects of the nozzle structure on the vapor volume percentage. The simulation findings demonstrate that when the input pressure is 20 MPa, the nozzle contraction angle α is 13.5°, and the expansion angle β is 60°, the vapor volume percentage from the nozzle can attain 95%. The cavitation cloud forms adjacent to the edge of the diffusion segment at the nozzle's output and disperses in the configuration of a vortex ring.

As the current angular cavitation nozzle, the swirling cavitation nozzle, and the self-oscillating pulse cavitation nozzle structures can only obtain good cavitation cleaning results under certain submerged conditions [16], under non-submerged conditions, the cavitation generated by the above nozzles mainly exists inside the nozzle. The cavitation effect rapidly diminishes as the jet exits the nozzle and enters the atmosphere. The energy produced during the collapse of the cavitation bubble is not entirely harnessed, leading to a limited effective target range of the jet. Furthermore, some large and critical parts cannot be immersed in water for cavitation jet treatment, thereby restricting the application of the cavitation jet in industrial production [17]. To solve this problem, researchers have successively proposed a variety of co-current artificial submersion cavitation nozzle structures, aiming to overcome the limitations of the submerged condition of the cavitation jet, increase the effective target range, and increase the applicability of the cavitation jet in industrial production. Soyama [18] designed a co-current overlapping nozzle structure, which allows shear cavitation to form at the annular interface where the internal nozzle's high-pressure water converges with the external nozzle's low-pressure water. The low-speed water jet replaces the static water in the traditional submerged jet, and this new type of artificial submerged nozzle initially achieves a cavitating jet without the need for submergence. Marcon et al. [19] obtained the cavitation performance and shot peening effect of a co-flow cavitation water jet nozzle via a cavitation jet striking force test, revealing that the maximum residual stress generated by the co-flow cavitation water jet nozzle striking on the 7075-t651 aluminum alloy target can reach 401 MPa. Vidvans et al. [20], to improve the cavitation performance of a co-current cavitation water jet nozzle, designed a co-current cavitation water jet nozzle structure with an organ pipe-type inner nozzle and characterized its cavitation strength by the bubble cloud visualization test and the residual stress of the target. The findings demonstrate that the cavitation strength of the nozzle exceeds that of the conventional cavitation nozzle by 61%, and the residual compressive stress is more significant than that of the conventional cavitation nozzle, which proves that the co-current cavitation water jet nozzle can achieve a better cavitation effect under the non-submerged condition. Wu et al. [21] investigated the viability of an artificially immersed cavitation jet for enhancing the 7075 aluminum alloy by examining the surface appearance and roughness of the target material at varying jet blasting velocities to ascertain the effects of the new jet blasting on its microstructure. The findings reveal that the shearing impact of high- and low-velocity jets can produce a bubble cloud when a cavitating jet is artificially immersed. It can overcome the conventional submerged cavitation jet target distance of insufficient defects and effectively impact the target surface plastic deformation to increase its strength for the application of artificial submerged cavitation jet target surface strengthening to provide technical support. Luo et al. [22] used ANSYS Fluent to numerically simulate the flow field of a co-current double-nozzle artificial submerged cavitation nozzle in order to investigate the cavitation shot peening ability of an artificial submerged nozzle. The influence of the internal nozzle geometry and inlet pressure on the vapor volume fraction was obtained. It was found that the inlet pressure was positively correlated with the vapor volume fraction. The convergence angle of the internal nozzle was negatively correlated with the vapor volume fraction, providing a reference for the structural design of the artificial submerged nozzle. Fan et al. [23] designed a double-current cavitation water jet nozzle structure to improve the cavitation jet's rock-breaking capability. The internal nozzle is a cavitation nozzle that can be filled with abrasive, and the external nozzle is a co-current low-speed jet nozzle structure. High-speed camera technology and target erosion experiments were used to investigate the rock-breaking performance of the nozzle under different jet pressures, and the effects of rock-breaking time and nozzle diameter on the cavitation performance were discussed. The results show that when the inlet pressure of the internal high-pressure cavitation nozzle is 50 MPa, the inlet pressure of the external co-current low-speed jet nozzle is 0.18 MPa; the low-speed and high-speed jets have the best effect of shear cavitation, which can erode rock bodies up to a depth of 600 μm. It provides a reference for applying co-current cavitation water jet nozzles in the field of rock breaking.

An artificially submerged cavitation jet nozzle structure comprises two inlet channels, one for high pressure and the other for low pressure; the flow structure is very complex and the jet is in the form of intense turbulence [24,25], which involves complex interphase interactions (cavitation bubble–cavitation bubble, cavitation bubble–liquid phase), so conventional experimental methods are challenging to capture the details in the vapor bubble motion and characterize the dynamic change pattern of the vapor cloud. Due to the swift advancement of computer technology and the emergence of high-resolution image display instruments, researchers have used computational fluid dynamics (CFD) software and high-speed camera technology for numerical simulation and real-time acquisition of the cavitation flow field, respectively, to realize the calculation and visualization of the flow characteristics of the cavitation jet flow field [26,27]. Sekyi et al. [28] analyzed the variation rule of the cavitation nozzle's bubble cloud based on CFD and high-speed camera technology and established the kinetic mathematical model of individual cavitation bubbles, and the comparison study of the grayscale pictures processed in MATLAB indicates that an increase in nozzle diameter and low pressure negatively influences the eroding action of the cavitating jet on the target. Du et al. [29] designed an artificially submerged cavitation nozzle based on the angular nozzle structural design principle. They performed CFD numerical simulations, which showed that the high-pressure water jet produces a significant cavitation effect at the jet boundary in the artificially created submerged environment, which can be maintained in the atmosphere. Li et al. [30] employed CFD techniques and impact tests to examine the effects of the self-made artificially immersed cavitating nozzle's geometrical factors, target distance, and inlet pressure on cavitation intensity. The data indicate that the shear effect formed between high-velocity and low-velocity is the core factor in producing cavitation incipient. The optimal cavitation capability of this nozzle occurs when the convergence angles α of the interior nozzle and β of the exterior nozzle are both 30°, with a target distance of 30 mm. Zheng [31] designed a two-stage series-connected double-chamber organ pipe cavitation nozzle structure. The cavitation capacity of the nozzle under different inlet pressures, diffusion angles, and diffusion part lengths was obtained using the CFD method. The outcomes indicate that the nozzle's diffusion angle exerts a more pronounced influence on cavitation capacity, velocity magnitude in the jet reflux region, and length of the isocore compared to inlet pressure and the length of the diffusion part. Onish et al. [32] integrated a CFD numerical simulation with rapid digital imaging to examine the cyclic dynamics of cavitation cloud creation, growth, and shedding produced by a cavitation nozzle. The results show that a pair of annular clouds composed of initial and subsequent newborn cavitation clouds are shed downstream in sequence. This process is cyclic reciprocity, and the Strouhal number, which dominates the shedding of the cavitation cloud, is stable at 0.21–0.29, which validates the precision of the computer-simulated outcomes. It has a particular value of reference for the improvement of the structural performance of the cavitation jet nozzle.

In summary, while certain researchers in this domain have conducted structural optimization, computational simulations, and rapid speed photography tests on artificially immersed cavitation nozzles, current research has not yet explored the structure of an artificially immersed cavitation nozzle featuring an organ pipe self-oscillating angular cavitation nozzle for the inner nozzle and conical–cylindrical combination nozzles for the outer nozzle. This study aims to eliminate the submerged condition limitation under the conventional cavitation water jet working condition and enhance the nozzle's cavitation capability. Based on the shear flow cavitation mechanism of the angular nozzle, the resonance principle of the organ pipe, and the principle of the jet pump, a new type of co-current cavitation water jet nozzle is designed. A numerical simulation of the nozzle's flow field is conducted using the CFD method, revealing the influence of inlet pressure on the vapor volume percentage, axial velocity, and radial velocity of the flow field. A comparative examination with the existing co-current cavitation nozzle structure to validate the efficacy of the homemade co-current cavitation water jet nozzle structure is made. This is with the aid of high-speed photographic equipment to capture the morphology and distribution of the artificial submerged cavitation jet cloud, thereby enhancing technical assistance for the advancement of theory and implementation of the artificial submerged cavitation water jet.

2. Materials and Methods

2.1. Parameters for Simulation Configuration

A numerical simulation of the water–water vapor two-phase flow field of a homemade co-current cavitation nozzle was carried out by using the CFD method and ANSYS Fluent 2022 R2 software. In the CFD computer simulation, water was chosen as the primary phase, with a density of 1000 kg/m^3 and a viscosity of 1×10^{-3} Pa·s, and water vapor was chosen as the secondary phase, with a density of 2.558×10^{-2} kg/m^3 and a viscosity of

1.26×10^{-5} Pa·s. The impact of pressure on the properties of the jet cavitation flow field was examined by varying the inlet pressure of the homemade co-current cavitation nozzle.

2.2. Structural Design of Co-Current Cavitation Water Jet Nozzle

2.2.1. Overall Structure

The homemade co-current cavitation water jet nozzle structure is schematically shown in Figure 1. As seen in Figure 1, the overall nozzle structure consists of two parts, the internal organ pipe self-oscillating angular cavitation nozzle and the external conical–cylindrical combination nozzle structure. The internal organ pipe self-oscillating angular cavitation nozzle is designed based on the angular nozzle shear flow cavitation mechanism and the organ pipe resonance principle and consists of an incident cavity, resonance cavity, convergent section, cylindrical section, and diffusion section; the external nozzle structure comprises a conical convergent portion and a cylindrical segment. The physical photo of the homemade co-current cavitation water jet nozzle structure is shown in Figure 2.

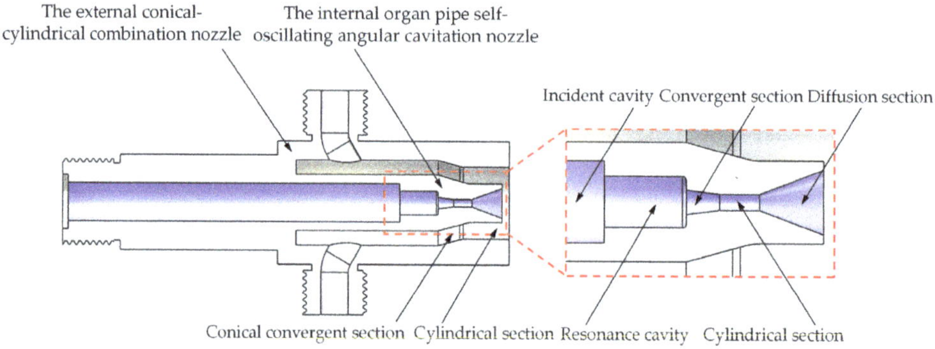

Figure 1. Schematic structure of homemade co-current cavitation water jet nozzle.

Figure 2. Homemade co-current cavitation water jet nozzle physical picture.

This study of a homemade combined co-current cavitation water jet nozzle has two advantages. Firstly, the internal organ pipe self-oscillation angular cavitation nozzle with the help of the organ pipe spray group resonance cavity of the feedback mechanism and amplification mechanism of the high-pressure water pulsation double modulation results in high-frequency pressure oscillation, which leads to the generation of cavitation. At the same time, the use of the angular cavitation nozzle outlet diffusion section shear effect of the nozzle outlet flow field to produce many vortices further enhances the role of the cavitation composite. Secondly, the conical convergent section of the external conical–cylindrical combined nozzle aids in the uniform introduction of the low-pressure jet, ensuring the wrapping effect of the peripheral flow beam on the internal high-speed jet. The cylindrical section structure can guide the external flow beam, promote the development of cavitation bubbles, ensure jet convergence, and significantly contribute to the mitigation of the reduction in axial velocity.

2.2.2. Structural Design of Internal Organ Pipe Self-Oscillating Angular Cavitation Nozzle

Figure 3 gives the internal organ pipe self-oscillating angular cavitation nozzle dimensions; the internal organ pipe self-oscillating angular cavitation nozzle is based on the angular nozzle shear flow cavitation mechanism and the organ pipe resonance principle, the traditional resonance cavity outlet cavity cylindrical contraction section is changed to a conical convergence section of the transition, and the angle of convergence α to choose the best recognized is 13.5° [23]. Only when the nozzle outlet cylindrical segment's diameter meets certain conditions can the pressure within the nozzle's flow field drop below the saturated vapor pressure, resulting in cavitation. Given a predetermined inlet pressure and flow rate, the nozzle's outlet diameter d can be as follows [33]:

$$d = \sqrt{\frac{4Q}{\sqrt{\frac{2p}{\rho}}\pi\mu}} \tag{1}$$

where Q denotes the internal organ pipe self-oscillating angular cavitation nozzle's flow rate (m^3/s); μ denotes the fluid flow factor; p denotes the internal organ pipe self-oscillating angular cavitation nozzle's inlet pressure (Pa); and ρ denotes the water's density (kg/m^3).

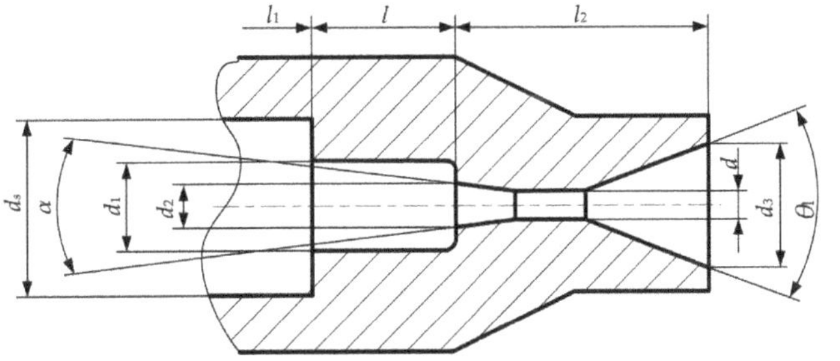

Figure 3. Dimensions of the internal organ pipe self-oscillating angular cavitation nozzle.

Substituting the actual working condition parameters, the nozzle's outlet diameter d can be obtained as 1.6 mm. According to the internationally commonly used angular cavitation water jet nozzle design in the proportional relationship between the nozzle diameter d and the length l_2, i.e., d:l_2 = 1:8 [13], then the internal organ pipe self-oscillation angular cavitation nozzle angular nozzle section length l_2 is taken as 12.8 mm.

The role of the resonant cavity of the internal cavitation nozzle in this study is to control the pressure gradient between the nozzle inlet cavity section and the outlet constriction section, to delay the position of cavitation occurrence, and to avert a substantial quantity of cavitation bubbles from obstructing the fluid flow pathway as a result of premature cavitation. On the other hand, the resonant cavity can match its intrinsic frequency and the critical self-excited structural frequency of the fluid to generate a self-oscillation phenomenon. The resonant cavity structure is a self-excited oscillatory amplifier consisting of length l and diameter d_1, where the inlet contraction section $(d_s/d_1)^2$ and outlet contraction section $(d_1/d_2)^2$ are the key parameters determining the cavitation effect of self-excited oscillation.

The internal organ pipe self-oscillating angular cavitation nozzle resonant cavity intrinsic frequency f is determined by the following equation [34]:

$$f = K_N \frac{a}{l} \tag{2}$$

where a denotes the altered wave speed within the resonant cavity of the organ pipe (m/s); l denotes the cavity length of the organ pipe-type resonant cavity (mm); and K_N denotes the modulus coefficient.

The modulus factor K_N is calculated as [35]

$$K_N = F\left\{N, \left(\frac{d_s}{d_1}\right)^2, \left(\frac{d_1}{d_2}\right)^2\right\} = \left\{\begin{array}{ll} \frac{2N-1}{4} & \left[\left(\frac{d_s}{d_1}\right)^2 \geq 1, \left(\frac{d_1}{d_2}\right)^2 \geq 1\right] \\ \frac{N}{2} & \left[\left(\frac{d_s}{d_1}\right)^2 \geq 1, \left(\frac{d_1}{d_2}\right)^2 \geq 1\right] \end{array}\right\} \quad (3)$$

where N denotes the coefficient of oscillation within the resonant cavity, $N = 1$; c denotes the value determined by the experiment; d_s denotes the incident cavity cross-section's diameter (mm); d_1 denotes the resonant cavity cross-section's diameter (mm); and d_2 denotes the conical convergence section at the exit of the resonant cavity's diameter (mm).

The critical self-excited structure frequency f^* of a self-oscillating cavitating jet is expressed in terms of the critical Strouhal number Sr^* of the nozzle as [34]

$$Sr^* = f^* \frac{d_2}{v} \quad (4)$$

where v denotes the nozzle outlet jet velocity (m/s).

Normally the Mach number Ma is expressed as [34]

$$Ma = \frac{v}{a} \quad (5)$$

Associative Equations (4) and (5) lead to

$$f^* = Sr^* \frac{Ma}{d_2} a \quad (6)$$

The organ pipe will generate the most robust resonance when the intrinsic frequency f of the resonant cavity of the internal organ pipe self-oscillating angular cavitation nozzle and the critical self-excited structure frequency f^* of the jet are equal. The joint Equations (2) and (6) can be obtained as follows:

$$\frac{l}{d_2} \approx \frac{K_N}{MaSr^*} \quad (7)$$

According to Crow et al. [36], the results of the jet test show that the conditions that can make the oscillating cavity produce strong resonance are a Mach number Ma in the range of 0.08 to 0.10 and when the critical Strouhal number Sr^* is equal to 0.3 or an integer multiple of 0.3 [34]. At this time, the nozzle cavitation effect is optimal. Therefore, in this study, when the Mach number Ma is taken as 0.08, the critical Strouhal number Sr^* is taken as 0.9, and the resonant cavity exit cone convergence section's diameter d_2 is taken as 2.4 mm, the resonant cavity's length is calculated to be $l = 8.3$ mm.

In this study, considering the high-pressure piping dimensions of the existing test-bed jet system, to achieve a better installation connection and flow stabilization, the internal organ pipe self-oscillating angular cavitation nozzle is taken to have an inlet cavity diameter d_s of 10 mm, and the inlet cavity length l_1 is extended to be 70 mm; according to $(d_s/d_1)^2 \geq 1$, $(d_1/d_2)^2 \geq 1$ and the optimal cross-section shrinkage ratio reasonable value range of 3.5–4.5 of the design criterion [37], the resonant cavity inlet diameter d_1 is calculated to be 5 mm; the expansion angle θ_1 is taken to be 40° considering the jet cavity morphology, axial length, and drag coefficient.

2.2.3. Structural Design of External Conical–Cylindrical Combination Nozzle

The external conical–cylindrical combination nozzle is manufactured according to the jet pump concept and the co-current overlapping cavitation nozzle design principle of artificial submergence. Through experimental studies, researchers found that the structural dimensions of the external nozzle of the artificially submerged cavitation nozzle have a crucial influence on the cavitation jet effect produced [38], so key parameters such as the axial distance H between the inner and outer nozzles, the outlet diameter D_0 of the outer nozzle, and the conical convergence angle θ_2 are defined by the structural parameters of the internal organ pipe self-oscillating angular cavitation nozzle.

The formula for calculating the outlet diameter D_0 of external conical–cylindrical combination nozzles [39,40] is as follows:

$$D_0 = \sqrt{\delta m_y} d_3 \quad (8)$$

where δ denotes the coefficient of nozzle jet area contraction, taking the value of 0.62; m_y denotes the ratio of inner and outer nozzle exit area, taking the value of 9.1; and d_3 denotes the internal nozzle outlet's diffusion section's cross-sectional diameter (mm).

The outer nozzle's optimal outlet length, L_{kopt}, was studied experimentally with different area ratios and operating pressures [39].

$$L_{kopt} = 1.02\sqrt{\delta m_y} d_3 \quad (9)$$

The conical convergence angle θ_2 of the external conical–cylindrical combination nozzle is taken as 30°, and the dimensions of the axial spacing H between the internal and external nozzle exits are taken as 1.5 mm. To reduce energy loss and disturbance during the introduction of low-pressure water flow, the double diversion port adopts an axisymmetric slant design. The angle θ_3 between the double diversion port and the axis line is 60°, and the diversion port's diameter is d_y = 6 mm.

2.3. *Method of Computation and Establishment of Boundary Conditions*

Given that the configuration of the homemade co-current cavitating water jet nozzle features co-current high- and low-pressure water ingress and acknowledging the complexity of the cavitating water jet as an unsteady flow, a high-efficiency pressure solver is employed. In the discrete solution of the specific pressure term equation, the PRESTO! format is selected. For the discrete solution of the volume fraction, the quick format is selected, and the second-order upwind format is used for all other equations; the convergence criterion is set as the residuals $R \leq 10^{-6}$ and the relaxation factor is adopted as the default. The coupled method of pressure and velocity is adopted as the coupled algorithm [41].

The high-pressure water inlet of the internal organ pipe self-oscillating angular cavitation nozzle and the two low-pressure water inlets of the external conical–cylindrical combination nozzle are set as pressure inlets, and the boundary of the flow field outside the outlet of the homemade co-current cavitation water jet nozzle is set as a pressure outlet.

3. Mathematical Pattern for Simulation Computations

3.1. *Multiphase Flow Model*

The internal flow field of the homemade co-current cavitating nozzle is a biphasic flow of water and water vapor. The water and water vapor phases of the medium penetrate and interact with each other. The two-phase medium is approximated to be a continuous medium. A large amount of vortex cavitation bubbles is discretely distributed, and the volume ratio of cavitation bubbles is less than 10% in the whole flow field. This study utilizes the mixture model as a multiphase flow model to simulate the complete flow field of the homemade co-current cavitating nozzle. Moreover, the cavitation model in the mixture model is chosen to perform the interphase mass conversion and make the numerical simulation results closer to reality. A cavitation water jet is a form of fluid motion that must satisfy the continuity, momentum, and second-phase volume fraction equation [42].

(1) The continuity equation for the water and water vapor phases is

$$\frac{\partial \rho_m}{\partial t} + \nabla \cdot (\rho_m v_m) = 0 \tag{10}$$

$$\rho_m = \sum_{k=1}^{2} \alpha_a \rho_a \tag{11}$$

where ρ_m denotes the hybrid-phase density (kg/m^3); v_{m} denotes the hybrid-phase velocity (m/s); t denotes time (s); α_a denotes the volume fraction of the water vapor phase; and ρ_a denotes the density of the water vapor phase (kg/m^3).

(2) The momentum equation for the water and water vapor phases is

$$\frac{\partial(\rho_m v_m)}{\partial t} + \nabla \cdot (\rho_m v_m v_m) = -\nabla p + \nabla \cdot \left[\mu_m\left(\nabla v_m + \nabla {v_m}^{\mathrm{T}}\right)\right] + \rho_m \vec{g} + F + \nabla \cdot \left(\sum_{k=1}^{2} \alpha_a \rho_a v_{dr,a} v_{dr,a}\right) \tag{12}$$

where p denotes the operational pressure (Pa); μ_m denotes the hybrid-phase viscosity (Pa·s); g denotes the acceleration of gravity (m/s^2); F denotes the volume force (Pa/m^3); and $v_{dr,a}$ denotes the water vapor phase's escape velocity (m/s).

(3) The water vapor volume percentage equation is

$$\frac{\partial(\alpha_a \rho_a)}{\partial t} + \nabla \cdot (\alpha_a \rho_a v_m) = -\nabla \cdot (\alpha_a \rho_a v_{dr,a}) + \sum_{a=1}^{n} m_{aw} - m_{wa} \tag{13}$$

where m_{aw} denotes the molecular interaction force from the water vapor phase to the water phase and m_{wa} denotes the molecular interaction force from the water phase to the water vapor phase.

3.2. Turbulence Model

The turbulence model directly affects the jet flow characteristics and the cavitation bubble morphology. Considering the substantial velocity disparity among the inside and outside flow fields of a co-current cavitation water jet nozzle, which results in the formation of a jet shear layer characterized by a high Reynolds number in both the inside and outside flow bundles, the RNG k-ε model, which is more accurate and adaptive in calculating the cavitation vortex structure, was adopted in the present study to solve for simulation [43].

(1) Based on the equation, the kinetic energy of turbulent flow k gives

$$\frac{\partial(\rho_m k)}{\partial t} + \frac{\partial(\rho_m k u_i)}{\partial x_i} = \frac{\partial}{\partial x_j}\left[\alpha_k(\mu_m + \mu_t)\frac{\partial k}{\partial x_j}\right] + G_k + \rho_m \varepsilon \tag{14}$$

where μ_t denotes the turbulent viscosity (Pa·s); G_k denotes the turbulent kinetic energy produced by the average velocity difference (m^2/s^2); and ε denotes the amount of loss of turbulent kinetic energy per unit of fluid (m^2/s^3).

(2) The equation for the dissipation rate ε is

$$\frac{\partial(\rho_m\varepsilon)}{\partial t}+\frac{\partial(\rho_m\varepsilon u_i)}{\partial x_i}=\frac{\partial}{\partial x_j}\left[\alpha_\varepsilon(\mu_m+\mu_t)\frac{\partial\varepsilon}{\partial x_j}\right]+C_{1\varepsilon}\frac{\varepsilon}{k}G_k-C_{2\varepsilon}\rho_m\frac{\varepsilon^2}{k} \quad (15)$$

where $C_{1\varepsilon}$ and $C_{2\varepsilon}$ denote the experimentally measured constant coefficients.

3.3. Cavitation Calculation Model

The Zwart–Gerber–Belamri cavitation model accurately represents the cavitation bubble shape in a highly turbulent cavitation flow environment, offering excellent computational stability and quick resolution. Consequently, this model is employed in this work for computational modeling of the homemade co-current cavitation water jet nozzle's flow field.

The model equation is [44]

When $P \le P_v$,

$$R_e = F_{vap}\frac{3\alpha_{nuc}(1-\alpha_v)\rho_v}{R_B}\sqrt{\frac{2}{3}\frac{(P_v-P)}{\rho_1}} \quad (16)$$

When $P > P_v$,

$$R_c = F_{cond}\frac{3\alpha_v\rho_v}{R_{nuc}}\sqrt{\frac{2}{3}\frac{(P-P_v)}{\rho_1}} \quad (17)$$

where P denotes the actual working condition surrounding pressure (Pa); P_v denotes the saturated pressure within the cavitation bubble (Pa); R_e and R_c denote the rate of water vapor phase creation and condensation; F_{vap} denotes the evaporation phase coefficient; F_{cond} denotes the condensation phase coefficient; α_{nuc} denotes the gas nucleus's volume fraction; R_B denotes the radius of the cavitation bubble (mm); and ρ_1 denotes the water phase's density (kg/m^3).

4. Simulation Results and Discussion

4.1. Finite Element Model

This study utilized Solidworks 2020 software to build a three-dimensional representation of the interior and exterior flow fields of the co-current cavitation water jet nozzle. The Fluent Meshing program was utilized to execute the tetra unstructured meshing for the nozzle's geometric model. The specific division method was as follows: firstly, Hexcore Octree tetrahedral unstructured mesh was used to divide the overall geometric model and secondly, BOI local mesh encryption was carried out for the nozzle outlet cylindrical section, as well as for the areas with complex flow characteristics, such as the intersection area of the inner and outer nozzles in the external flow field, to accurately capture the purpose of the small-scale vortices and air bubbles. Figure 4 illustrates the finite element model.

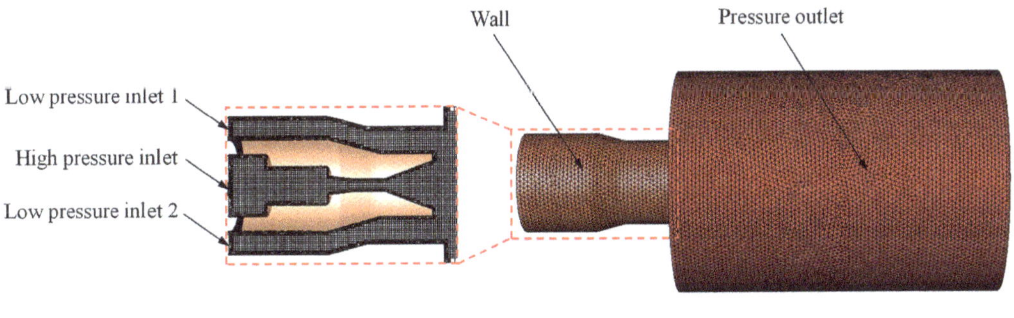

Figure 4. Finite element model.

4.2. Mesh Irrelevance Test

The level of detail of the mesh is essential for the precision of numerical simulation calculations; therefore, in practical fluid dynamics calculation applications, by properly encrypting the mesh, the numerical simulation solution will gradually converge to the final accurate solution. However, the increasing number of meshes will seriously consume computer resources, corresponding to the rising time cost. Therefore, the finite element model needs a mesh irrelevance test to improve computational efficiency while maintaining accuracy. This study investigates the highest vapor volume percentage inside the flow field for five different mesh numbers with an external nozzle input pressure of 3 MPa and an internal cavitation nozzle input pressure of 10 MPa for the co-current cavitation water jet nozzle, and the outcomes are demonstrated in Table 1 after the computation reaches the steady state.

Table 1. Flow field mesh irrelevance test results.

Number of Meshes/Pcs	Maximum Vapor Volume Percentage/%
1,049,513	92.0
1,299,513	94.6
1,549,513	95.2
1,799,513	95.2
2,049,513	95.1

Figure 5 illustrates the variation curves of the maximal vapor volume percentage within the nozzle flow field for different mesh densities. According to the indication in Figure 5, when the quantity of meshes is fewer than 1,549,513, there is a notable disparity existing in the calculated maximal vapor volume percentage within the nozzle flow field; when the quantity of meshes is higher than or equivalent to 1,549,513, the continuing increase in the number of meshes on the calculation of the vapor volume percentage of the results of the difference is negligible. Considering the computational efficiency and accuracy of numerical simulation, the number of mesh cells of the nozzle model is determined to be 1,549,513.

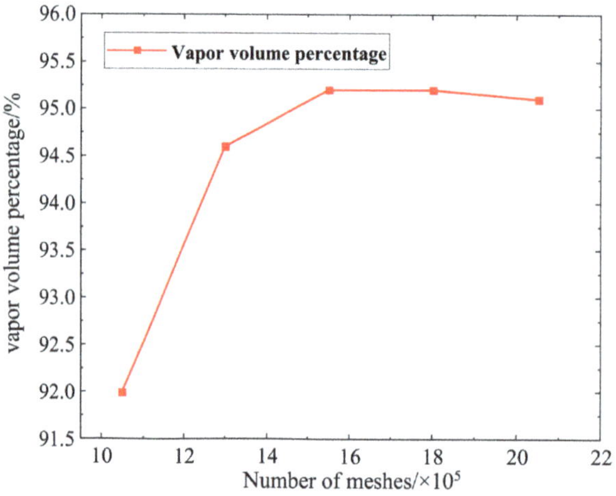

Figure 5. Variation curve of maximum vapor volume percentage for different numbers of meshes.

4.3. Effect of Internal Nozzle Inlet Pressure on Maximum Vapor Volume Percentage

The pressure at the input of the inner nozzle of the homemade co-current cavitation water jet nozzle provides energy for the creation of jet cavitation incipient conditions, and when the inlet pressures of the two low-pressure inlets of the external conical–cylindrical combination nozzle are kept constant, the vapor volume percentage within the overall flow farm and the jet velocity vary with alterations in the input pressure of the internal cavitation nozzle. Figure 6 illustrates the distribution of the highest vapor volume percentage inside the flow farm of the co-current cavitating water jet nozzle at an external nozzle input pressure of 3 MPa and internal nozzle input pressures of 10, 15, 20, 25, and 30 MPa.

As demonstrated in Figure 6, the vapor volume of the homemade co-current cavitation water jet nozzle has an axisymmetric distribution, and the vapor volume percentage gradually rises with the augmentation of the input pressure of the internal cavitating nozzle. Specifically, when water jets via the cylindrical section of the internal cavitation nozzle and enter the position of the wall of the outlet diffusion section, cavitation occurs due to the pressure gradient between the proximal wall portion of the outlet diffusion section and the central jet. When high-pressure water is expelled via the internal nozzle output to create a cavitating jet, the external low-pressure jet wraps the internal high-speed cavitating jet and enters the atmosphere. Owing to the significant velocity differential among the interior and exterior jets, a violent exchange of shear momentum occurs between the jets, causing the fluid within the jet shear region to form a vortex structure under the action of viscous forces and reverse pressure differences. The pressure at the center of the vortex continuously decreases. When the pressure decreases below the saturated vapor pressure of the environment, cavitation will occur and bubbles will be generated, thus promoting the cavitation effect. The tiny bubbles generated in the boundary layer will grow and move downstream with the jet. During the development of the movement, the bubbles will continuously combine with the primary bubbles to achieve a good jet cavitation effect under non-submerged conditions. This is consistent with the findings of Lei [27] and others that artificial submerged nozzle cavitation initially occurs at the exit cylindrical section and diffusion section, and the leading cause of cavitation is the shear interaction between the internal fast jet and the peripheral flow.

Figure 6 also illustrates the cavitation effect of the co-current cavitation water jet nozzle, which exhibits an upward trend with the input pressure of the internal nozzle. The greater the input pressure, the greater the pressure and velocity gradients among the interior high-pressure jet with the exterior low-pressure jet, the more significant the cavitation effect, and the greater the vapor volume percentage. When the input pressures are 10, 15, 20, 25, and 30 MPa, the maximal vapor volume percentages are 95.2%, 95.6%, 95.7%, 95.8%, and 96.4%, respectively, which proves that the homemade co-current cavitation water jet nozzle can produce a good cavitation effect. It is pertinent to note that when the inlet pressure of the internal nozzle rises, the momentum exchange among the interior high-pressure jet with the exterior low-pressure jet in the shear boundary layer becomes more intense, causing the length and range of the cavitation cloud in the flow field to expand. The size of the cavitation cloud penetrates the entire external flow field, proving that the cavitation bubbles produced by the homemade co-current cavitation nozzle can be effectively maintained in the atmosphere, eliminating the application limitation that the cavitation jet can only obtain better results in a submerged environment.

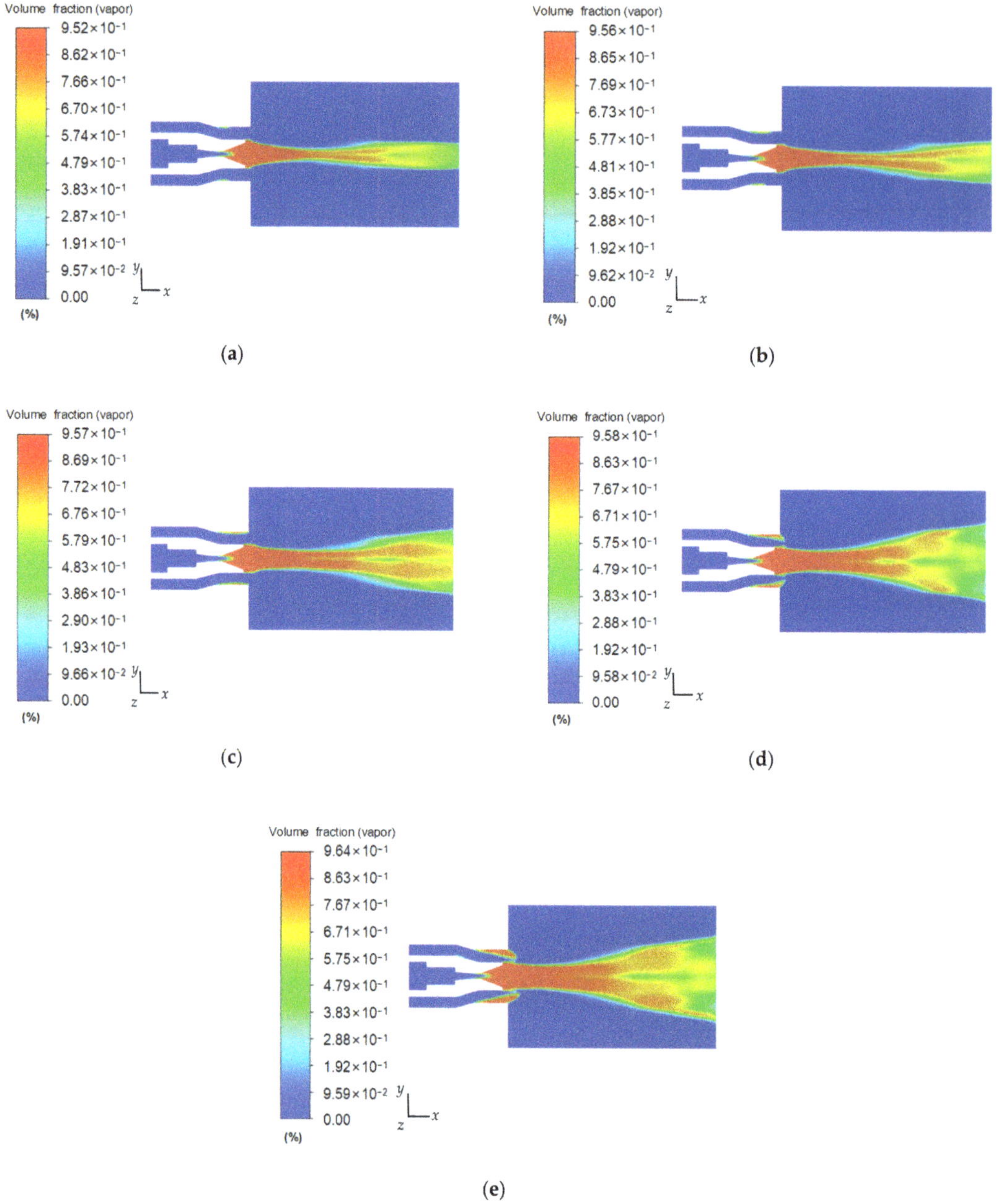

Figure 6. Cloud diagram of maximum vapor volume percentage distributions at (**a**) 10 MPa, (**b**) 15 MPa, (**c**) 20 MPa, (**d**) 25 MPa, and (**e**) 30 MPa.

Figure 6 also demonstrates that the jet undergoes a significant energy exchange with the atmosphere during its downstream movement, gradually destabilizing its coherence. Due to the movement of the jet and changes in ambient pressure, the radial dimensions of the jet's cavitation region decrease at a specific downstream position. As the jet continues to move downstream for a certain distance, the diffusion of the internal and outer jet bodies becomes more and more pronounced and the cavitation bubble group collapses over a wide area. When the input pressure of the internal nozzle exceeds or equals 20 MPa, a distinct cloud shedding can be seen downstream of the external flow field. The greater the inlet pressure of the internal nozzle, the more intense the entrainment of the external low-speed jet by the internal high-speed jet, and cavitation will also occur at the position of the cylindrical section of the external conical–cylindrical combination nozzle outlet, further promoting the generation of cavitation bubbles. This corresponds with the numerical simulation findings of Liu et al. [45], who found that the external nozzle outlet of an artificially submerged nozzle also produces a cavitation effect. The annular jet of the external nozzle can efficiently enlarge the low-pressure region afterward to the outlet of the internal nozzle, effectively expanding the outlet wall of the cavitation nozzle and enhancing the shear effect between the jets. This develops a low-pressure vortex ring upon the jet's interaction with the atmosphere, hence enhancing the cavitation effect.

4.4. *Influence of Input Pressure of Internal Nozzle on Axial Velocity*

Figure 7 gives a cloud diagram of the axial velocity distribution in the flow farm for the co-current cavitation water jet nozzle with the external nozzle input pressure of 3 MPa and the internal nozzle input pressures of 10, 15, 20, 25, and 30 MPa, respectively.

Figure 7 illustrates that the axial velocity distribution for the flow field in the home-made co-current cavitation water jet nozzle exhibits similarity across varying input pressures, the axial velocity distribution cloud map is symmetrical, and the maximum axial velocity of the flow field gradually rises with the augmentation of the inlet pressure of the internal nozzle. Specifically, the low-pressure peripheral flow first accelerates through the conical convergent section of the external conical cylindrical combination nozzle structure. Then, it intersects with the high-pressure cavitating jet at the position of the diffusion section at the outlet of the internal organ pipe self-oscillation angular cavitation nozzle. The low-pressure peripheral flow, on the one hand, plays a role in slowing down the high-pressure cavitation jet into the atmospheric environment of the "package" role; on the other hand, it plays a role in the high-pressure cavitation jet to continue to provide a low-velocity environment of the body of water so as to ensure the cavitation phenomenon produced by the "shear" effect, so that the main body of the jet flow rate by the external conical–cylindrical combination of the nozzle outlet into the atmosphere can maintain a certain distance. As the jet continues to move toward the atmosphere, the internal and external flow beams of the volume of suction and doping phenomenon intensified and the peripheral low-pressure flow beams showed a gradual contraction trend. The isocore velocity of the jet downstream appears to be a "necked" phenomenon, which is due to the vortex structure formed by the significant shear effect between the fluids, making the flow field appear to be a velocity component parallel to the y–z plane. Due to the atmospheric environment's continuous action, the jet core's velocity decays and the pairing and merging process between the vortices intensifies, causing the shear stress between the fluids to weaken gradually and the momentum exchange within the shear layer to become more violent. When the external nozzle input pressure is 3 MPa and the internal nozzle input pressures are 10, 15, 20, 25, and 30 MPa, the maximum axial velocities in the nozzle flow field are 148, 181, 208, 235, and 257 m/s, respectively.

Subsequently, as illustrated in Figure 7, when the internal jet traverses the exit cylindrical section of the internal cavitation nozzle, the jet axial velocity reaches the maximum; when the jet continues to move to the exit diffusion section of the internal cavitation nozzle, due to the shear effect caused by the velocity gradient between the inner and outer flow beams, it results in significant fluctuations in the jet velocity between the interior and exterior flow beams, and the axial velocity begins to decrease. This is consistent with Luo et al. [22], who found that the axial velocity of the flow field of an artificially submerged cavitation jet nozzle shows a significant stratification phenomenon due to the shear effect, and the axial velocity of the flow field is the highest along the position of the jet axis and progressively diminishes along the axis to the two sides.

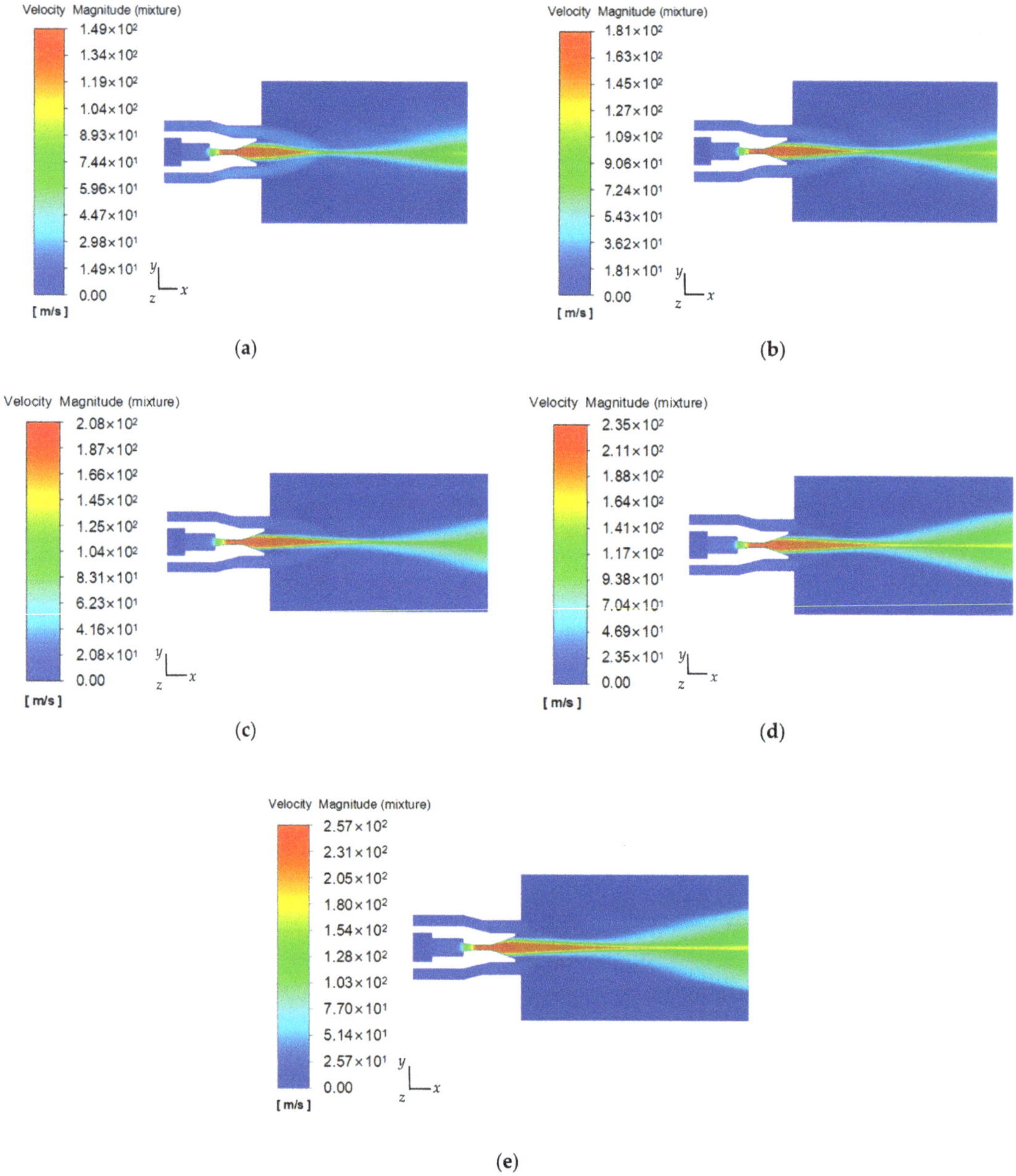

Figure 7. Cloud diagram of axial velocity distribution at different inlet pressures of (**a**) 10 MPa, (**b**) 15 MPa, (**c**) 20 MPa, (**d**) 25 MPa, and (**e**) 30 MPa.

4.5. Effect of Inlet Pressure of Internal Nozzle on Radial Velocity

Figure 8 shows the radial velocity distribution curves of five sections with x = 10, 15, 20, 25, and 30 mm away from the outlet of the internal cavitation nozzle along the jet direction when the inlet pressure of the external nozzle is 3 MPa and the inlet pressure of the internal nozzle is 10, 15, 20, 25, and 30 MPa, respectively.

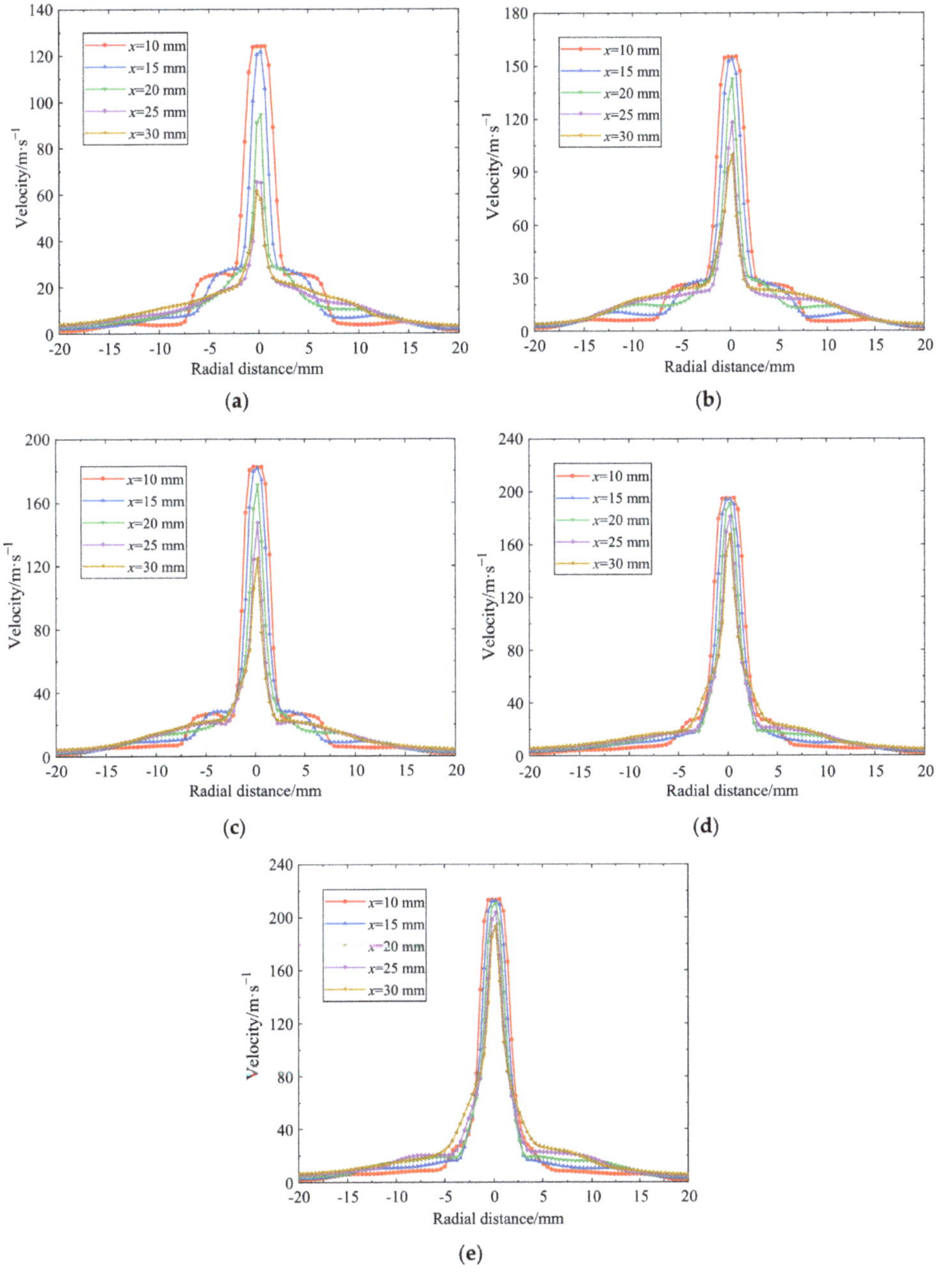

Figure 8. Radial distribution of jet velocity at different inlet pressures of (**a**) 10 MPa, (**b**) 15 MPa, (**c**) 20 MPa, (**d**) 25 MPa, and (**e**) 30 MPa.

As shown in Figure 8, the radial velocity distribution curves of flow fields at different sections exhibit good symmetry, and the jet velocity gradually decreases along the radial direction. At section x = 10 mm, which is closest to the nozzle outlet, the section is in the initial stage of the jet, exhibiting good convergence performance. The jet has a large velocity gradient in this section, and the shear effect between fluids is very strong, which is conducive to the formation of vortex structures and provides a prerequisite for the occurrence of cavitation phenomena. When the jet moves to the x = 30 mm section, the velocity gradient in the section is significantly reduced, and the shear effect is significantly weakened, indicating that the effect of artificial submerged cavitation is negatively correlated with the target distance.

Figure 8 also illustrates that when the input pressure of the internal organ tube self-oscillating angular cavitation nozzle increases, the maximum radial velocity of each radial section is significantly increased and the velocity gradient at the remote section (x = 25, 30 mm) is significantly increased, which is conducive to enhancing the shear effect between the inner and outer flow beams. As the input pressure of the inner nozzle rises, the core velocity of the jet increases from 140 m/s to 240 m/s, which intensifies the mixing and shear effect of the inner and outer fluids and is more conducive to the formation of vortex structures under negative pressure. The internal high-speed jet near the axis will move outward in the radial direction, and the external low-speed jet will move inward in the radial direction, thus forming a state in which the external low-speed jet wraps the internal high-speed jet, delaying the divergence of the jet and increasing the energy and length of the jet beam, which is conducive to maintaining the cavitation effect and erosion performance of the cavitating jet in the atmospheric environment. This aligns with the conclusion of Liu et al. [46] that the low-speed water flow around the periphery wraps the high-speed jet inside at the initial stage of the artificially submerged nozzle jet, which can delay the jet divergence and increase the length of the jet beam.

4.6. Visualization Test to Verify Cavitation Jet

Figure 9a illustrates the visualization test system of the homemade co-current cavitation water jet nozzle. As can be seen from Figure 9a, the system primarily comprises a water storage tank, a low-pressure plunger pump, a horizontal three-plunger high-pressure pump, high-pressure pipes, a homemade co-current cavitation water jet nozzle, a high-speed camera, a strobe light source, a computer, and a visualization box. During the experiment, the external conical–cylindrical combination nozzle of the homemade co-current cavitation water jet nozzle had an inlet pressure of 3.0 MPa, and the internal organ pipe self-oscillating angular cavitation nozzle inlet pressure was selected to be 25 MPa. To fully capture the cavitation cloud's dynamic development, the camera was positioned directly opposite the water jet during filming, and a black background cloth was placed on the visual box wall opposite the camera to improve light absorption. The frame rate of the high-speed camera was set to 4000 fps, the shooting interval was 0.25 ms, and the exposure time was 198 μs. Dimensional calibration was performed to facilitate subsequent quantitative analysis of the shooting results. The high-velocity camera recorded real-time photos of the shape of the high-speed gas–liquid two-phase fluid cavitation cloud injected into the visualization box by a homemade co-current cavitation water jet nozzle. A physical photograph of the system is shown in Figure 9b.

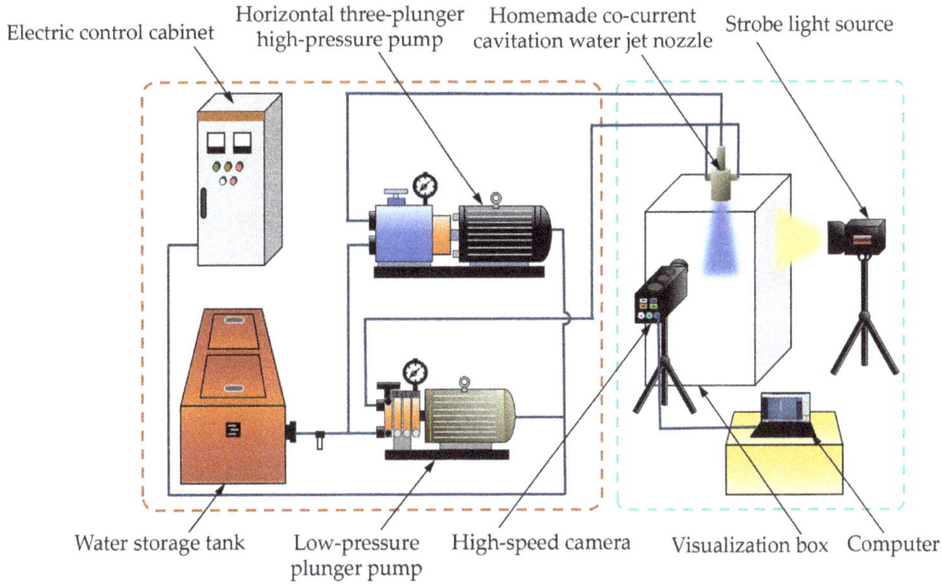

(**a**)

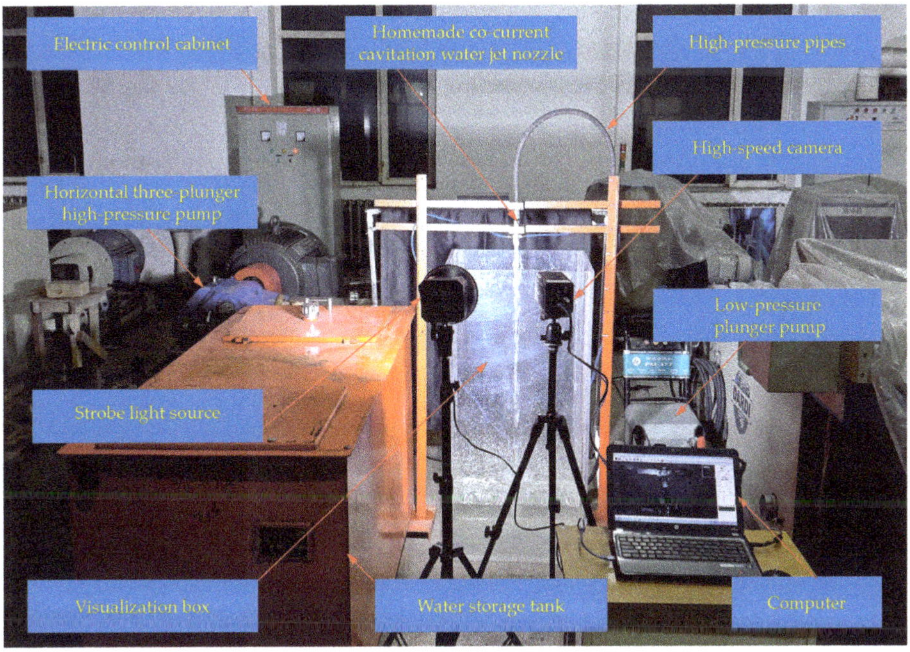

(**b**)

Figure 9. Visualization test system diagram of co-current cavitation water jet nozzle. (**a**) Visualization of system diagram. (**b**) Visualization of system physical photos.

The jet generated by the homemade co-current cavitation water jet nozzle is a free jet discharged into the air, distinguishing it from the traditional submerged cavitation jet introduced into stagnant water. The inner and outer jets of the co-current cavitation water jet are on the same axis and have significant differences in speed and pressure. This causes the inner and outer jets to shear against each other and exchange much energy. Disturbed by external flow beams, conventional image analysis and processing techniques for visualizing cavitation clouds, such as the binary image method and the frame difference method [47], cannot clearly and accurately analyze images of cavitation clouds. Therefore, for the detection of cavitation under non-submerged conditions, researchers have proposed the pseudo-color image processing technique, which has been applied to the cavitation detection of centrifugal pumps and has become a common means of fault diagnosis for centrifugal pumps [48]. Pseudo-color image processing is an image processing technique that assigns specific colors to different gray value regions through specific mathematical relationships, thereby converting grayscale images into color images. The purpose is to improve the discernment of image details and enhance the visualization effect of the image through the color expression of the data. This study used the pseudo-color image processing technique to not only visually reveal the macroscopic structural characteristics of the non-submerged cavitating jet but also to capture the subtle dynamic changes in the non-submerged cavitating cloud. The specific processing method is a spatial domain gray-color transformation method based on a pseudo-color image processing technique. First, the original image is processed in grayscale to obtain a grayscale map. Then, the grayscale map is subdivided into multiple subregions. Each subregion is independently transformed in the red, green, and blue (RGB) color transformations, with the three-color transformation processes being independent of each other, allowing for the fine control of each color, thereby achieving a precise enhancement of image details and obtaining the dynamic change law of the cavitation cloud of a homemade co-current cavitation water jet nozzle in non-submerged conditions.

Figure 10 shows the original images, grayscale images, and pseudo-color images of the cavitation cloud's dynamic development in the co-current cavitation water jet during two consecutive periods, with the external nozzle inlet pressure at 3.0 MPa and the internal nozzle inlet pressure at 25 MPa. As seen in the original images of Figure 10a, the overall shape of the external flow of the co-current cavitation water jet is conical and divergent. The presence of the external flow has a better binding effect on the internal jet. The jet continues to move downstream, maintaining a certain width against the black background. The grayscale images in Figure 10b illustrate that the external low-speed flow of the co-current cavitation water jet encases the internal high-speed jet emanating from the nozzle exit, subsequently dispersing into the atmosphere. A distinct velocity gradient exists between the internal and external flow beams, exacerbating the turbulent shear of the fluid and facilitating the progression of cavitation. The grayscale image presents the luminous irregular white cavitation cloud on a black background, characterized by extensive coverage and density. The pseudo-color images in Figure 10c demonstrate that the black line segments, extending from left to right, delineate the cavitation clouds during the first and second consecutive periods, encapsulating the entire cavitation process from birth to growth, shedding, and collapse within the wireframe. Specifically, at t = 0 ms, the primary cavitation bubbles converge at the outlet of the homemade co-current cavitation water jet nozzle. Since the vortex structure has yet to develop in this area, the cavitation bubble size is relatively small. As the shear force increases, the cavitation bubble enters the development stage and the size and density of the cavitation cloud gradually increase. The cavitation cloud further expands to its maximum width at t = 0.5–1.5 ms. As the turbulence intensifies between the inner and outer flow beams, the combined action of turbulence, coherent vortices, and atmospheric drag affects the cavitation cloud. The end of the cavitation cloud becomes accumulated and asymmetric, eventually leading to a gradual decrease in the size of the cavitation bubble and the shedding and collapse of the cavitation cloud. At the same time, a new upstream cavitation cloud cluster is born and begins to develop. At t = 2 ms, the cavitation cloud of the previous period almost wholly disappears and the cavitation cloud of the next period develops rapidly. The cavitation cloud of the next period will continue the development process of birth–growth–shedding–collapse of the previous period, which is similar to the evolution law of the cavitation cloud obtained by Cui et al. [49] in a visual experiment of the cavitation water jet. The visualization test results indicate that the cavitation bubble development process of the non-submerged cavitation jet generated by the homemade co-current cavitation water jet nozzle exhibits notable continuity, periodicity, and regularity, resulting in an effective cavitation effect and confirming the validity of the numerical simulation results.

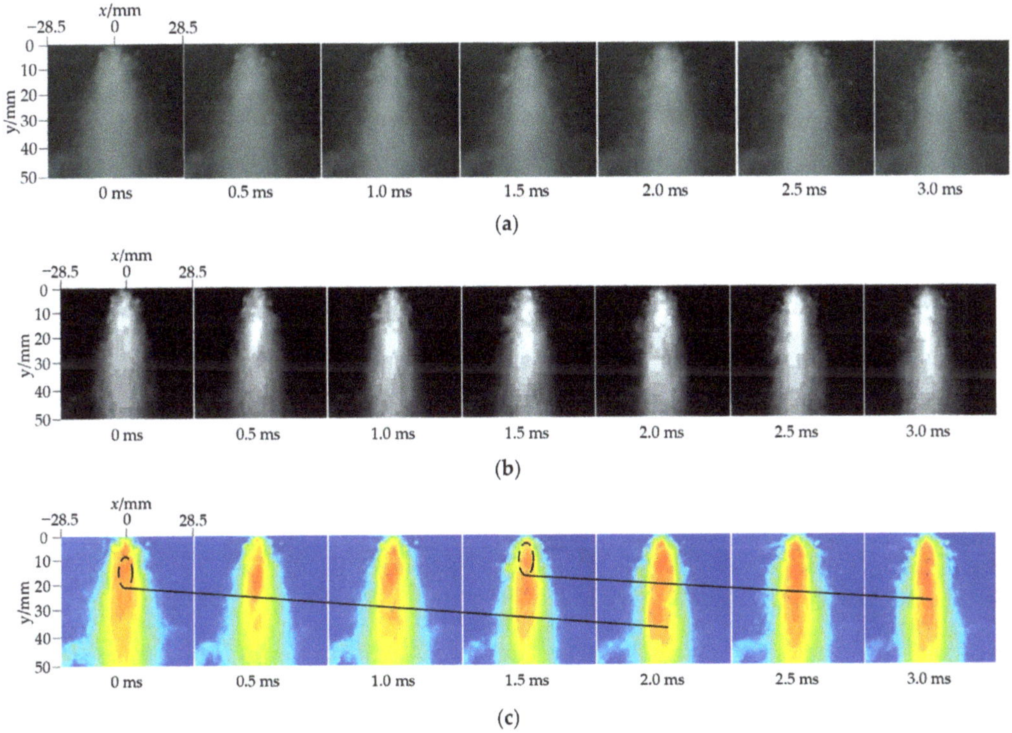

Figure 10. Dynamic evolution process diagram of cavitation cloud. (**a**) Original images. (**b**) Grayscale images. (**c**) Pseudo-color images.

5. Conclusions

This study designed a homemade co-current cavitation water jet nozzle structure, which consists of an internal organ pipe self-oscillating angular cavitation nozzle and an external conical–cylindrical combined nozzle, to enhance the cavitation ability of the cavitating jet in the atmospheric environment (non-submerged condition). The gas–liquid two-phase flow field of the nozzle was solved using the CFD method, and it was compared with the existing artificial submerged cavitating nozzle to verify the rationality of the homemade co-current cavitation water jet nozzle design.

The volume fraction of vapor generated by the co-current cavitation water jet nozzle is distributed axially symmetrically, which has a positive correlation with the internal nozzle's inlet pressure. When the external nozzle input pressure is 3 MPa and the internal nozzle input pressures are 10, 15, 20, 25, and 30 MPa, the vapor volume percentages generated by the nozzle are 95.2%, 95.6%, 95.7%, 95.9%, and 96.4%, respectively, which shows that the homemade co-current cavitation water jet nozzle has a good cavitation performance.

The pseudo-color images of the cavitation cloud, generated by the homemade co-current cavitating water jet nozzle, reveal a significant periodic dynamic change rule following the application of the spatial domain gray-color transformation method. One change period is 1.5 ms, which indicates that the nozzle can generate a stable cavitating jet in a non-submerged condition, verifying the accuracy of the CFD numerical simulation results.

Author Contributions: Conceptualization, methodology, validation, formal analysis, investigation, resources, writing—original draft preparation, visualization, supervision, C.G.; methodology, supervision, software, validation, investigation, data curation, H.S.; conceptualization, investigation, resources, writing—review and editing, project administration, funding acquisition, X.D.; investigation, visualization, Y.J. All authors have read and agreed to the published version of the manuscript.

Funding: This research was supported by the National Natural Science Foundation of China (no. 52075151) and the Natural Science Foundation of Heilongjiang Province (no. LH2020E123).

Data Availability Statement: The data presented in this study are available upon request from the corresponding author. As the data in this paper belong to the National Natural Science Foundation of China, it involves related privacy and is not owned by individuals.

Conflicts of Interest: The authors declare no conflicts of interest.

References

1. Kan, K.; Binama, M.; Chen, H.X.; Zheng, Y.; Zhou, D.Q.; Su, W.T.; Muhirwa, A. Pump as turbine cavitation performance for both conventional and reverse operating modes: A review. *Renew. Sustain. Energy Rev.* **2022**, *168*, 112786. [CrossRef]
2. Long, Y.; An, C.; Zhu, R.S.; Chen, J.P. Research on hydrodynamics of high velocity regions in a water-jet pump based on experimental and numerical calculations at different cavitation conditions. *Phys. Fluids* **2021**, *33*, 045124. [CrossRef]
3. Xu, S.J.; Wang, J.; Cai, B.H.; Cheng, H.Y.; Ji, B.; Zhang, Z.T.; Long, X.P. Investigation on cavitation initiation in jet pump cavitation reactors with special emphasis on two mechanisms of cavitation initiation. *Phys. Fluids* **2022**, *34*, 013308. [CrossRef]
4. Usta, O.; Köksal, C.S.; Korkut, E. A systematic study into the cavitation erosion test for marine propeller materials by cavitating jet technique. *Ocean Eng.* **2023**, *284*, 115252. [CrossRef]
5. Li, Z.F. Criteria for jet cavitation and cavitation jet drilling. *Int. J. Rock Mech. Min. Sci.* **2014**, *71*, 204–207. [CrossRef]
6. Zhong, X.; Dong, J.M.; Liu, M.S.; Meng, R.X.; Li, S.; Pan, X.X. Experimental study on ship fouling cleaning by ultrasonic-enhanced submerged cavitation jet: A preliminary study. *Ocean Eng.* **2022**, *258*, 111844. [CrossRef]
7. Dong, J.M.; Li, S.; Meng, R.X.; Zhong, X.; Pan, X.X. Research on Cavitation Characteristics of Two-Throat Nozzle Submerged Jet. *Appl. Sci.* **2022**, *12*, 536. [CrossRef]
8. Soyama, H. Cavitation Peening: A Review. *Metals* **2020**, *10*, 270. [CrossRef]
9. Soyama, H.; Korsunsky, A.M. A critical comparative review of cavitation peening and other surface peening methods. *J. Mater. Process. Technol.* **2022**, *305*, 117586. [CrossRef]
10. Wang, Y.; Zhu, C.; Li, F.Z.; Yu, C.; Chen, S.J.; Wu, J.Y. Processing characteristics and loading principle on micro blanking of cavitation jet. *Int. J. Adv. Manuf. Technol.* **2024**, *133*, 1317–1329. [CrossRef]
11. Dong, X.; Li, J.L.; Li, Z.; Jin, X.M. Numerical simulation of flow field in cavitation water jet nozzle. *J. Heilongjiang Univ. Sci. Technol.* **2018**, *28*, 560–565.
12. Yang, Z.G.; Yi, S.J.; Zhao, S.X.; Shi, W. Influence of the inlet pressure on impulse characteristics of aluminum samples. *Surf. Rev. Lett.* **2021**, *28*, 2150110. [CrossRef]
13. Soyama, H. Effect of nozzle geometry on a standard cavitation erosion test using a cavitating jet. *Wear* **2013**, *297*, 895–902. [CrossRef]
14. Panda, A.; Anisimov, V.M.; Anisimov, V.V.; Pandova, I.; Klymenko, A.; Ermakov, P. Cavitation nozzles with expansion chamber. *MM Sci. J.* **2022**, *2022*, 6020–6025. [CrossRef]
15. Dong, W.Q.; Yao, L.G.; Luo, W.L. Numerical Simulation of Flow Field of Submerged Angular Cavitation Nozzle. *Appl. Sci.* **2023**, *13*, 613. [CrossRef]
16. Yang, M.G.; Xiao, S.N.; Kang, C.; Wang, Y.L. Effect of geometrical parameters on submerged cavitation jet discharged from profiled central-body nozzle. *Chin. J. Mech. Eng. (Engl. Ed.)* **2013**, *26*, 476–482. [CrossRef]

17. Cao, Z.P.; Liu, H.L.; Xie, C.X. Cavitation and Cleaning Characteristics of New Annular Cavitation Nozzles. *Hydraul. Pneum.* **2020**, *2*, 136–144.
18. Soyama, H. Introduction of compressive residual stress using a cavitating jet in air. *J. Eng. Mater. Technol. ASME* **2004**, *126*, 123–128. [CrossRef]
19. Marcon, A.; Melkote, S.N.; Yoda, M. Effect of nozzle size scaling in co-flow water cavitation jet peening. *J. Manuf. Process.* **2018**, *31*, 372–381. [CrossRef]
20. Vidvans, A.N.; Melkote, S.N.; Sanders, D.G. Enhancing Cavitation Intensity in Co-Flow Water Cavitation Peening With Organ Pipe Nozzles. *J. Eng. Mater. Technol. ASME* **2021**, *143*, 071010. [CrossRef]
21. Wu, S.H.; Yang, Y.F.; Shi, W.D.; Wang, G.W.; Wu, X.L.; Wu, R. Performance of 7075 Aluminum Alloy Strengthened by Artificially Submerged Cavitation Water Jet Peening. *Surf. Technol.* **2024**, *53*, 190–199. [CrossRef]
22. Luo, Y.; Zang, J.Y.; Zheng, H.X. Flow Field and Gas Field Distribution of Non-Submerged Cavitation Water Jet Based on Dual-Nozzle with Concentric Configuration. *Water* **2023**, *15*, 2904. [CrossRef]
23. Fan, C.X.; Li, D.; Kang, Y.; Zhang, H.T. Effect of low-speed waterjet pressure on the rock-breaking performance of unsubmerged cavitating abrasive waterjet. *Pet. Sci.* **2024**, *21*, 2638–2649. [CrossRef]
24. Straka, W.A.; Meyer, R.S.; Fontaine, A.A.; Welz, J.P. Cavitation inception in quiescent and co-flow nozzle jets. *J. Hydrodyn. Ser. B* **2010**, *22*, 771–777. [CrossRef]
25. Marcon, A.; Melkote, S.N.; Castle, J.; Sanders, D.G.; Yoda, M. Effect of jet velocity in co-flow water cavitation jet peening. *Wear* **2016**, *360*, 38–50. [CrossRef]
26. Yang, Q.; Xu, H.; Li, Y.G.; Zhang, W.H.; Wei, Y.J.; Wang, C. Topology and cavitation number characteristics of the gaseous jet-induced tail cavity under co-flow. *Phys. Fluids* **2022**, *34*, 013303. [CrossRef]
27. Lei, C.C.; Deng, S.S.; Guan, J.F.; Zhang, S.X.; Guo, G.D.; Zhang, T.F.; Guo, L.H. Flow Field Numerical Simulation of Artificial Submerged Cavitating Water Jet Based on Fluent. In Proceedings of the International Conference on Computer Science and Technology (CST), Shenzhen, China, 8–10 January 2017; pp. 1142–1150.
28. Sekyi-Ansah, J.; Wang, Y.; Tan, Z.R.; Zhu, J.; Li, F.Z. The Dynamic Evolution of Cavitation Vacuolar Cloud with High-Speed Camera. *Arab. J. Sci. Eng.* **2020**, *45*, 4907–4919. [CrossRef]
29. Du, P.; Lv, S.S. Research on the performance of manually flooded cavitation nozzles. *Mech. Eng. Autom.* **2022**, *10*, 1–3, 6.
30. Li, H.A.; Chen, J.W.; Guo, J.; Zhu, H.; Lin, Y.; Ge, H. A Study of Cavitation Erosion in Artificial Submerged Water Jets. *Appl. Sci.* **2024**, *14*, 4804. [CrossRef]
31. Zheng, P. CFD simulation of dual-cavity self-resonating cavitating nozzle. *Sadhana-Acad. Proc. Eng. Sci.* **2024**, *49*, 125. [CrossRef]
32. Onishi, T.; Li, K.Z.; Ji, H.; Peng, G.Y. Shedding of Cavitation Clouds in an Orifice Nozzle. *Fluids* **2024**, *9*, 156. [CrossRef]
33. Wang, P.F.; Han, H.; Liu, R.H.; Gao, R.Z.; Wu, G.G. Effect of outlet diameter on atomization characteristics and dust reduction performance of X-swirl pressure nozzle. *Process Saf. Environ. Prot.* **2020**, *137*, 340–351. [CrossRef]
34. Li, G.S.; Shen, Z.H. *Theory and Application of Self-Oscillating Cavitation Jet*; China University of Petroleum Press: Beijing, China, 2008.
35. Wang, X.L.; Kang, Y.; Zhang, M.D.; Yuan, M.; Li, D. The Effects of the Downstream Contraction Ratio of Organ-Pipe Nozzle on the Pressure Oscillations of Self-Resonating Waterjets. *Energies* **2018**, *11*, 3137. [CrossRef]
36. Crow, S.C.; Champagne, F.H. Orderly structure in jet turbulence. *J. Fluid Mech.* **1971**, *48*, 547. [CrossRef]
37. Li, G.S.; Shen, Z.H. Experimental study of the effect of submerged self-oscillating cavitating jets on rock under normal pressure. *J. East China Pet. Inst.* **1987**, *11*, 12–22.
38. Hu, S.G.; Zhu, M.Z. Current status of cavitation water jet research and its application. *J. East China Univ. Technol.* **1996**, *18*, 1–8.
39. Lu, H.Q. *Theory and Application of Jet Pump Technology*; Water Resources and Electric Power Press: Beijing, China, 1989; p. 316.
40. Zhang, H.Y.; Zou, D.H.; Yang, X.L.; Mou, J.G.; Zhou, Q.W.; Xu, M.S. Liquid-Gas Jet Pump: A Review. *Energies* **2022**, *15*, 6978. [CrossRef]
41. Dong, X.; Guo, C.H.; Peng, D.Q.; Jiang, Y. Numerical Simulation of Air-Water-Flake Graphite Triple-Phase Flow Field in a Homemade Double-Nozzle Jet Micro-Bubble Generator. *Minerals* **2024**, *14*, 533. [CrossRef]
42. Mangela, D.B.; Raj, A.K.; Mohanan, K.P.; Jayaraj, S. Computational analysis of two-phase flow in a converging-diverging propulsion nozzle suitable for water transport. *Sadhana-Acad. Proc. Eng. Sci.* **2022**, *47*, 63. [CrossRef]
43. Fluent, A. *Ansys Fluent Theory Guide*; Ansys Inc.: Canonsburg, PA, USA, 2020; pp. 51–53.
44. Paul, D.; Agarwal, H.; Ponangi, B.R. CFD analysis of two-phase cavitating flow in a centrifugal pump with an inducer. *Heat Transf.* **2020**, *49*, 3854–3881. [CrossRef]
45. Liu, H.L.; Cao, Z.P.; Xie, C.X.; Chen, G.P.; Li, D.F.; Wang, J.W. Research on erosion characteristics of a novel cavitation nozzle under nonsubmerged condition. *Proc. Inst. Mech. Eng. Part C J. Mech. Eng. Sci.* **2021**, *235*, 988–998. [CrossRef]
46. Liu, B.X.; Wang, H.F.; Chen, Z.G. Simulation and study on non-submerged cavitating jet from a novel composite nozzle. *Min. Mach.* **2018**, *46*, 58–62. [CrossRef]
47. Yang, Y.F.; Li, W.; Shi, W.D.; Zhou, L.; Zhang, W.Q. Experimental Study on the Unsteady Characteristics and the Impact Performance of a High-Pressure Submerged Cavitation Jet. *Shock Vib.* **2020**, *2020*, 1701843. [CrossRef]

48. Goel, A.K.; Naikan, V.N.A. Exploring the Diagnostic Potential of Infrared Thermography for Experimental Assessment of Cavitation and Air Entrainment-induced Faults in Centrifugal Pumps. *J. Appl. Fluid Mech.* **2024**, *17*, 352–369. [CrossRef]
49. Cui, Y.Y.; Zhao, M.J.; Ding, Q.M.; Cheng, B. Study on Dynamic Evolution and Erosion Characteristics of Cavitation Clouds in Submerged Cavitating Water Jets. *J. Mar. Sci. Eng.* **2024**, *12*, 641. [CrossRef]

applied sciences

MDPI

Review

Research Progress and Engineering Applications of Viscous Fluid Mechanics

Jianjun Peng [1,2,*], Run Feng [1], Meng Xue [1], Erhao Zhou [1], Junhua Wang [1,2], Zhidan Zhong [1,2] and Xiangchen Ku [1,2,*]

[1] School of Mechanical and Electrical Engineering, Henan University of Science and Technology, Luoyang 471003, China; fr@stu.haust.edu.cn (R.F.); xm@stu.haust.edu.cn (M.X.); zeh@stu.haust.edu.cn (E.Z.); wangjh@haust.edu.cn (J.W.); zzd@haust.edu.cn (Z.Z.)
[2] Collaborative Innovation Center of Henan Province for High-End Bearing, Luoyang 471003, China
* Correspondence: pjj@haust.edu.cn (J.P.); kuxch@126.com (X.K.)

Abstract: This paper systematically reviews the development of viscous fluid mechanics and expounds the current research status of viscous fluid mechanics from the aspects of permeability experiments and research on viscous fluids, particle distribution and movement, droplet impact, modern computer technology simulation, etc. This paper summarizes the application status of viscous fluid mechanics theory in aerospace, fluid simulation, bioengineering, pipeline transportation, and other engineering fields, analyzes the influence of multiphysical field coupling, complex flow, and new materials on viscous fluid mechanics, and puts forward its development trend in algorithms, simulation, biomedicine, aerospace, and other fields.

Keywords: viscous fluids; experiments; simulations; research advances

1. Introduction

Viscous fluids are actual fluids with viscous properties, which are manifested by the nature of the fluid preventing its own flow, and this viscosity is caused by the interaction forces between the fluid molecules. As an important branch of fluid mechanics, viscous fluid mechanics studies the behavior and properties of fluids under the influence of viscous effects [1–3]. The most classical study is Newton's assumption that the shear stress inside the fluid is proportional to the velocity gradient, thus introducing the concept of kinetic viscosity, which is applicable to many fluids in practical applications, such as water and air [4,5]. The formula and schematic diagram are shown below (Figure 1).

$$F = \mu \frac{d_x}{d_y} A, \tag{1}$$

However, in some engineering and natural phenomena, the viscous effects of fluids may significantly affect the kinematic properties of fluids, such as viscous hysteresis phenomena, the deformation and diffusion behaviors of fluids, etc. [6–9]. The study of viscous fluids is not limited to traditional Newtonian fluid models but also includes non-Newtonian fluids such as plastic, elastic, and viscoelastic fluids, which exhibit complex nonlinear behavior at high strain rates or high stresses. In addition to viscous fluids, viscous fluid theory also applies to fluids that are not viscous in nature (such as air, water, ethanol) which exhibit high viscosity when flowing in confined spaces (such as through membrane pores, porous materials, droplets hitting surfaces, etc.). Other relevant formulas and criteria for viscous fluid motion include the Froude criterion, Euler equation, N-S equation, Reynolds criterion, laminar flow, turbulence, turbulent flow, boundary layer theory, etc. [10,11].

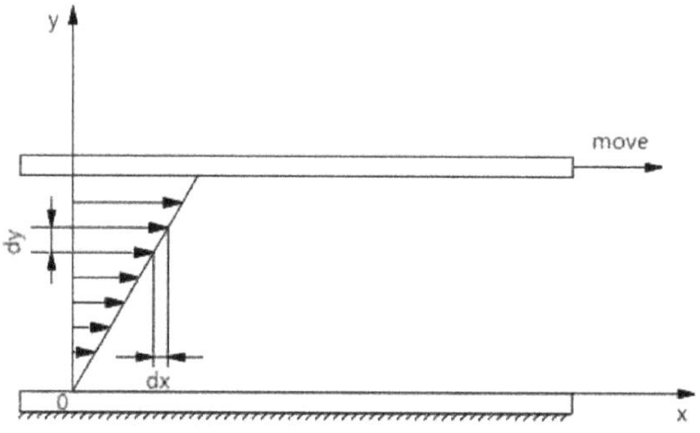

Figure 1. Velocity gradient diagram of liquid interlayer motion. The lower plate in the channel is fixed, and the upper plate shifts to the right to drive the liquid movement. The proportional coefficient in the formula is called the dynamic viscosity coefficient, or the internal friction coefficient, the force direction of the viscous force F is the negative direction of the X-axis in the figure, and A is the unit area of the liquid.

In a wide range of fields of scientific exploration and engineering, many problems involve the permeability of viscous fluids moving on the surface or inside various materials and their specific modalities. For example, in the field of aircraft design, an in-depth understanding of the turbulence characteristics on aerodynamic surfaces and the potential impact of air viscosity on aircraft performance is essential [12,13]; in river engineering practice, an accurate assessment of the erosion of water flow on the riverbed and riverbank structures is equally indispensable [14]. In the pipeline transportation of chemical viscous liquids such as oil and liquefied natural gas, it is necessary to fully consider the potential impact of the viscosity of these liquids on the transportation efficiency [15]. To solve these problems, it is indispensable to use experiments and theoretical calculations to accurately determine various key parameters of viscous fluids [16]. In this paper, the basic theory, mathematical models, and practical applications of viscous fluid dynamics are systematically explored. This article summarizes the recent progress in the direction of experimental exploration and simulation research on the state of motion of viscous fluids, including the current problems and limitations. Meanwhile, the future development trend is envisioned, aiming to provide guidance and reference for research in the field of viscous fluid kinematic states.

2. Advances in the Study of the State of Motion of Viscous Fluids

Here are a few common ways to investigate the properties of viscous liquids and how they have evolved.

2.1. Experimentation and Research

2.1.1. Experiments and Studies on the Permeability of Media to Viscous Fluids

Experiments and studies on the permeability of media to viscous fluids usually focus on investigating the permeability properties of fluids in porous media and their dynamic behavioral properties. Such studies have important applications in petroleum engineering, geology, civil engineering, and other fields [17–19]. The following are some common related contents and methods: (1) Porous media and permeability: porous media can be rocks, soils, rocks, or other materials in petroleum reservoirs. Their pore structure and connectivity determine the permeability of the media to the viscous fluid. (2) Permeability experimental methods: (i) the pressure-driven method (conventional method) involves applying a pressure difference to push the fluid through the porous medium, measuring the rate of fluid passage and pressure changes to calculate the permeability; (ii) the constant pressure method involves maintaining a constant pressure difference, measuring the rate of fluid permeation and permeability; and (iii) the resistance method involves determining the permeability properties by measuring the resistance of the fluid to passage through the porous medium. (3) The influence of the permeability of viscous fluids: the permeability properties of viscous fluids (e.g., petroleum, polymer solutions, etc.) are usually different from those of nonviscous fluids because viscosity affects the flow rate and distribution of the fluid. (4) Experimental equipment and techniques: permeability determination equipment (e.g., permeameters, permeability test benches, etc.) is used, usually in conjunction with pressure transducers, flow meters, and other instruments for accurate measurements. (5) Numerical simulation: Computational fluid dynamics (CFD) and numerical simulation methods are used to simulate the permeability behavior of viscous fluid in porous media and predict the permeability and fluid distribution of the medium to the viscous fluid [20–23]. These studies are important for optimizing oil extraction, groundwater resource management, environmental pollution prevention, and other fields and can help engineers and scientists to better understand and control the behavior of viscous fluids in complex media [24].

Currently, Darcy's law and Forchheimer's law are the main laws for determining the permeability of viscous fluids. Darcy's law describes the relationship between the rate of permeation of a viscous fluid in a porous medium and the gradient of osmotic pressure under steady-state seepage conditions [25]. However, this law is mainly used for the determination of macroscopic permeability k under constant conditions, and it is questionable whether the fine channel changes under microscopic conditions affect this law. Forchheimer's law as a model for describing non-Darcy's flow is based on the overall average of the porous media model and does not take into account the specific effect of the microstructure inside the medium, which may limit its accuracy in some delicate porous media [26,27]. Both of these problems have been explored by researchers worldwide.

In this regard, Prof. Howard A. Stone et al. [28] of Princeton University carried out an experimental study on the pressure-driven passage of viscous fluids through a superelastic porous membrane. They fabricated their own experimental setup, which is schematically shown in Figure 2. The study focused on the permeability of thin PDMS membranes under convective conditions, and the experimental conclusions pointed out that macroscopic deformation can significantly affect the microstructure of soft materials and their transport properties.

Stafie, N, et al. [29] investigated the effect of polydimethylsiloxane (PDMS) crosslinking on the permeation properties of polyacrylonitrile (PAN)/PDMS nanofiltration (NF) composite membranes and found results that were inconsistent with the results of the swelling of the dense, This may be due to the lower pore intrusion rate of the composite film prepared at 10/2 compared to 10/1, or due to the heterogeneity of the silicone network. Delli, ML, (Delli, Mohana L.) et al. [30] investigated the change in permeability in a viscous material, natural gas-containing hydrate sediments, by a series of experiments, using different initial water saturations to reach hydrate saturations of up to 45% and measuring the corresponding fluid permeability during steady-state flow, which gradually decreases as the hydrate saturation increases.

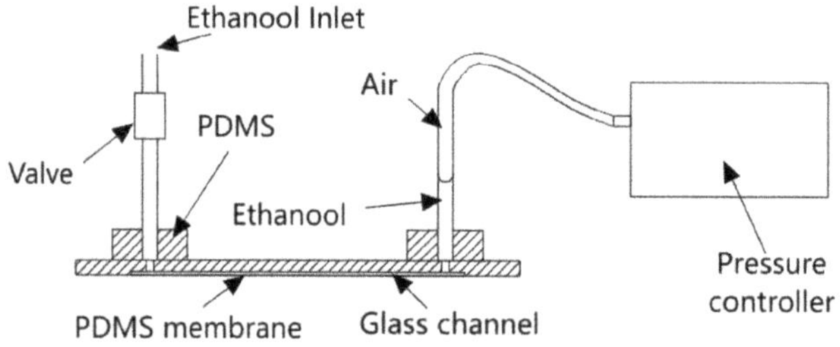

Figure 2. The section diagram of the experimental device which used alcohol and air to explore the permeability of thin PDMS film under convective conditions. The glass channel under the device is 5 mm wide, 0.5 mm deep, and 45 mm long, and a PDMS film with a thickness of 60–200 μm is attached to the lower plate surface to form a closed channel. The two pipelines importing ethanol solution and air are fixed by thick PDMS blocks. The valve of the ethanol pipeline on the left controls the input and exit of ethanol, and the air pipeline on the right is connected with the pressure control pump to adjust the pressure in the channel. Finally, the position of the semilunar surface of the PDMS membrane over time is recorded by a high-speed camera during the experiment.

Wang Shifang et al. [31] studied the spherical flow of Newtonian fluids in porous media based on fractal theory and technology. Three-dimensional spherical porous media percolation is widely found in hydrocarbon reservoirs. When only a small part of the reservoir is opened, the flow area has spherical radial flow, that is, the fluid flows from the center of the outward wellbore. The fractal permeability model of spherical porous media percolation is shown in Figure 3. The experiment reveals the physical mechanism affecting the spherical permeability and verifies the correctness of the fractal model of spherical permeability.

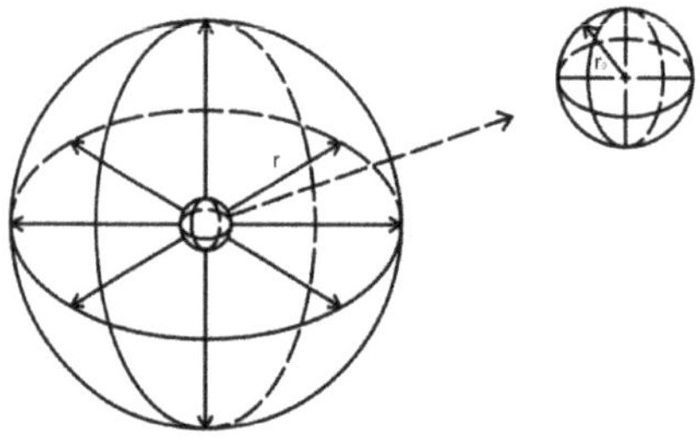

Figure 3. Seepage radius diagram of spherical porous media. In the figure, r is the radial distance from a point in the reservoir to the center of the well, and r_0 is the radius of the well. The results show that the spherical permeability increases with the increase in pore surface integral shape dimension and porosity. It decreases with the increase in tortuosity fractal dimension and with increasing radial distance [31].

Wang Mang et al. [32] conducted an experimental study of the effect of different fluids and clay minerals on the permeability of rocks; the conclusions of their study showed that the nature of the viscous fluid itself has an influence on the macroscopic permeability of the permeable material, and the physical and chemical relationship between the fluid and the permeable material should also be considered along with the liquid permeability. The experimental device of media permeability to liquid designed by their research is shown in Figure 4. The experimental device provides a new experimental method for studying the permeability of viscous liquid.

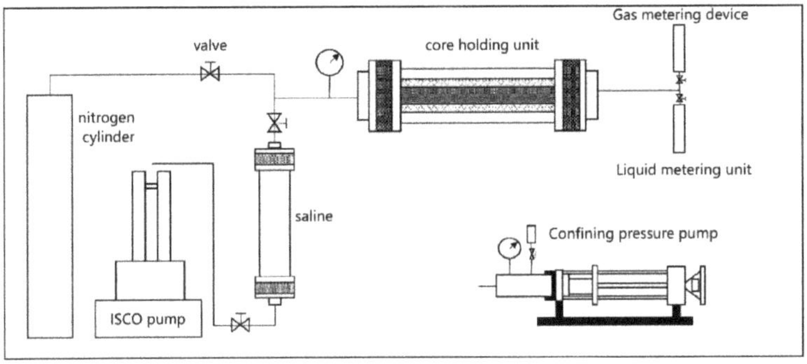

Figure 4. Schematic diagram of permeability measuring device. The gas and liquid metering device in the device is used to help determine the target permeability. When the experimental device measures the gas permeability, a high-pressure nitrogen cylinder and a core holder are connected to the gas metering device. To measure the permeability of brine, the liquid injection ISCO pump, an intermediate container filled with brine, and a core holder are connected to the liquid metering device. Firstly, nitrogen was used to measure the permeability of all sandstone test blocks, and the device was kept in a dry state during the whole process. After gas measurement, salt water was used to measure the permeability. After decompressing the piston at the bottom of the sealed tank, the test block was removed from the tank and weighed, and the permeability was determined by calculating saturation using porosity and brine density [32].

Sun Weitao et al. [33] analyzed the complex fluid model of a tight reservoir and its applicability, using glycerol and methane gas for the pore structure of the tight reservoir at a micro- and nano-scale, and analyzed the theoretical basis and scope of application of the complex fluid model of the tight reservoir in detail. Their study concluded that the Cauchy correction model is a correction model used when the fluid does not meet the hydrodynamic characteristics of the continuous medium, and the Forchheimer model is a correction model that can be used when the fluid turbulence phenomenon is serious and its fluid flow still meets the hydrodynamic characteristics of the continuous medium. Ding Liang et al. [34], in order to study the permeability characteristics of non-homogeneous porous media, using high-temperature alloy-powder-sintered porous media as the experimental substrate, with pure air, water, and ethanol as the circulation medium, conducted an experimental study of the fluid through the porous media plate driving force and its movement characteristics of the relationship between the non-homogeneous porous media in the state of the single-phase flow of permeability characteristics. The experiments resulted in an obvious non-Darcy phenomenon, in which the viscous interaction between different fluids and the microporous solid wall is different, which may lead to the driving force required by high-viscous fluids being smaller than that of low-viscous fluids, resulting in inconsistency with the permeability determination of Darcy's theorem.

In the above experiments on the study of Cauchy's criterion and Forchheimer's law for the permeability of viscous liquids, the conclusions generally agree that Forchheimer's law is applicable to the case of non-Darcy flow, which is widely applicable, but it is relatively difficult to determine the parameters; Cauchy's law of hydrodynamics has great difficulty in describing complex fluids and in obtaining experimental data, and Cauchy's law will not be effective in the case of a change in microstructure due to deformation. With Cauchy's law of hydrodynamics, it is very difficult to describe complex fluids and to obtain experimental data, and Cauchy's law is invalid in the case of microstructures changed by deformation. The experimental method and experimental setup of the above experiments can provide a good idea for the subsequent related research, which is of great significance to promote the development of research on the permeability of viscous fluids.

2.1.2. Experiments and Studies on Droplet Impacts in Viscous Fluids

The study of viscous fluid droplet impact is an important branch in the field of fluid mechanics and surface phenomena, which involves the dynamic behaviors and physical processes of droplets in contact with solid or liquid surfaces under different conditions. These studies not only have far-reaching impacts on basic science but also have important implications in engineering applications. The dynamic process of droplet impact involves the deformation, expansion, and sputtering of droplets [35,36]. As a droplet approaches a surface, its shape is affected by surface tension, viscous, and inertial forces, which in turn affect the post-impact response. For example, higher surface tension and viscosity cause droplets to be more difficult to expand, which may cause droplet buckling or sputtering, while lower surface tension may contribute to droplets expanding more easily and forming a thin film. The phenomena of energy transfer and dissipation during droplet impact are also crucial [37]. Upon impact, energy is transferred between the droplet and the contacting surface, and some of this energy is either dissipated as heat or converted into the kinetic energy of the droplet, affecting the subsequent behavior and morphological evolution of the droplet. The understanding of these processes is particularly critical for engineering applications such as inkjet printing, coating technologies, and microfluidic system design [38,39]. The study of viscous fluid droplet impingement is also of direct practical relevance for solving problems in a variety of real-life and industrial applications. For example, in medical devices, understanding the expansion and adhesion behavior of droplets on various surfaces can help design more efficient fluid dispensing systems; in coating technology, controlling droplet impingement and expansion can improve the uniformity and quality of coatings [40]. Overall, the study of viscous fluid droplet impingement not only deepens our understanding of fluid dynamics and surface phenomena but also provides an important foundation and guidance for modern engineering technologies and applications.

The droplet impact velocity, angle, and impact energy have important effects on the experimental results. Too high or too low impact energy may lead to the atypical behavior of droplets or unrepeatable experimental results [41]. If the droplet cannot be fully wetted on contact with a solid surface or if there is a large contact angle, this will affect the deformation and rebound behavior of the droplet, making it difficult for the experimental results to accurately reflect the theoretical model [42]. Also, the Weber number of the liquid can greatly affect the collision results of viscous droplets [43].

With the development of technology and the increase in the demand for precise control, high-frame-rate high-speed video cameras for viscous droplet collision experiments to detect the experimental collision process, simulation software to simulate the collision process, and other methods of scientific research and investigation have been gradually and widely used [44–46]. For example, in the study of the impact of the Weber number on droplet collision, experimental controllable variables such as substrate elasticity, impact velocity, liquid viscosity, and liquid material can be systematically changed, and then, high-speed cameras can be used to clearly observe the phenomena of the diffusion, rebound, retraction, and wetting of droplets with different viscosities after impacting different substrates. By observing the experimental images, we can further study the influence of liquid viscosity on the critical rebound Weber number, maximum spreading, and contact time and speculate the mathematical model of the change in the Weber number before and after the collision [47–49]. In this line of thinking, Liu Hailong et al. [50], based on high-speed micro-digital camera technology and the establishment of the finite element numerical model of droplet impact on a superhydrophobic wall, carried out experiments to study the dynamic behavior of Newtonian fluid droplets with different viscosities (0.9–27.7 m Pa·s) impacting on a superhydrophobic wall (the static contact angle is 158°).

Figure 5 shows the experimental process of the impact of Newtonian fluid droplets with different viscosity on the superhydrophobic wall captured by the high-speed camera when the We number is 28.2. It can be seen that the performance of the high-speed camera can observe a very clear droplet collision process.

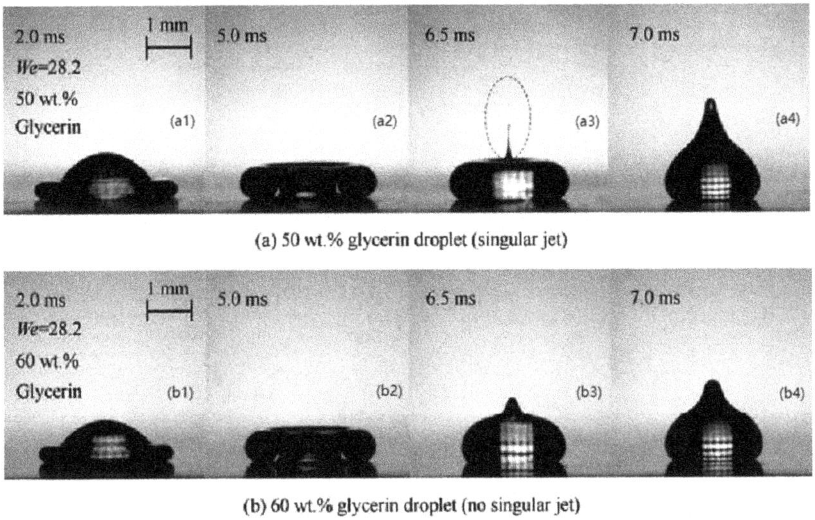

Figure 5. Snapshots of Newtonian fluid droplets with different viscosities impacting on the superhydrophobic surface at *We* = 28.2 (all of the scale bars represent 1 mm): (**a1–a4**) show the process of a 50 wt.% glycerol droplet impacting the superhydrophobic wall, and (**b1–b4**) show the process image of the initial state, 5.0 ms, 6.5 ms, and 7.0 ms, respectively, after the process of a 60 wt.% glycerol droplet impacting the superhydrophobic wall [50].

On the issue that the impact velocity, angle, and impact energy of droplets have important effects on the experimental results, Chengyao Wang et al. [51] investigated the impact kinetic behaviors of water, anhydrous ethanol, and low-surface-energy droplets of different viscosities on hydrophobic surfaces through experimental visualization and explored and analyzed the effects of the three influencing factors, namely, the impact velocity, the surface wettability, and the droplet viscosity, on the impact kinetics of viscous droplets. The impact dynamics of viscous droplets were explored and analyzed. Qian Wenwei et al. [52] studied the impact process of viscous liquid and dense particle jets against a wall with a background of gasifier gasification technology and concluded that when the particle size is small, the dense particle impact flow shows a particle film morphology similar to the liquid impact; the higher the solid content rate, the lower the particle rebound probability and the more homogeneous the velocity and energy of the particles, which makes it easier to present the particle film morphology and other conclusions. In order to analyze the process of collision aggregation of unequally large droplets on walls with different wettability, Xing Lei et al. [53] prepared unequally large droplets with a volume ratio of 1:2 and carried out collision experiments in which small droplets on substrates with different wettability hit large droplets from different contact angles using high-speed camera technology to analyze the kinetic behaviors of droplet collision aggregation. The experimental device is shown in Figure 6, which provides a new idea of device construction for droplet collision experiments.

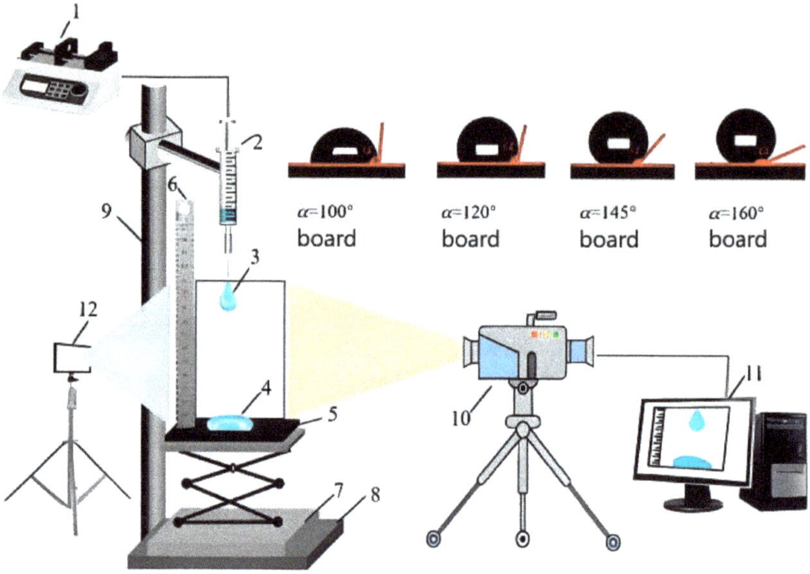

Figure 6. Experimental system installation diagram. In the specific experiment, the lifting table surface is first controlled to fall, and the distance between the needle and the substrate is 3 mm (determined by the scale). The distilled water is controlled by a microinjection system and gathers into droplets at the needle point, breaking the quasi-static equilibrium and falling freely onto the substrate. Then, the table rises by 3 mm, and the second drop and the first drop collide and coalesce, making a large drop twice the volume of a single drop. After the ascending table reaches the preset falling height h, small droplets are released from the tip of the needle in the same vertical position to ensure that the collision droplet is centered on the fixed droplet. The experiment is repeated four times for each group. The whole process is monitored in real time by high-speed cameras and computers, taking images of the impact [53]. 1—microfluidic injection system; 2—2 mL syringe; 3—impingement droplet; 4—solidified droplet; 5—substrate; 6—scale; 7—high-precision three-direction elevator; 8—optical platform; 9—large-stroke elevator; 10—high-speed camera; 11—computer; 12—light source.

There exists the problem that the droplets cannot be completely wetted when contacting the solid surface, or there exists a large contact angle, which affects the deformation and rebound behavior of the droplets, making it difficult for the experimental results to accurately reflect the theoretical model. Zong Shaoqiang et al. [54] used a high-speed shadow imaging technique to investigate the evolution of viscous droplets (glycerol aqueous solution) impacting the drying and pre-wetting mesh surface to form liquid fingers and fragmentation and concluded that an increase in droplet viscosity and a decrease in mesh width would inhibit the complete penetration of droplets into the drying mesh surface and so on. Xing Lei et al. [55] used a high-speed camera to set up an experimental device to study the different changes in droplets impacting plates with different angles and humidity at different We numbers, and their experimental results provide some reference bases for complex fluid motion to a certain extent.

In addition to some droplet collision deposition and flight problems, researchers used viscous liquids of different materials and sprinkler heads of different specifications to simulate the spraying and collision process of viscoelastic fluid through experiments and observed the influence of viscoelasticity caused by the presence of polymers in the solution on the droplet deposition process and the influence of sprinkler heads on the atomization rupture of the solution with high accuracy [56,57].

In the above experiments, the construction of the experimental setup and the collection and prediction of the results are mostly based on the new century high-speed camera, which can instantly turn the droplet collision into thousands of frames for the researcher to study the mechanism of droplet collision in a more detailed and accurate way. Therefore, high-speed cameras play an important role in the study of viscous liquids and the motion of some other fluids.

2.1.3. Experiments and Studies on the Flat-Plate Gap Method

The flat-plate gap method for viscous fluids is an important experimental method for investigating the properties of viscous fluids flowing between two parallel solid surfaces. This traditional experiment not only reveals the behavior of the fluid at the microscopic scale but also measures important parameters such as its rheological properties and viscosity [58]. In the experiments of the flat-plate gap method for viscous fluids, it is first necessary to prepare two parallel solid surfaces, usually flat plates or pipe walls. The flatness of these surfaces has an important effect on the experimental results, so high-precision experimental equipment is usually used [59,60]. Next, an appropriate sample of viscous fluid is selected for the experiment. The choice of viscous fluid can be determined according to the specific purpose and requirements of the experiment [61], e.g., fluid samples with different viscosities and different rheological properties can be selected. While the experiment is in progress, a small gap, usually between a few micrometers and a few millimeters, is set between two flat plates. The size of this gap will directly affect the flow characteristics of the fluid therein. The fluid can be guided to flow in the gap by applying some pressure or by applying an electric field (for galvanic fluids) [62,63]. During the experiments, the flow data of the fluid in the gap are obtained by means of appropriate measuring devices such as pressure transducers, flowmeters, or laser measuring devices. The commonly measured parameters include the flow velocity distribution, shear stress distribution, and pressure distribution of the fluid [64–66].

The measurement of these data is important for inferring the viscosity, flow resistance, and flow behavior of viscous fluids in confined spaces, but this experimental method has the problems that the control of experimental conditions requires high-precision equipment and operating techniques, and the stability of experimental environments is highly required. Therefore, Prausová, H, et al. [67] investigated the laminar, turbulent, and transitional flow regimes of a viscous fluid in a narrow gap between two substrates with a parallel distance of 2 mm and obtained the static and stagnant pressure ratio distributions along the axis of the channel using both optical and gas-dynamics methods. Christensen, AH, et al. [68] developed a mathematical model to investigate the flow behavior of viscous fluid in a narrow gap between two rectangular elastomeric plates with a distance of 2 mm. The model is based on low-Reynolds-number fluid dynamics and linear plate theory, and it is concluded that the outflow volume of a viscous fluid increases linearly with increasing pressure. Lin, JR, et al. [69] investigated the combined effect of the properties of viscous liquids subjected to pressure as well as non-Newtonian coupled stresses by developing a mathematical model of piezoelectric-viscous dependence and non-Newtonian coupled stresses in a wide parallel rectangular-plate extruded membrane based on Stokes microcontinuum medium theory and incorporating factors related to the variation in viscosity with pressure. Vijayakumar, B, et al. [70] designed a set of mathematical models based on the Cartesian coordinate system for the effects of piezo-viscous dependence and non-Newtonian coupling stress on the performance of cylindrical-perforated-plate extrusion-film bearings. The digital model is shown in Figure 7.

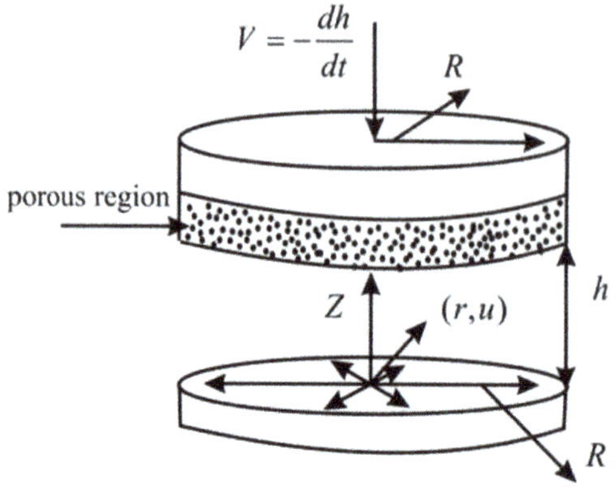

Figure 7. Systematic diagram of squeeze-film bearing with porous circular plate. The digital model assumes that the velocity of the upper plate in the negative direction of the Z-axis of the Cartesian coordinate system drives the collision and extrusion between the parallel circular perforated plate and the fixed lower plate and predicts the bearing characteristics of the extrusion die of the parallel circular perforated plate. The pressure results under different viscosity–pressure parameters are numerically calculated and compared with the isoviscous coupling stress and Newtonian lubricant. The results show that the pressure of the piezo-viscous non-Newtonian fluid is significantly higher than that of the isoviscous Newtonian fluid and the isoviscous non-Newtonian fluid [70].

Neufeld, JA, et al. [71] studied the diffusive motion patterns between a thin film of viscous fluid and a thin elastomeric sheet on a plate by extrusion using a permeable horizontal plate and showed that an air gap forms underneath the viscous fluid as it is extruded through the plate until the fluid re-flows onto the plate. For very light or stiff elastic sheets, the air gap contacts may extend well beyond the fluid discharge zone. Zamanov, AD, et al. [72] envisioned a fluid viscoelastic model consisting of a viscoelastic plate and a half-plane filled with a viscous fluid, which is shown schematically in Figure 8. Using the exact equations of viscoelastic dynamics for describing the motion of the plate and the linearized N-S equations for describing the fluid flow, the problem of forced vibration of the system under plane strain was investigated.

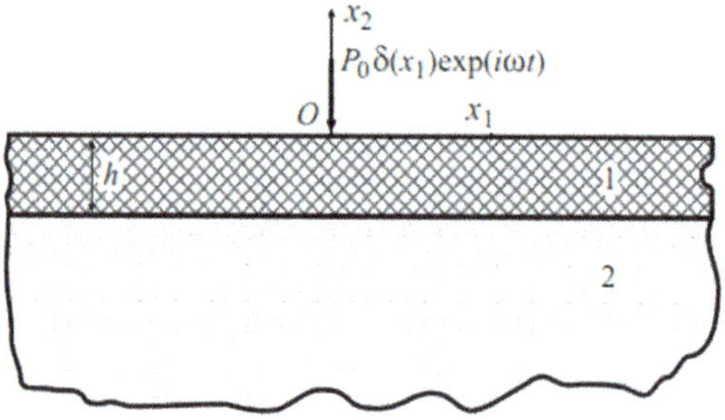

Figure 8. Schematic of a hydroviscoelastic system: 1—viscoelastic plate; 2—viscous fluid. Based on the Cartesian coordinate system $Ox_1x_2x_3$ on the surface of the upper plate, a hydroviscoelastic system consisting of a plate of viscoelastic material and a half-plane filled with compressible viscous fluid is established. In this system of coordinates, the plate occupies the area $\{|x_1| < \infty, -\infty < x_2 < 0, |x_3| < \infty\}$ and the fluid occupies the area $\{|x_1| < \infty, -\infty < x_2 < -h, |x_3| < \infty\}$, where h is the thickness of the plate. It is assumed that the system is subjected to a normal harmonic force of amplitude p, uniformly distributed on axis Ox_3. The plane strain state is generated in the plate, and the liquid plane flow occurs in the plane Ox_1x_2 [72].

As can be seen from the above researchers' studies on the motion of viscous fluids between plate gaps, this type of experiment is complicated by the experimental setup, so much so that most of the researchers use digital models for predictive studies. In terms of scientific research, this approach helps to gain a deeper understanding of the flow properties of viscous fluids at the microscopic scale, explore the complex rheological behavior of fluids, and save experimental costs. In engineering applications, this digital model prediction method can provide important data support for designing viscous fluid lubrication systems, optimizing fluid transport pipelines, and so on.

2.1.4. Experiments and Studies on the Oscillation Method

The viscosity of a fluid is measured or its state of motion is studied using an oscillating-disk- or oscillating-cylinder-type geometry or by external shock conduction, which is inferred from the frequency of the oscillations and the response of the oscillator. The basic principle is to use external excitation to cause oscillations within or on the surface of the fluid and to reveal the fluid's hydrodynamic behavior and oscillatory modes by observing and analyzing the oscillatory phenomena [73,74]. This method is commonly used to study polymer solutions or complex liquid systems containing additives [75]. As a research tool, the viscous liquid oscillation method, although it has an important role in fluid dynamics and engineering applications, has some potential problems and challenges, which include complex fluid behavior and nonlinear effects, difficulties in ensuring the control and stability of the experimental conditions, the large amount of resultant data requiring complex data processing algorithms and technical support, and the lack of a precise interpretation for the phenomena of liquid oscillations that have been observed and recorded, which makes the exact interpretation and physical significance of the different oscillation modes difficult to analyze [76–78].

The results of oscillations of single geometries in the experiments of the oscillation law of viscous fluids often do not allow a clear determination of the exercises between the two. Nuriev, AN, et al. [79] studied the low-amplitude and high-frequency oscillations induced by the reciprocating motion of ellipsoidal cylinders with different axial ratios in viscous fluids based on the law of harmonics, and the limits of the asymptotic theory were determined through the observation of oscillations of the surface of the liquid, which is more diverse than the study of the oscillatory behavior of individual geometries in viscous liquids. In addition to the study of viscous liquids using geometries probed into the liquid to produce displacements or oscillations, the use of acoustic waves to impart oscillations to the liquid is a very worthwhile research method. In view of this problem, Kremer, J et al. [80] implemented non-contact positioning suspension of viscous liquids such as silicone oil and propanol based on acoustic suspension, and placed them under the same ambient pressure to measure the surface tension and viscosity of each liquid. The construction diagram of the experimental device is shown in Figure 9.

Brenn, G, et al. [81], based on the transient deformation induced by oscillation, built an experimental setup and predicted and analyzed the small-amplitude axisymmetric-shape oscillations of a viscoelastic droplet suspended in a gas and improved the Jeffreys model to obtain a new characteristic equation for the complex frequency, which quantifies the effects of the two time scales, stress relaxation and deformation delay time, on the oscillatory behavior of the droplet. Due to the traditional experimental research method of building physical platforms having many inconveniences, such as the stability being difficult to control and the experimental results being difficult to obtain, many researchers use computer technology to digitize the physical environment of the object of study in order to control the variables, and the digital simulation makes it easy to extract the results. Feng, L, et al. [82] took into account the emergence of severe precession and frequency shifts in experiments in the measurement of the surface tension of liquid introducing non-negligible measurement errors in the determination, so the effects of the oscillation amplitude of viscous droplets and the viscous effect on the droplet oscillation were investigated by three-dimensional modeling simulation. The modeling and simulation are shown in Figures 10 and 11.

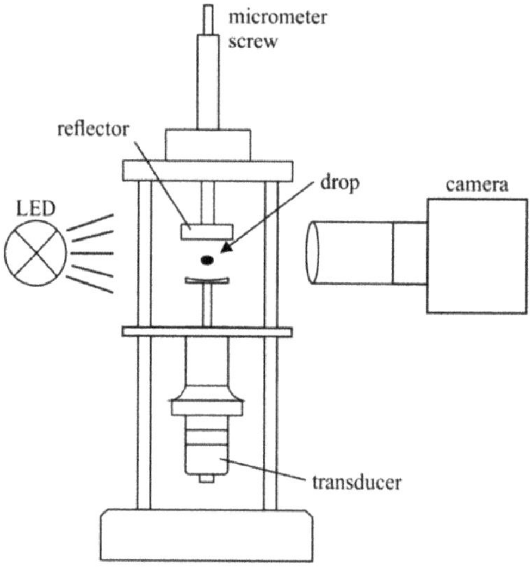

Figure 9. Acoustic levitation system. Below the device is the acoustic levitator, which is mainly composed of a reflector and a piezoelectric transducer. The suspension voltage is controlled by the amplifier, and the ultrasonic wave is reflected by the reflector to ensure that the droplet is in a relatively stable suspension state in the air. The surface tension and viscosity of the acoustically suspended droplets are measured by the direct amplitude modulation of the transducer-driven signal. A slow-motion high-speed camera is used to record the experimental process and observe whether the droplet rotates during the measurement. LED lights provide enough light to clearly observe and record the suspension and oscillation of small droplets [80].

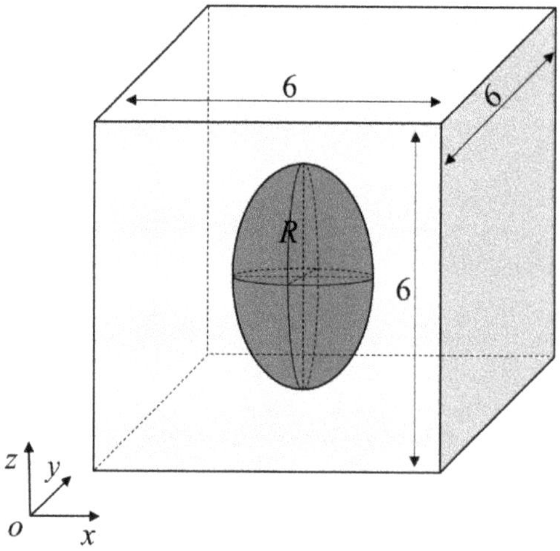

Figure 10. Schematic diagram of an oscillating droplet. The computational domain of the model is a $6 \times 6 \times 6$ cube with a drop of radius R = 1 placed in the center [82].

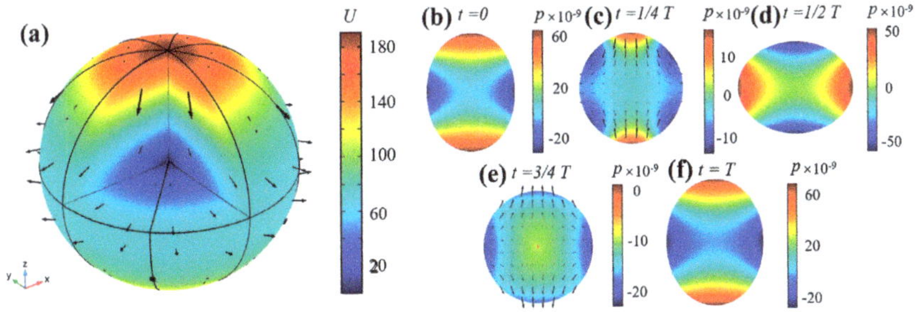

Figure 11. Typical oscillation behaviors of droplet with We = 1×10^{-5} and $\varepsilon = 0.2$. (**a**) Three-dimensional view of the flow field of an oscillating droplet. (**b–f**) Variations in flow and hydrodynamic pressure field of droplet in zox slice. Figure (**b**) is the initial stage of the simulation, (**c**) is the state of 1/4 period, (**d**) is the state diagram of 1/2 period, (**e**) is the state diagram of 3/4 period, and (**f**) is the final state [82].

The research methods introduced above have mainly covered three research methods, i.e., the traditional contact oscillation method, the non-contact oscillation method, and the digital simulation method. Each of the three methods has its own advantages and shortcomings, and the following Table 1 makes a simple comparison between these methods for the readers to refer to.

Table 1. Comparison of research methods for viscous fluids.

Method	Cost	Accuracy	Time Consumption	Future Development Potential
Conventional contact shock method	High	Relatively low	High	Low
Non-contact oscillation method	High	Moderate	High	Relatively low
Digital simulation method	Low	High	Low	High

When conducting relevant experiments, researchers should consider factors such as money, time, and target results for research exploration. However, I think that with the development of computers, simulation will be the mainstay of research methods in the future. The table showing the comparison between the traditional physical experiment method and the computer simulation method is also of reference significance for the previous several summaries.

2.2. An Investigation of the Simulation of the State of Motion of Viscous Fluids by Modern Computer Technology

The phenomenon of viscous liquid fluid exists widely in nature, daily life, and industrial production, and since ancient times, scientists have never stopped investigating the process of viscous fluid motion. In the 21st century, with the significant improvement of computer computing power, the application of computer and general software to simulation analysis of fluid dynamics, a new fluid motion research method, has become popular [83,84]. This kind of software adopts numerical computation methods programmed to directly solve the flow master equations (e.g., Euler's equations or Navier–Stokes equations) in order to reveal the laws of various flow phenomena. They integrate the knowledge of various disciplines such as mathematics, physics, computer science, fluid mechanics, 3D modeling, scientific visualization, etc., and the simulation methods are mainly classified into mesh and meshless methods [85,86]. At present, the mainstream fluid simulation and meshing software and algorithm programs include ANSYS Fluent, MATLAB, FEM, MST, Flo EFD, AICFD, Polyflow, RealFlow, AcuSolve, XFlow, PowerFLOW, VirtualFlow, Flowmaster, etc. These fluid simulation software and algorithms cover a wide range of applications from 2D model simulation to 3D model simulation [87–89]. Scientific researchers simulate and analyze the motion process of fluids based on mathematical models or three-dimensional physical models, which are more efficient, accurate, and intuitive compared to the pure mathematical computation method of predicting fluid motion in the last century, which provides a certain theoretical basis for the judgment of the motion phenomena of viscous fluids and has become an important method of research on fluid-related motion [90–92].

2.2.1. Computerized Algorithmic Solution of Viscous Fluids

Currently, many scientists use computers to compile algorithmic programs to help them find the relationship between various variables from the huge amount of experimental data and at the same time set the program to iterate autonomously through the computer, and the computational speed far exceeds that of the original handwritten calculations. Mentzoni et al. [93] proposed an oscillatory flow condition relevant to marine applications based on the Navier–Stokes equation. A two-dimensional viscous flow field model for porous-plate hydrodynamics was developed and analytically solved using MATLAB, and the accuracy of the model was experimentally verified. Kyung Hyun Ahn et al. [94] investigated the particulate system of solid particles suspended in a viscoelastic medium based on a novel numerical algorithm of the lattice Boltzmann method (LBM). Zhu Liang et al. [95] used the Giesekus model of viscoelastic fluid and spherical rigid particles as the research object and used the quasi-constant algorithm based on the relative motion model to investigate the distribution characteristics of the lateral lift force of the particles in the channel; the computational model is shown in Figure 12.

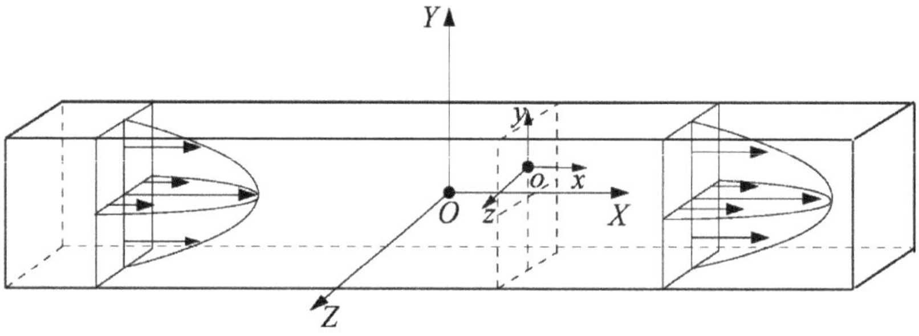

Figure 12. Schematic diagram of calculation model. In the figure, a single spherical particle moves in a square straight tube with a section side length H. In a fixed coordinate system O − XYZ, the flow field is unsteady. When the particle moves to a stable state driven by the fluid, its translation velocity and rotation velocity remain unchanged. When the relative motion model is used, the particles are placed in several transverse positions of the channel section, and the coordinate system o − xyz is established with the geometric center of the particles as the origin, and the particles are translated at the same speed. In this translation coordinate system o − xyz, the particle motion can be regarded as having only rotation velocity and no translation velocity; the movement velocity of the channel wall and the particle translation velocity are reversed. In the translational coordinate system o − xyz, the whole flow field is transformed into a fixed boundary and relatively steady flow, which can be solved numerically by using a fixed grid [95].

Hanqing Li et al. [96], for Oldroyd-B-type viscoelastic fluid, used the lattice Boltzmann method (LBM) to obtain the flow field distributions of the viscoelastic fluid in the dilatational and constrictive flow paths (the flow paths are shown in Figure 13) and numerically analyzed the settling process and characterization of the particles in dilatational and constrictive flows based on the point-source particle model.

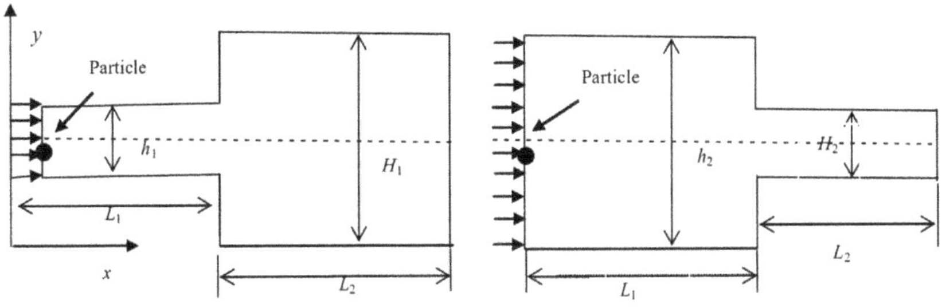

Figure 13. Schematic diagram of geometric model of channel expansion and contraction. In the figure, the left is a two-dimensional expanding flow channel geometry model, and the right is a two-dimensional shrinking flow channel geometry model. The black dot in the figure represents the starting position of particle movement, and the dimensions of each geometry in the figure are L_1 = 120 mm, L_2 = 120 mm, h_1 = 40 mm, H_1 = 120 mm, h_2 = 120 mm, and H_2 = 40 mm [96].

Different from the forward algorithm which is based on using known conditions to obtain unknown results, Tsepelev, IA, [97], with the final state of the target fluid being highly viscous and incompressible, used algorithms to deduce the unknown initial state of the fluid in reverse, using OpenFOAM software based on linear algebra to develop a new iterative algorithm in reverse, with the computational error and input data error on the opposite side. Xu, XY, et al. [98] improved the SPH algorithm in order to simulate the transient free-surface flow of viscous and viscoelastic fluids, which improved the accuracy and stability of the algorithm. It can accurately and stably simulate the flow and oscillating pressure field at the free surface of viscous and viscoelastic fluids, which provides a more efficient method for the study of numerical algorithms for viscous fluids.

2.2.2. Simulation of Viscous Fluid Motion by Computer Techniques

In addition to the use of advanced computer technology for the numerical calculation of the model of the fluid, we can also use simulation software to model the target fluid environment so that the resulting results contain static and dynamic processes and a series of numerical values and plots, which can be more intuitive for researchers to understand the process of fluid motion [99]. The basic solution process is as follows: Firstly, the continuous fluid dynamics equations (e.g., the Navier–Stokes equations) are discretized into discrete points or cells using finite difference, finite volume, and finite element methods, which are common discretization methods, so that they can be processed by computers [100,101]. This is followed by the selection of appropriate time integration methods (e.g., explicit or implicit methods, Runge–Kutta methods, etc.) that allow the computer to solve the fluid dynamics evolution over time. These methods take into account stability and accuracy, especially for nonlinear and unsteady flows. The fluid region then needs to be divided into discrete cells or grids according to the geometry and boundary conditions of the fluid domain, and the grid density or local encryption, grid type selection, etc., are set according to the size and needs of the model [102–104]. An example is shown in Figure 14, which provides a simple reference for your understanding. The figure is a two-dimensional model of a pipeline. After setting the entrance, exit, and wall surface, dividing the grid, and setting various physical fields and load conditions, simulation can be performed.

Finally, appropriate iterative solvers (e.g., conjugate gradient method, stability iteration method, etc.) are to be used to solve the discretized fluid dynamics equations. These methods usually take into account the nonlinear, viscous, and inertial properties of the fluid [105]. Of course, the correct treatment of boundary conditions is crucial when computing fluid dynamics. This includes the setting of velocities, pressures, or other physical quantities at the boundary of the fluid flow to ensure the accuracy and physical soundness of the numerical solution [106]. For larger models, more complex structures, or simulation environments for fluid simulation, it is also necessary to reasonably allocate the number of computer running cores for the solution in order to prevent insufficient arithmetic power leading to the inability to give simulation results after a large number of time iterations [107].

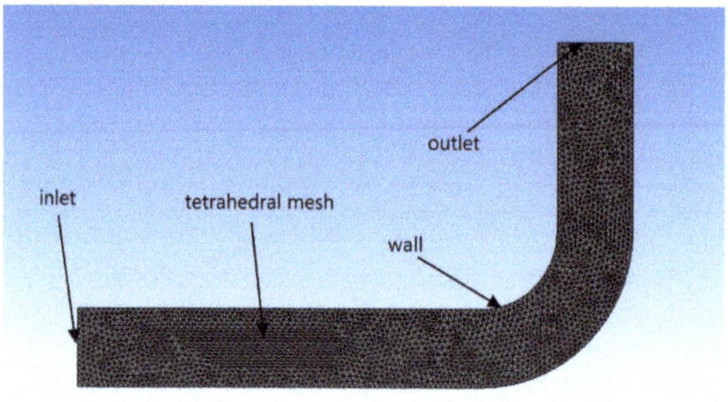

Figure 14. Grid layout and boundary condition setting of liquid-conveying pipeline.

As far as simulation technology in the study of viscous fluid motion is concerned, Xu Baiping and others [108] used the transient finite element method (FEM) combined with the mesh superposition technique (MST) with high-viscosity Newtonian fluid as the object of study, using numerical simulation methods to study the mixing motion process of viscous fluid in a cavity under the action of triangular rotor mixing with different staggered column angles, and the simulation image is shown in Figure 15.

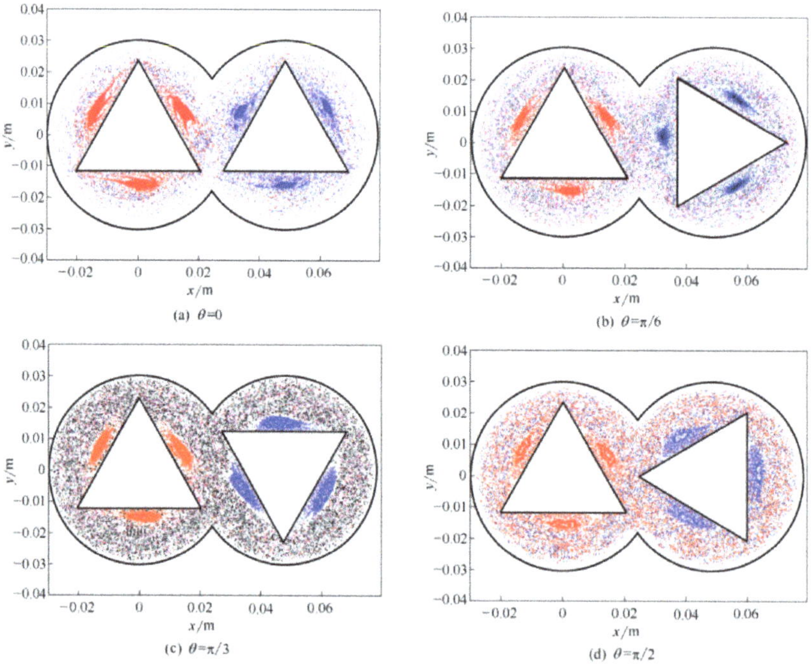

Figure 15. Simulation diagram of liquid particle movement mixing at different staggered angles; subgraphs (**a**–**d**) are the simulation images at the dislocation angles 0, $\pi/6$, $\pi/3$, and $\pi/2$, respectively. The red and blue particles in the picture represent the viscous liquid that enters on both sides when stirred [108].

Liang Yuanfei et al. [109] investigated the turbulence damping rate of viscoelastic fluid for wall turbulence of up to 18.9% by building a three-dimensional computational basin for simulation operations through OpenFOAM, and the results of the power spectral density showed that the energy attenuation of the viscoelastic flow was slower than that of Newtonian fluid. Kaiyuan Shi et al. [110] proposed a fully nonlinear potential flow method, the spectral coupled boundary element method (SCBEM), to address the problem of limited fluid computational resources. Compared with the traditional BEM method, the algorithm speed is greatly improved by using the region decomposition method. The numerical results also confirm the validity and accuracy of the simulation results. Zhu Yafei et al. [111] took the binary viscous fluid as the main research object, visualized and analyzed the flow diffusion behavior of two viscous fluids, hemolysate and hematoma, through the multiphysics field simulation software COMSOL Multiphysics, and improved and designed medical drainage tubes according to the visualization results. Shen Yang et al. [112] established a relative motion model to understand the effect of the shear thinning effect on the particle motion in viscoelastic fluid; a simulation of the mesh division and the vector diagrams are shown in Figures 16 and 17. The results show that under the influence of a shear thinning effect of different Weber numbers, the elasticity of the viscoelastic fluid is reduced, leading to the movement of particles to the outer wall.

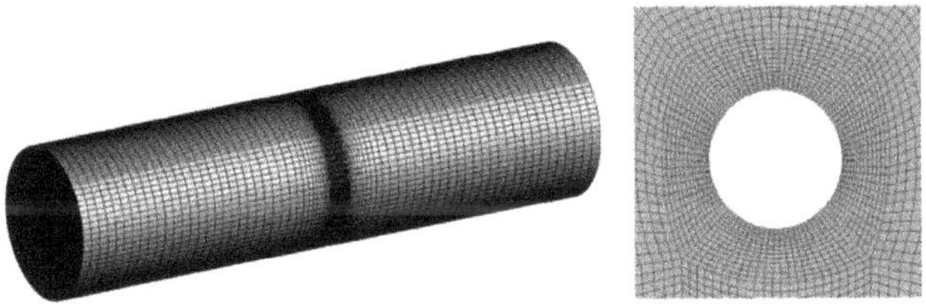

Figure 16. Meshing around pipes and particles [112].

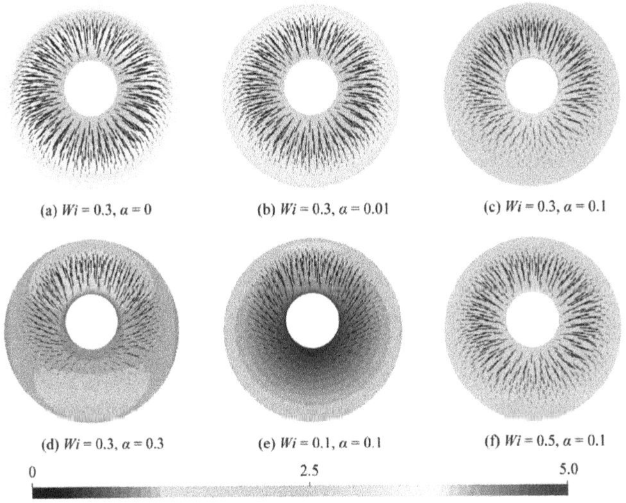

Figure 17. The cloud image of the first normal stress difference coefficient and velocity vector of the x = 0 section at r+ = 0.1. Figure (**a–d**) are the simulation vector diagrams when Wi = 0.3 with a flow factor α = 0, 0.01, 0.1, and 0.3; Figure (**e**,**f**) are the simulation vector diagrams when Wi = 0.1 and 0.5 with flow factor α = 0.3 [112].

Qiu, LC, et al. [113] applied projection-based incompressible smooth particle hydrodynamics (ISPH) to the simulation of the deformation process of a viscous droplet, and the simulation results showed that projection-based ISPH with the particle displacement technique can be used to simulate the deformation process of a viscous droplet stably and accurately.

In general, the visualization of the simulation technology to make the original only rely on the formula deduction and imagination of the viscous fluid exploration and understanding of the process has become a vivid image and more intuitive, very powerful in promoting the development of fluid mechanics. However, the computational power of the computer is a major obstacle to the simulation, and the future development of viscous fluid exploration will inevitably accompany the development of computing power and progress.

At the end of this section, this paper provides a comparison of the characteristics between algorithms and simulations for researchers' reference. Algorithms and simulations each have their own unique strengths and limitations, as shown in Table 2, and an understanding of their characteristics and applicability scenarios can help in choosing the most appropriate tools and methods for solving viscous-fluid-related engineering or scientific problems.

Table 2. Comparative table of advantages and limitations of algorithms and simulations.

Method	Precision of Mathematical Results	Visualization	Time Cost	Complexity Adaptability
Arithmetic	High	Mediocre	Relatively low	Low
Emulate	Relatively low	First-rate	High	High

3. Application Areas of Engineering

The theory of viscous fluid dynamics (or simply viscous fluid theory) plays a crucial role in the field of engineering. The theory not only provides engineers with tools to deeply understand and predict the behavior of various fluids but also directly contributes to the development of several industries in design, optimization, and innovation. In aerospace engineering, the theory of viscous hydrodynamics is of crucial importance for designing the aerodynamic shape of aircraft and optimizing airfoils [114,115]. Through the in-depth study of viscous fluid dynamics, engineers are able to gain a deep understanding of aerodynamic effects such as aerodynamic drag and lift generation mechanisms. This knowledge helps to optimize the aerodynamic shape of the vehicle, reduce drag, improve flight efficiency, reduce fuel consumption, and thus extend range [116–118]. In automotive engineering, viscous fluid dynamics theory is widely used to improve the aerodynamic performance of vehicles. By gaining a deeper understanding of the complexity of air flow during vehicle operation, engineers can design and optimize body shapes, mirrors, and other streamlined components to reduce air resistance and improve the fuel efficiency and driving stability of vehicles [119,120]. In marine engineering, the theory of viscous hydrodynamics is used to study the hydrodynamic performance of marine structures (e.g., platforms, buoys, etc.). By analyzing the motion characteristics of water flow on the surface of structures, engineers can optimize the design to improve the stability and wind and wave resistance of the structures to ensure that they can operate safely in extreme marine environments [121–123]. In the oil, gas, and chemical industries, the theory of viscous fluid mechanics plays an important role in the design and optimization of piping systems. Engineers need to have an in-depth understanding of the flow characteristics of fluids within pipelines to minimize energy loss, control pressure drops, avoid pipeline clogging, and ensure the stable and efficient transportation of fluids [124–126]. In biomedical engineering, the theory of viscous fluid dynamics is important for studying the flow of blood in the cardiovascular system.

For example, in the simulation model of heart valves, viscous fluid dynamics can simulate the flow of blood over the valves and evaluate the hydrodynamic properties of the valves when they open and close, which can help optimize the design and surgical plan of artificial heart valves [127–129]. In industrial production, the theory of viscous hydrodynamics is used to improve the efficiency and quality of coating technology and printing engineering. By precisely controlling the droplet impaction, expansion, and drying processes, engineers can improve the uniformity and durability of coatings, which in turn affects the final quality and appearance of the product. This theory has a wide range of applications in fields such as electronics and packaging materials [130,131]. In environmental engineering, the theory of viscous hydrodynamics is applied to the design of wastewater treatment systems and water management programs. Through a deeper understanding of the complex behavior of water flow in the treatment process, engineers can optimize the design of treatment facilities, improve the treatment efficiency, and reduce the impact on the environment, thus protecting water resources and the ecological environment [132–134]. As shown in Figure 18, this paper provides a schematic diagram of engineering application for readers' reference, which includes some research directions and the applications of related technologies involved in viscous fluid mechanics in the engineering field. The applications in aerospace, fluid simulation, bioengineering, pipeline transportation, and other engineering fields in the figure correspond to the above.

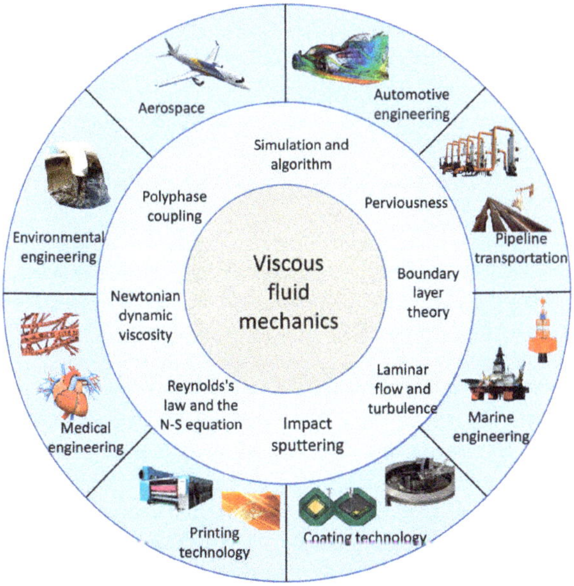

Figure 18. Viscous fluid mechanics engineering applications.

In summary, the wide application of viscous fluid dynamics theory in the field of engineering not only provides an effective means to solve complex fluid dynamics problems but also greatly promotes the technological progress and innovation in various industries. With the continuous development of science and technology and increasingly diversified engineering needs, the research and application of viscous fluid dynamics will continue to play a pivotal role in various fields, providing engineers with a solid theoretical foundation and powerful technical tools to help them cope with a variety of challenges and promote the continuous development of engineering technology.

4. Future Trends and Directions

Viscous fluid mechanics, as an important branch of fluid mechanics, has made remarkable progress in the past decades and has been widely used in several engineering and scientific fields. The future research and development outlook can be viewed from the following aspects (Figure 19):

(1) Future research in viscous fluid mechanics will pay more attention to the coupling of multiple physical fields. The physical fields and their effects that need attention include the following: (1) Thermal field: Temperature changes can significantly affect the viscosity of the fluid; usually, a temperature increase will lead to a decrease in the viscosity of the fluid, while a decrease in temperature will increase the viscosity. This coupling is particularly important in high-temperature lubrication and polymer fluid handling. (2) Chemical reactions: Chemical reactions can change the composition of a fluid and thus affect its viscosity. For example, cross-linking reactions in polymer fluids can lead to nonlinear changes in viscosity, which is critical for the processing and control of viscoelastic materials. (3) Electromagnetic fields: Magnetic fluids can produce morphological changes and controlled flow behavior under an applied magnetic field, such as electromagnetically driven fluids in liquid metals. This coupling can have applications in liquid metal metallurgy, aerospace technology, and biomedical sensors. (4) Solid mechanics: in fluids in contact with solid walls, the morphology and roughness of the solid surface can significantly affect the friction and adhesion behavior of the fluid.(5)Acoustic effects: Acoustic fields can affect the viscoelasticity and flow properties of fluids, for example, in the ultrasonic acoustic field for microfluidic manipulation and in biomedical imaging. This coupling helps to control the mixing and stirring process of microfluidics; not only limited to the above physical fields, or even using two or more kinds of physical fields coupled together, numerical simulation or complex experimental methods can be used in studies and analysis in order to add a comprehensive, in-depth understanding of complex engineering phenomena and natural phenomena. This multiphysical field coupling research method is expected to provide new ideas and solutions for innovations in the fields of energy conversion, environmental protection, and biomedicine. I believe that the optimization trends in this area are as follows: (1) for complex systems and phenomena, such as biofluids, multiphase flows, and non-Newtonian fluids, multiscale modeling and simulation methods should be improved to capture the details in different spatial and temporal scales; (2) for numerical simulation, new numerical simulation techniques and algorithms such as mixed-element methods, implicit-coupling methods, and adaptive mesh techniques should be improved to facilitate the numerical modeling and analysis of complex fluid systems;(3)experimental techniques, such as microfluidic experimental platforms and advanced imaging techniques, should be improved to further improve the precision of experimental device components and optimize real-time sensing techniques, which can more accurately validate theoretical models and numerical predictions.

(2) One of the future research priorities will be the in-depth understanding and control of complex flow phenomena. Viscous fluid complex flow phenomena include the following: (1) non-Newtonian fluid flow: non-Newtonian fluid flow is not only affected by the velocity gradient but may also be affected by the non-uniformity and time-dependence of the flow field, resulting in complex flow behaviors, such as shear thinning and shear thickening phenomena; (2) multiphase flow: this refers to the simultaneous presence of two or more substances (e.g., gases, liquids, or solid particles) of the flow system, with a focus on gas–liquid, liquid–liquid, solid–liquid, and other combinations of flow, such as foam flow, bubble flow, the movement of

particles in suspension, etc.; (3) biological fluid flow: the flow mechanism of blood flow in living organisms and fluid exchange across the walls between cells is very complex, and in the future, it may be possible to simulate blood flow, intercellular fluid exchange, and other biophysical fluid dynamics phenomena, which will aid in the diagnosis of diseases, the research and development of new drugs, and bioengineering; and (4) complex coupled flow: complex coupled flow mainly refers to those flow phenomena that are not only affected by the flow characteristics of the fluid itself but are also affected by the interaction between the fluid and the solid, the fluid and the electric field or magnetic field, etc., for example, electromagnetic fluids (such as plasma), magnetic fluids (such as magnetorheological fluids), and complex fluid systems that interact with particles. I think there are several optimization trends in this area: (1) data-driven method: using machine learning and data mining techniques to process a large amount of experimental data and numerical simulation data, such as the bird flocking algorithm and the whale algorithm in machine learning, to discover the patterns and laws in the flow and optimize the parameters and boundary conditions of the numerical model to improve the accuracy and efficiency of the simulation results; (2) experimental simulation synergistic method: comparing and verifying the experimental data with the numerical simulation results and continuously improving the numerical model to enhance its applicability and realism in complex flow systems; In addition to this, we can also can deepen the research on microscale flow, boundary layer control, wake phenomena, etc. By deepening the understanding and control of these complex flow phenomena, we can provide strong support for the innovation and development of related fields.

(3) The development of viscous fluid mechanics will also be closely related to the application of new materials. This area currently has significant trends in the following areas: (1) Porous and complex media: with the deepening of the study of fluid mechanics in porous media, new porous materials (such as 3D printed structures) and complex media (such as nanomaterials) could be developed, and these materials may exhibit nonlinear, non-Newtonian flow properties, requiring new numerical simulation and experimental techniques to understand their flow behavior, for example, the permeability of the material, the surface tension and viscosity of liquid nanomaterials, etc. (2) Biomaterials: The field of biofluid dynamics has an increasing demand for research on biomaterials (e.g., soft tissues, cell membranes, etc.), and the characteristics of these materials may lead to complex flow behaviors, which need to be combined with experimental and simulation methods to conduct an in-depth study. In this regard, I'm more optimistic about the potential of nanotechnology in the study of viscous fluids, and we need to actively explore how to utilize new technologies such as nanotechnology to change the viscosity and surface tension of fluids and develop new fluid control and regulation technologies to meet different engineering needs and challenges.

As a rapidly developing field, future research in viscous fluid dynamics will continue to explore the nature of complex fluid phenomena in depth. By combining advanced computational techniques and interdisciplinary research methods with highly sophisticated camera technology, control technology, detection technology, etc., viscous fluid dynamics is expected to promote applications and innovations in many fields, such as engineering, science, and medicine, and provide more accurate and effective theoretical and methodological support for solving practical problems.

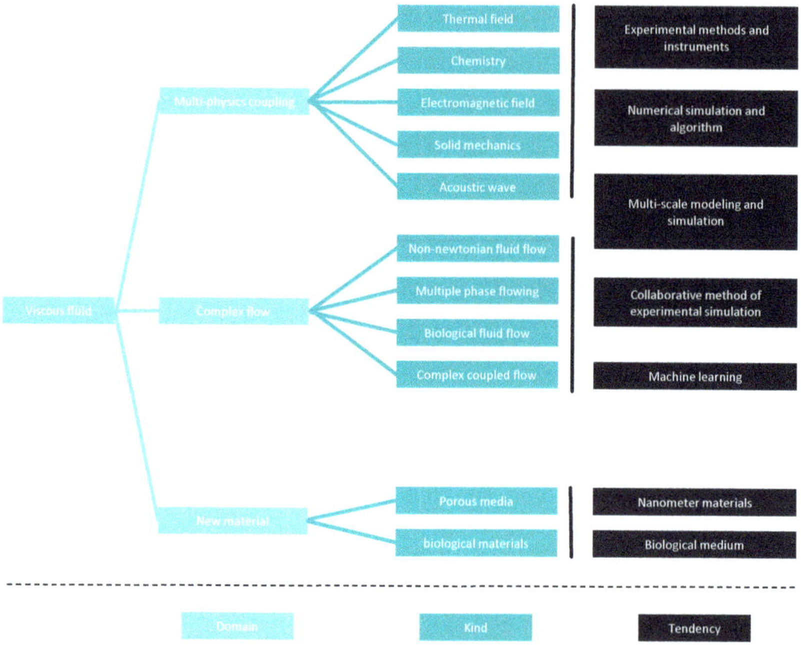

Figure 19. Research direction and trend of viscous fluid mechanics.

Author Contributions: Conceptualization, J.P. and X.K.; methodology, J.P.; software, R.F.; validation, J.P., X.K. and R.F.; formal analysis, E.Z.; investigation, R.F.; resources, J.W.; data curation, M.X.; writing—original draft preparation, R.F.; writing—review and editing, R.F.; visualization, M.X.; supervision, J.P.; project administration, J.P. and X.K.; funding acquisition, Z.Z. and J.P. All authors have read and agreed to the published version of the manuscript.

Funding: This research was funded by the Key Research Project of Universities of Henan Province (22A460014), the Tribology Science Foundation of the State Key Laboratory of Tribology of Advanced Equipment (skltkf22b12), Key R&D projects in Henan Province (231111222900), and the joint fund (industrial sector) of R&D projects in Henan Province (225101610003).

Data Availability Statement: No new data were created or analyzed in this study.

Conflicts of Interest: The authors declare no conflict of interest.

References

1. Romatschke, P. New developments in relativistic viscous hydrodynamics. *Int. J. Mod. Phys. E* **2010**, *19*, 1–53. [CrossRef]
2. Phillips, T.N.; Roberts, G.W. Lattice Boltzmann models for non-Newtonian flows. *IMA J. Appl. Math.* **2011**, *76*, 790–816. [CrossRef]
3. Aharonov, E.; Rothman, D.H. Non-Newtonian flow (through porous media): A lattice-Boltzmann method. *Geophys. Res. Lett.* **1993**, *20*, 679–682. [CrossRef]
4. Hu, H.; Zhang, G.; Zhao, T. Overview of computational fluid dynamics and discrete element coupling methods. In Proceedings of the 5th China Symposium on Hydraulic and Hydropower Geotechnical Mechanics and Engineering, Yichang, China, 17–21 September 2014; p. 4.
5. Liu, Y.-Z. Overview of dynamic research methods for liquid-filled rigid bodies. *Mech. Pract.* **2015**, *37*, 397–400.
6. Silva, D.P.; Coelho, R.C.; Pagonabarraga, I.; Succi, S.; da Gama, M.M.T.; Araújo, N.A. Lattice Boltzmann simulation of deformable fluid-filled bodies: Progress and perspectives. *Soft Matter* **2024**, *20*, 2419–2441. [CrossRef] [PubMed]
7. Coussot, P. Yield stress fluid flows: A review of experimental data. *J. Non-Newton. Fluid Mech.* **2014**, *211*, 31–49. [CrossRef]
8. Tatsumi, T. Fluid Mechanics in the turn of the century. *Fluid Dyn. Res.* **1999**, *24*, 315–319. [CrossRef]
9. Rozenman, G.G.; Ullinger, F.; Zimmermann, M.; Efremov, M.A.; Shemer, L.; Schleich, W.P.; Arie, A. Observation of a phase space horizon with surface gravity water waves. *Commun. Phys.* **2024**, *7*, 165. [CrossRef]

10. Otaru, A. Review on processing and fluid transport in porous metals with a focus on bottleneck structures. *Met. Mater. Int.* **2020**, *26*, 510–525. [CrossRef]
11. Crespí-Llorens, D.; Vicente, P.; Viedma, A. Generalized Reynolds number and viscosity definitions for non-Newtonian fluid flow in ducts of non-uniform cross-section. *Exp. Therm. Fluid Sci.* **2015**, *64*, 125–133. [CrossRef]
12. Delplace, F.; Leuliet, J. Generalized Reynolds number for the flow of power law fluids in cylindrical ducts of arbitrary cross-section. *Chem. Eng. J. Biochem. Eng. J.* **1995**, *56*, 33–37. [CrossRef]
13. Abell, M.; Ames, W. Symmetry reduction of Reynolds equation and applications to film lubrication. *J. Appl. Mech.* **1992**, *59*, 206–210. [CrossRef]
14. Metzner, A.; Reed, J. Flow of non-newtonian fluids—Correlation of the laminar, transition, and turbulent-flow regions. *Aiche J.* **1955**, *1*, 434–440. [CrossRef]
15. Chhabra, R.P.; Richardson, J.F. *Non-Newtonian Flow and Applied Rheology: Engineering Applications*; Butterworth-Heinemann: Oxford, UK, 2011.
16. Garcia, A.; Vicente, P.G.; Viedma, A. Experimental study of heat transfer enhancement with wire coil inserts in laminar-transition-turbulent regimes at different Prandtl numbers. *Int. J. Heat Mass Transf.* **2005**, *48*, 4640–4651. [CrossRef]
17. Ni, H.; Zhang, L.; Xiong, X.; Mao, Z.; Wang, J. Supercritical fluids at subduction zones: Evidence, formation condition, and physicochemical properties. *Earth-Sci. Rev.* **2017**, *167*, 62–71. [CrossRef]
18. Adam, J.; Locmelis, M.; Afonso, J.C.; Rushmer, T.; Fiorentini, M.L. The capacity of hydrous fluids to transport and fractionate incompatible elements and metals within the Earth's mantle. *Geochem. Geophys. Geosystems* **2014**, *15*, 2241–2253. [CrossRef]
19. Bai, J.; Wang, T.; Zou, L.; Li, P. Large Eddy Simulation of compressible multi-media viscous Fluid and Turbulence. *Explos. Shock. Waves* **2010**, *30*, 262–268.
20. Khan, Z.A.; Ishaq, M.; Ghani, U.; Ullah, R.; Iqbal, S. Analysis of magnetohydrodynamics flow of a generalized viscous fluid through a porous medium with Prabhakar-like fractional model with generalized thermal transport. *Adv. Mech. Eng.* **2022**, *14*, 16878132221132949. [CrossRef]
21. Elnaqeeb, T.; Shah, N.A.; Mirza, I.A. Natural convection flows of carbon nanotubes nanofluids with Prabhakar-like thermal transport. *Math. Methods Appl. Sci.* 2020, *early view*. [CrossRef]
22. Hemalatha, R.; Kameswaran, P.K.; Murthy, P. Effect of nanoparticle shape on the mixed convective transport over a vertical cylinder in a non-Darcy porous medium. *Int. Commun. Heat Mass Transf.* **2022**, *133*, 105962. [CrossRef]
23. Makris, N.; Dargush, G.; Constantinou, M. Dynamic analysis of generalized viscoelastic fluids. *J. Eng. Mech.* **1993**, *119*, 1663–1679. [CrossRef]
24. Awasthi, M.; Tamsir, M. Viscous Corrections for the Viscous Potential Flow Analysis of Magneto-Hydrodynamic Kelvin-Helmholtz Instability through Porous Media. *J. Porous Media* **2013**, *16*, 663–676. [CrossRef]
25. Craster, R.V.; Matar, O.K. Dynamics and stability of thin liquid films. *Rev. Mod. Phys.* **2009**, *81*, 1131–1198. [CrossRef]
26. Pérez-Ràfols, F.; Wall, P.; Almqvist, A. On compressible and piezo-viscous flow in thin porous media. *Proc. R. Soc. A Math. Phys. Eng. Sci.* **2018**, *474*, 20170601. [CrossRef]
27. Jegatheeswaran, S.; Ein-Mozaffari, F.; Wu, J. Laminar mixing of non-Newtonian fluids in static mixers: Process intensification perspective. *Rev. Chem. Eng.* **2020**, *36*, 423–436. [CrossRef]
28. Song, R.; Stone, H.A.; Jensen, K.H.; Lee, J. Pressure-driven flow across a hyperelastic porous membrane. *J. Fluid Mech.* **2019**, *871*, 742–754. [CrossRef]
29. Stafie, N.; Stamatialis, D.; Wessling, M. Effect of PDMS cross-linking degree on the permeation performance of PAN/PDMS composite nanofiltration membranes. *Sep. Purif. Technol.* **2005**, *45*, 220–231. [CrossRef]
30. Delli, M.L.; Grozic, J.L. Experimental determination of permeability of porous media in the presence of gas hydrates. *J. Pet. Sci. Eng.* **2014**, *120*, 1–9. [CrossRef]
31. Wang, S.; Wu, T.; Xia, K. Fractal permeability model of spherical porous media. *J. Cent. China Norm. Univ. (Nat. Sci. Ed.)* **2020**, *54*, 45–49. [CrossRef]
32. Wang, M.; Li, X.; Yang, X.; Liu, H.; Zhang, Z.; Xu, D.; Liu, Z.-J. Experimental study on the influence of different fluids and clay minerals on rock permeability. *Prog. Geophys.* **2022**, *37*, 275–283.
33. Sun, W.; Xiong, F.; Cao, H.; Yang, Z.; Lu, M. Complex fluid model and its applicability analysis in tight reservoir. *Geophys. Prospect. Pet.* **2021**, *60*, 136–148.
34. Ding, L.; Ma, J.; Wang, J. Experimental study on permeability characteristics of sintered porous media. *Q. J. Mech.* **2012**, *33*, 375–381. [CrossRef]
35. Kowal, K.N. Viscous banding instabilities: Non-porous viscous fingering. *J. Fluid Mech.* **2021**, *926*, A4. [CrossRef]
36. Graeber, G.; Martin Kieliger, O.B.; Schutzius, T.M.; Poulikakos, D. 3D-printed surface architecture enhancing superhydrophobicity and viscous droplet repellency. *ACS Appl. Mater. Interfaces* **2018**, *10*, 43275–43281. [CrossRef]
37. Bartolo, D.; Josserand, C.; Bonn, D. Retraction dynamics of aqueous drops upon impact on non-wetting surfaces. *J. Fluid Mech.* **2005**, *545*, 329–338. [CrossRef]

38. Bird, J.C.; Dhiman, R.; Kwon, H.-M.; Varanasi, K.K. Reducing the contact time of a bouncing drop. *Nature* **2013**, *503*, 385–388. [CrossRef]
39. Hao, C.; Zhou, Y.; Zhou, X.; Che, L.; Chu, B.; Wang, Z. Dynamic control of droplet jumping by tailoring nanoparticle concentrations. *Appl. Phys. Lett.* **2016**, *109*, 021601. [CrossRef]
40. Bartolo, D.; Boudaoud, A.; Narcy, G.; Bonn, D. Dynamics of non-Newtonian droplets. *Phys. Rev. Lett.* **2007**, *99*, 174502. [CrossRef]
41. Aksoy, Y.T.; Eneren, P.; Koos, E.; Vetrano, M.R. Spreading of a droplet impacting on a smooth flat surface: How liquid viscosity influences the maximum spreading time and spreading ratio. *Phys. Fluids* **2022**, *34*, 042106. [CrossRef]
42. Eckert, M. Ludwig Prandtl and the growth of fluid mechanics in Germany. *Comptes Rendus Mécanique* **2017**, *345*, 467–476. [CrossRef]
43. Bartram, E.; Goldsmith, H.; Mason, S. Particle motions in non-newtonian media: III. *Furth. Obs. Elasticoviscous Fluids. Rheol. Acta* **1975**, *14*, 776–782. [CrossRef]
44. Patil, A. Laminar flow heat transfer and pressure drop characteristics of power-law fluids inside tubes with varying width twisted tape inserts. *J. Heat Transf.* **2000**, *122*, 143–149. [CrossRef]
45. Carillo, S.; Jordan, P.M. On the structure of isothermal acoustic shocks under classical and artificial viscosity laws: Selected case studies. *Meccanica* **2023**, *58*, 1121–1139. [CrossRef]
46. Finotello, G.; De, S.; Vrouwenvelder, J.C.; Padding, J.T.; Buist, K.A.; Jongsma, A.; Innings, F.; Kuipers, J. Experimental investigation of non-Newtonian droplet collisions: The role of extensional viscosity. *Exp. Fluids* **2018**, *59*, 113. [CrossRef]
47. Lin, M.; Vo, Q.; Mitra, S.; Tran, T. Viscous droplet impingement on soft substrates. *Soft Matter* **2022**, *18*, 5474–5482. [CrossRef]
48. Yang, L.; Liu, X.; Wang, J.; Zhang, P. An Experimental Study on Complete Droplet Rebound from Soft Surfaces: Critical Weber Numbers, Maximum Spreading, and Contact Time. *Langmuir* **2024**, *40*, 2165–2173. [CrossRef] [PubMed]
49. Chang, T.-B.; Chen, R.-H. Experimental investigation into deposition/splashing behavior of droplets impacting vibrating surface. *Adv. Mech. Eng.* **2017**, *9*, 1687814017730004. [CrossRef]
50. Liu, H.; Wang, J.; Shen, X.; Zheng, N.; Wang, J. Experiment and simulation of viscous fluid droplet impacting singular jet on superhydrophobic wall. *Chin. Surf. Eng.* **2023**, *36*, 124–134.
51. Li, J.; Zhao, C.; Wang, C. Experimental study on the dynamics of droplet impacting on solid surface. *Microfluid. Nanofluidics* **2023**, *27*, 69. [CrossRef]
52. Qian, W.W. Experimental Research and Numerical Simulation of Viscous Liquid and Dense Particle Jet Collision Process. Master's Thesis, East China University of Science and Technology, Shanghai, China, 2016.
53. Xing, L.; Zhang, J.; Jiang, M.; Zhao, L.; Guan, S. Experimental study on collision coalescence process of unequal large droplets on different wettability walls. *Chem. Eng.* **2024**, *52*, 43–48.
54. Zong, S.; Xu, L.; Hao, J. Experimental study on droplet impact of viscous Newtonian fluid on dry or pre-wet mesh. *Chin. J. Mech. Mech.* **2024**, *56*, 101–111.
55. Xing, L.; Li, J.; Jiang, M.; Zhao, L. Experimental study of droplet collision with different wettability inclined plate. *J. Eng. Thermophys.* **2024**, *45*, 2001–2011.
56. Jung, S.; Hoath, S.D.; Hutchings, I.M. The role of viscoelasticity in drop impact and spreading for inkjet printing of polymer solution on a wettable surface. *Microfluid. Nanofluidics* **2013**, *14*, 163–169. [CrossRef]
57. Jung, S.; Hoath, S.D.; Martin, G.D.; Hutchings, I.M. Experimental study of atomization patterns produced by the oblique collision of two viscoelastic liquid jets. *J. Non-Newton. Fluid Mech.* **2011**, *166*, 297–306. [CrossRef]
58. Fürst, J.; Straka, P.; Příhoda, J.; Šimurda, D. Comparison of several models of the laminar/turbulent transition. *EPJ Web Conf.* **2013**, *45*, 01032. [CrossRef]
59. Langtry, R.B.; Menter, F.R. Correlation-based transition modeling for unstructured parallelized computational fluid dynamics codes. *AIAA J.* **2009**, *47*, 2894–2906. [CrossRef]
60. Mehendale, S.; Jacobi, A.; Shah, R. Fluid flow and heat transfer at micro-and meso-scales with application to heat exchanger design. *Appl. Mech. Rev.* **2000**, *53*, 175–193. [CrossRef]
61. Hou, G.; Wang, J.; Layton, A. Numerical methods for fluid-structure interaction—A review. *Commun. Comput. Phys.* **2012**, *12*, 337–377. [CrossRef]
62. Dowell, E.H.; Hall, K.C. Modeling of fluid-structure interaction. *Annu. Rev. Fluid Mech.* **2001**, *33*, 445–490. [CrossRef]
63. Pihler-Puzović, D.; Illien, P.; Heil, M.; Juel, A. Suppression of Complex Fingerlike Patterns at the Interface between Air and a Viscous Fluid by Elastic Membranes. *Phys. Rev. Lett.* **2012**, *108*, 074502. [CrossRef]
64. Naduvinamani, N.; Hiremath, P.; Gurubasavaraj, G. Effect of surface roughness on the couple-stress squeeze film between a sphere and a flat plate. *Tribol. Int.* **2005**, *38*, 451–458. [CrossRef]
65. Lopes Filho, M.C.; Nussenzveig Lopes, H.J.; Titi, E.S.; Zang, A. Approximation of 2D Euler equations by the second-grade fluid equations with Dirichlet boundary conditions. *J. Math. Fluid Mech.* **2015**, *17*, 327–340. [CrossRef]
66. Chae, D.; Constantin, P.; Wu, J. Inviscid models generalizing the two-dimensional Euler and the surface quasi-geostrophic equations. *Arch. Ration. Mech. Anal.* **2011**, *202*, 35–62. [CrossRef]

67. Prausová, H.; Bublík, O.; Vimmr, J.; Luxa, M.; Hála, J. Clearance gap flow: Simulations by discontinuous Galerkin method and experiments. *EPJ Web Conf.* **2015**, *92*, 02073. [CrossRef]
68. Christensen, A.H.; Jensen, K.H. Viscous flow in a slit between two elastic plates. *Phys. Rev. Fluids* **2020**, *5*, 044101. [CrossRef]
69. Lin, J.-R.; Chu, L.-M.; Li, W.-L.; Lu, R.-F. Combined effects of piezo-viscous dependency and non-Newtonian couple stresses in wide parallel-plate squeeze-film characteristics. *Tribol. Int.* **2011**, *44*, 1598–1602. [CrossRef]
70. Vijayakumar, B.; Kesavan, S. Effects of pressure distribution on parallel circular porous plates with combined effect of piezo-viscous dependency and non-Newtonian couple stress fluid. *J. Phys. Conf. Ser.* **2018**, *1000*, 012021. [CrossRef]
71. Box, F.; Neufeld, J.A.; Woods, A.W. On the dynamics of a thin viscous film spreading between a permeable horizontal plate and an elastic sheet. *J. Fluid Mech.* **2018**, *841*, 989–1011. [CrossRef]
72. Zamanov, A.; Ismailov, M.; Akbarov, S. The Effect of Viscosity of a Fluid on the Frequency Response of a Viscoelastic Plate Loaded by This Fluid. *Mech. Compos. Mater.* **2018**, *54*, 41–52. [CrossRef]
73. Aureli, M.; Basaran, M.; Porfiri, M. Nonlinear finite amplitude vibrations of sharp-edged beams in viscous fluids. *J. Sound Vib.* **2012**, *331*, 1624–1654. [CrossRef]
74. Aureli, M.; Porfiri, M. Low frequency and large amplitude oscillations of cantilevers in viscous fluids. *Appl. Phys. Lett.* **2010**, *96*, 164102. [CrossRef]
75. Justesen, P. A numerical study of oscillating flow around a circular cylinder. *J. Fluid Mech.* **1991**, *222*, 157–196. [CrossRef]
76. Apfel, R.E.; Tian, Y.; Jankovsky, J.; Shi, T.; Chen, X.; Holt, R.G.; Trinh, E.; Croonquist, A.; Thornton, K.C.; Sacco Jr, A. Free oscillations and surfactant studies of superdeformed drops in microgravity. *Phys. Rev. Lett.* **1997**, *78*, 1912. [CrossRef]
77. Kitahata, H.; Tanaka, R.; Koyano, Y.; Matsumoto, S.; Nishinari, K.; Watanabe, T.; Hasegawa, K.; Kanagawa, T.; Kaneko, A.; Abe, Y. Oscillation of a rotating levitated droplet: Analysis with a mechanical model. *Phys. Rev. E* **2015**, *92*, 062904. [CrossRef] [PubMed]
78. Trinh, E.; Zwern, A.; Wang, T. An experimental study of small-amplitude drop oscillations in immiscible liquid systems. *J. Fluid Mech.* **1982**, *115*, 453–474. [CrossRef]
79. Nuriev, A.N.; Egorov, A.G.; Kamalutdinov, A.M. Hydrodynamic forces acting on the elliptic cylinder performing high-frequency low-amplitude multi-harmonic oscillations in a viscous fluid. *J. Fluid Mech.* **2021**, *913*, A40. [CrossRef]
80. Kremer, J.; Kilzer, A.; Petermann, M. Simultaneous measurement of surface tension and viscosity using freely decaying oscillations of acoustically levitated droplets. *Rev. Sci. Instrum.* **2018**, *89*, 015109. [CrossRef] [PubMed]
81. Brenn, G.; Teichtmeister, S. Linear shape oscillations and polymeric time scales of viscoelastic drops. *J. Fluid Mech.* **2013**, *733*, 504–527. [CrossRef]
82. Feng, L.; Shi, W.-Y. Viscous Effect on the Frequency Shift of an Oscillating-Rotating Droplet. *Microgravity Sci. Technol.* **2023**, *35*, 27. [CrossRef]
83. Ghaffari, A.; Hashemabadi, S.H.; Ashtiani, M. A review on the simulation and modeling of magnetorheological fluids. *J. Intell. Mater. Syst. Struct.* **2015**, *26*, 881–904. [CrossRef]
84. Chae, D. On the a priori estimates for the Euler, the Navier-Stokes and the quasi-geostrophic equations. *arXiv* **2007**, arXiv:0711.3403. [CrossRef]
85. Chowdhury, M.; Fester, V.G. Modeling pressure losses for Newtonian and non-Newtonian laminar and turbulent flow in long square edged orifices. *Chem. Eng. Res. Des.* **2012**, *90*, 863–869. [CrossRef]
86. Rahman, M.; Tachie, M. Effects of Reynolds number on turbulent characteristics of surface jet. In Proceedings of the Fluids Engineering Division Summer Meeting, Washington, DC, USA, 10–14 July 2016. V01BT20A002.
87. Wang, F.Z.; Animasaun, I.; Muhammad, T.; Okoya, S. Recent advancements in fluid dynamics: Drag reduction, lift generation, computational fluid dynamics, turbulence modelling, and multiphase flow. *Arab. J. Sci. Eng.* **2024**, *49*, 10237–10249. [CrossRef]
88. Bakhtyar, R.; Razmi, A.M.; Barry, D.A.; Yeganeh-Bakhtiary, A.; Zou, Q.-P. Air–water two-phase flow modeling of turbulent surf and swash zone wave motions. *Adv. Water Resour.* **2010**, *33*, 1560–1574. [CrossRef]
89. Guerrero, J.; Sanguineti, M.; Wittkowski, K. CFD study of the impact of variable cant angle winglets on total drag reduction. *Aerospace* **2018**, *5*, 126. [CrossRef]
90. Jeon, J.; Lee, J.; Vinuesa, R.; Kim, S.J. Residual-based physics-informed transfer learning: A hybrid method for accelerating long-term cfd simulations via deep learning. *Int. J. Heat Mass Transf.* **2024**, *220*, 124900. [CrossRef]
91. Kadivar, M.; Tormey, D.; McGranaghan, G. A comparison of RANS Models used for CFD prediction of turbulent flow and heat transfer in rough and smooth channels. *Int. J. Thermofluids* **2023**, *20*, 100399. [CrossRef]
92. Van Hirtum, A.; Grandchamp, X.; Cisonni, J. Reynolds number dependence of near field vortex motion downstream from an asymmetrical nozzle. *Mech. Res. Commun.* **2012**, *44*, 47–50. [CrossRef]
93. Wang, Z.; Mao, B.; Zhu, R.; Bai, X.; Han, X. Study on oscillating flow model of fractional Maxwell viscoelastic colloidal damper. *J. Ordnance Eng.* **2020**, *41*, 984–995.
94. Lee, Y.K.; Ahn, K.H. A novel lattice Boltzmann method for the dynamics of rigid particles suspended in a viscoelastic medium. *J. Non-Newton. Fluid Mech.* **2017**, *244*, 75–84. [CrossRef]

95. Zhu, L.; Wang, Q.; Qian, J.; Xue, Z. Mechanical characteristics of particle migration with viscoelastic fluid in a square tube. *Light Ind. Mach.* **2019**, *37*, 39–44.
96. Li, H.; Li, Y.; Zhou, W.J.; He, L. Lattice Boltzmann simulation analysis of particle settlement in viscoelastic fluid expansion and contraction flow. *Q. J. Mech.* **2020**, *41*, 319–328. [CrossRef]
97. Tsepelev, I. Iterative algorithm for solving the retrospective problem of thermal convection in a viscous fluid. *Fluid Dyn.* **2011**, *46*, 835–842. [CrossRef]
98. Xu, X.; Deng, X.-L. An improved weakly compressible SPH method for simulating free surface flows of viscous and viscoelastic fluids. *Comput. Phys. Commun.* **2016**, *201*, 43–62. [CrossRef]
99. Moulinec, C.; Violeau, D.; Laurence, D.; Lee, E.; Stansby, P.; Xu, R. Comparisons of Weakly Compressible and Truly Incompressible Algorithms for the SPH Mesh Free Simulation with Polyhedral Meshes. *J. Comput. Phys.* **2008**, *227*, 8417–8436.
100. Li, G.; Ma, X.; Zhang, B.; Xu, H. An integrated smoothed particle hydrodynamics method for numerical simulation of the droplet impacting with heat transfer. *Eng. Anal. Bound. Elem.* **2021**, *124*, 1–13. [CrossRef]
101. Saintillan, D. Rheology of active fluids. *Annu. Rev. Fluid Mech.* **2018**, *50*, 563–592. [CrossRef]
102. Pudasaini, S.P.; Mergili, M. A multi-phase mass flow model. *J. Geophys. Res. Earth Surf.* **2019**, *124*, 2920–2942. [CrossRef]
103. Tian, F.-B.; Dai, H.; Luo, H.; Doyle, J.F.; Rousseau, B. Fluid–structure interaction involving large deformations: 3D simulations and applications to biological systems. *J. Comput. Phys.* **2014**, *258*, 451–469. [CrossRef]
104. Sun, P.-N.; Le Touze, D.; Oger, G.; Zhang, A.-M. An accurate FSI-SPH modeling of challenging fluid-structure interaction problems in two and three dimensions. *Ocean Eng.* **2021**, *221*, 108552. [CrossRef]
105. Atangana, A.; Baleanu, D. New fractional derivatives with nonlocal and non-singular kernel: Theory and application to heat transfer model. *arXiv* **2016**, arXiv:1602.03408. [CrossRef]
106. Balduzzi, F.; Bianchini, A.; Maleci, R.; Ferrara, G.; Ferrari, L. Critical issues in the CFD simulation of Darrieus wind turbines. *Renew. Energy* **2016**, *85*, 419–435. [CrossRef]
107. Khayyer, A.; Gotoh, H.; Falahaty, H.; Shimizu, Y. An enhanced ISPH–SPH coupled method for simulation of incompressible fluid–elastic structure interactions. *Comput. Phys. Commun.* **2018**, *232*, 139–164. [CrossRef]
108. Xu, B.; Liu, B.; Liu, Y.; Tan, S.; Du, Y.; Liu, C. Numerical simulation of distributed mixing under the action of different staggered angular triangular rotors. *J. Chem. Eng.* **2020**, *71*, 3545–3555.
109. Liang, Y.-f. Numerical Simulation of Viscoelastic Fluid turbulence drag Reduction. Master's Thesis, Guangzhou University, Guangdong, China, 2022.
110. Shi, K.; Zhu, R.; Xu, D. A spectral coupled boundary element method for the simulation of nonlinear surface gravity waves. *J. Ocean. Eng. Sci.* **2023**, *in press*. [CrossRef]
111. Zhu, Y. Visual Analysis and Application of viscous liquid mutual diffusion. Master's Thesis, North China University of Science and Technology, Tangshan, China, 2021.
112. Shen, Y.; Wang, Q.; Liu, T.-J. Effect of shear thinning rheological properties on particle migration in microchannels. *Appl. Math. Mech.* **2024**, *45*, 637–650.
113. Qiu, L.-C. Numerical simulation of deformation process of viscous liquid drop based on the incompressible smoothed particle hydrodynamics. *Acta Phys. Sin.* **2013**, *62*, 124702.
114. Frezzotti, M.L.; Ferrando, S. The chemical behavior of fluids released during deep subduction based on fluid inclusions. *Am. Mineral.* **2015**, *100*, 352–377. [CrossRef]
115. Buttazzo, G.; Frediani, A.; Frediani, A.; Montanari, G. Best wing system: An exact solution of the Prandtl's problem. In *Variational Analysis and Aerospace Engineering*; Springer: New York, NY, USA, 2009; pp. 183–211.
116. Eckert, M. Theory from wind tunnels: Empirical roots of twentieth century fluid dynamics. *Centaurus* **2008**, *50*, 233–253. [CrossRef]
117. Sekundov, A. On the boundary conditions for differential turbulence models. *Fluid Dyn.* **2012**, *47*, 20–25. [CrossRef]
118. Singla, A.; Ray, B. Effects of surface topography on low Reynolds number droplet/bubble flow through a constricted passage. *Phys. Fluids* **2021**, *33*, 011301. [CrossRef]
119. Eskandari, A.; Ahmadi Arpanahi, R.; Daneh-Dezfuli, A.; Mohammadi, B.; Hosseini Hashemi, S. Investigation of vibration of the nano rotating blade coupled with viscous fluid medium by considering the nonlocal elastic theory. *J. Low Freq. Noise Vib. Act. Control.* **2024**, *43*, 863–879. [CrossRef]
120. Zhang, D.; Zheng, W.; Ma, Q.; Yang, P. Numerical simulation of viscous flow around rowing based on FLUENT. In Proceedings of the 2009 ISECS International Colloquium on Computing, Communication, Control, and Management, Sanya, China, 8–9 August 2009; pp. 429–433.
121. Yang, H.-y.; Pan, M. Engineering research in fluid power: A review. *J. Zhejiang Univ. Sci. A* **2015**, *16*, 427–442. [CrossRef]
122. Zhang, Z.-r.; Li, B.-q.; Zhao, F. The comparison of turbulence models applied to viscous ship flow computation. *J. Hydrodyn.* **2004**, *19*, 637–642.

123. Nadeem, S.; Waqar, H.; Akhtar, S.; Zidan, A.M.; Almutairi, S.; Ghazwani, H.A.S.; Kbiri Alaoui, M.; Tarek El-Waked, M. Mathematical assessment of convection and diffusion analysis for a non-circular duct flow with viscous dissipation: Application of physiology. *Symmetry* **2022**, *14*, 1536. [CrossRef]
124. Li, X.-L.; Fu, D.-X.; Ma, Y.-W.; Liang, X. Direct numerical simulation of compressible turbulent flows. *Acta Mech. Sin.* **2010**, *26*, 795–806. [CrossRef]
125. Sharp, N.S.; Neuscamman, S.; Warhaft, Z. Effects of large-scale free stream turbulence on a turbulent boundary layer. *Phys. Fluids* **2009**, *21*, 095105. [CrossRef]
126. Bhatti, M.; Ellahi, R.; Zeeshan, A. Study of variable magnetic field on the peristaltic flow of Jeffrey fluid in a non-uniform rectangular duct having compliant walls. *J. Mol. Liq.* **2016**, *222*, 101–108. [CrossRef]
127. Elbaz, S.; Gat, A. Dynamics of viscous liquid within a closed elastic cylinder subject to external forces with application to soft robotics. *J. Fluid Mech.* **2014**, *758*, 221–237. [CrossRef]
128. Tchomeni, B.X.; Alugongo, A. Modelling and numerical simulation of vibrations induced by mixed faults of a rotor system immersed in an incompressible viscous fluid. *Adv. Mech. Eng.* **2018**, *10*, 1687814018819341. [CrossRef]
129. Gushchin, V.; Matyushin, P. Classification of the stratified fluid flows regimes around a square cylinder. *AIP Conf. Proc.* **2015**, *1684*, 100002.
130. Belotserkovskii, O.M.; Gushchin, V.A.E.; Kon'shin, V.N. The splitting method for investigating flows of a stratified liquid with a free surface. *USSR Comput. Math. Math. Phys.* **1987**, *27*, 181–191. [CrossRef]
131. Amouroux, S.C.; Heider, D.; Gillespie, J.W., Jr. Permeability estimation of nano-porous membranes for nonwetting fluids. *J. Porous Media* **2010**, *13*, 319–329. [CrossRef]
132. Raupp, S.M.; Schmitt, M.; Walz, A.-L.; Diehm, R.; Hummel, H.; Scharfer, P.; Schabel, W. Slot die stripe coating of low viscous fluids. *J. Coat. Technol. Res.* **2018**, *15*, 899–911. [CrossRef]
133. Stone, H.A. Interfaces: In fluid mechanics and across disciplines. *J. Fluid Mech.* **2010**, *645*, 1–25. [CrossRef]
134. Ajdari, A.; Bocquet, L. Giant Amplification of Interfacially Driven Transport by Hydrodynamic Slip:Diffusio-Osmosis and Beyond. *Phys. Rev. Lett.* **2006**, *96*, 186102. [CrossRef] [PubMed]

fluids

MDPI

Article

Analysis of the Evolution of Shock Waves When a Muzzle Jet Impacts a Constrained Moving Body

Zijie Li and Hao Wang *

School of Energy and Power Engineering, Nanjing University of Science and Technology, Nanjing 210016, China; lizijie@njust.edu.cn
* Correspondence: wanghao1960@126.com

Abstract: The evolution form of the flow field generated via the impact of a muzzle jet on a constrained moving body changes from the state of fully free-spatial development to that of constrained development, and it involves the problem of interference, owing to the spatiotemporal coupling of various kinds of shock waves and vortices. Against this backdrop, the authors use the dynamic mesh method to establish two models for simulating the flow field and exploring the mechanism of development and the characteristics of propagation of disturbances induced via the shock waves as the muzzle jet impacts a constrained moving body. The results show that the muzzle jet exhibited a circumferentially asymmetric shape under the influence of the constrained track. The shock wave leaned towards the upper part of the muzzle, and its speed of propagation above the muzzle was higher than that below the muzzle. Meanwhile, the vortex that should have been present below the muzzle disappeared, and it was replaced with a separation line. Changes in the pressure of the flow field and important parameters of the moving body also became more complex due to the influence of the constrained track.

Keywords: shock wave; vortex; muzzle jet; constrained track; dynamic mesh

1. Introduction

The traditional muzzle has a completely circular shape, or it is equipped in such devices as muzzle brakes of varying shapes [1,2]. However, a recently developed launch device was equipped with a long track in front of the muzzle for the non-destructive recovery of a projectile during the operation of a rocket track [3,4]. This leads to a unique form of motion for the projectile. That is, the movement of the projectile changes from high-speed motion in an open, full space to constrained high-speed motion. This renders the characteristics of transient development of the moving shock wave in this novel muzzle jet significantly different from those in the traditional muzzle jet.

The muzzle jet poses a series of aerodynamic problems in which such complex flows as shock waves/shock wave interference, a strong moving shock wave, and strong vortices represent the major phenomena of flow in the flow field [5–7]. The shock wave in the conventional muzzle develops in a free field in full space, and it is spherically symmetric in three dimensions (3D). Compared with the traditional muzzle jet, the projectile moves forward on the track as propelled by this novel type of muzzle jet to form a special moving boundary at a right angle with the track, where this constrains the development of the muzzle jet in free space. The muzzle jet subsequently exhibits highly 3D, circumferential, and asymmetric characteristics that lead to the formation of multiple types of vortices [8]. The development of the muzzle jet thus becomes complex under the interference of the track. It is, therefore, important to analyze the complex phenomena of flow in muzzle jets with restricted boundaries to inform scientific research in the area and solve key technical problems [9,10].

A considerable amount of research has been devoted to the muzzle jet, and it has led to the establishment of a relatively systematic theory. The characteristics of the transient evolution of shock waves caused by interference with the muzzle device, and the impact of the muzzle jet on a high-speed moving body in full space, have long been popular subjects of research in this area [11,12].

Schmidt et al. [13,14] measured the characteristic parameters of a set of muzzle brake structures by using a 20-mm small-caliber gun, and they obtained clear images of the muzzle jet. They analyzed the characteristics and flow mechanism of the muzzle jet, and they observed that the coupling between the flow fields along the axial direction was very strong but that along the transverse direction was very weak.

Zeqing et al. [15,16] conducted a systematic study of the flow field of a 7.62-mm small-caliber launcher with a cylindrical muzzle device impacting a high-speed projectile in full space. The results showed that the vorticity in the flow field was mainly concentrated at the boundary of the jet, and a large number of small-scale vortices appeared downstream of the Mach disk. As the bottle-shaped shock wave contracted, the vortex ring above the three wave points detached from the boundary of the jet and developed independently. Instability subsequently occurred at the boundary, and this led to the formation of a large number of large-scale vortices directly downstream of the Mach disk. The jet inside the muzzle was much more complex under the influence of the muzzle device (cylindrical nozzle).

Huanhao et al. [17,18] used large-eddy simulations to examine the characteristics of the flow field during the interaction between a supersonic mainstream and a transverse jet, and they used the results to derive the law of evolution of the morphology of the 3D shock wave. The leading edge of the jet formed a bow-shaped shock wave and a separated 'λ' shock wave, owing to the strong inverse pressure gradient formed via the interaction between the supersonic mainstream and the transverse jet. In addition, the interaction between the strong compression-induced shock wave, the shear layer, and the horseshoe vortex accelerated instability in the flow field, and this resulted in the formation of a complex multi-scale structure of the downstream flow. Therefore, the complex shock wave system in the flow field of the supersonic transverse jet plays an important role in inducing turbulence in it, as well as strengthening the mixing between the jet and the mainstream.

Zijie [19,20] identified the evolution characteristics of the shock wave of muzzle jets in case of an initial interference with the jet. The velocity of the shock wave increased in the muzzle jet subjected to this interference, and it led to more pronounced wrapping of the moving projectile. This led to a strengthening of the effect of the shock wave at the bottom of the muzzle jet, and it hindered the generation of the Mach disk.

Zijie, Wang Hao [21] conducted experiments and numerical simulations on a 300 mm caliber balanced gun and obtained the development process and laws of a muzzle jet in near-Earth space. Research has found that muzzle shock waves reflect off the ground, forming shock waves with opposite propagation directions. The shock wave generated via ground reflection intersects with the muzzle shock wave, forming a new wavefront. A vortex was generated near the new wavefront, and the shape of the shock wave at the bottom of the projectile was also disrupted, exacerbating the initial disturbance of the moving projectile.

The above brief summary of the literature shows that researchers have extensively examined the characteristics of the flow field of a muzzle jet impinging on a moving body in a free state and half-space from the perspectives of mechanisms, numerical simulation methods, and visualization experiments. But few studies have investigated the development and evolution of shock waves and vortices when the muzzle jet impacts a constrained moving body. This is a new scientific problem that has important guiding significance for solving the key technical difficulties in the engineering application of integrated gun rocket sled launch systems. Therefore, it is necessary to conduct research on the development and evolution mechanisms of various forms of shock waves and vortices in the flow field of jet impact-constrained moving bodies. Accordingly, the authors of this study establish a representative model of simulation with the FLUENT software 2019R1, which utilizes a k–ε turbulence model and combines it with dynamic grid technology. At the same time, we compile the six degrees of freedom of the moving body for the secondary development of the software to examine the evolution characteristics of the shock wave of a supersonic jet as it impacts a constrained high-speed moving body. We identify the process of transient evolution of shock waves in the flow field by combining theoretical calculations, simulations, and experiments.

The remainder of this article is structured as follows: Section 2 provides a detailed introduction to the physical model, the model of the mesh, the dynamic mesh method, boundary conditions, and the methods of solution. We also report the results of tests in that section. Section 3 discusses the evolution characteristics of the shock wave when a supersonic jet impacts a constrained moving body, including the structure of the muzzle jet, the evolution of the shock wave, the generation and evolution of vortices, the transient changes in the pressure at key points, and the differences in the key parameters of the moving body. Section 4 summarizes the conclusions of this study.

2. Physical Model and Mesh Model

2.1. Physical Model

To clearly represent the development and evolution of shock waves when the muzzle jet impacts a constrained moving body, where this is different from cases involving traditional muzzle jets, we established two models of numerical simulation: one without a constrained track in front of the muzzle (Case 1) and the other with it (Case 2). Except for the presence/absence of the constrained track, all other parameters of the two models were identical. We focus here on the structures and sizes of the inlet of the jet and the moving body. Figure 1a–c show the overall structural diagram of the proposed model, and schematic diagrams of the models without the constrained track (Case 1) and with it (Case 2), respectively. The model included a constrained track, a tube, a moving body, and support plates. Its dimensions were set according to the parameters of the corresponding experiments conducted by the authors, and its size-related parameters are shown in Table 1. The diameter of the muzzle Dm was 65 mm, and the length of the moving body Lb was 200 mm. The size of the entire computational domain Lx was 5.7 m along the direction of motion of the body, which moved over a distance, Li, of 5 m. The lengths Ly and Lz of the computational domain were 2.5 m. Due to its structural design, the new muzzle of the launch device was not completely circular but had a shape that conformed to that of the constrained track. The moving body also had a corresponding shape. The cross-sectional shape of the track is rectangular. The height is 200 mm, and the width is 25 mm. In this article, numerical simulations were conducted on half of the physical model with a width of 12.5 mm. The length fully satisfies the length of the physical model in the direction of projectile motion.

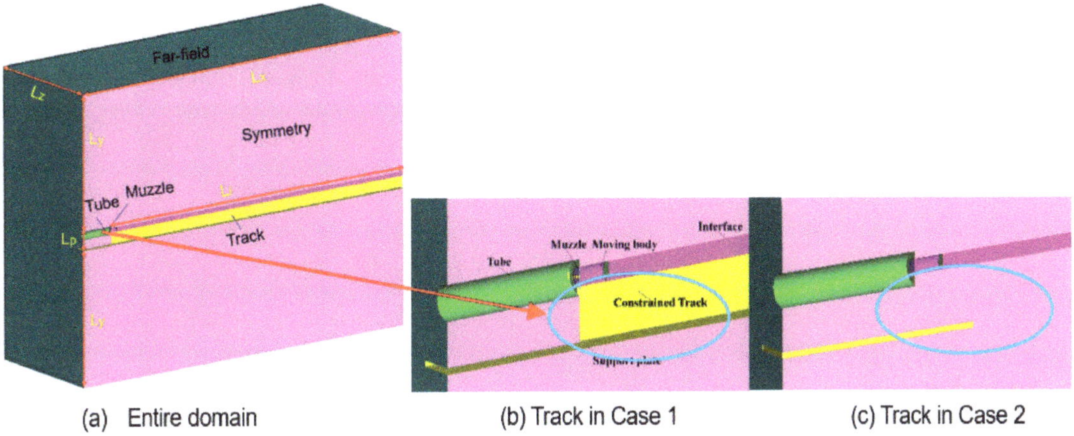

Figure 1. Simulated physical structures of the computational domain and the track.

Table 1. The size-related parameters of the model.

Parameter	Value
Lx	5.7 m
Ly	2.5 m
Lz	2.5 m
Dm	65 mm
Lb	200 mm
Lp	255 mm
Li	5.0 m

2.2. Meshing Method

To reduce the cost of calculations and the time needed to perform them, we conducted simulations on only half of the models considered here (as shown in Figure 2). The computational domain was divided into approximately 9.8 million structured grids. When simulating a complex flow field, it is often difficult to generate a single, high-quality structural grid. Moreover, the simulations required the dynamic grid method, which made it necessary to divide the entire computational domain into subdomains of stationary and dynamic grids for processing. The interface was used for numerical exchange.

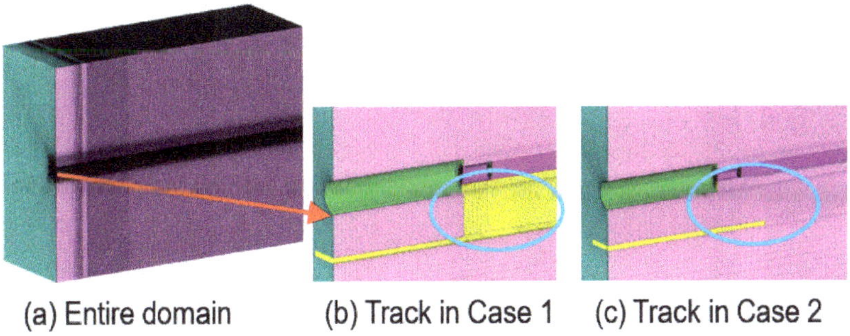

Figure 2. Schematic diagram of the model of the grid.

2.3. Dynamic Mesh Method

We needed to use dynamic grid technology to handle the changes in the grid caused by the motion of the body during the calculations. The dynamic grid method based on three algorithms was used to calculate the requisite adjustments to the internal nodes of the grid: the layering, smoothing, and remeshing algorithms. Because we assumed that the body moved only in one direction, it was appropriate to use the layering method to simulate its motion. The layering method mainly uses a pair of interfaces for data transfer. The interface is mainly used to handle the data transfer between region interfaces in multi-region computing models. Therefore, in order to keep the computing domain smooth in multi-region computing models, interfaces need to be created at the boundaries of mutual contact. Meanwhile, this method requires setting the segmentation and merging factors of the mesh. The mesh behind the body shifted forward along the x-axis due to the motion of the projectile. A layer of the mesh near the muzzle was thus elongated and had a side length of hs. The part of the mesh that exceeded the boundary of the computational domain was truncated, and the length of the layer of mesh near the boundary of the computational domain was hc. Assuming that the ideal size of the mesh was hi, we divided it into two parts when the size of the newly generated mesh satisfied h > (1 + Cs) hi. When the size of the mesh satisfied the condition h > Cshi, the grid was merged into a single grid. In this example, hi = 2.0 mm, Cs = 0.4, and Cl = 0.2.

2.4. Boundary Conditions and Solution Methods

We simulated the model of a supersonic jet impinging on a high-speed moving body in Fluent software 2019 R1. Figure 3 shows the types of boundaries used in the simulations. The outlet of pressure was set as the boundary condition of the domain of calculation, while the tube and track were assumed to be non-sliding insulated walls. The moving body was assumed to adhere to the boundary condition of the moving wall, and it was controlled via a user-defined function (UDF) with six degrees of freedom (6 DOFs). The initial velocity was set to 500 m/s according to the results of prior experiments. The inlet of pressure was set as the boundary condition for the muzzle, and it was 45 MPa, according to the experiments. The environmental pressure was 101,325 Pa, and the temperature was 300 K.

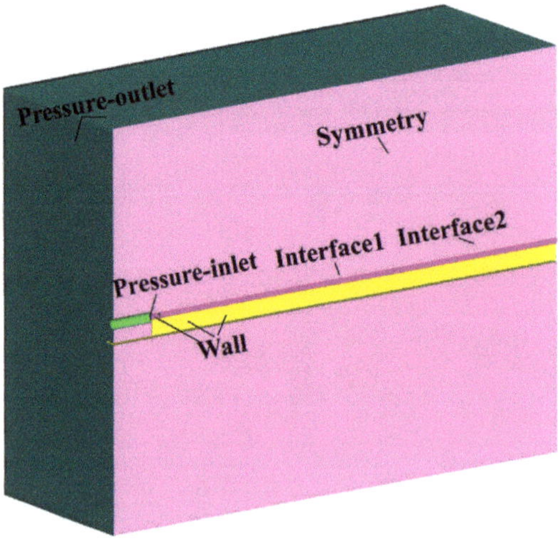

Figure 3. Boundary conditions of the model of the muzzle jet.

We processed the motion of the body in the block-based division of the mesh by using the finite volume method in combination with the structural dynamic method. The Navier–Stokes (N–S) equation solver based on density correction and the k-ε realizable model of turbulence was used. The inviscid convective flux was split by using the Roe-FDS scheme, and the implicit scheme was used with the time-marching method to accelerate the convergence of the numerical calculations. The equation of flow control was discretized using the second-order upwind scheme, and the material was assumed to be an ideal gas.

2.5. Test Verification

These simulations were based on a small-scale experiment conducted by the authors. Due to constraints imposed by the experimental equipment and the conditions of implementation, the constrained track was short, such that it might have skewed the results of the experiment. The experimental platform is the main jet generator, which ignites the gunpowder inside the gun barrel through an ignition device, generating high-temperature and high-pressure gas to launch the moving body. The PEPE KD2002L-02 free field pressure sensor is used, with a measurement range adjustable from 1–10 MPa, to monitor the pressure changes in key characteristic points in front of the muzzle over time. The signal is collected using the DEWE2-A13 transient data recorder produced by DEWETRON Industrial Equipment Company in Austria. Figure 4 is the experimental platform involved in this article; the schematic diagram of the experimental plan and the measuring device are shown in Figures 5 and 6.

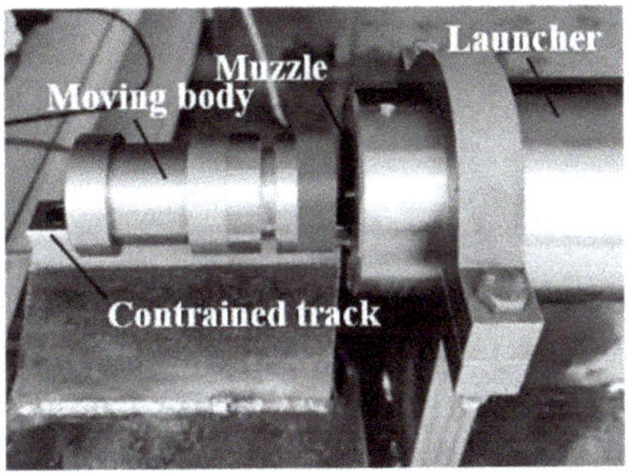

Figure 4. Experiment platform of the nozzle of the engine used for the rocket track test.

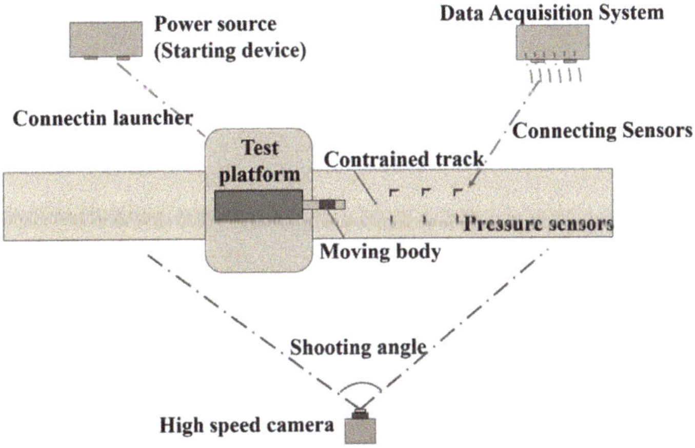

Figure 5. The schematic diagram of the experimental plan.

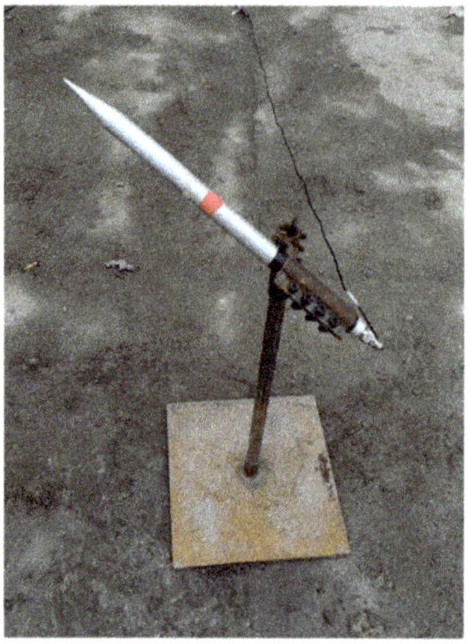

Figure 6. The measuring device of the free field pressure sensor.

We verified the experiments based on numerical simulations, and subsequently mainly analyzed the model of numerical simulation in this study. Figure 7 compares the changes in pressure over time between the experiment and the simulation at the key monitoring point (1500, 30, and 10) mm, and it is evident from it they were consistent. Therefore, the method of simulation and the mesh used in this article could accurately represent the characteristics of the muzzle jet.

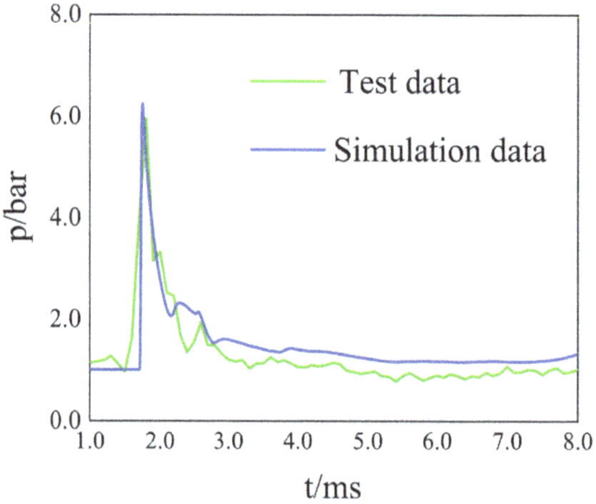

Figure 7. Comparison of pressure between the results of the experiments and simulations.

2.6. Grid Independence Verification

We provide three sets of grids, which are defined as coarse mesh, fine mesh, and dense mesh, with 6 million, 9.8 million, and 12 million, respectively. The numerical simulation results of different grid models considered for the thrust variation on the bottom of the moving body in Case 1 are shown in Figure 8. As the figure shows, the results obtained using fine mesh and dense mesh are very close, with an error of less than 1%. On the contrary, there are significant differences in the results obtained from the coarse grid model.

Therefore, it can be concluded that, when using a fine mesh model, the simulation results are independent of the mesh size.

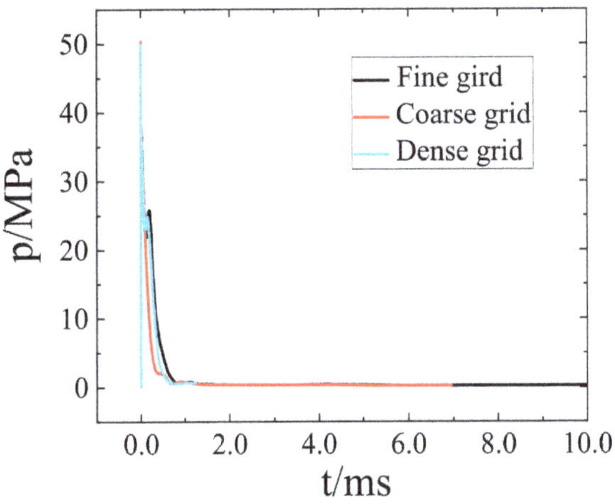

Figure 8. The thrust variation on the bottom of the moving body for grid independence verification.

3. Analysis of Results

We aim to analyze the evolution of shock waves in the muzzle jet under different boundary conditions. The process of interest includes the development of shock waves in the muzzle, the structure of vortices generated by them, and the flight attitude of the moving body. In view of this, we examine the characteristics of the development of flow in the muzzle jet on a symmetric plane, the development and evolution of vortex structures in the flow field, changes in pressure at symmetric monitoring points, and the key parameters of motion of the body in the two models considered here.

3.1. Analysis of Muzzle Jet

We monitored the contours of pressure at different times to compare the models of the muzzle jet with and without constrained tracks. We do not provide more details on the conventional muzzle jet because past studies have already reported a sufficient amount of research on it [2,22]. Figures 9 and 10 show the contours of pressure on the plane of symmetry in Case 1 and Case 2 from 0.5–5 ms. It is clear from them that the muzzle jet differed significantly in the two models.

Figure 9 shows the preliminary structure of the flow field of the muzzle in Case 1. The muzzle jet emitted high-temperature and high-pressure gas, which then rapidly expanded. As its pressure decreased, the velocity of the gas increased. It subsequently spread and pushed the air around the muzzle to form a spherical shock wave. The muzzle jet had a roughly circumferential symmetric structure because the muzzle was not a complete circle. The muzzle jet developed much more quickly at the junction of the trajectory of the muzzle, where this was also the local minimum cross-section. However, the development of the flow field of gas was insufficient, such that it influenced only a small portion of the area. Nonetheless, typical muzzle jets remain adequately representative.

The shock wave at the muzzle continued to develop over time, and the scope of its impact gradually expanded. Figure 9b,c show that the muzzle jet developed gradually, and its asymmetry decreased over time. A complete wave system consisting of a multi-level shock wave, which was composed of coronal shock waves, reflected shock waves, Mach disks, and overlapping discontinuities, was formed around the muzzle. It is clear from Figure 9d–f that the muzzle jet developed thoroughly, and it had an axisymmetric shape.

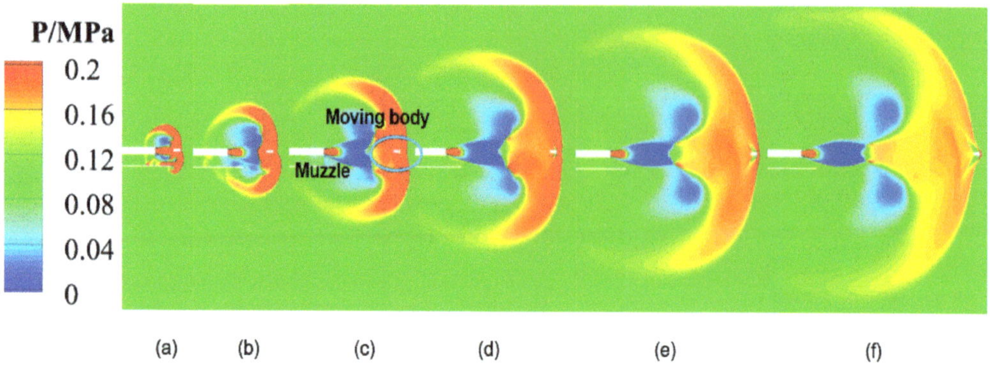

Figure 9. Contours of pressure of the muzzle jet on the plane of symmetry over time in Case 1: (**a**) t = 0.5 ms; (**b**) t = 1.0 ms; (**c**) t = 2.0 ms; (**d**) t = 3.0 ms; (**e**) t = 4.0 ms; (**f**) t = 5.0 ms.

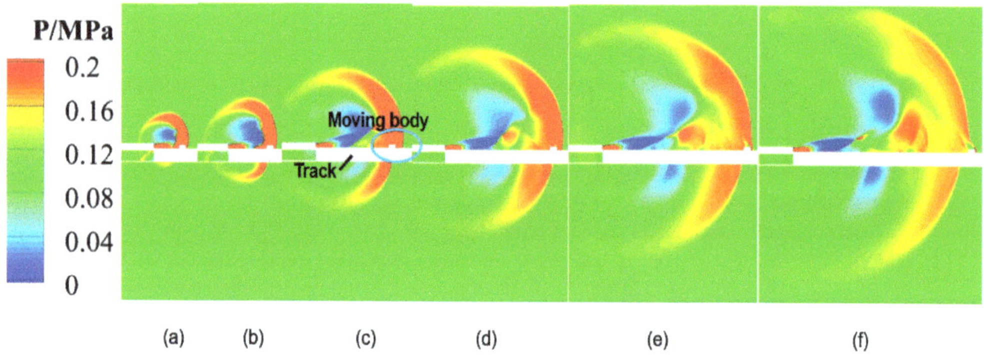

Figure 10. Contours of pressure of the muzzle jet on the plane of symmetry over time in Case 2: (**a**) t = 0.5 ms; (**b**) t = 1.0 ms; (**c**) t = 2.0 ms; (**d**) t = 3.0 ms; (**e**) t = 4.0 ms; (**f**) t = 5.0 ms.

The flow fields of the two models exhibited certain differences, owing to their different tracks. The track cut-off space for the downward development of the muzzle jet in Case 2, causing it to twist and deform while exhibiting a highly asymmetric axial shape. The flow field was no longer a spherical wavefront, and this hindered its development. Therefore, the shock waves needed to bypass the track in order to continue propagating. They were reflected at this point such that this caused another shock wave to form in the opposite direction of propagation, thereby forming a corresponding wavefront along the track. The track prevented the shock wave from propagating downwards, and it was reflected and propagated upwards to form an irregularly shaped wavefront. On the whole, it developed in 3D space in the area above the track.

Figure 10a shows that high-temperature and high-pressure gas was ejected from the muzzle along the constrained track, and it influenced only the area above the track, owing to the insufficient development of the muzzle jet. Figure 10b shows that the shock wave influenced the lower area of the constrained track over time, and the body moved forward on the track. It is clear from Figure 10c,d that the coronal shock wave in front of the moving body was significantly asymmetric, and its speed of propagation in the upper region was significantly higher than that in the lower region. The Mach disk was also truncated by the track, such that it was incompletely formed. Figure 10e,f show that the muzzle jet had fully developed, with a prominent decrease in its asymmetry. The development of the flow field was dependent on its previous states, and it featured asymmetric coronal shock waves, reflected shock waves, multi-level shock waves composed of Mach disks, and a complete wave system of overlapping discontinuities that had formed around the muzzle.

3.2. Analysis of Vortices

To further compare the muzzle jets with differently constrained boundaries, we analyzed the development of vortex structures in the range of 1.0–4.0 ms. The presence of the track hindered the development and propagation of the shock waves in the muzzle, changed their direction of propagation, and generated lateral airflow that led to changes in the vortex structure in the muzzle jet. The vortex increased the energy consumption and diffusion of the flow field, and it influenced the distributions of its pressure and velocity. The eddy currents moved, deformed, merged, paired, and ruptured, and remained far downstream as their strength gradually decreased. The change in the vortical structure was a phenomenon unique to the muzzle jet with constrained boundaries, and it was not observed in the traditional muzzle jet. We thus focus here on the generation and evolution of vortices in the flow field.

Figures 11 and 12 show the evolution of vortices in the symmetric muzzle jet in the two cases considered here over time. Figure 11a shows that two symmetrical vortices were formed in the conventional muzzle jet over time. These vortices were located in the low-pressure zone, with the streamline directed from the center of the vortex to outside it. This means that the spiral point was stable and difficult to dissipate. The vortices developed steadily over time.

In comparison with the above, it is evident that the vortical structure in Case 2 differed significantly from that in Case 1 due to interference by the constrained track, support plate, and test platform. The vortex above the muzzle in Case 2 was similar to that in Case 1, and it was located in the low-pressure zone. Its streamline pointed outward from the center of the vortex, where this showed that the spiral point was stable, and it was difficult to dissipate. No vortex was formed in the area below the track, and the separation line was generated instead. This is because the track hindered the development of the muzzle jet above it and changed its pattern of development. The vortex is a rotating flow formed in a fluid, which can cause energy loss, fluid instability, and turbulence in the flow field. Therefore, eliminating vortices is of great significance for optimizing fluid motion and improving flow field stability. If the vortex structure in the fluid is easily dissipated, the vortex intensity in the flow field will be smaller, and the vortex dissipation rate will be higher. In this article, in models without constrained tracks, vortices are relatively stable, while in models with constrained tracks, vortices are relatively unstable. This indicates that constraining the orbit leads to a more unstable flow field, making the flow more complex and energy dissipation severe.

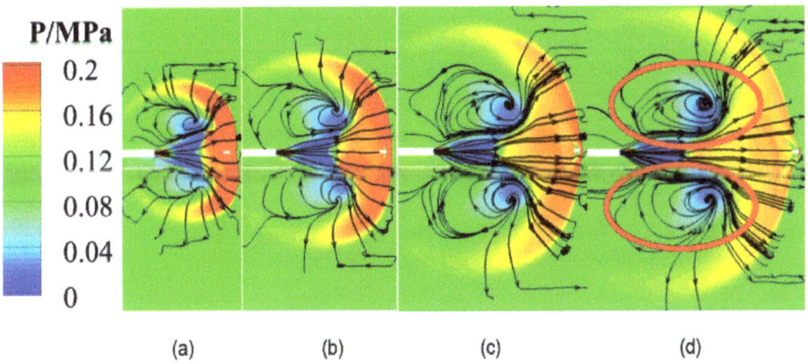

Figure 11. Contours of vortices of the muzzle jet on the plane of symmetry over time in Case 1: (**a**) t = 1.0 ms; (**b**) t = 2.0 ms; (**c**) t = 3.0 ms; (**d**) t = 4.0 ms.

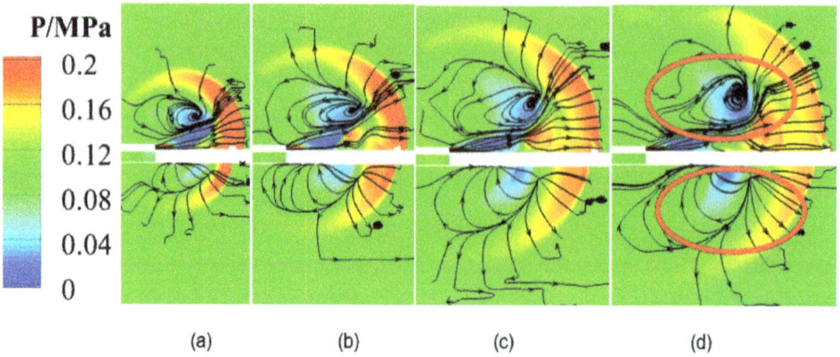

Figure 12. Contours of vortices of the muzzle jet on the plane of symmetry over time in Case 2: (**a**) t = 1.0 ms; (**b**) t = 2.0 ms; (**c**) t = 3.0 ms; (**d**) t = 4.0 ms.

3.3. Analysis of Pressure Changes

We also monitored changes in pressure at key points over time to illustrate the impact of the constrained track on the muzzle jet. Figure 13 provides a schematic diagram of the locations of the monitoring points and Table 2 shows their specific locations.

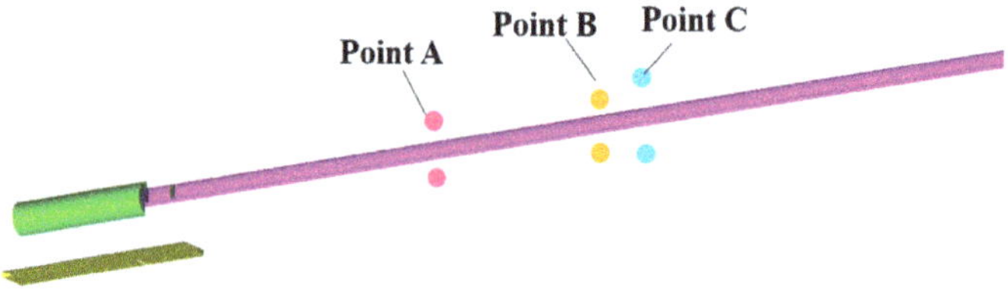

Figure 13. Distribution of the monitoring points on the track.

Table 2. Locations of the monitoring points.

Point	Location Parameter (x, y, z)/mm
A-above A-below	(1900, 30, 100) (1900, 30, −100)
B-above B-below	(3000, 60, 100) (3000, 60, −100)
C-above C-below	(3500, 70, 200) (3500, 70, −200)

Figure 14 shows a comparative analysis of changes in the pressure at the monitoring points considered. It is clear from it that the changes in pressure, which should have remained the same as before, now differed due to the influence of the track. When the shock wave did not propagate to a given monitoring point, the pressure at this point was assumed to be the environmental pressure. When the wavefront of the shock wave propagated to this point, the pressure at it rose. The gas continued to expand, and pressure at the relevant monitoring point decreased and finally became stable.

The shock wave is a complicated system containing alternating expansion- and compression-induced waves, where this makes it difficult to identify changes in the pressure. A comparison of the pressure between the upper and lower symmetrical points in Case 1 showed that the changes in pressure at these points were almost identical. The reason for the difference is to control a single variable; that is, although there is no constrained track, the shape of the muzzle still maintains the same shape as in Case 2, which is not a complete circle. Therefore, it is reasonable for the pressure to have been slightly different at the upper and lower points of symmetry.

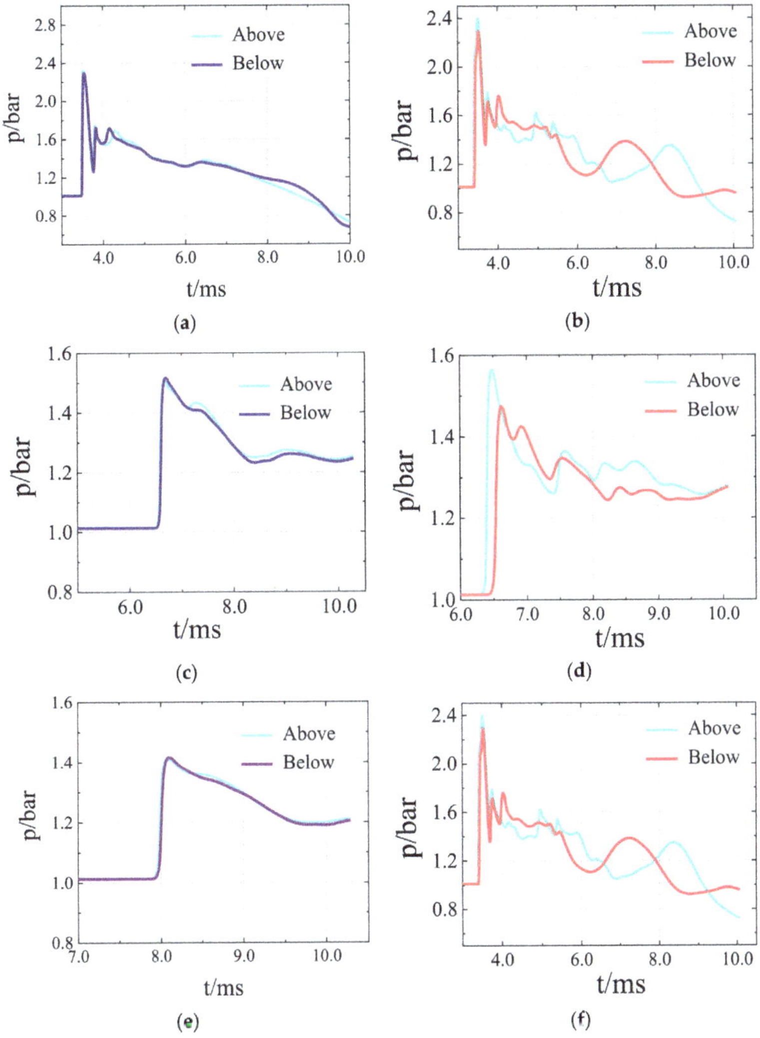

Figure 14. Pressure changes at monitoring points: (**a**) Point A for Case 1; (**b**) Point A for Case 2; (**c**) Point B for Case 1; (**d**) Point B for Case 2; (**e**) Point C for Case 1; (**f**) Point C for Case 2.

Figure 14a,b show that the trends of changes in pressure remained similar between the cases. The body moved at about 500 m/s. The shock wave moved to this point at approximately 3.8 ms, and the disturbance in pressure was observed immediately. It rapidly increased and then gradually decreased. Due to the influence of the track in Case 2, the muzzle jet deviated upwards, and this led to a higher maximum pressure at A-above than that at A-below. The pattern of development of the shock waves also became more complex, because of which the severe fluctuations in pressure during the descending stage were also significantly different from those in Case 1.

Similarly, it is clear from Figure 14c,d and Figure 14e,f that the shock wave moved to this characteristic point at approximately 6.5 ms. The change in pressure at the upper monitoring point occurred earlier than at the lower monitoring point, indicating that the shock wave tended to develop above the muzzle. Meanwhile, the maximum pressure at the upper monitoring point was higher than that at the lower monitoring point. The pattern of development of the shock waves was also relatively more complex, such that the pressure oscillated severely during the descending phase.

In addition, Figure 15 provides a comparison of the changes in pressure over time between the cases at the same monitoring point. It is clear from it that the two cases had similarities, as well as differences. The speed of propagation of the shock wave in Case 2 was higher than that in Case 1 in all Figures, indicating that the presence of the constraining track had increased the speed of propagation of the shock wave. The figures show that the variation in pressure at the characteristic point in Case 1 was relatively smooth and singular, while that in Case 2 was complex. This proves that the constrained track increased the complexity of the development of the shock wave.

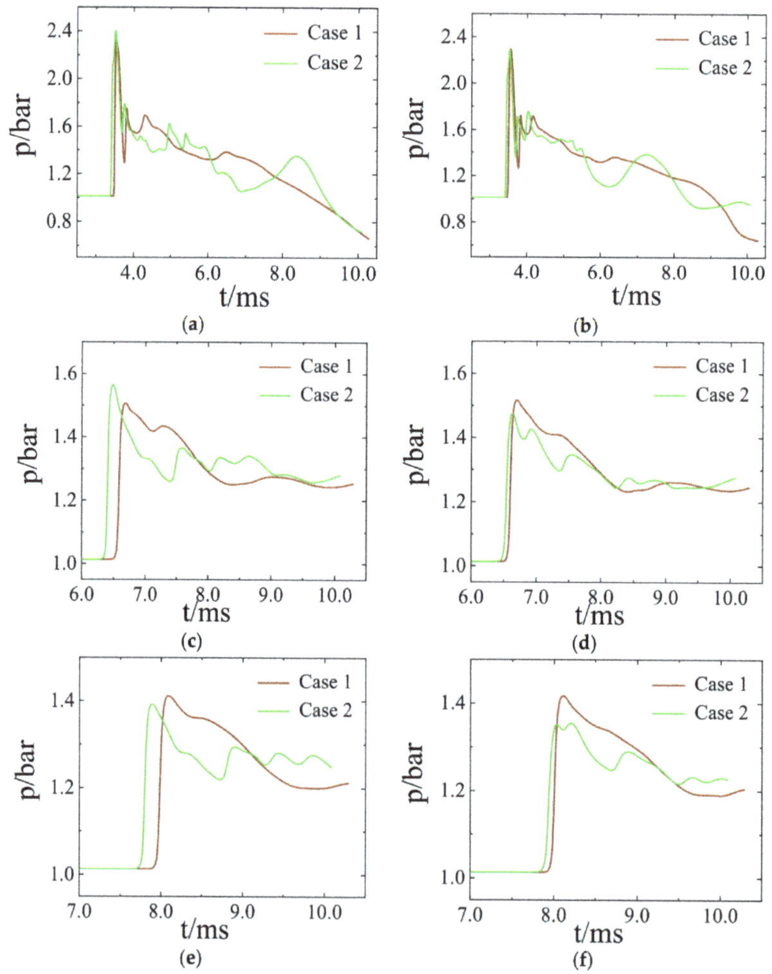

Figure 15. Pressure changes at monitoring points: (**a**) A—above; (**b**) A—below; (**c**) B—above; (**d**) B—below; (**e**) C—above; (**f**) C—below.

Specifically, Figure 15a–e show that changes in the initial pressure in Case 2 occurred much earlier than in Case 1, indicating that the speed of propagation of the shock wave in Case 2 was significantly higher than that in Case 1. These three points were all located above the muzzle, and this verifies that the constrained track caused the shock wave to develop above the muzzle.

It is clear from Figure 15b,d,f that the initial pressure in Case 2 changed slightly earlier than in Case 1, indicating that the speed of propagation of the shock wave in Case 2 was slightly higher than that in Case 1. These three points were all located below the muzzle, and this shows that the constrained track caused the speed of propagation of the shock wave in the lower region to still be higher than that in the model without the constrained track.

3.4. Analysis of Key Parameters of Moving Body

Figure 16 compares the key parameters of motion of the body. Figure 16a shows the Mach number of the moving body, while Figure 16b shows the drag on it. It is clear from Figure 16a that the Mach number of the moving body in Case 2 was higher than that in Case 1. The Mach numbers of the moving body in the two cases were very similar close to the muzzle, owing to the dominant influence of the thrust of the muzzle jet. The body gradually moved away from the muzzle, whose effect of thrust rapidly decayed. The Mach number of the moving body was more significantly influenced due to interference via the track in this case. In conjunction with the above results, it is evident that the shock waves developed more quickly in Case 2, and the Mach number of the moving body was also higher than in Case 1.

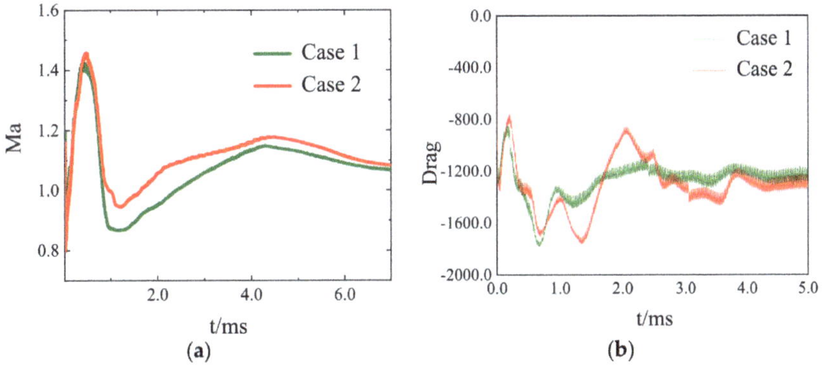

Figure 16. Mach number and drag force of the moving body: (**a**) Mach number; (**b**) drag force.

Figure 16b shows that fluctuations in resistance in Case 2 were significantly more complex than in Case 1. Because the thrust had the major influence on the muzzle jet close to the muzzle, the values of resistance of the moving body in both cases were similar. The body gradually moved away from the muzzle, and the effect of the thrust of the muzzle decayed rapidly. The change in the resistance of the moving body was significantly influenced by its speed and the track. It is also evident that the change in resistance in Case 1 was relatively smooth because the moving body flew freely in fully open space. On the contrary, the form of development of the shock wave was significantly influenced in Case 2, and it led to significant fluctuations in the resistance of the moving body.

4. Conclusions

In this study, the authors established two models of simulation to investigate the impact of a muzzle jet on a constrained moving body based on dynamic grid technology, with the aim of discussing the unique development and evolution of the shock waves and vortices. We drew the following conclusions:

(a) The presence of the constrained track led to the twisting and deformation of the shock wave of the spherical muzzle, where this should have been circumferentially symmetric. The overall shock wave leaned towards the upper part of the center of the muzzle, and its speed of development above the muzzle was much higher than that below it.

(b) The influence of the constrained track on the muzzle jet led to the generation of lateral airflow that influenced the morphology of the vortices. The vortices were supposed to be symmetrical above and below the muzzle. However, the vortex above the muzzle became asymmetrical, and that below it disappeared. Only a separation line was formed below the muzzle.

(c) The characteristics of development of the complex shock waves of the muzzle, owing to the constrained track, were notably different from those in its absence, with more pronounced fluctuations in the pressure at the monitoring points and the drag force on the moving body. The Mach number of the moving body was also higher than that in the model without the track.

Author Contributions: Z.L. performed the data analysis, conducted the formal analysis, and wrote the manuscript. H.W. performed the revision of the paper. All authors have read and agreed to the published version of the manuscript.

Funding: The evolution mechanism and control method of shock waves in a near ground constrained moving body under the impact of a dual bore jet, funded by the National Youth Science Foundation of China with grant numbers 124022682024.

Data Availability Statement: All data and models that support the findings of this study are available from the corresponding author upon reasonable request. All data and models generated or used during the study appear in the submitted article.

Acknowledgments: We thank Yonggang Yu for help with the written discussions in this paper.

Conflicts of Interest: The authors declare no conflicts of interest.

References

1. Jiang, S.; Wang, H.; Lin, C.; Wang, J. Establishment and simulation of two-dimensional interior ballistics model of large caliber Davis gun considering the axial movement of tubular charge clusters. *Acta Armamentarii* **2016**, *37*, 1941–1948. [CrossRef]
2. Wang, G.; Cheng, C.; Zhang, X.; Huang, X. Numerical simulation and analysis of muzzle flow during a rarefaction wave gun firing. *Propellants Explos. Pyrotech.* **2021**, *46*, 45–56. [CrossRef]
3. Goto, K.; Kato, Y.; Ishihara, K.; Matsuoka, K.; Kasahara, J.; Matsuo, A.; Funaki, I.; Nakata, D.; Higashino, K.; Tanatsugu, N. Thrust validation of rotating detonation engine system by moving rocket track test. *J. Propul. Power* **2020**, *37*, 419–425. [CrossRef]
4. Zhang, G.; Kim, H.D. Numerical simulation of shock wave and con tact surface propagation in micro shock tubes. *J. Mech. Sci. Technol.* **2015**, *29*, 1689–1696. [CrossRef]
5. Li, Z.J.; Wang, H. Effect of Precursor Flow Field of Muzzle on the Combustion Gas Jet Flow of Gun Propellant. *Chin. J. Energetic Mater.* **2017**, *25*, 282–290. [CrossRef]
6. Teramoto, S. Large-Eddy Simulation of Transitional Boundary Layer with Impinging Shock Wave. *AIAA J.* **2015**, *43*, 2354–2363. [CrossRef]
7. Mäkinen, T.; Pertilä, P. Shooter localization and bullet trajectory, caliber, and speed estimation based on detected firing sounds. *Appl. Acous.* **2010**, *71*, 902–913. [CrossRef]
8. Jones, A.R.; Medina, A.; Spooner, H.; Mulleners, K. Characterizing a burst leading-edge vortex on a rotating flat plate wing. *Exp. Fluids* **2016**, *57*, 1–16. [CrossRef]

9. Moumen, A.; Grossen, J.; Ndindabahizi, I.; Gallant, J.; Hendrick, P. Visualization and analysis of muzzle flow fields using the background-oriented schlieren technique. *J. Vis.* **2020**, *23*, 409–423. [CrossRef]
10. Florio, L.A. Effect of vent opening area and arrangement on gas flow field as gas propelled cylinder exits a flow tube. *Meccanica* **2010**, *45*, 475–501. [CrossRef]
11. Souverein, L.J.; Dupont, P.; Debiève, J.-F.; Dussauge, J.-P.; van Oudheusden, B.W.; Scarano, F. Effect of Interaction Strength on Unsteadiness in Shock-Wave-Induced Separations. *AIAA J.* **2010**, *48*, 1480–1493. [CrossRef]
12. Lees, L.; Reeves, B.L. Supersonic Separated and Reattaching Laminar Flows: I. General Theory and Application to Adiabatic Boundary Layer-Shock Wave Interactions. *AIAA J.* **2015**, *2*, 1907–1920. [CrossRef]
13. Schmidt, E.M.; Fansler, K.S.; Shear, D.D. Trajectory Perturbations of Fin-Stabilized Projectiles Due to Muzzle Blast. *J. Spacecr. Rocket.* **2015**, *14*, 339–344. [CrossRef]
14. Schmidt, E.M.; Shear, D. Optical measurements of muzzle blast. *AIAA J.* **2012**, *13*, 1086–1091. [CrossRef]
15. Guo, Z.Q.; Jiang, X.H.; Wang, Y. Parallel numerical simulation of muzzle reacting flow. *Chin. J. Comput. Mech.* **2013**, *30*, 111–116+123. [CrossRef]
16. Guo, Z.; Qiao, H.; Jiang, X. Numerical simulation on the characteristics of flow field of the muzzle of embedded aircraft gun. *Acta Armamentarii* **2017**, *38*, 2373–2378. [CrossRef]
17. Zhang, H.; Xiao, Y.; Zheng, C.; Chen, Z.; Xu, D.; Zhu, S.; Song, W. Characteristics of three-dimensional flow field structure of supersonic transverse jet. *J. Propul. Technol.* **2023**, *44*, 30–39. [CrossRef]
18. Zhang, H.; Chen, Z.; Jiang, X.; Li, H. Investigations on the exterior flow field and the efficiency of the muzzle brake. *J. Mech. Sci. Tech.* **2013**, *27*, 95–101. [CrossRef]
19. Li, Z.; Wang, H. Analysis of Ground Distance Effect on Hypervelocity Jet Flow and Stability of Projectile in Half-space. *AIP Adv.* **2022**, *12*, 045013. [CrossRef]
20. Li, Z.; Wang, H. Mechanism of Evolution of Shock Wave of Muzzle Jet under Initial Interference and Its Simplified Model. *Aerospace* **2024**, *11*, 381. [CrossRef]
21. Li, Z.; Wang, H.; Chen, J. Ground effects on the hypervelocity jet flow and the stability of projectile. *Eng. Appl. Comput. Fluid Mech.* **2018**, *12*, 375–384. [CrossRef]
22. Zhao, X.Y.; Zhou, K.D.; He, L.; Lu, Y.; Wang, J.; Zheng, Q. Numerical Simulation and Experiment on Impulse Noise in a Small Caliber Rifle with Muzzle Brake. *Shock Vib.* **2019**, *26*, 5938034. [CrossRef]

fluids

MDPI

Article

Enhancement of Aerodynamic Performance of Two Adjacent H-Darrieus Turbines Using a Dual-Rotor Configuration

Douha Boulla [1,*], Saïf ed-Dîn Fertahi [1,*], Maryam Bernatchou [2], Abderrahim Samaouali [1] and Asmae Arbaoui [1]

1 "Thermodynamics and Energy" Research Team, Energy Research Center, Physics Department, Faculty of Science, Mohammed V University in Rabat, 4 Avenue Ibn Batouta, BP 1014, Rabat 10000, Morocco; abderrahimsamaouali@yahoo.fr (A.S.); arbaoui_asmae@yahoo.fr (A.A.)

2 Physics Department, Team of Modeling and Simulation in Mechanics and Energetics (MSME), Faculty of Sciences, Mohammed V University in Rabat, 4 Avenue Ibn Batouta, BP 1014, Rabat 10000, Morocco; bernatchou.maryam@gmail.com

* Correspondence: douhaboulla@gmail.com (D.B.); fertahi.sayfdin@gmail.com (S.e.-D.F.); Tel.: +212-689-803-781 (D.B.)

Abstract: Improvements in the aerodynamic performance of the H-Darrieus turbine are crucial to address future energy requirements. This work aims to optimize the behavior of two adjacent turbines through the addition of a dual H-Darrieus rotor. The first rotor is composed of three NACA 0021 blades, while the second comprises a single Eppler 420 blade. This study focuses on 2D CFD simulation based on the solution of the unsteady Reynolds-averaged Navier–Stokes (URANS) equations, using the sliding mesh method and $k-\omega$ SST turbulence model. The simulation results indicate a 17% improvement in the efficiency of the two turbines integrating dual rotors, compared to the two isolated turbines, for $\alpha = 0°$. Moreover, the power coefficient (C_p) reaches maximum values of 0.49, 0.42, and 0.40 for angles of attack of 30°, 25°, and 20°, respectively, at TSR = 2.51. Conversely, the selection of an optimal angle of attack allows the efficiency of the two H-Darrieus turbines to be increased. It is also shown by the results that the effect of stagnation is reduced and lift is maximized when the optimum distance between two adjacent turbines is chosen. Moreover, the overall aerodynamic performance of the system is enhanced by the potential of a dual-rotor configuration, and the wake between the two turbines is disrupted, which can result in a decrease in energy production within wind farms.

Keywords: CFD; h-Darrieus; aerodynamic performance; dual rotors; two turbines; sliding mesh

1. Introduction

To address the rising global demand for electricity, renewable energies are currently occupying an important place in energy politics, as they are environmentally friendly and help to reduce emissions of greenhouse gases. Among these energy sources, particular attention has been given to wind turbine technologies providing ecological and economic advantages due to their aerodynamic function, geometric design, and installation [1]. These wind turbines are classified into two main categories: horizontal-axis wind turbines (HAWT) and vertical-axis wind turbines (VAWT). There is currently growing interest in vertical-axis wind turbines for several reasons: they are better suited to turbulent winds, they require less infrastructure for orientation, they operate more silently, and their architecture makes them an attractive solution for energy production. These vertical-axis wind turbines (VAWTs) are mainly divided into two types: Darrieus and Savonius rotors. The Savonius rotor has a low power output, despite its simple design and high self-starting

 https://doi.org/10.3390/fluids10090239

capability. On the other hand, the Darrieus rotor has high power and low self-starting capability [2].

Previous research has focused on improving the aerodynamic performance of the H-Darrieus rotor. Among these studies, we mention that of Eltayeb et al. [3], who improved the aerodynamic performance of urban wind systems integrated with solar shafts, using Plain Flaps (PFs) and Gurney Flaps (GFs) to modify NACA 0015 aerodynamic profiles through a CFD simulation and the URANS $k-\omega$ SST model. Their results show that the PF configuration of 0.6 c at 10° reduced negative torque, improved self-starting capability, and achieved the best power coefficient (C_P) of 0.46 at TSR = 2.5, as well as a 9.8% increase over the NACA 0015 airfoil. Mohamed et al. carried out an analysis of vertical-axis turbines (VAWTs) using 2D CFD simulation, on 20 different airfoils, with the aim of maximizing torque and power coefficients [4]. Ferreira et al. [5] used the PIV method to visualize the flow on the upper surface of the airfoil for three TSR values. The analysis described separation vortices at the leading edge and released vorticity at the trailing edge. These results were used to assess the aerodynamic performance of the H-Darrieus rotor. Castelli et al. [6] presented a CFD analysis of a Darrieus vertical-axis micro-wind turbine. The validation was obtained in a wind tunnel through a systematic comparison between numerical simulations and experimental data. In Revanche, a statistical analysis of the y^+ parameter was used to determine the optimal size of mesh elements next to the profiles and to optimize the accuracy of numerical performance prediction. Mohamed et al. [7] analyzed the performance of a three-bladed Darrieus turbine, adopting 20 profile types from various families and using CFD simulations based on the Reynolds-averaged Navier–Stokes equations. The results showed that the power coefficient of the turbine equipped with the LS (1)-0413 airfoil increased by 16% compared with the performance obtained with the NACA 0021 airfoil. El Baz et al. [8] investigated the effect of using a slotted profile versus the NACA 0018 profile on the performance and starting characteristics of the H-Darrieus rotor using CFD simulation. In this case, the analysis of the turbine's aerodynamic behavior showed a 15% higher optimization compared with standard blades. Ma et al. [9] developed an automatic high-solidity airfoil optimization system to improve Darrieus turbine performance. Their results illustrate that this system provides a proven technique for generating optimal and suitable turbine airfoils to achieve robust energy efficiency. Ghaffari et al. [10] examined the impact of different cavity parameters (dimples) on the aerodynamic performance of the H-Darrieus rotor, employing computational fluid dynamics (CFD) and the $k-\omega$ SST (shear stress transport) turbulence model. Their simulation results showed that wind turbine efficiency was improved by 18% and 25% when the cavity profile with an optimum TSR (TSR = 2.6) was used.

Regarding the use of dual rotors to improve VAWT aerodynamic performance, we cite Shen et al. [11], who used a model of a double Darrieus vertical-axis wind turbine (DD-VAWT) to increase the power coefficient while reducing the maximum instantaneous torque coefficient. The results of their CFD simulation reveal a significant improvement in the performance of the DD-VAWT at low TSR. Cheng et al. [12] proposed a two-Darrieus-turbine design (DDWTs) using a genetic expression programming (GEP) model. Their results indicate that this GEP model shows high accuracy, enabling an improvement in energy production of 7.5%. Ghafoorian et al. [13] examined the performance of the Darrieus vertical-axis wind turbine by integrating a Darrieus–Savonius hybrid rotor. Their simulation results show that the hybrid rotor performs better at a low TSR of 64% of the power coefficient, while the Darrieus rotor performs better at a high TSR. Abdel-razak et al. [14] studied the distance between turbine centers, configuration angles, rotation directions, and stall angle using the Taguchi optimization method in order to improve the performance of three Darrieus–Savonius hybrid turbines. The hybrid rotor can maintain performance equivalent to that of the isolated rotor up to a TSR of 4.1. To improve performance at low wind speeds, Kumar et al. [15] proposed a model of an adaptive hybrid Darrieus turbine (AHDT) based on the Reynolds-averaged Navier–Stokes (RANS) equations. Their objective was to analyze the impact of the Darrieus rotor diameter (DR) on the Savonius rotor (DT). The results of their CFD simulation indicate that an optimum DR/DT ratio of value 3 maximizes the power coefficient of the Darrieus rotor. Ghobadian et al. [16] proposed a new double-row wind turbine, combining J-shaped and conventional blades, using CFD simulation and Taguchi's method to exploit both drag and lift forces. The turbine achieved a better maximum power coefficient value of 0.52 at low TSR. These results show that the adoption of a Darrieus-type double-row hybrid turbine optimizes the efficiency of vertical-axis wind turbines.

Several studies have addressed the optimization of Darrieus wind turbine efficiency. We cite Dessoky et al. [17], who analyzed the aerodynamic performance of two-blade Darrieus turbines with NACA 0021 airfoils, separated by a distance. Their results show that the second turbine achieves better performance if the first turbine operates at a high TSR. Omar Sherif et al. [18] showed that using several Darrieus turbines in near proximity to each other increases energy production and improves wake valorization. The results of their numerical CFD simulation, adopting the Actuator Line Method (ALM), showed that the ALM is a reliable, economical tool for accurately reproducing the flow field between vertical-axis wind turbines. On the other hand, Mao et al. [19] proposed two-turbine configurations with a spacing no lower than the turbine diameter. Their CFD results show an increase in the power coefficient with the number of turbines reaching 0.588 for a TSR of 3.0. Shaheen et al. [20] were interested in analyzing a Darrieus multi-turbine design with the aim of developing efficient vertical-axis wind farms. The CFD results of multi-turbine clusters enable the conception of an average power coefficient up to 30% higher than that of an isolated turbine. Li Zou et al. [21] presented two turbines aligned at variable distances. However, numerical simulations were conducted using the sliding mesh method coupled with Large Eddy Simulation (LES). The comparison with isolated turbines revealed a significant improvement in the power coefficient for both turbines. Mishra et al. [22] used CFD simulation and the Taguchi method, combined with analysis of variance (ANOVA), to identify the parameters influencing the starting performance of adjacent rotors. The data from their simulation indicate that downstream rotor 2 started faster than upstream rotor 1, showing an ability to self-start at an angular velocity of 57 rad/s and at a lower TSR.

The previously cited works have focused on improving the performance of the H-Darrieus wind turbine, either by using dual rotors or by combining several turbines. In our study, we evaluated the performance of an H-Darrieus wind turbine using both dual rotors and two turbines, which represents the originality of our work. Two-dimensional CFD simulations have been carried out on the two adjacent wind turbines, each comprising dual rotors with a different number and type of aerodynamic profiles. The first rotor, situated in the interior of the H-Darrieus wind turbine, is equipped with three blades of the NACA 0021 profile. The second, situated on the exterior, upstream of the first, is equipped with a single blade of the Eppler 420 profile, which was chosen to be added to the wind turbine due to its remarkable performance [23]. This second rotor acts as a curtain, minimizing the negative impact on the flow around the interior rotor, reducing mechanical complexity, and decreasing aerodynamic drag during rotation. The objective of this analysis is to increase the energy yield of two adjacent H-Darrieus wind turbines incorporating a double rotor system, with the aim of improving turbine efficiency and overall field production. These two turbines are separated by different distances, namely 1.6D, 2.4D, 3.2D, 4D, and 4.8D, in order to determine the optimal distance that minimizes performance losses due to interaction effects. The analysis also includes variations in certain geometric parameters such as chord, angle of attack, and azimuth angle. The CFD simulation model was validated by comparing it to the work of Castelli et al. [24] and Islam et al. [25], through a comparative analysis of power coefficients as a function of TSR. To this end, the results were visualized through the power coefficient, moment coefficient, drag, and lift, as well as the magnitude velocity distribution, pressure, and turbulent kinetic energy. In addition, these analyses confirm that a dual-rotor turbine optimizes system performance compared with an isolated configuration. This improvement also contributes to enhancing wind distribution and stability between two adjacent turbines.

2. Methodology

2.1. Geometrical Model

A three-bladed H-Darrieus VAWT device has been adopted in our study, based on the configuration proposed by Castelli et al. [24], with the aim of assessing the reliability of the results obtained by CFD simulation. The study domain proposes a model composed of two adjacent turbines to which we have added another rotor to improve their performance, as illustrated in Figure 1. Table 1 shows the geometric characteristics of the H-Darrieus dual rotors. Each turbine integrates dual rotors, the first of which is composed of three NACA0021-type airfoils with a blade chord length $c = 0.0858$ m, a diameter (L) of 1.030 m, a height (h) of 1.456 m, and a solidity (σ) of 0.5. In contrast, the second rotor is composed of a single Eppler 420-type airfoil, presenting a blade chord length $c = 0.30$ m and a diameter (D) of 1.25 m; the trailing edge of the blade has been modified as illustrated in Figure 2e by connecting the top and bottom surfaces of the three airfoils with by a straight line of 0.38 mm, to avoid the generation of mesh elements. The study considers blade azimuth angles θ equal to 45°, 90°, 135°, 180°, and 225° for the first rotor and 60°, 120°, 180°, 240°, and 300° for the second rotor to enhance the performance of the two H-Darrieus vertical-axis aerodynamic turbines, accelerating airflow, increasing stability, reducing drag, improving wind interaction between the two turbines, and minimizing energy losses. The approach of using the Eppler 420 airfoil in the second rotor has been chosen because of its adequate advantages over H-Darrieus turbines, in particular, its low thickness and better efficiency at different velocities, enabling lift (C_l) and drag (C_d) coefficients to be maintained even when the angle of attack α is increased [26].

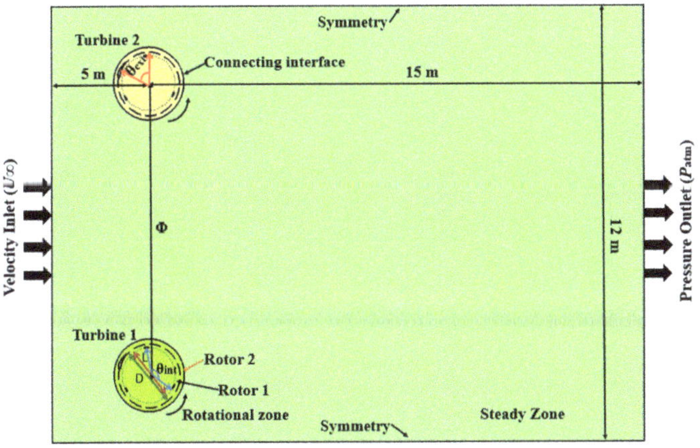

Figure 1. Configuration model of turbines with H-Darrieus dual rotors and the characteristics of the domain.

Table 1. Main geometric characteristics of the H-Darrieus dual rotors.

Characteristics	Rotor 1	Rotor 2
Number of blades	3	1
Blade airfoil	NACA 0021	Eppler 420
Diameter (m)	1.030	1.25
Chord (m)	0.0858	0.30
Solidity	0.5	0.5
Height (m)	1	1

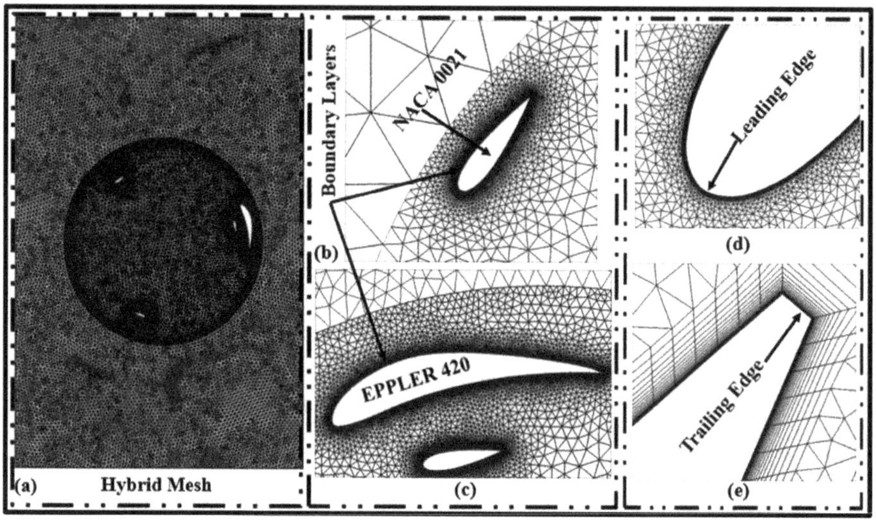

Figure 2. (**a**) Hybrid mesh using quadrilateral and triangular elements. (**b**) Region near the NACA. 0021 blade. (**c**) Region near the EPPLER 420 blade. (**d**) Leading edge of a NACA 0021 airfoil. (**e**) Trailing edge of a NACA 0021 airfoil.

2.2. Computational Domain and Boundary Conditions

The present study used CFD simulation to evaluate the aerodynamic performance of the turbines while analyzing the air flow around the dual rotors. Here, 2D rather than 3D modeling was adopted because of its optimized computational cost, and this allows for obtaining sufficient results with an acceptable error, provided that the three-dimensional phenomena have little influence on the studied quantities. As the average velocity or the pressure varies little according to the ignored dimension of space, the 2D simulation becomes relevant and does not affect the accuracy of the solution [27]. The computational domain is shown in Figure 1 and is separated into two zones: a rotating zone in the form of a cylinder that includes the three blades of the first rotor and a single blade of the second rotor of the two turbines, and a stationary zone in the form of a rectangle delimited by interfaces and generated by ANSYS R1, 2022 software [28]. At the inlet, the length is 5 m from the center of the turbine, and at the outlet, it is 15 m. The width is fixed at 12 m. In addition, the boundary conditions are expressed as follows: free stream velocity flow U_∞ = 9 m/s at the inlet and atmospheric pressure at the outlet. Symmetry boundary conditions are applied to the upper and lower parts of the domain. The fluid employed in this work is air with a density (ρ) of 1.225 kg/m^3, at a temperature of 300 K, a dynamic viscosity (μ) of 1.7894×10^{-5} kg/m.s, and a turbulence intensity of 5%. The sliding mesh (SM) technique was exploited to simulate the rotational motion of the H-Darrieus dual-rotor-equipped turbine in the unsteady regime [29,30]. Non-slip interaction conditions were used at the interfaces and blade surfaces of the dual rotors [31].

2.3. Mesh Generation and Grid Independence

To achieve accurate results, it is necessary to focus on the spatial discretization of the computational domain. Figure 2 shows the mesh distribution around the dual rotors, including the two airfoils (NACA 0021 and Eppler 420). However, a hybrid mesh (quadrilateral and triangular) has been used throughout the entire surface of the dual rotors for the two H-Darrieus turbines, using ANSYS Workbench R1, 2022 software. Specifically, finely meshed quadrilateral structured elements are used carefully for boundary layer meshing, and in rotating and stationary regions, unstructured triangular grids are adopted for the system because of their capacity to stabilize flow abruptness and minimize the number of mesh elements. On the other hand, a well-refined and highly accurate mesh was selected for the domain to capture the complex behaviors of the aerodynamic flow around the Darrieus turbines and ensure a balance between mesh quality near the walls and solution reliability, while managing computational costs [32].

The $k - \omega$ SST turbulence model was adopted, which suggests a y+ range between 1 and 5. In our study, we fixed y+ at 1 to obtain adequate accuracy and estimate the first near-wall layer [33]. In addition, the mesh boundary layer was induced with 25 cell layers and a cell growth ratio of 1.2 for both rotors. Reynolds numbers of around Re = 6.3×10^5 and Re = 7.7×10^5 were calculated for this work based on the diameters of the first and second H-Darrieus rotors, respectively. The mesh is particularly fine near the walls, as can be observed in Figure 2. Furthermore, the major difference between these mesh types lies in the size of the grid around the blades. The minimum grid thicknesses are 2.4×10^{-4} m, 1.1×10^{-4} m, and 3.7×10^{-5} m, equivalent to coarse, medium, and fine meshes, respectively, in order to ensure that flow phenomena are appropriately modeled.

Figure 3 illustrates the variation in the power coefficient (C_P) as a function of the number of elements for a tip speed ratio (TSR) of 2.51. In addition, when analyzing evolutions in C_P, a difference was found between a medium mesh and a fine mesh, with a maximum variation of 0.92%. A sensitivity evaluation was conducted to determine the independent grid for a 2D turbine, using the following five meshes: 123,000, 347,500, 482,123, 653,200, and 750,140 elements. It can be observed that the power coefficient increases with the number of mesh elements from 123,000 to 653,200 elements and then becomes almost constant above 653,200 elements. Consequently, at the value of 653,200 elements, the coefficient stabilizes, and the solution becomes grid-independent from this value onwards.

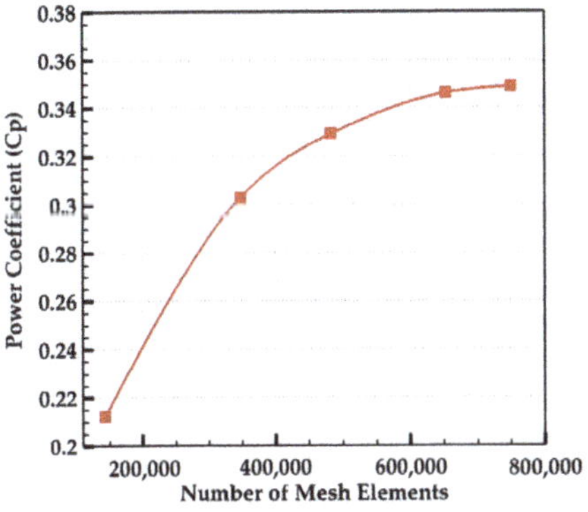

Figure 3. Variation in C_P with number of mesh elements.

2.4. Solver Settings

The present work adopted CFD numerical simulation to study the aerodynamic flow around the dual-rotor-containing turbine, using the URANS approach for 2D incompressible flow with Ansys Fluent software. This approach offers higher efficiency and lower computational cost than other models. In addition, a $k-\omega$ SST turbulence model was used to solve the Navier–Stokes equations; this model is better suited to analyzing air flow around dual H-Darrieus turbine rotors due to its performance in solving complex boundary layer problems. The sliding mesh technique was used to simulate the rotational domain of the H-Darrieus wind turbine, while ensuring better resolution of the dynamics and minimizing computation time [34]. The SIMPLE algorithm was applied to solve the velocity and pressure coupling, associated with a second-order upwind scheme for the spatial discretization of the URANS equations and the first-order scheme for the temporal discretization, in order to obtain a more reliable prediction and to achieve accurate CFD simulation of velocity, pressure, momentum, kinetic energy, and turbulence. In the CFD studies applied to the H-Darrieus turbine, the classical method used to define the time step was based on the fixed rotation velocity (ω) expressed in (rpm) and calculated a time step (Δt) that captured an elementary degree of rotation of the turbine blades. On the other hand, the following time steps corresponding to 1° of elementary blade rotation were used: 5.87, 4.87, 4.29, 3.96, 3.84, 3.22, and 3.03 $\times 10^{-4}$ s. This value of 1 degree was adopted on the basis of the simulations carried out [24,35]. An internal number of iterations (100) was used in order to maintain the residual convergence criterion for each physical time step was set to 10^{-4}. In addition, a Y + value of 1 was chosen to ensure optimal accuracy of the viscous boundary layer near the wall.

3. URANS Equations

3.1. Mass and Momentum Equations

The fluid around the airfoils is considered incompressible and unstable, and is described by the URANS equations, conservation of mass and momentum, utilizing the sliding mesh technique and $k-\omega$ SST turbulence model. These equations are represented as follows [36,37]:

The URANS mass conservation equation:

$$\frac{\partial \overline{u}_i}{\partial x_i} = 0 \tag{1}$$

The URANS momentum conservation equation:

$$\frac{\partial(\rho\overline{u}_i)}{\partial t} + \frac{\partial(\rho\overline{u}_i\overline{u}_j)}{\partial x_j} = -\frac{\partial \overline{p}}{\partial x_i} + \frac{\partial}{\partial x_j}\left[\mu_{eff}\left(\frac{\partial \overline{u}_i}{\partial x_j} + \frac{\partial \overline{u}_j}{\partial x_i}\right)\right] \tag{2}$$

The effective viscosity:

$$\mu_{eff} = \mu + \mu_t \tag{3}$$

where ρ represents the fluid density, $\overline{u}_i$ denotes the time-averaged velocity in the i th direction, x_i is the coordinate in the i th direction, $\overline{p}$ indicates the time-averaged pressure, μ_t is the turbulent viscosity, and μ is the molecular viscosity.

3.2. Sliding Mesh Technique

The sliding mesh technique was employed to accurately simulate the interaction between the stationary and rotating fluid zones of the H-Darrieus rotor. This method requires an unsteady (transient) computation, generating time-dependent results rather than time-averaged values. Despite this, a difference in meshing at the interface between the turbine and the tunnel was explained by the fact that each area is meshed independently according to its geometric and simulation constraints. To ensure the continuity of physical quantities, particularly velocity and pressure, between the two domains, the solver performs data interpolation at the interface, which can introduce a slight loss of accuracy. Despite this difference, good results were observed, provided that a fine, good-quality, and well-adapted mesh is selected. The integral form of the equation can be expressed as follows [38]:

$$\frac{d}{dt}\left(\int_V \rho\phi dV\right) + \int_{\partial V} \rho\Phi\left(\vec{u} - \vec{u}_g\right)d\vec{A} = \int_{\partial V} \Gamma\nabla\Phi d\vec{A} + \int_V S_\phi dV \tag{4}$$

where ρ is the air density, u is the flow velocity, u_g is the mesh moving velocity, Γ is the diffusion coefficient, and S_ϕ is the source term of Φ.

The time derivative term in Equation (4) is formulated as shown in Equation (5) below:

$$\frac{d}{dt}\left(\int_V \rho\phi dV\right) = \frac{\left[(\rho\Phi V)^{n+1} - (\rho\Phi V)^n\right]}{\Delta t} \tag{5}$$

where n and $(n + 1)$ denote the quantity at the current time level tn and the next time at $tn + 1$, respectively. The volume of the temporal level $(n + 1)$, V^{n+1}, is calculated from Equation (6) [39].

$$V^{n+1} = V^n + \frac{dV}{dt}\Delta t \tag{6}$$

where $\frac{dV}{dT}$ is the volume time derivative.

The temporal rate of change in cell volume is zero, and Equation (6) can be simplified to Equation (7) [39].

$$V^{n+1} = V^n \tag{7}$$

Equation (5) is therefore transformed into Equation (8) [40].

$$\frac{d}{dt}\left(\int_V \rho\phi dV\right) = \frac{\left[(\rho\Phi)^{n+1} - (\rho\Phi)^n\right]V^n}{\Delta t} \tag{8}$$

3.3. SST K-Omega Turbulence Model

To calculate the flow in the near-wall region, the $k - \omega$ shear stress transport (SST) turbulence model was used. This hybrid model combines two models: the turbulent kinetic energy k and the specific dissipation rate ω, which are obtained from the transport Equations (9) and (10), respectively, according to the proposals of [40,41]:

$$\frac{\partial(\rho k)}{\partial t} + \frac{\partial(\rho k\overline{u}_j)}{\partial x_j} = \frac{\partial}{\partial x_j}\left((\mu + \sigma_K\mu_t)\frac{\partial k}{\partial x_j}\right) + P_k - \beta^*\rho\omega k \tag{9}$$

$$\frac{\partial(\rho\omega)}{\partial t} + \frac{\partial(\rho\omega\overline{u}_j)}{\partial x_j} = \frac{\partial}{\partial x_j}\left((\mu + \sigma_\omega\mu_t)\frac{\partial\omega}{\partial x_j}\right) + \frac{\gamma\rho}{\mu_t}P_k - \beta\rho\omega^2 \tag{10}$$

The constants σ_K, σ_ω, β^*, γ, and β are model coefficients, and P_k represents the turbulence production.

The turbulent viscosity is defined using Equation (11) detailed in [42].

$$\mu_t = \frac{\rho k}{\omega}\frac{1}{max[1/\alpha^*, SF_2/a_1\omega]} \tag{11}$$

where S is the magnitude of the strain rate and α^* is defined in Equation (12).

$$\alpha^* = \alpha^*_\infty \left(\frac{\alpha^*_0 + Re_t/R_k}{1 + Re_{t/} R_k} \right) \quad (12)$$

where $Re_t = \frac{\rho k}{\mu\omega}$; $R_k = 6$; $\alpha^*_0 = \frac{\beta_i}{3}$; $\beta_i = 0.072$; and when the Reynolds number of the $k-\omega$ model is high, $\alpha^* = \alpha^*_\infty = 1$ [43].

$$F_2 = tanh\left(\varnothing_2^2\right) \quad (13)$$

and

$$\phi_2 = max\left(2\frac{\sqrt{k}}{0.09\omega k}, \frac{500\mu}{\rho y^2 \omega}\right) \quad (14)$$

where y is the distance to the next surface.

3.4. Performance Indicators

The tip speed ratio (TSR) presents the ratio between the product of the ω angular velocity and the R radius of the turbine and the free stream velocity flow designated by U_∞, and it is represented by the following equation, Equation (15) [44,45]:

$$TSR = \frac{\omega R}{U_\infty} \quad (15)$$

The simulation is carried out for tip speed ratios (TSRs) between 1.7 and 3.3. This is achieved by fixing the incoming free stream velocity flow at (9 m/s) while varying the azimuth angle rotor (θ°) and angle of attack (α°) for the dual rotors, as shown in Equation (15).

Several papers focus on numerical and experimental analyses aimed at optimizing and approximating the power coefficient (C_P) to the Betz limit of 60%. In addition, the torque coefficient (C_m) is a coefficient of performance that measures the rotor's ability to transform incoming wind energy into mechanical torque [46,47].

The performance indicator can be represented by the following equations, Equations (16) and (17) [48].

$$C_m = \frac{M_t}{0,5\rho U_\infty^2 AR} \quad (16)$$

$$C_P = \frac{P_t}{0,5\rho U_\infty^3 A} = TSR \times C_m \quad (17)$$

A is the section covered by the wind, M_t is the rotor torque indicator, and P_t is power.

4. Results and Discussion

4.1. Validation

Validation of our 2D CFD simulation model is carried out by comparing the results obtained with those of the experimental study and CFD simulation carried out by Castelli et al. [24], and also with the CFD simulation results of Islam et al. [25]. The studies by Castelli and Islam are both based on a vertical-axis wind turbine (VAWT) with an H-Darrieus rotor.

The geometric conception is represented by a wind turbine equipped with three blades of aerodynamic profile NACA 0021 and having different characteristics, including a chord length of $c = 0.0858$ m, a diameter of 1.030 m, a height of 1.456 m, and a strength of $\sigma = 0.5$, with a free stream velocity flow of $U_\infty = 9$ m/s and a turbulence intensity of 5%. The study is carried out using the sliding mesh method and the $k-\omega$ SST turbulence model, which is distinguished by a superior ability to characterize airflow separation.

Figure 4 reveals the variation in the power coefficient (C_P) as a function of different TSR values, namely 1.70, 2.04, 2.32, 2.51, 2.6, 3.09, and 3.3, for our study and that of Castelli (experimental and CFD simulation) and Islam (CFD simulation). The results are compared using the relative error (RE) method, which consolidates the power coefficient ($C_{P_{exp,i}}$) obtained experimentally from Castelli et al. [24] and the power coefficient ($C_{P_{sim,i}}$) determined by numerical simulations. This is expressed by the following equation, Equation (18) [28]:

$$RE = \frac{1}{n}\sum_{i=1}^{n} \frac{\left|C_{P_{exp,i}} - C_{P_{sim,i}}\right|}{C_{P_{exp,i}}} \times 100 \quad (18)$$

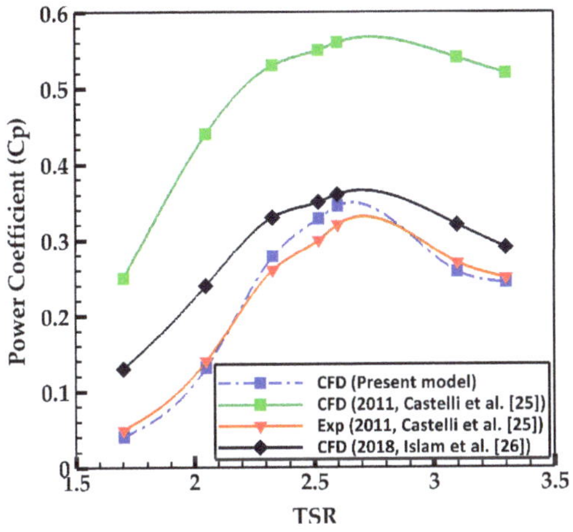

Figure 4. Validation model: CFD results vs. experimental data.

Furthermore, Castelli et al. [24] indicate a relative error of 20% for CFD simulation, while our CFD study presents a relative error of 7% for $C_P = 0.34$ at TSR = 2.6, and Islam et al. [25] reveal a relative error of 16%. We can deduce that the C_P values we obtained are closer to those obtained by Castelli using an experiment. We therefore concluded that our 2D CFD configuration is appropriate for realizing a parametric study with the aim of improving the aerodynamic performance of the two turbines integrating dual rotors.

4.2. Parametric Study Analysis

4.2.1. Analysis of Power and Torque Coefficients

This work highlights the potential of the two H-Darrieus turbines in dual rotors and in two different aerodynamic profiles, in order to reduce energy losses and maximize the aerodynamic performances of the two turbines.

Figure 5a illustrates the variation in the coefficient of performance as a function of TSR, comparing the dual H-Darrieus rotors of the turbine. From this figure, we can observe that the coefficient of performance (C_P) takes on maximum values equal to 0.34 and 0.38 for a tip speed ratio (TSR) equal to 2.6 for the dual rotors. Then, we observe a decrease in power coefficients when TSR values are between 3.09 and 3.3 for α and θ equal to 0°. On the other hand, the second rotor of the Eppler 420 airfoil has maximum C_P, reflecting that it offers around 12% improvement over the NACA 0021 airfoil in turbine performance. Other studies also confirm the use of a dual-rotor wind turbine to optimize energy efficiency at a low tip speed ratio (TSR). In addition, this configuration offers a better capacity to predict an optimal power coefficient, which is a significant advantage for aerodynamic conception and performance maximization [11,12].

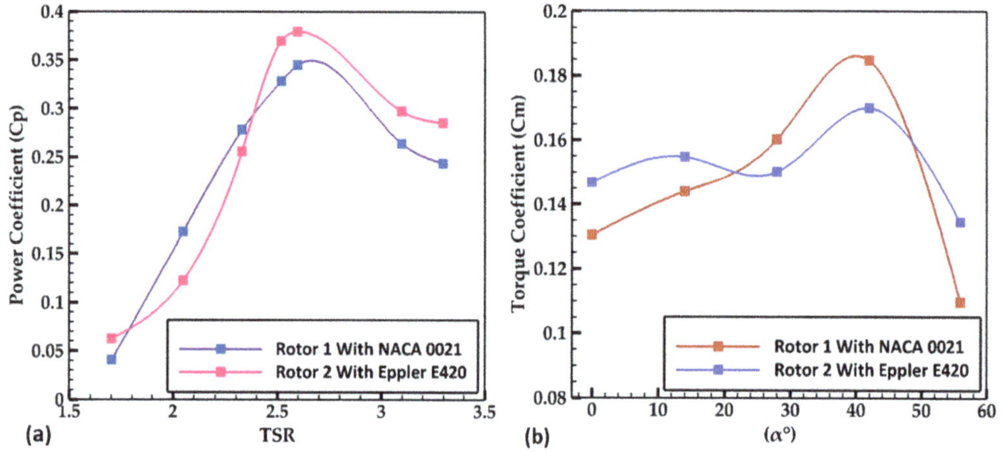

Figure 5. Comparative analysis of a dual-rotor turbine: (**a**) power coefficient as a function of TSR, (**b**) torque coefficient as a function of angle of attack.

A comparative analysis of the moment coefficient (C_m) as a function of angle of attack is shown in Figure 5b for the dual rotors including the NACA 0021 and Eppler 420 airfoils. For each of these profiles, moment coefficient values are presented for four angles of attack, namely $\alpha = 0°$, $\alpha = 14°$, $\alpha = 28°$, $\alpha = 42°$, and $\alpha = 56°$, and at an optimum TSR equal to 2.51. For the NACA 0021 profile, the moment coefficient (C_m) increases with increasing angle of attack throughout the interval, taking a maximum value equal to 0.19 for $\alpha = 42°$ and a minimum value of 0.11 when $\alpha = 56°$. On the other hand, for the Eppler 420 profile, the C_m values reveal a different variation from that observed for the NACA 0021 profile. The maximum moment coefficient of $C_m = 0.17$ is observed at $\alpha = 42°$, while its minimum value reaches 0.13 and is found at $\alpha = 56°$. In this case, the values of the moment coefficient C_m initially increase with the angle of attack until $\alpha = 14°$ is reached and then decrease slightly at $\alpha = 28°$, after which they increase to attain their maximum at $\alpha = 42°$ and then begin to decrease until $\alpha = 56°$ is reached.

For the same values of α and for a TSR equal to 2.51 for the two profiles, we emphasize a difference in rotor performance for a certain value of angle of attack. As an example, the Eppler 420 profile shows an initial increase in C_m, followed by a decrease at $\alpha = 28°$ before a further increase to reach maximum performance and then a decrease for high angles of attack due to flow separation, which causes a rapid decrease in the moment coefficient and results in a loss of stability around the profile. This requires specific angles of attack to optimize energy recovery. For the NACA 0021 profile, C_m generally increases for all α values, reaching its maximum optimization, and then decreases more steeply than the Eppler 420 profile for angles of attack that are excessively high. This difference highlights the influence of angle of attack on rotor performance. In any case, the two profiles reach the same maximum value of C_m.

The impact of choosing a second rotor with a single Eppler 420 profile blade lies mainly in reducing mechanical complexity, manufacturing costs, and aerodynamic drag during rotation, which can improve efficiency, especially at low velocities. This rotor has been moved upstream of the first rotor, which is located inside the turbine, to ensure optimal balance and easy start-up. However, we can conclude that the second rotor gave rise to the first rotor, with the aim of achieving a more efficient H-Darrieus configuration in order to maintain sustainable energy productivity. This configuration far surpasses conventional Savonius–Darrieus hybrid designs, as well as the single-rotor H-Darrieus design [49,50].

4.2.2. Effect of Dual-Rotor Integration on the Performance of Two Turbines

The parametric study focuses on optimizing the aerodynamic performance of the two H-Darrieus turbines, and this is achieved by focusing on the effect of the dual rotors with different types and numbers of airfoils on the efficiency of the H-Darrieus turbines, operating at a free stream velocity flow $U_\infty = 9$ m/s, and applying the sliding mesh method.

A dual-rotor configuration was considered in order to reduce the effects of aerodynamic interaction between two adjacent synchronized turbines, which mainly influence each other through the wind turbulence generated between them. These interactions, particularly through the wake effect, can significantly reduce the performance of turbines located close to each other. The first rotor, located inside the turbine, has three blades. The second, located outside, has a single blade. The second rotor not only optimizes energy recovery, but also generates a less concentrated and therefore less perturbing wake around the first rotor downstream.

Figure 6 shows the variation in the power coefficient as a function of the tip speed ratio (TSR) achieved for two turbines with dual rotors; one has three NACA 0021 airfoil blades with a chord length equal to 0.0858 m, while the other has a single Eppler 420 airfoil blade with a chord length of 0.30 m at $\theta = 0°$, with two isolated turbines comprising only the first rotor with a fixed angle of attack of 0°. The dual-rotor turbines have a variable angle of attack with values of 0°, 15°, 20°, 25°, and 30° associated with the dual rotors. This figure shows that C_P takes on the maximum values for tip speed ratios (TSRs) equal to 2.51 and 2.6, for all angles of attack. Beyond this, for TSR values between 3.09 and 3.3, power coefficients tend to decrease. Analytically, the minimum power coefficient value of 0.31 is observed at TSR = 2.6 and $\alpha = 0°$ for both isolated turbines. In contrast, for the dual-rotor turbines, the maximum power coefficient value of 0.49 is observed for $\alpha = 30°$. Thereafter, C_P decreases slightly, reaching 0.42 for $\alpha = 25°$ and then $C_P = 0.40$ for $\alpha = 20°$ at TSR = 2.51. Finally, when the angle of attack equals 15° and 0°, the power coefficient reaches its maximum value of 0.37 and 0.36, respectively, at TSR = 2.6.

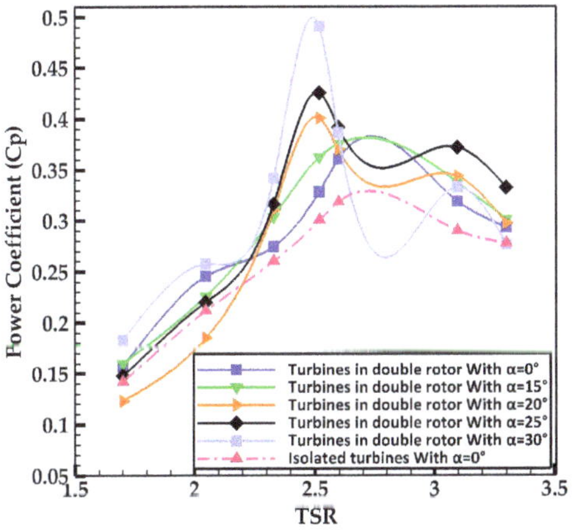

Figure 6. Comparative study of the power coefficient versus TSR for different angles of attack.

Previous studies based on 2D URANS simulations have confirmed that interactions between several H-Darrieus turbines can generate an optimization of the global performance compared with an isolated turbine. These improvements depend in particular on optimal management of the angle of attack and the number and spacing of turbines [19]. On the other hand, these results show that dual-rotor turbines achieve higher efficiency even when operating at a fixed angle of attack of 0°. This visual illustration of the results highlights a physically consistent relationship between rotor number and turbine performance, which is associated with the ability to extract maximum power [51]. Furthermore, the global efficiency of the two turbines incorporating dual rotors was improved by 17%, compared with the two isolated turbines for $\alpha = 0°$. In addition, a suitable arrangement of the airfoils and an optimum overlap of the angle of attack between the dual rotors are capable of improving the energy efficiency and enhancing the performance of the H-Darrieus turbine.

We conclude that the choice of a dual-rotor configuration including an external rotor incorporating a single high-performance Eppler 420 blade enables it to act as a curtain, creating a less dense wake and thus limiting its negative impact on the flow around the internal rotor. This configuration helps reduce parasitic effects, particularly performance losses and turbulence between the rotors, while modifying the flow between them, optimizing and improving the overall performance coefficient of the turbine.

Although the rotors rotate independently from a single generator, technical solutions such as a flywheel, clutch system, or converter allow the movements of the dual rotors to be combined, ensuring continuous generator operation. This optimized coupling promotes more stable electricity production.

4.2.3. Effect of Distance Between Two Adjacent Turbines on Wind Turbine Performance

Figure 7 illustrates the evolution of lift (C_l) and drag (C_d) coefficients as a function of different distances between two adjacent turbines integrating dual rotors, at an optimum TSR equal to 2.51 and for an angle of attack $\alpha = 42°$. This figure indicates that a low distance between turbines can result in a reduction in lift. In contrast, C_l values increase with distance, reaching maximum values of 1.34 when the distance (Φ) is 4D. Then, when the distance (Φ) is 4.8D, we observe an increase in drag coefficient (C_d) of 0.64, while lift decreases by 0.94. This observation may highlight the fact that this distance is less efficient, as it generates more drag, which adversely affects the performance and stability of the two airfoils. Furthermore, the lift coefficient (C_l) generates a higher aerodynamic efficiency than the drag capacity (C_d), and therefore, an optimal distance of 4D might be more suitable when the two turbines are applied in the wind direction, while ensuring optimal lift for each airfoil and avoiding vortices, to maintain robust energy efficiency within a wind farm. On the other hand, an insufficient distance between two turbines allows the airflow to be directed towards the second turbine, disturbing the wake produced by the first, which can have a negative influence on the turbine's efficiency and lift. However, to minimize performance losses, it is preferable to maintain an optimal distance between turbines so that the wind has time to regenerate between them, contributing to greater efficiency. Furthermore, too great or too small a distance can adversely affect the energy density of wind turbines. It is therefore necessary to find a balance between aerodynamic performance and optimal position management.

Figure 8 compares the total pressure distribution for different adjacent distances between the two H-Darrieus turbines, which include two separate rotors: the first with profile NACA 0021 having a chord length of 0.0858 m, and the second rotor with profile Eppler 420 having a chord length of 0.30 m, at TSR = 2.51 and $\theta = 0°$. The values of these distances, according to Figure 7, are 1.6 D, 2.4 D, and 3.2 D, representing low, medium, and high efficiency, respectively. This distance selection enables us to observe the distances' effect on adjacent turbines. CFD simulation results emphasize the presence of a system with two adjacent turbines in the same direction as the wind. For a mean-to-optimum-value distance of 3.2D, this system generates a high and favorable total pressure distribution in the wake zone. On the other hand, in the case of a dual-rotor system with a reduced distance of 1.6 D, the pressure generated by the airflow decreases slightly. This provokes flow field blockage, resulting in the formation of larger vortices in the wake zone and adversely affecting turbine performance. In conclusion, the dual-rotor configuration improves the aerodynamic performance of the turbines thanks to an arrangement based on optimum spacing between the turbines. Nevertheless, turbines located in the center of the configuration (Φ = 1.6D) suffer a reduction in efficiency due to the formation of a low-pressure zone above the blades that reduces their ability to produce more energy. In addition, the contours reveal that high pressure upstream of the airfoils generates vortices on the upper surface, helping to generate power and minimize energy losses. However, the second rotor of the Eppler 420 profile has a positive effect, improving overall performance and enhancing the lift (Figure 7) of the first rotor. Consequently, pressure field analysis is essential to assess the impact of the distance between the two turbines and optimize the efficiency of the first rotor, in particular by integrating a second rotor. Evaluating the aerodynamic performance of the H-Darrieus wind turbine is crucially dependent on the choice of a configuration with an optimum distance adapted to the velocity gradient distributions around the dual-rotor turbine.

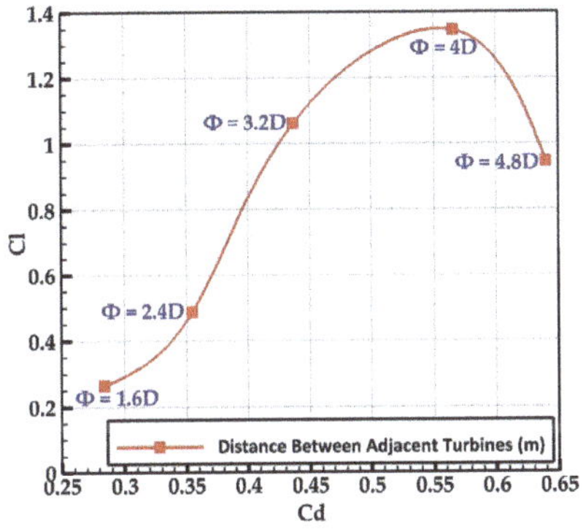

Figure 7. Variation in drag and lift coefficients as a function of different distances (Φ) between adjacent turbines.

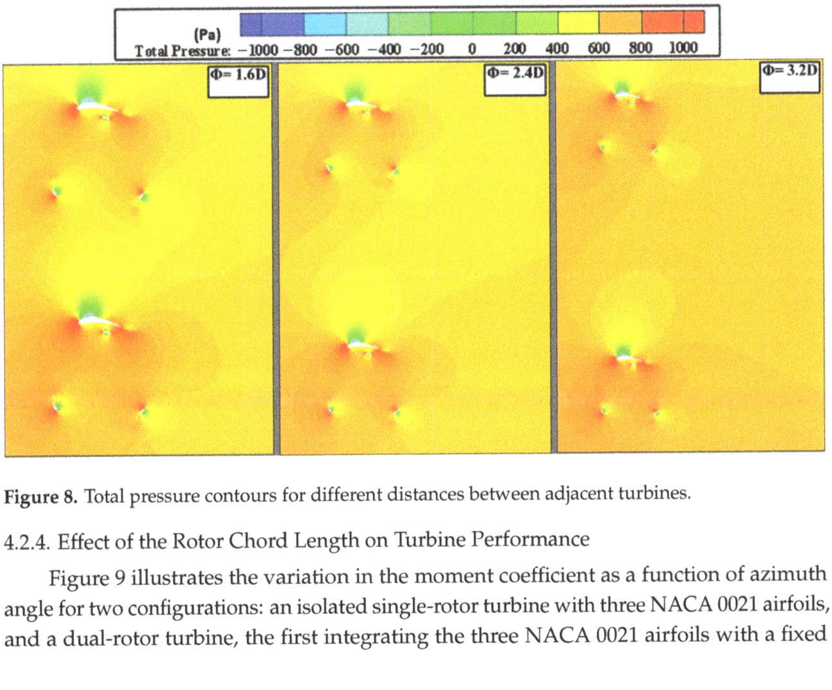

Figure 8. Total pressure contours for different distances between adjacent turbines.

4.2.4. Effect of the Rotor Chord Length on Turbine Performance

Figure 9 illustrates the variation in the moment coefficient as a function of azimuth angle for two configurations: an isolated single-rotor turbine with three NACA 0021 airfoils, and a dual-rotor turbine, the first integrating the three NACA 0021 airfoils with a fixed chord length equal to 0.0858 m and the second containing a single Eppler 420 airfoil with different chord length values of 0.20, 0.30, and 0.40 m at TSR = 2.51. The aim is to find the ideal H-Darrieus configuration offering stable performance. The results in this figure show that for the isolated turbine with a single rotor, the moment coefficient reaches maximum values of 0.11, 0.03, and 0.02 at azimuth angles of 90°, 180°, and 270°, respectively, while minimum values are found at θ of 45°, 135°, 225°, and 315°. Furthermore, for a dual-rotor configuration with C = 0.20 m, the maximum moment coefficient is observed at 0.13 and 0.04 for θ equal to 90° and 180°, before decreasing at angles of 45°, 135°, 225°, and 270°. Furthermore, for the C = 0.30 m dual-rotor turbine, the moment coefficient peaks at values of 0.16, 0.06, and 0.05 at azimuth angles equal to 90°, 180°, and 270°, respectively. The coefficient then decreases progressively at angles of 45°, 135°, 225°, and 315°. On the other hand, the dual-rotor configuration at C = 0.40 m has the highest C_m values of 0.18, 0.08, and 0.07 for azimuth angles equal to 90°, 180°, and 270°, respectively. We additionally observe that C_m takes on negative values for azimuth angles equal to 45°, 135°, 225°, and 315° due to flow separation and vortex formation. We can consider that the moment coefficient increases with increasing chord length in the case of the second rotor. This indicates that the dual-rotor turbine, with a higher chord length value, is capable of generating greater efficiency and demonstrating maximum performance compared to a single H-Darrieus turbine. Furthermore, it can be summarized that the dual-rotor configuration tends to offer better stability and more precise control, even if it requires more complex management compared to a single rotor, which can present limited power and torque as well as certain disadvantages depending on the application.

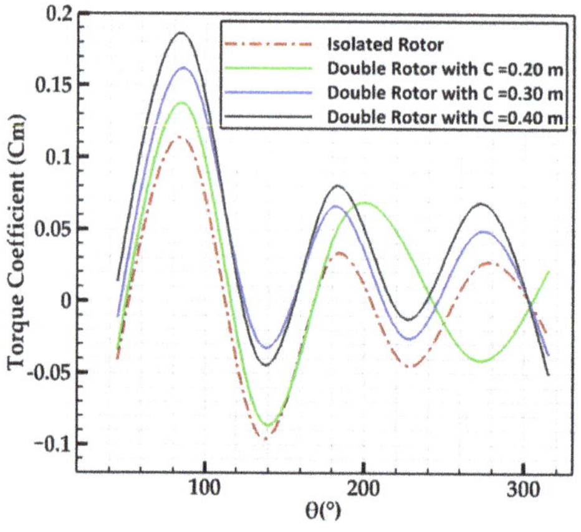

Figure 9. Evolution of the torque coefficient as a function of azimuth angle, for an insulated rotor configuration and a dual-rotor configuration with various chord lengths of the Eppler 420 profile.

The distribution of turbulent kinetic energy contours around a dual-rotor turbine is illustrated in Figure 10 to study the effect of different chord lengths of 0.2 m, 0.3 m, and 0.4 m of the second rotor (Eppler 420) on the first rotor, and more precisely on the aerodynamic performance of the H-Darrieus turbine at TSR = 2.51 for α = 15°. Analysis of these contours shows a vortex flow structure around the blades that increases with increasing chord length. The observation also indicates a high turbulent intensity enveloping the Eppler 420 blade for C = 0.4 m, which affects the performance of all three NACA 0021 blades, and this configuration valorizes a maximization of the kinetic energy transferred to the last ones [14]. For C = 0.3 m, a moderate intensity was visualized, signifying a high availability of the flow energy quantity by improving the aerodynamic performance of the H-Darrieus turbine. In conclusion, the turbulent kinetic energy contours serve to show the optimum configuration based on the chord length values most useful for the conceptualization of dual-rotor wind turbines, that is, where the flow carries the most turbulent energy. In summary, choosing a minimum chord value for an Eppler profile can generate a low improvement compared to a configuration with an optimum chord length of C = 0.3 m to C = 0.4 m, while still enhancing the torque and power applied to the blades (Figure 9). In addition, it is essential to note that once the chord length surpasses these optimum values, the flow around the blades becomes excessively dense, which can cause mechanical stresses to develop on the first rotor, increasing turbine instabilities, leading to configuration disruptions and inducing efficiency losses.

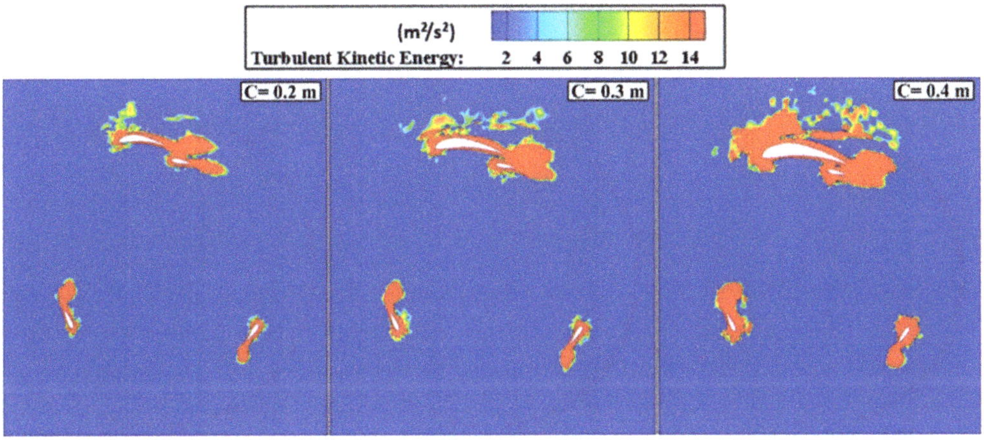

Figure 10. Turbulent kinetic energy contours for a dual-rotor system at different chord lengths of the Eppler 420 profile.

4.2.5. Velocity Contour Analysis

The variance in velocity magnitude is illustrated in the contours of Figure 11 for several azimuth angles θ, namely 45°, 90°, 135°, 180°, and 225° for the first rotor and 60°, 120°, 180°, 240°, and 300° for the second rotor at the optimum TSR of 2.51 and α = 0°. The velocity field is divided into two regions: one where the wind is in front (upstream) of the turbine and the other where the wind is behind (downstream) the blades. The results show that the velocity increases upstream and downstream of the airfoils, while it decreases in the area of the upper and lower surfaces, particularly for θ equal to 45°, 60°, 90°, 120°, 180°, 240°, 225°, and 300°. However, the rotation of the second Eppler 420 profile as it enters the downstream zone generates an accumulation of vortices, resulting in a reduction in power output. Furthermore, for angles θ equal to 135° and 180°, the velocity decreases slightly at the leading and trailing edges of the Eppler 420 profile, while increasing at the edges of the three NACA 0021 profiles. This shows that the second rotor makes a positive contribution to flow stability around the first rotor when the two blades of each rotor are very close to each other. In this case, the vortices generated around the Eppler 420 profile can interact with the other profiles, maximizing turbine torque and power. This interaction produces an acceleration of the flow, promoting a prompt start-up and enhancing the turbine's durability. In this respect, the velocity field can evolve towards an optimal limit thanks to the interaction between several rotors. These interactions can generate regions of overpressure or underpressure, thus modifying the stagnation zone and improving the energy output of the H-Darrieus turbine.

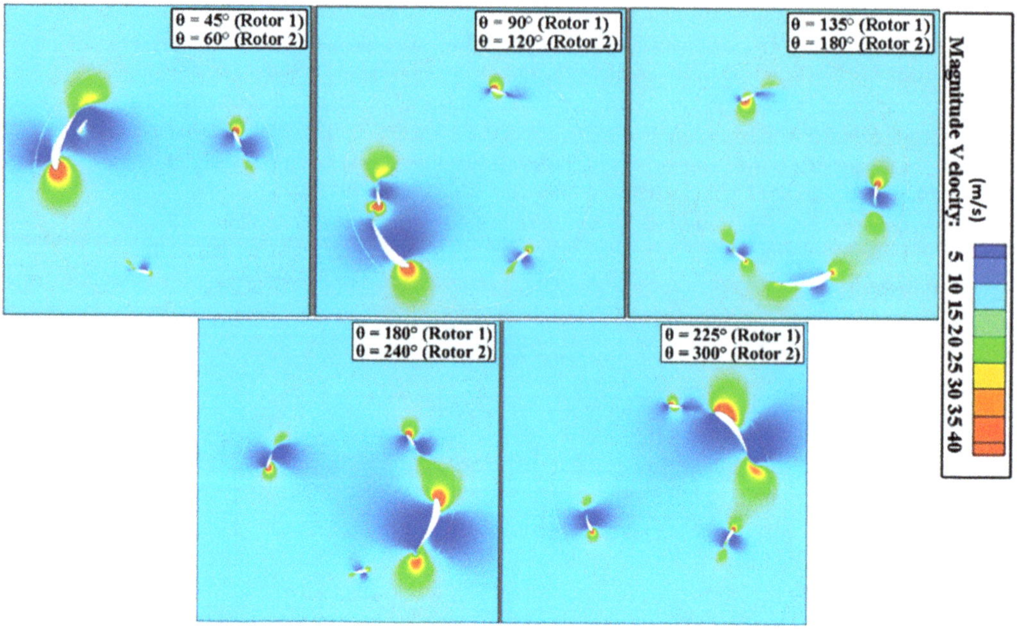

Figure 11. Magnitude velocity contours for dual-rotor turbine at different azimuth angles.

5. Conclusions

The aim of this article is to improve the aerodynamic performance of two adjacent H-Darrieus turbines by adding another rotor to the turbines. Our innovative configuration consists of two turbines separated by a distance and includes dual rotors with Eppler 420 and NACA 0021 airfoils. A 2D CFD simulation is performed using calculations based on the solution of the unsteady Reynolds-averaged Navier–Stokes equations (URANS), adopting the sliding mesh method and the $k - \omega$ SST turbulence model. Validation of this simulation model was carried out for the isolated H-Darrieus turbine, based on the evolution of the power coefficient (C_P) as a function of TSR. This validation analysis showed a good correlation between our results and those obtained by Castelli et al. [24].

The results of our study have shown that a dual-rotor turbine incorporating high-performance airfoils outperforms the two isolated single-rotor turbines. This is obtained through higher power output and better energy capture. The main CFD results obtained are presented below:

- A 12% increase in the coefficient of performance C_P at TSR = 2.6 was observed when a second Eppler 420 airfoil rotor was added to the first three-bladed NACA 0021 airfoil rotor.
- The efficiency of the two turbines incorporating dual rotors was improved by 17% compared with the two isolated turbines for $\alpha = 0°$. In addition, the choice of an optimum angle of attack for the dual-rotor airfoils improves energy efficiency and maximizes the performance of the two H-Darrieus turbines by attenuating wake disturbance.
- Using an optimum distance (Φ) of 4D between two adjacent turbines ensures efficient lift for each turbine, maintains robust energy efficiency, and demonstrates maximum performance within a wind farm.
- Increasing the chord length of the second rotor Eppler 420 blade by C = 0.4 m results in a higher moment coefficient, which translates into higher efficiency generation for H-Darrieus turbines.

- Analysis of the flow field reveals stable flow around the dual rotors of the H-Darrieus turbine as a function of the optimum parameter of the two airfoils. In addition, the second rotor contributes positively to improving the flow around the first rotor, thereby enhancing turbine efficiency and reducing wake size, enabling higher flow quality to be achieved. This configuration could be considered for future applications. On the other hand, choosing the optimum distance between two turbines incorporating dual rotors enhances their aerodynamic performance.

This study highlights the potential of dual-rotor H-Darrieus wind turbines equipped with high-performance airfoils to improve performance and increase the lift of the two adjacent turbines, offering high prospects for energy production, flow stabilization, and reduced head losses.

Future work will include the integration of artificial intelligence into the H-Darrieus turbine to improve efficiency and reliability, considering the installation of vertical-axis turbines (VAWTs) in hydrodynamic environments, and the study of the influence of a new 3D design on the self-starting capacity, with the aim of enhancing the system's sustainability and improving the performance of VAWTs.

Author Contributions: D.B.: Conceptualization, Methodology, Validation, Investigation, Resources, Writing—Original Draft, Software, Visualization, Formal Analysis, and Writing—Review and Editing. S.e.-D.F.: Methodology, Software, Validation, Investigation, Resources, Writing, Funding Acquisition, Supervision, and Visualization. M.B.: Methodology, Validation, Investigation, Resources, Writing, Supervision, and Visualization. A.S.: Methodology, Validation, Investigation, Resources, and Visualization. A.A.: Investigation, Validation, and Visualization. All authors have read and agreed to the published version of the manuscript.

Funding: This research received no external funding.

Data Availability Statement: In this study, no new data were produced. The analysis was carried out using existing publicly accessible data and previously published literature. Researchers seeking to obtain the data utilized in this study can consult the sources and references cited in the manuscript.

Acknowledgments: The corresponding author, Boulla Douha, wishes to take this opportunity to extend heartfelt thanks to the editor and reviewers for their invaluable feedback, which has significantly enhanced the quality of this scientific contribution.

Conflicts of Interest: The authors declare no conflicts of interest.

Abbreviations

2D	Two-Dimensional
CFD	Computational Fluid Dynamics
NACA	National Advisory Committee for Aeronautics
SST	Shear Stress Transport
TSR	Tip Speed Ratio
TSST	Turbulent Scale Simulation Transport Model
VAWT	Vertical-Axis Wind Turbine
SM	Sliding Mesh
RANS	Steady Reynolds-Averaged Navier–Stokes
HAWT	Horizontal-Axis Wind Turbine

μ	Dynamic Viscosity (kg/(m.s.))
ω	Darrieus Angular Velocity (rad/s)
ρ	Fluid Density (kg/m^3)
α	Angle of Attack (degrees)

$k-\omega$	SST Shear Stress Transport Turbulence Model
$k-\varepsilon$	Turbulence Model
A	Area Swept by Darrieus (m^2)
R	Darrieus Radius (m)
Re	Reynolds Number
C_d	Drag Coefficient
C_l	Lift Coefficient
C_m	Torque Coefficient
Cp	Power Coefficient
g	Acceleration Due to Gravity Vector (m/s^2)
P	Pressure (Pa)
U_∞	Free Stream Velocity (m/s)
y+	Dimensionless Wall Distance
θ	Azimuth Angle (degrees)
Φ	Distance between Adjacent Turbines (m)
Δt	Time Step (s)

References

1. Tasneem, Z.; Al Noman, A.; Das, S.K.; Saha, D.K.; Islam, M.R.; Ali, M.F.; R Badal, M.F.; Ahamed, M.H.; Moyeen, S.I.; Alam, F. An analytical review on the evaluation of wind resource and wind turbine for urban application: Prospect and challenges. *Dev. Built Environ.* **2020**, *4*, 100033. [CrossRef]
2. Hau, E. *Wind Turbines: Fundamentals, Technologies, Application, Economics*; Springer: Berlin/Heidelberg, Germany, 2006.
3. Wallaaldin, A.E.; Jarupula, S. Performance enhancement of Darrieus wind turbines using Plain Flap and Gurney Flap configurations: A CFD analysis. *Results Eng.* **2024**, *24*, 103400. [CrossRef]
4. Mohamed, M.H. Performance investigation of H-rotor Darrieus turbine with new airfoil shapes. *Energy* **2012**, *47*, 522–530. [CrossRef]
5. Carlos, S.F.; Gijs, V.K.; Gerard, V.B.; Fulvio, S. Visualization by PIV of dynamic stall on a vertical axis wind turbine. *Exp. Fluids* **2009**, *46*, 97–108.
6. Raciti, C.M.; Pavesi, G.; Battisti, L.; Benini, E.; Ardizzon, G. Modeling strategy and numerical validation for a Darrieus vertical axis micro-wind turbine. In Proceedings of the ASME 2010 International Mechanical Engineering Congress and Exposition, Vancouver, BC, Canada, 12–18 November 2010; p. IMECE2010-39548.
7. Mohamed, M.H.; Dessoky, A.; Faris, A. Blade shape effect on the behavior of the H-rotor Darrieus wind turbine: Performance investigation and force analysis. *Energy* **2019**, *179*, 1217–1234. [CrossRef]
8. Mohamed, O.S.; Ibrahim, A.; Etman, A.K.; Abdelfatah, A.A.; Elbaz, A.M.R. Numerical investigation of Darrieus wind turbine with slotted airfoil blades. *Energy Convers. Manag. X* **2020**, *5*, 100026. [CrossRef]
9. Ma, N.; Lei, H.; Han, Z.; Zhou, D.; Bao, Y.; Zhang, K. Airfoil optimization to improve power performance of a high-solidity vertical axis wind turbine at a moderate tip speed ratio. *Energy* **2018**, *150*, 236–252. [CrossRef]
10. Sobhani, E.; Ghaffari, M.; Maghrebi, M.J. Numerical investigation of dimple effects on Darrieus vertical axis wind turbine. *Energy* **2017**, *133*, 231–241. [CrossRef]
11. Shen, Z.; Gong, S.; Zu, H.; Guo, W. Multi-objective optimization study on the performance of double Darrieus hybrid vertical axis wind turbine based on DOE-RSM and MOPSO-MODM. *Energy* **2024**, *299*, 131406. [CrossRef]
12. Cheng, B.; Du, J.; Yao, Y. Power prediction formula for blade design and optimization of Dual Darrieus Wind Turbines based on Taguchi Method and Genetic Expression Programming model. *Renew. Energy* **2022**, *192*, 583–605. [CrossRef]
13. Farzad, G.; Seyed, R.M.; Mohammad, E.; Mehdi, M. A systematic investigation on the hybrid Darrieus-Savonius vertical axis wind turbine aerodynamic performance and self-starting capability improvement by installing a curtain. *Next Energy* **2025**, *6*, 100203.
14. Mahmoud, H.; Abdel, R.; Mohamed, E.; Shinichi, O.; Hamdy, H. Optimization of the power performance for three hybrid Darrieus-Savonius rotors based on the Taguchi method. *J. Phys. Conf. Ser.* **2024**, *2857*, 012011.
15. Mohan Kumar, P.; Surya, M.R.; Sivalingam, K.; Lim, T.-C.; Ramakrishna, S.; Wei, H. Computational Optimization of Adaptive Hybrid Darrieus Turbine: Part 1. *Fluids* **2019**, *4*, 90. [CrossRef]
16. Mirzaeian, H.; Ghobadian, B.; Mirhosseini, M. Performance Analysis and Optimization of Dual-Row Vertical Axis Wind Turbines with Innovative Hybrid Blades. *Energy Sci. Eng.* **2025**, *13*, 3171–3184. [CrossRef]

17. Amgad, D.; Pradip, Z.; Thorsten, L.; Ewald, K. Numerical Investigations of Two Darrieus Turbine Rotors Placed One Behind the Other with Respect to Wind Direction. In Proceedings of the Thirteenth International Conference of Fluid Dynamics, Cairo, Egypt, 21–22 December 2018.
18. Omar, S.M.; Pier, F.M.; Francesco, P.; Francesco, B.; Alessandro, B. Is the Actuator line method able to reproduce the interaction between closely-spaced Darrieus rotors a critical assessment on wind and hydrokinetic turbines. *Energy Convers. Manag.* **2023**, *293*, 117473. [CrossRef]
19. Tian, W.; Ni, X.; Li, B.; Yang, G.; Mao, Z. Improving the efficiency of Darrieus turbines through a gear-like turbine layout. *Energy* **2023**, *267*, 126580. [CrossRef]
20. Mohammed, S.; Shaaban, A. Efficient clusters and patterned farms for Darrieus wind turbine. *Sustain. Energy Technol. Assess.* **2017**, *19*, 125–135.
21. Jin, G.; Zong, Z.; Jiang, Y.; Zou, L. Aerodynamic analysis of side-by-side placed twin vertical-axis wind turbines. *Ocean Eng.* **2020**, *209*, 107296. [CrossRef]
22. Esmaeel, F.; Rakesh, M.; Frankie, F.J.; Hossein, F. Optimization and analysis of self-starting capabilities of vertical axis wind turbine pairs: A CFD-Taguchi approach. *Ocean Eng.* **2024**, *302*, 117614.
23. Fertahi, S.e.-D.; Belhadad, T.; Kanna, A.; Samaouali, A.; Kadiri, I. Assessment of fairing geometry effects on H-Darrieus hydroturbine performance using 2D URANS CFD simulations. *Energy Convers. Manag.* **2023**, *293*, 117434. [CrossRef]
24. Castelli, M.R.; Englaro, A.; Benini, E. The Darrieus wind turbine: Proposal for a new performance 8prediction model based on CFD. *Energy* **2011**, *36*, 4919–4934. [CrossRef]
25. Islam, H.; Mohamed, M. Aerodynamic performance enhancements of H-rotor darrieus wind turbine. *Energy* **2018**, *142*, 531–545. [CrossRef]
26. Chandra, S.; Tyagi, R. Study of Eppler 423 Airfoil with Gurney Flap and Vortex Generators. *Adv. Aerosp. Sci. Technol.* **2020**, *5*, 51001. [CrossRef]
27. Didane, D.H.; Rosly, N.; Zulkafli, M.F.; Shamsudin, S.S. Numerical investigation of a novel contra-rotating vertical axis wind turbine. *Sustain. Energy Technol. Assess.* **2019**, *31*, 43–53. [CrossRef]
28. Fluent, A. *Ansys Fluent Theory Guide*; ANSYS Inc.: Canonsburg, PA, USA, 2020; pp. 0001–1026.
29. Mahdi, Z.; Mohammad, J.M.; Seyed, R.V. Starting torque improvement using J-shaped straight-bladed Darrieus vertical axis wind turbine by means of numerical simulation. *Renew. Energy* **2016**, *95*, 109–126.
30. Wilberforce, T.; Alaswad, A. Performance analysis of a vertical axis wind tur bine using computational fluid dynamics. *Energy* **2023**, *263*, 125892. [CrossRef]
31. Asr, M.T.; Nezhad, E.Z.; Mustapha, F.; Wiriadidjaja, S. Study on start-up characteristics of H-Darrieus vertical axis wind turbines comprising NACA 4-digit series blade airfoils. *Energy* **2016**, *112*, 528–537. [CrossRef]
32. Balduzzi, F.; Bianchini, A.; Ferrara, G.; Ferrari, L. Dimensionless numbers for the assessment of mesh and timestep requirements in CFD simulations of Darrieus wind turbines. *Energy* **2016**, *97*, 246–261. [CrossRef]
33. Marsh, P.; Ranmuthugala, D.; Penesis, I.; Thomas, G. The influence of turbulence model and two and three-dimensional domain selection on the simulated performance characteristics of vertical axis tidal turbines. *Renew. Energy* **2017**, *105*, 106–116. [CrossRef]
34. Tescione, G.; Ragni, D.; He, C.; Simão, F.C.J.; Van Bussel, G.J.W. Near wake flow analysis of a vertical axis wind turbine by stereoscopic particle image velocimetry. *Renew. Energy* **2014**, *70*, 47–61. [CrossRef]
35. Boulla, D.; Fertahi, S.e.-D.; Bernatchou, M.; Samaouali, A. CFD Numerical Simulation Study of Aerodynamic Performance of H Darrieus Rotor. *CFD Lett.* **2025**, *17*, 117. [CrossRef]
36. Menter, F.R. Two-equation eddy-viscosity turbulence models for engineering applications. *AIAA J.* **1994**, *32*, 1598–1605. [CrossRef]
37. Marini, M.; Aristide, M.; Antonio, S. Performance of vertical axis wind turbines with different shapes. *J. Wind Eng. Ind. Aerodyn.* **1992**, *39*, 83–93. [CrossRef]
38. Boulla, D.; Fertahi, S.e.-D.; Bernatchou, M.; Samaouali, A.; Ajani, M.; Boussaq, C. Enhancement of aerodynamic performance of H-Darrieus rotor using wraparound fairing system: A 2D CFD study. *Int. J. Renew. Energy Dev.* **2025**, *14*, 1058–1071. [CrossRef]
39. Versteeg, H.K.; Malalasekera, W. *An Introduction to Computational Fluid Dynamics: The Finite Volume Method*; Pearson Education: London, UK, 2007.
40. Hoseinzadeh, S.; Amin, B.; Seyed, M.M.; Ali, S.; Stephan, H. A detailed experimental airfoil performance investigation using an equipped wind tunnel. *Flow Meas. Instrum.* **2020**, *72*, 101717. [CrossRef]
41. Vennell, R. Exceeding the betz limit with tidal turbines. *Renew. Energy* **2013**, *55*, 277–285. [CrossRef]
42. Wang, S.; Ingham, D.B.; Ma, L.; Pourkashanian, M.; Tao, Z. Numerical Investigation on dynamic stall of low Reynolds number flow around oscillating airfoils. *Comput. Fluids* **2010**, *39*, 1529–1541. [CrossRef]
43. Hao, W.; Ding, Q.; Li, C. Optimal performance of adaptive flap on flow separation control. *Comput. Fluids* **2019**, *179*, 437–448. [CrossRef]
44. Willmott, C.J.; Ackleson, S.G.; Davis, R.E.; Feddema, J.J.; Klink, K.M.; Legates, D.R.; O'donnell, J.; Rowe, C.M. Statistics for the evaluation and comparison of models. *Oceans* **1985**, *90*, 8995–9005. [CrossRef]

45. Baizhuma, Z.; Kim, T.; Son, C. Numerical method to predict ice accretion shapes and performance penalties for rotating vertical axis wind turbines under icing conditions. *J. Wind Eng. Ind. Aerodyn.* **2021**, *216*, 104708. [CrossRef]
46. Ahmad, M.; Shahzad, A.; Akram, F.; Ahmad, F.; Shah, S.I.A. Design optimization of Double Darrieus hybrid vertical axis wind turbine. *Ocean Eng.* **2022**, *254*, 111171. [CrossRef]
47. Paulo, A.S.F.; Silva, P.; Tsoutsanis, A.; Antoniadis, F. Simple multiple reference frame for high order solution of hovering rotors with and without ground effect. *Aerosp. Sci. Technol.* **2021**, *111*, 10651. [CrossRef]
48. Fertahi, S.e.-D.; Samaouali, A.; Kadiri, I. CFD comparison of 2D and 3D aerodynamics in H-Darrieus prototype wake. *E-Prime-Adv. Electr. Eng. Electron. Energy* **2023**, *4*, 100178. [CrossRef]
49. Zhang, W.; Cao, Y.; Qian, Z.; Wang, J.; Zhu, Y.; Yang, Y.; Wang, Y.; Wu, G. Research on Aerodynamic Performance of Asynchronous-Hybrid Dual-Rotor Vertical Axis Wind Turbines. *Energies* **2024**, *17*, 4424. [CrossRef]
50. Huang, H.; Li, J.; Li, G. Improving the self-starting and operating characteristics of vertical axis wind turbine by changing center distance in part of blades. *J. Build. Eng.* **2023**, *68*, 105974. [CrossRef]
51. Yang, Y.; Cao, Y.; Qian, Z.; Wang, J.; Zhu, Y.; Chen, X.; Zhang, W.; Wang, Y.; Wu, G.; Chen, S. A Study on the Effect of Turbulence Intensity on Dual Vertical-Axis Wind Turbine Aerodynamic Performance. *Energies* **2024**, *17*, 4124. [CrossRef]

applied sciences

MDPI

Article

Study on the Evolution Patterns of Cavitation Clouds in Friction-Shear Cavitating Water Jets

Xing Dong [1,*], Yun Jiang [2], Chenhao Guo [2] and Lu Chang [1]

[1] School of Mechanical Engineering, Heilongjiang University of Science and Technology, Harbin 150022, China; changlu1932@163.com
[2] School of Safety Engineering, Heilongjiang University of Science and Technology, Harbin 150022, China; jiangyun011010@163.com (Y.J.); 18145974766@163.com (C.G.)
* Correspondence: dongxingwrh@163.com; Tel.: +86-13069897538

Abstract: Current cavitating water jet technology for mineral liberation predominantly relies on the micro-jet impact generated by bubble collapse. Consequently, conventional nozzle designs often overlook the shear effects on mineral particles within the internal flow path. Moreover, the cavitation cloud evolution mechanisms in nozzles operating on this innovative principle remain insufficiently explored. This study systematically evaluates the cavitation performance of an innovatively designed cavitating jet nozzle with friction-shear effects (CJN-FSE), whose optimized internal structure enhances the interlayer shear and stripping effects crucial for the liberation of layered minerals. Utilizing high-speed imaging, we visualized submerged friction-shear cavitating water jets and systematically investigated the dynamic evolution patterns of cavitation clouds under jet pressures ranging from 15 to 35 MPa. The results demonstrate that the nozzle achieves effective cavitation, with jet pressure exerting a significant influence on the morphology and evolution of the cavitation clouds. As the jet pressure increased from 15 to 35 MPa, the cloud length, width, and average shedding distance increased by 37.05%, 45.79%, and 211.25%, respectively. The mean box-counting dimension of the cloud contour rose from 1.029 to 1.074, while the shedding frequency decreased from 1360 to 640 Hz. Within the 15–25 MPa range, the clouds showed periodic evolution, with each cycle comprising four stages: inception, development, shedding, and collapse. At 30 MPa, mutual interference between adjacent clouds emerged, leading to unsteady shedding behavior. This study thereby reveals the influence of jet pressure on the dynamic evolution patterns and unsteady shedding mechanisms of the clouds. It provides a theoretical and experimental basis for subsequent research into the nozzle's application in liberating layered minerals and proposes a new design paradigm for cavitation nozzles tailored to the mechanical properties of specific minerals.

Keywords: cavitating water jet; nozzle; cavitation cloud; high-speed imaging; shedding frequency

1. Introduction

Cavitation is a phase-change phenomenon induced by local pressure variations within a liquid [1–3]. The immense energy released instantaneously during bubble collapse represents its most valuable engineering characteristic [4,5]. This is because the asymmetric collapse of bubbles near a solid boundary generates high-speed micro-jets directed toward the wall, thereby imparting significant mechanical effects on the material surface [5–7]. Consequently, cavitation technology has found substantial application in fields that leverage this intense mechanical action, including efficient cleaning, surface derusting, material comminution, and material surface strengthening [8–12].

As a significant engineering implementation of the cavitation effect, cavitating jet technology actively induces bubble generation through specially designed nozzle structures and harnesses the intense impact effects resulting from bubble collapse to enhance the jet's performance [13]. As research on cavitating water jet technology deepens, its potential for mineral comminution and liberation has come into sharper focus. Guo et al. [14] conducted multiple comminution experiments on mica using abrasive cavitating jets and discovered that the cavitation effect significantly improves energy utilization efficiency. Dvorsky et al. [15] performed theoretical and experimental studies on the pure vapor cavitation mechanism in a novel water jet mill (WJM), finding that cavitation collapse induced by extreme velocity gradients can efficiently break down micron-sized silicon particles to the nanoscale. Sen et al. [16] conducted grinding tests on magnetite using a customized water jet experimental device equipped with a cavitation chamber, confirming that the growth and collapse of cavitation bubbles effectively promote particle refinement, representing a key mechanism for enhancing comminution efficiency. Galecki et al. [17] performed batch closed-circuit comminution tests on mono-sized coal samples using a self-designed cavitation chamber, observing that cavitation aids particle breakage, with backpressure-free conditions favoring refinement and the presence of backpressure broadening the product size distribution. Furthermore, Galecki et al. [18] investigated the effects of feed size and operating pressure on particle comminution efficiency, finding that coarser bituminous coal is more easily comminuted, and increasing pressure not only yields finer products but also alters particle morphology from flaky to blocky. These studies collectively corroborate the effectiveness of the cavitation effect in mineral comminution, revealing the promoting role of the energy instantaneously released during bubble collapse in particle liberation. However, this body of research primarily focuses on utilizing the impact energy for overall particle fragmentation and generally overlooks the shear effects exerted by the nozzle structure on mineral particles during liberation. Particularly for layered-structure minerals, efficient liberation relies not only on impact damage from bubble collapse but more critically on the directional application of shear force along weak interlayer interfaces to achieve efficient delamination [19]. Most current nozzle designs overlook this crucial mechanism. Therefore, it is essential to systematically optimize cavitation nozzle structures to synergistically excite both the necessary shear force and the cavitation effect. This optimization should be tailored to the mechanical properties and liberation behavior of layered minerals. Such an approach is of significant importance for enhancing the applicability and engineering maturity of cavitating jet technology in mineral processing.

The cavitating jet nozzle serves as the core component for applying cavitating jets to mineral comminution. Its structural parameters significantly influence the cavitation intensity of the jet and the stability of cavitation cloud development [20–22], thereby determining the ultimate liberation and comminution efficiency for minerals. Consequently, optimizing the nozzle structure for specific application scenarios is a critical pathway to enhancing technical efficacy. In recent years, scholars have conducted extensive research on the relationship between nozzle structure and cavitation performance using advanced numerical and experimental methods. Hutli et al. [23] combined high-speed imaging with copper specimen erosion tests and found that the nozzle geometric structure has the most significant influence on jet performance. Liu et al. [24] compared the cavitation effects of cylindrical, organ-pipe, and converging-diverging nozzles, indicating that cylindrical nozzles generate the most cavitation clouds, whereas the contraction truncation angles in converging-diverging nozzles inhibit the full development of cavitation clouds. Kamisaka et al. [25] proposed, through aluminum specimen erosion tests, that cavitation cloud lifetime is a key parameter affecting erosion intensity and found that nozzles with orifices near the throat exit and medium-length guide tubes exhibit optimal erosion performance. Cai et al. [26] demonstrated through erosion tests on aluminum specimens and sandstone that nozzles with a whistle-shaped outlet and a shorter throat can generate cavitation clouds with larger volume, lower shedding frequency, longer effective range, and stronger erosion capability. Prasetya et al. [27] employed X-ray phase-contrast imaging to reveal that proportional scaling of the nozzle size affects the diameter of the initial cavitation bubbles within the nozzle, with the bubble size showing a positive correlation with the nozzle dimensions. Wang et al. [28] found that organ pipe nozzles with a diffuser section at the outlet can enhance shear vortex intensity, thereby improving cavitation effects. Guo et al. [29] designed a dual-nozzle co-flow cavitating jet and confirmed its favorable cavitation performance, capable of forming cavitation clouds with distinct periodic variation patterns. Ullas et al. [30] compared Venturi tubes with different throat lengths using multi-modal measurement techniques and found that throat length significantly influences the cavitation shedding mechanism and collapse intensity. Yang et al. [31] conducted high-speed imaging tests on nozzles with divergence angles of 40°, 80°, and 120°, finding that the nozzle with an 80° divergence angle produced cavitation clouds with superior performance in both concentration and distribution range. Vidvans et al. [32] compared the performance of organ pipe nozzles (with Strouhal numbers of 0.14, 0.28, and 0.42) and long straight-hole nozzles in co-flow cavitating jet peening, indicating that the organ-pipe nozzle with St = 0.28 achieved the optimal cavitation intensity and peening efficacy. While these studies have laid a crucial foundation for understanding cavitation mechanisms and nozzle optimization, it is important to note that the majority focus on traditional applications such as material erosion and surface cleaning. The predominant design objectives are to enhance impact strength, increase erosion area, or intensify local damage—effects that primarily leverage the shock waves and micro-jets from bubble collapse. In contrast, research on cavitation nozzles specifically designed for the liberation of layered minerals remains scarce. Given that the efficient liberation of such minerals relies more heavily on shear-induced delamination at bedding interfaces than on impact fragmentation alone, there is an urgent need to develop specialized nozzle structures capable of synergistically inducing cavitation effects and enhancing shear actions, tailored to the mechanics of interlayer delamination.

In summary, despite the significant potential of cavitating jet technology in mineral processing and the considerable progress made in understanding nozzle structures and cavitation performance, research on nozzles specifically for liberating layered minerals remains in its infancy. Existing designs predominantly focus on enhancing the impact effect from bubble collapse and have paid insufficient attention to the shear effects experienced by mineral particles within the nozzle. To address this gap, the cavitation performance of a novel cavitating jet nozzle with friction-shear effects (CJN-FSE) is investigated. The CJN-FSE features a specially designed internal flow path to intensify the shear-induced liberation of mineral particles. Using high-speed imaging, we systematically examined the dynamic evolution patterns of cavitation clouds generated by the nozzle under jet pressures ranging from 15 to 35 MPa. The observed characteristics were compared with those of traditional nozzles reported in the literature to identify common cavitation behaviors. This work elucidates the influence of jet pressure on the cavitation dynamics of the CJN-FSE, providing a theoretical and experimental basis for its application in liberating layered minerals and proposing a new design strategy for the efficient liberation of such minerals.

2. Materials and Methods

2.1. Cavitating Jet Nozzle with Friction-Shear Effects

The disintegration process of layered mineral particles under cavitating jet action involves complex multi-physics dynamics. The primary mechanisms include interlayer delamination induced by pressure release; micro-jet impact from cavitation bubble collapse; water-wedge effects (i.e., the wedging action resulting from fluid penetration into the interlayer cracks or fissures of layered minerals) at solid–liquid interfaces; and the combined action of friction-shear coupling effects among mineral particles and between particles and nozzle walls [19]. This study focuses on a friction-shear cavitating water jet, which is a submerged jet generated by an independently designed CJN-FSE. This nozzle is specifically engineered to leverage these multiple physical mechanisms for mineral liberation. Its core innovation lies in the active intensification of friction-shear coupling effects, which synergistically enhances interlayer delamination and cavitation-induced damage. The structure of the CJN-FSE, shown in Figure 1, incorporates the following key functional features:

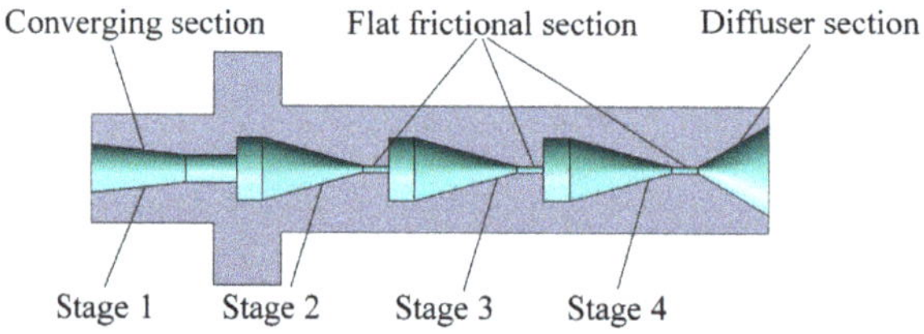

Figure 1. Internal channel geometry of the series-connected cavitating jet nozzle with friction-shear effects (CJN-FSE).

(1) Fluid acceleration and kinetic energy impartment. According to the continuity equation for incompressible flow:

$$Q = A_i \cdot v_i = A_o \cdot v_o \quad (1)$$

where Q is the volumetric flow rate, A_i is the inlet cross-sectional area of the converging section, A_o is the outlet cross-sectional area of the converging section, v_i is the average flow velocity at the inlet, and v_o is the average flow velocity at the outlet. The first nozzle stage, featuring a converging section, induces a significant reduction in the flow path cross-sectional area along the flow direction ($A_o < A_i$). According to Equation (1), this results in the outlet average velocity v_o significantly exceeding the inlet velocity v_i under a constant flow rate Q, thereby conferring high initial kinetic energy to the fluid.

(2) Intensification of shear stress within the fluid, enhancing friction-shear coupling effects. Stages 2 to 4 each incorporate flat frictional sections within their internal flow paths. For a Newtonian fluid (assuming no-slip boundary conditions), the shear stress τ within the fluid satisfies Equation [33]:

$$\tau = \mu \gamma \quad (2)$$

where μ is the dynamic viscosity of the fluid and γ represents the velocity gradient. The confined channel height (700–600 μm) in these flat frictional sections induces a significant velocity gradient, yielding high shear stress. This intensified shear stress constitutes the core driving force of the friction-shear coupling effects. Consequently, as layered mineral particles traverse these regions, they undergo intense friction and shear, promoting effective interlayer delamination.

(3) Active induction of cavitation. Stage 4 incorporates a diffuser section. The geometric expansion of this diffuser is designed to generate a substantial pressure drop in the flow field, creating the low-pressure conditions required for cavitation inception. The propensity for cavitation is characterized by the cavitation number, σ [34–36]:

$$\sigma = \frac{p_2 - p_v}{p_1 - p_2} \cong \frac{p_2}{p_1} \quad (3)$$

where p_1 is the jet pressure (at the nozzle inlet), p_2 is the ambient pressure (the static water pressure inside the cavitation tank at the nozzle outlet), and p_v is the saturated vapor pressure of water at room temperature. Cavitation occurs when σ falls below a critical value σ_c. The diffuser section is designed to reduce the ambient pressure p_2. This decrease in p_2 lowers the cavitation number σ, thereby inducing cavitation. Studies indicate [31,37] that optimal cavitation effects are achieved with divergence angles within 40–60°. The divergence angle θ of conventional nozzles is defined by:

$$\tan\frac{\theta}{2} = \frac{r_o - r_i}{L_k} \quad (4)$$

where r_o is the equivalent outlet radius, r_i is the equivalent inlet radius, and L_k is the diffuser length. Based on Equation (4), the equivalent divergence angles for the present nozzle were calculated to be 60° in the height direction and 47.27° in the width direction, both of which fall within the recommended effective range.

Unlike traditional cavitating jets, which rely primarily on shear layers between the jet and the surrounding fluid, the core characteristic of the friction-shear cavitating water jet lies in its internal flow path structure. This structure is designed to actively enhance solid–liquid friction, thereby inducing intense shear stress within the fluid. This "friction-shear" coupling effect acts directly on mineral particles, providing the necessary mechanical force for interlayer delamination. Thus, the term "friction-shear" denotes an active design mechanism where solid–liquid friction is harnessed to intensify shear forces, with the ultimate goal of enhancing mineral liberation efficiency.

2.2. Test System and Experimental Setup

The cavitating jet visualization test system, as schematically and physically depicted in Figure 2, consisted of a water reservoir, a high-pressure pump, a control console, the CJN-FSE, a high-speed camera system, a light source with a diffuser panel, a computer, and a cavitation tank. The system was driven by a CHC16/50 S-L triplex plunger pump (UDOR S.p.A., Rubiera, Italy), with a rated pressure of 50 MPa and a maximum flow rate of 16 L/min. The high-speed imaging was captured by a Y4-S1 camera (Integrated Design Tools, Pasadena Inc., Pasadena, CA, USA), which was equipped with an AF Nikkor 50 mm f/1.8D fixed-focal-length lens (Nikon Corporation, Tokyo, Japan); this camera system offers a maximum sampling frequency of 130,000 fps and a maximum image resolution of 1024 × 1024 pixels. A 300 W LED white surface light source with a color temperature of 6500 K, combined with a diffuser panel, served as the illumination system to ensure a uniform light field in the imaging area. An overflow port was installed on the wall of the cavitation tank to maintain a constant liquid level during experiments, thereby ensuring a stable confining pressure at the nozzle outlet. All high-pressure experimental procedures strictly adhered to safety protocols. The rated working pressure of all pressure-bearing components, including the high-pressure pump, pipelines, and the nozzle, significantly exceeded the maximum experimental value of 35 MPa. Furthermore, the system was outfitted with safety valves and other overpressure protection devices to guarantee a safe and controllable experimental process.

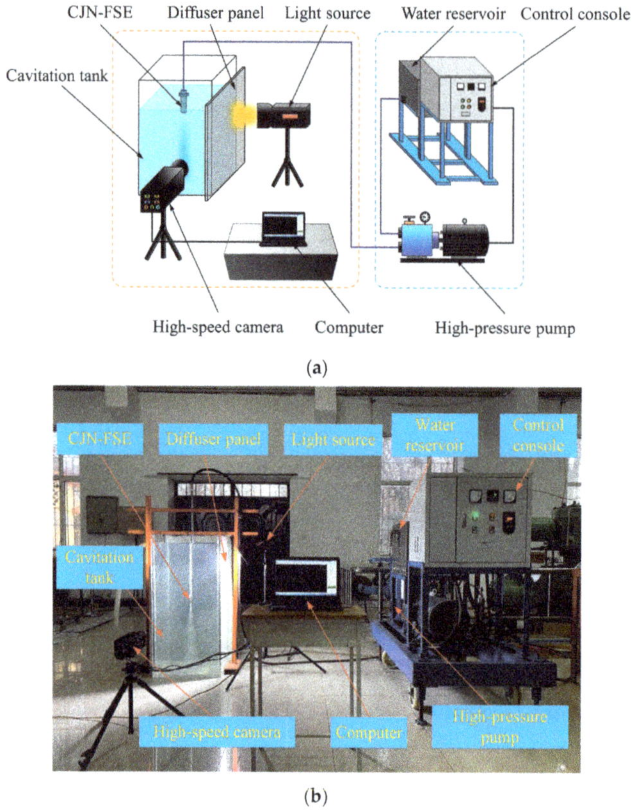

Figure 2. Cavitating jet visualization test system: (**a**) Schematic diagram of the visualization system; (**b**) Physical prototype of the visualization system.

2.3. Test Procedure

The experimental procedure comprised the following steps. First, the cavitation tank was filled until water overflowed, maintaining a dynamic equilibrium at the liquid surface to ensure a stable confining pressure environment. The assembly comprising the high-pressure hose, nozzle holder, and nozzle was then rigidly mounted using adjustable brackets, ensuring the nozzle axis was vertically aligned and the submerged depth at the outlet was 500 mm. The high-speed camera was positioned 130 cm from the nozzle, orientated to center the nozzle exit within the field of view while capturing the full evolution of the cavitation clouds. Spatial calibration was performed using a standard scale ruler, after which all component positions were secured.

The high-speed camera was configured with the following parameters: a sampling frequency of 5000 fps, an exposure time of 198 μs, an aperture of f/4, and a capture resolution of 800 × 600 pixels. This frame rate was selected because preliminary observations indicated a maximum cavitation cloud shedding frequency below 2000 Hz, thereby satisfying the Nyquist sampling criterion. The exposure time balanced sufficient light intake and image clarity, while the resolution adequately met the requirements for both spatial calibration accuracy and resolving key cavitation cloud features at the specified high frame rate. Under this configuration, the depth of field was estimated to exceed 10 cm, amply covering the primary region of cavitation cloud evolution. Furthermore, as the imaging field was located in the central area of the lens and geometric distortion at an f/4 aperture was negligible, lens distortion was not considered a significant factor.

A step-loading protocol was employed during testing. The jet pressure was increased at a rate of 0.5 MPa/s until reaching the target pressures of 15, 20, 25, 30, and 35 MPa. At each target pressure, data acquisition commenced only after the system had maintained stable operation for 30 s. For every jet pressure condition, the high-speed camera recorded the dynamic cavitation cloud behavior over a duration of 0.2 s (equivalent to 1000 frames), which was sufficient to capture numerous complete evolution cycles for robust statistical analysis.

To verify the reliability of the experimental results, three independent replicate tests were conducted for each operating condition. For each replicate, the system was completely restarted and the full procedure was followed to mitigate potential influences from prior initial conditions. A comparison of the average cavitation cloud length, width, and shedding frequency from the three replicates revealed that all relative deviations were within 10%, confirming the good reliability and repeatability of the data.

Finally, key flow parameters were calculated based on the experimental settings. The volumetric flow rate Q was determined for each jet pressure p_j. Subsequently, the Reynolds number Re and the cavitation number σ, the latter based on the nozzle outlet pressure, were derived. The corresponding values for all test conditions are summarized in Table 1.

Table 1. Operation parameters under different jet pressures.

p_j (MPa)	Q (L/s)	Re	σ
15	0.158	1.860×10^5	0.333×10^{-3}
20	0.182	2.143×10^5	0.250×10^{-3}
25	0.204	2.402×10^5	0.200×10^{-3}
30	0.223	2.626×10^5	0.167×10^{-3}
35	0.241	2.838×10^5	0.143×10^{-3}

Spatial calibration was performed using a standard scale ruler with 1 mm graduations. The ruler was carefully positioned within the plane of the jet centerline and adjusted to be parallel to the camera's imaging sensor to minimize parallax error, ensuring coverage of the entire cavitation cloud evolution region. A 16.0 cm segment on the ruler was selected as a reference length. Using Image Pro Plus 6.0 software, the pixel width corresponding to this physical length was measured repeatedly across multiple images, yielding an average value of 400 pixels. The spatial conversion coefficient, η, was then determined as follows [38]:

$$\eta = \frac{d_o}{d_i} \tag{5}$$

where d_o is the actual physical dimension, and d_i represents the corresponding number of pixels in the image. Substituting the values into Equation (5) gives η = 0.400 mm/pixel, meaning each pixel represents a physical area of 0.400 mm $\times$ 0.400 mm.

Given the high-speed camera sensor pixel size of 13.68 µm, the optical magnification factor was calculated to be approximately 0.0342. An uncertainty analysis was conducted for the spatial measurements. The primary sources of uncertainty were identified as the residual parallelism deviation between the scale ruler and the imaging plane, and the pixel edge interpretation error during image analysis. The combined standard uncertainty from these sources was estimated to be $\pm$1.0 pixel. Contributions from optical distortion in the central field of view and the manufacturing tolerance of the scale ruler were deemed negligible. Consequently, the absolute standard uncertainty for spatial measurements is $\pm$0.40 mm. This uncertainty is substantially smaller than the characteristic changes in cavitation cloud dimensions observed across the tested pressure range, which confirms that the measurement error does not compromise the reliability of the main findings.

3. Results

3.1. Image Processing Methods and Procedure

To accurately analyze the dynamic evolution of the cavitation clouds, a systematic image processing workflow was implemented. The key steps involved were grayscale conversion, time-averaged analysis, contour extraction, and the Frame Difference Method (FDM). The specific procedures for these key techniques are detailed below.

Cavitation clouds are opaque and reflective [39], yielding high-contrast optical signatures in high-speed imaging. Taking advantage of this, the clouds were captured using side-lighting illumination, appearing as high-intensity white regions in the raw images. To analyze their dynamics accurately, grayscale image processing was applied. This conversion transforms original three-channel RGB images into single-channel grayscale images, which removes chromatic interference and enhances the contrast between the cavitation clouds and the background, thereby improving identification accuracy. The conversion was performed using the standard weighted average formula:

$$\boldsymbol{I}_{\text{gray}} = 0.299\boldsymbol{R} + 0.587\boldsymbol{G} + 0.114\boldsymbol{B} \tag{6}$$

where $\boldsymbol{R}$, $\boldsymbol{G}$, $\boldsymbol{B}$ are the pixel values of the red, green, and blue channels, respectively, and $\boldsymbol{I}_{\text{gray}}$ is the converted grayscale value, ranging from 0 to 255.

Time-averaged analysis was employed to extract the steady-state features and characterize the time-averaged morphology of the cavitation clouds by performing statistical averaging over the image sequence in the temporal dimension [31]. It is mathematically defined as

$$\boldsymbol{I}_{\text{mean}} = \frac{1}{N}\sum_{\text{t}=1}^{N} \boldsymbol{I}_{\text{t}} \tag{7}$$

where $\boldsymbol{I}_{\text{mean}}$ is the time-averaged grayscale matrix, $\boldsymbol{I}_{\text{t}}$ denotes the grayscale matrix at frame t, and N represents the number of images within the computational time domain. (N = 1000, corresponding to 0.2 s). This technique effectively suppresses interference from transient fluctuations, thereby revealing the underlying time-averaged cloud structure.

To enable comparative analysis across images, the time-averaged grayscale images were normalized to the [0, 1] interval using the following transformation [31]:

$$\boldsymbol{I}_{\text{norm}} = \frac{\boldsymbol{I}_{\text{t}} - \min(\boldsymbol{I}_{\text{mean}})}{\max(\boldsymbol{I}_{\text{mean}}) - \min(\boldsymbol{I}_{\text{mean}})} \tag{8}$$

where $\boldsymbol{I}_{\text{norm}}$ is the normalized time-averaged grayscale matrix. For subsequent contour map generation, pixels with normalized intensities in the range [0, 0.1] were clamped to a value of 0.1. This threshold was applied because this low-intensity range predominantly corresponds to background optical noise (e.g., from water and sensor dark current) rather than the cavitation cloud signal. Suppressing this noise facilitates a clearer visualization of the cavitation cloud morphology.

Extracting the binary contour of the cavitation clouds from the grayscale images was a prerequisite for the fractal dimension calculation. This was achieved via image binarization using a fixed global threshold method in MATLAB R2023b:

$$BW_{(x,y)} = \begin{cases} 1 & \text{if } \boldsymbol{I}_{\text{mean}(x,y)} \geq T \\ 0 & \text{otherwise} \end{cases} \tag{9}$$

where $I_{\text{mean}(x,y)}$ is the value in the time-averaged grayscale matrix at row x, column y; T is the binarization threshold, manually selected between 0.2 and 0.3 to ensure consistent contour extraction across all pressure conditions. The Canny edge detection algorithm was then applied to the binary images to obtain a single-pixel-width contour curve. Finally, all contour images were rescaled to a standard size of 1024 × 1024 pixels to maintain dimensional consistency for subsequent analysis.

The FDM was used to analyze cloud dynamics. It operates by computing the grayscale difference between consecutive frames, producing a difference matrix:

$$D_{\text{t}} = I_{\text{t}} - I_{\text{t}-1} \tag{10}$$

where D_{t} is the grayscale difference matrix, and $I_{\text{t}-1}$ represents the grayscale matrix at frame t − 1. In the resulting matrix, positive and negative values correspond to pixels that have brightened or darkened, respectively, between frames. This directly captures the dynamic evolution processes of the cavitation clouds [39].

3.2. Dynamic Behavior of Cavitation Clouds Under Constant Jet Pressure

Figure 3 illustrates the dynamic evolution of cavitation clouds within a 3 ms timeframe at 20 MPa jet pressure. The image sequence was acquired at equal time intervals (0.2 ms), with the abscissa and ordinate representing temporal and spatial dimensions, respectively. Analysis of grayscale characteristics in sequential images enables clearer identification of morphological evolution patterns at millisecond timescales. As illustrated in Figure 3, under a jet pressure of 20 MPa, cavitation clouds near the nozzle exit exhibit distinct periodic dynamic evolution. A complete cycle comprises four phases: inception, development, shedding, and collapse. Time-resolved imaging shows that at 0.2 ms, Cavitation Cloud A has nucleated within the wake region of preceding clouds. At 0.4 ms, the preceding cloud sheds from Cloud A, which then enters the development phase (0.4–1.4 ms). During this phase, the length and width of Cloud A increase progressively by 255.56% and 59.52%, respectively. During 1.4–1.6 ms, cavitation cloud A sheds from cloud B and subsequently enters the collapse phase (1.6–2.6 ms), exhibiting two characteristic evolutionary behaviors: significant decay in axial velocity at the cloud tail and rapid attenuation in density/length of the cavitation cloud. Complete collapse occurs at 2.8 ms. Temporal analysis further indicates that subsequent clouds B, C, and D nucleate at 1.0, 1.8, and 2.6 ms, respectively, each following the same evolutionary phases as Cloud A.

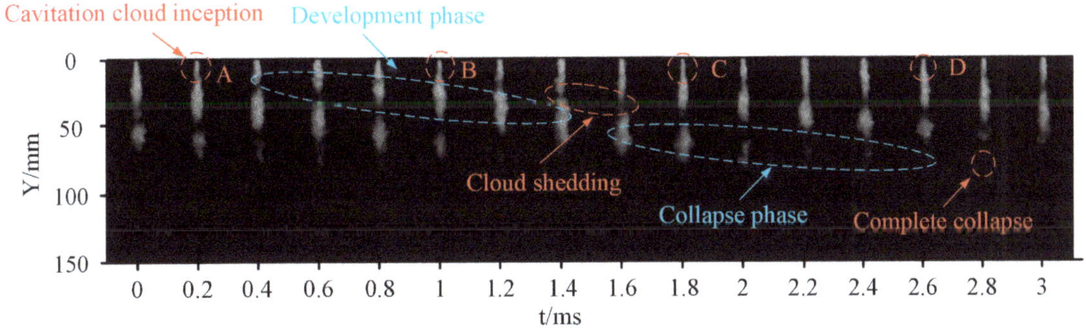

Figure 3. Time-resolved evolution of the cavitating jet at 20 MPa jet pressure. The labeled A, B, C, and D in the figure are all newly generated cavitation clouds.

This periodic evolution is fundamentally driven by the cyclic generation, development, and shedding of vortices in the shear layer. Upon exiting the nozzle, the high-speed water jet establishes a strong velocity gradient against the surrounding stagnant water, triggering shear layer instabilities that lead to the formation of discrete vortex structures. The pressure at the core of these vortices drops, as described by Bernoulli's principle. When the core pressure falls below the local saturated vapor pressure, cavitation occurs, giving rise to the visible cavitation clouds. Once the vortices reach a critical scale, they shed, and the attached cavitation clouds consequently enter the shedding and collapse stages, thereby initiating a new cycle of vortex generation. The diffuser section of the nozzle effectively intensifies this vortex generation process, resulting in the production of a stable, periodic sequence of cavitation clouds. This observed iterative periodicity, demonstrating clear temporal self-similarity across the inception, development, shedding, and collapse phases at 20 MPa, validates the intrinsic stability of the cavitating jet pulsations generated by the CJN-FSE. These characteristics are consistent with the periodic cavitation cloud evolution documented for other nozzle types [2,34,36,40–43], collectively confirming that the CJN-FSE produces stable and effective cavitation.

3.3. Characteristics of Cavitation Clouds Under Varying Jet Pressures

3.3.1. Morphological Features of Cavitation Clouds

This method decouples interference from periodic cloud evolution and plunger pump pressure fluctuations, enabling accurate reconstruction of time-averaged morphological features. This paper performs temporal averaging on cavitation cloud images under varying jet pressures. The temporally averaged grayscale images of cavitation clouds over 0.2 s durations at 15–35 MPa jet pressures are illustrated in Figure 4. In Figure 4b, the exclusion of pixels with grayscale values in the range of 0 to 0.1 (normalized to 0 to 1 scale) eliminates interference from low-intensity regions on cavitation cloud morphology in contour maps. Statistical analysis of cloud dimensions using Image Pro Plus 6.0 software revealed a 45.79% increase in width and a 37.05% increase in length as jet pressure increased from 15 to 35 MPa. The variation curves of cavitation cloud length and width versus jet pressure are shown in Figure 5. It can be seen from Figure 5 that both the length and width of the cavitation cloud increase with increasing jet pressure, but the rate of increase diminishes as the jet pressure rises.

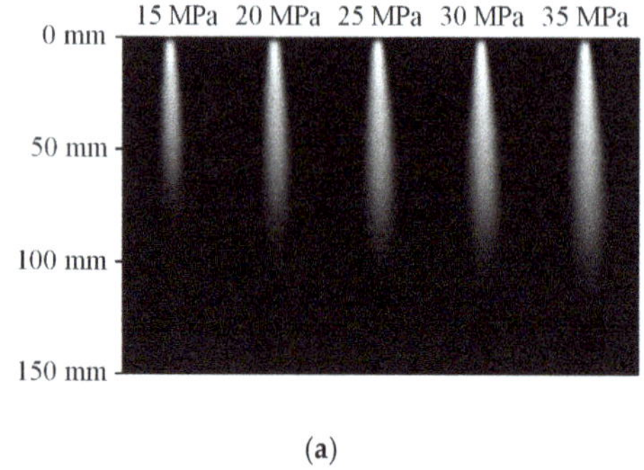

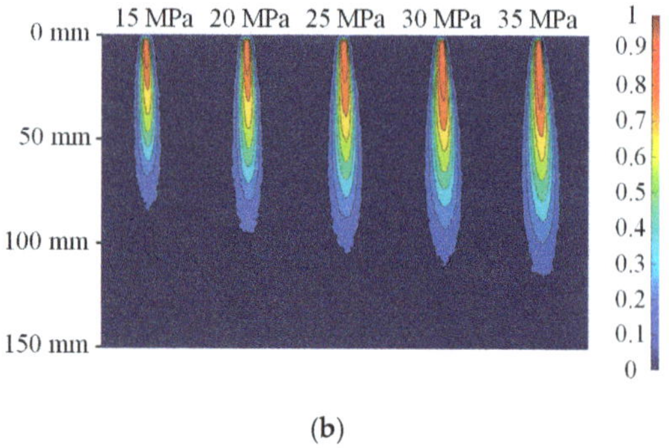

Figure 4. Cavitation clouds under varying jet pressures: (**a**) Time-averaged grayscale images; (**b**) Contour maps of time-averaged grayscale.

The fractal dimension serves as an effective metric for quantifying the complexity and irregularity of geometric shapes, finding extensive application across engineering disciplines [44]. For contour curves, it provides a quantitative measure of morphological complexity. Among various computational methods, the box-counting dimension is particularly prevalent for analyzing complex curves, owing to its conceptual and computational simplicity and straightforward implementation. Moreover, the cavitation cloud contour exhibits self-similarity, a characteristic for which the box-counting dimension is well-suited. The method is also computationally efficient and consistent, making it ideal for processing the large volumes of binary image data generated from transient contours under different pressures, thereby facilitating robust statistical comparison. For these reasons, the box-counting dimension was selected in this study to characterize the contour complexity of the cavitation clouds. The box-counting dimension D_F of a cavitation cloud contour curve F is defined as follows [45,46]:

$$\boldsymbol{D}_F = -\lim_{r \to 0} \frac{\log(N_r)}{\log r} \tag{11}$$

where r denotes the side length of the square boxes (grid cells) used to cover the plane containing curve F, and N_r represents the minimum number of boxes required to completely cover curve F.

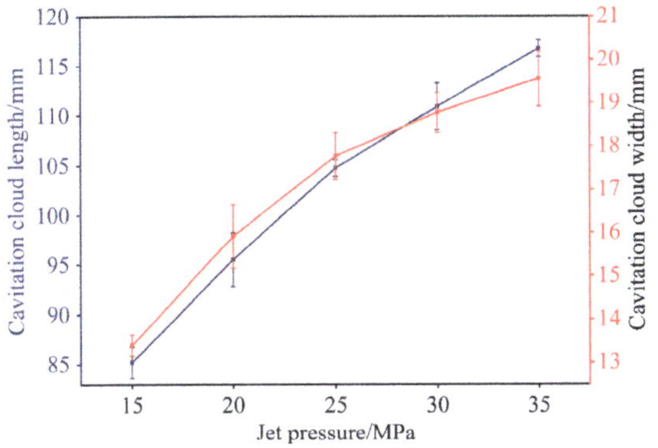

Figure 5. Variation curves of cavitation cloud dimensions versus jet pressure. Data points represent the mean values from three independent repeated tests (n = 3), and error bars indicate the standard deviation (SD).

The box-counting dimension was computed using a geometric sequence of box sizes, with side lengths doubling from $2^0 = 1$ pixel to $2^{10} = 1024$ pixels, resulting in 11 distinct scales. A linear regression was then performed on the data plotted in a double logarithmic coordinate system, where $\log_2(1/r)$ and $\log_2(N_r)$ represented the abscissa and ordinate, respectively. The slope of the resulting best-fit line provides the estimate for the box-counting dimension, D_F. Figure 6 presents a representative cavitation cloud contour at 35 MPa and its corresponding double logarithmic plot. As shown in Figure 6b, the strong linear correlation observed in the plot indicates scale invariance of the contour across the analyzed range of scales, thereby confirming its fractal nature and validating the application of the box-counting method in this study.

However, cavitation clouds exhibit highly dynamic evolution, and their instantaneous morphology, influenced by flow field disturbances, possesses significant randomness. Consequently, the box-counting dimension of the cavitation cloud contour curve at a single moment cannot comprehensively and stably characterize the overall complexity of the cavitation cloud under a specific jet pressure. To overcome the limitations of single-phase analysis and enable effective comparison of cavitation cloud contour complexity under different jet pressures, this study calculates the box-counting dimension for the contour curves of all cavitation clouds captured within the 0.2 s sampling period under the same jet pressure. The arithmetic mean of these values is then used as the characteristic parameter representing the contour complexity of the cavitation cloud under that operating condition. This method effectively suppresses transient random fluctuations through statistical averaging in the time domain, making the calculation results more reflective of the inherent complexity of the cavitation cloud structure. It thereby provides a stable and statistically significant quantitative metric for comparisons under different jet conditions.

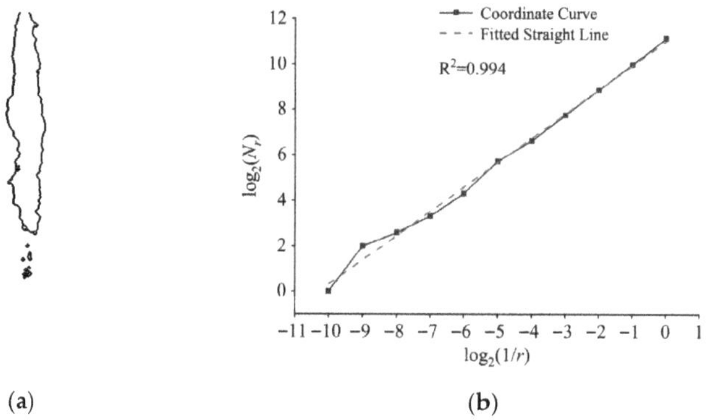

Figure 6. Contour curve of the cavitation cloud and its double logarithmic plot for box-counting dimension calculation at 35 MPa jet pressure: (**a**) Contour curve of the cavitation cloud, (**b**) Double logarithmic plot for box-counting dimension.

Applying this method, the mean box-counting dimension of the cavitation cloud contour was calculated across the jet pressure range of 15–35 MPa. Figure 7 plots this mean value against jet pressure, with the corresponding statistical data provided in Table 2. It is noteworthy that the calculated fractal dimensions exhibited a relatively large standard deviation, originating from the inherently transient and turbulent nature of the cavitation clouds, which causes significant random fluctuations in their instantaneous morphology. Nevertheless, the mean value derived from a statistical sample of 1000 frames robustly represents the time-averaged contour complexity. As illustrated in Figure 7, the mean box-counting dimension demonstrates a clear monotonic increase from 1.029 to 1.074 as the jet pressure rises from 15 MPa to 35 MPa. This upward trend signifies a transition in the cloud's morphology: from a relatively smooth and simple contour at lower pressures to an increasingly fragmented and intricate one at higher pressures. The proliferation of small-scale structures and the growing irregularity of the boundary directly contribute to this increase in morphological complexity.

From a fluid mechanics standpoint, an increase in jet pressure intensifies the shear interaction at the jet boundary, enhancing flow instability and vortex generation, which in turn leads to a more complex cavitation cloud morphology and a correspondingly higher box-counting dimension of the cloud contour. Simultaneously, the elevated pressure promotes the activation and growth of cavitation nuclei, contributing to the observed increases in cloud length and width. These observations collectively demonstrate that the nozzle design, particularly the diffuser section, effectively creates low-pressure zones and enhances shear layer development, thereby confirming its efficacy in cavitation generation. The strengthening of cavitation effects with rising jet pressure is expected to enhance the comminution efficiency for mineral particles. However, a diminishing trend in the growth rates of cloud dimensions and contour complexity is observed as pressure exceeds a certain level. This signifies that the enhancement of cavitation by jet pressure is subject to the law of diminishing marginal returns; specifically, the incremental gain in cavitation intensity per unit pressure increase becomes smaller at higher operating pressures. Therefore, optimizing the jet pressure for mineral disintegration applications necessitates a trade-off between achieving high comminution efficiency and managing energy consumption. Employing excessively high jet pressures is not only energy-intensive but also provides progressively smaller improvements in disintegration performance.

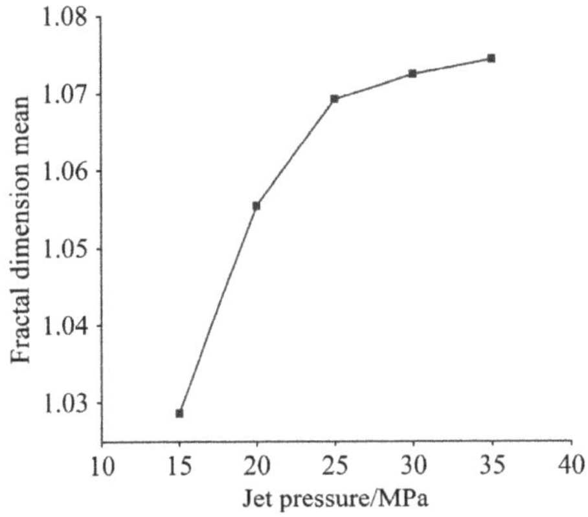

Figure 7. Variation curve of the mean box-counting dimension of the cavitation cloud contour versus jet pressure.

Table 2. Statistical values of the box-counting dimension for the cavitation cloud contour under different jet pressures.

p_j (MPa)	$D_{F\text{mean}}$	SD	R^2_{mean}	n
15	1.0287	0.0243	0.9954	1000
20	1.0555	0.0258	0.9975	1000
25	1.0692	0.0202	0.9983	1000
30	1.0725	0.0208	0.9980	1000
35	1.0744	0.0196	0.9967	1000

3.3.2. Dynamic Evolution of Cavitation Clouds Under Varying Jet Pressures

Figure 8 presents grayscale images of cavitation clouds over a 3 ms period at jet pressures of 15–35 MPa. The blue dashed line marks the shedding characteristic plane. The vertical distance between the shedding plane and the nozzle exit defines the average shedding distance. Statistical measurements show a 211.25% increase in average shedding distance as jet pressure rises from 15 to 35 MPa. Figure 9 plots this distance versus jet pressure, revealing a nonlinear growth trend. Notably, at 35 MPa, the average shedding distance increases markedly compared to lower-pressure conditions.

As shown in Figure 8, at 15 MPa, the nozzle produces periodic cavitation clouds, though of limited axial extent. Increasing the jet pressure to 20 MPa elevates the nozzle outlet flow rate and velocity, which intensifies shear-induced vortex generation. This process yields larger and more energetic vortex structures, establishing a more extensive low-pressure region that, in turn, exacerbates cavitation, promotes volumetric expansion, and leads to a substantial elongation of the clouds. Under this condition, the four characteristic stages of a full cycle—inception, development, shedding, and collapse—are distinctly visible. At 25 MPa, the jet maintains a similar periodic evolution but with heightened cavitation intensity near the nozzle exit. Concurrently, the shedding plane moves downstream, the clouds attain larger dimensions prior to shedding, and the overall jet length is extended. A critical transition occurs beyond 25 MPa, where a combination of reduced cloud tail velocity and increased pre-shedding size triggers interfacial fusion between a shedding cloud and its successor, as evidenced in Figure 8c (1–1.2 ms) and Figure 8d (0.2–0.4 ms). This coalescence becomes more pronounced at higher pressures. At 30 MPa, the temporal overlap between clouds intensifies; the preceding cloud has not fully collapsed or shed before the subsequent cloud rapidly generates and develops, resulting in significant spatial overlap and mutual interference. This interaction disrupts the inherent periodicity of the vortex-shedding process, preventing a fraction of the clouds from completing a full shedding-collapse cycle and thus triggering a transition from steady to unsteady shedding behavior. The non-shedding clouds subsequently undergo a gradual collapse, as seen in Figure 8d (1.2–1.4 ms) and Figure 8e (1.8–2 ms). At 35 MPa, this unsteady shedding is markedly pronounced, which considerably increases the average shedding distance.

Comparative studies indicate that the observed cloud merging between shedding and adjacent cavitation clouds in this work also occurs in clouds generated by other nozzles. For instance, Zhu et al. [35] documented the merging of shed clouds with primary clouds, noting that increased jet pressure promotes coalescence, consistent with our findings. Furthermore, observations by Cui et al. [40] on cylindrical cavitating nozzles at 20 MPa revealed cloud merging within 0.6 to 0.7 ms. Similarly, observations by Hutli et al. [47] on conical diffuser cavitation nozzles at 26.7 MPa jet pressure revealed interface fusion between adjacent cavitation clouds within the 0.276 to 0.354 ms timeframe.

At jet pressures $\geq$ 30 MPa, cavitation clouds exhibit unsteady shedding behavior. To clearly distinguish clouds across cycles and analyze their dynamics at elevated pressures, we applied the FDM to process grayscale image sequences. The FDM detects motion regions by comparing pixel-wise variations between consecutive frames [31,48].

Figure 10 presents the FDM maps of the cavitation clouds under the five jet pressures, where adjacent dashed lines demarcate identical evolution cycles. Based on the principles of FDM, the translational motion of a cloud along the jet axis induces concurrent positive and negative grayscale fluctuations in the image sequence. Therefore, the FDM maps capture a superposition of the genuine cavitation processes (bubble development and collapse) and the apparent motion due to cloud displacement.

To distinguish true collapse from pure translation, an area-based quantitative criterion was established. During pure translational motion without significant phase change, the area of the grayscale disappearance zone and the emergence zone should be approximately equal. Conversely, a collapse event, characterized by a net reduction in cloud volume, manifests as a grayscale disappearance zone that is significantly larger than the emergence zone. The specific criterion was defined as follows: a cloud was classified as undergoing pure translation if the ratio of the disappearance zone area to the emergence zone area fell within 0.8–1.2. A ratio exceeding 1.5 for several consecutive frames was identified as a collapse event. These threshold values were determined through statistical analysis of multiple characteristic evolution sequences and were found to effectively discriminate between the two physical processes.

Figure 10 reveals that cavitating jets under all five pressures exhibit similar periodic evolution within individual cycles. As jet pressure increases, the FDM maps display intensified hollowing effects at their central regions, indicating heightened cloud density that persistently occupies the jet core. Consequently, grayscale fluctuations from cloud inception and migration concentrate along jet peripheries. When jet pressure is below 30 MPa, mutual interference between cavitation clouds is relatively small, resulting in distinct periodic characteristics in the developmental changes of cavitation clouds across cycles. When jet pressure reaches 30 MPa, interference between cavitation clouds occurs within the 0.4–1 ms timeframe. At 35 MPa jet pressure, this mutual interference becomes significantly more pronounced, leading to significant interference with cavitation cloud collapse during the 1.2 to 1.8 ms interval.

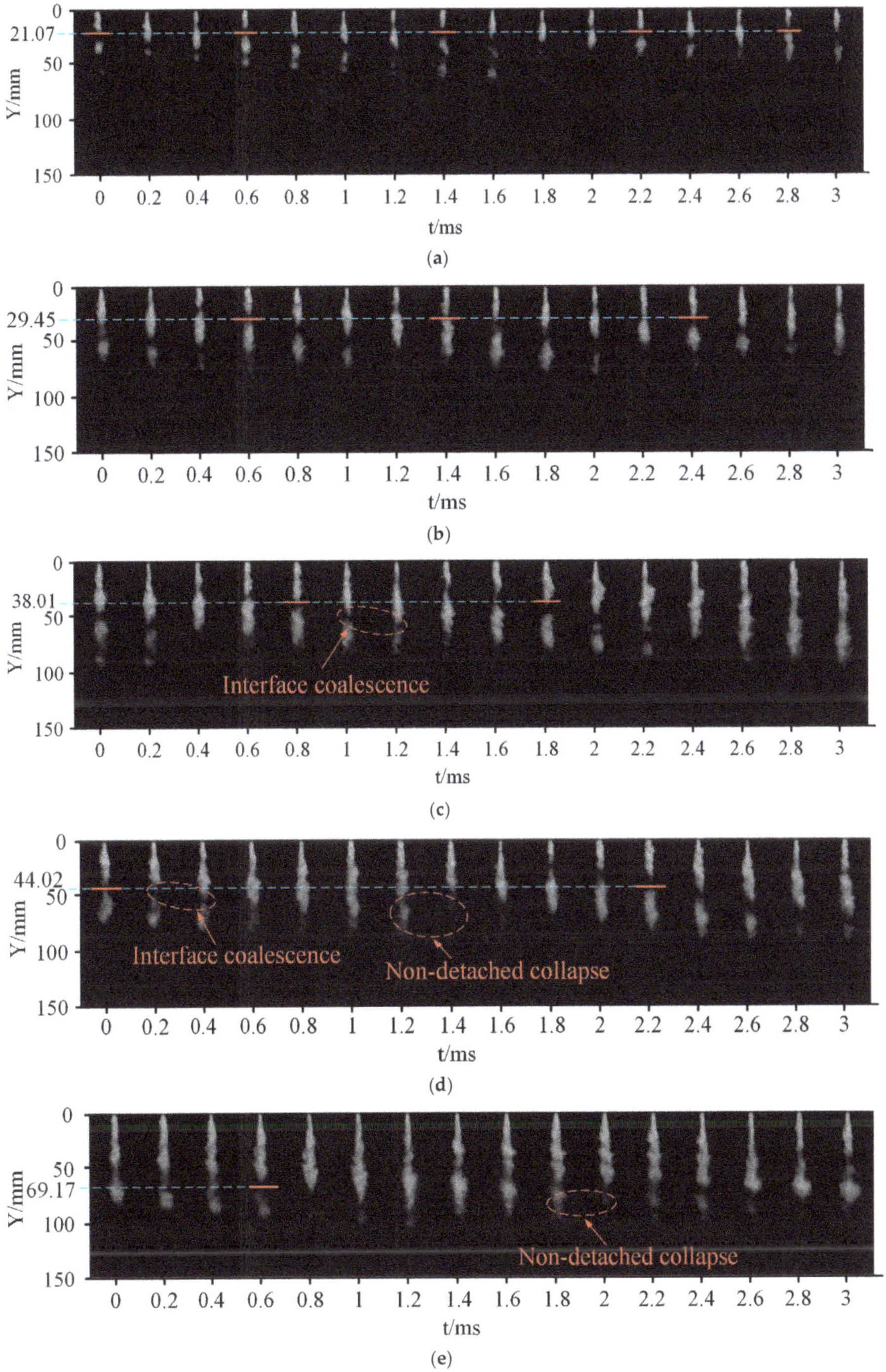

Figure 8. Grayscale images of cavitation clouds under varying jet pressures: (**a**) 15 MPa; (**b**) 20 MPa; (**c**) 25 MPa; (**d**) 30 MPa; (**e**) 35 MPa. The red line segments designate the shedding positions of the cavitation clouds, while the blue dashed line represents the characteristic plane of cavitation cloud shedding.

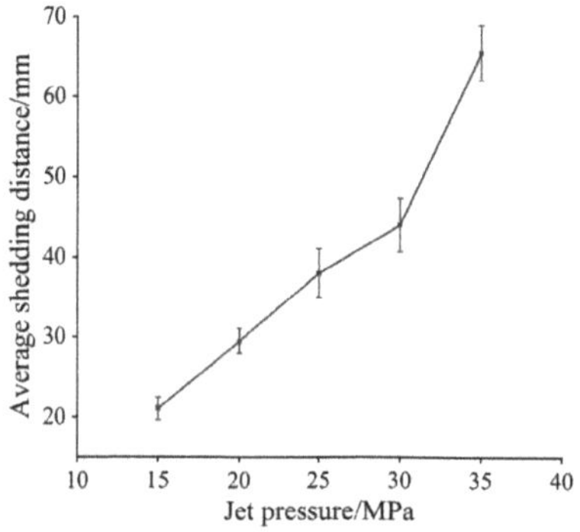

Figure 9. Variation curve of average shedding distance versus jet pressure. Data points represent the mean values from three independent repeated tests (n = 3), and error bars indicate the standard deviation (SD).

The core design objective of the CJN-FSE is to intensify friction-shear coupling effects on mineral particles through internal flow paths while synergistically inducing cavitation to enhance disintegration efficiency. The observed dynamic characteristics of cavitation clouds—particularly their periodicity, stability, and spatiotemporal collapse distribution—are critical for achieving this synergy. At 15–25 MPa, cavitation clouds exhibit clear, stable periodic shedding. This periodicity implies a predictable spatiotemporal distribution of bubble collapses. Combined with the nozzle's internally designed high-shear zones and stable periodic micro-jet impacts from cavitation collapse, this theoretically enables uniform and efficient interlayer delamination of layered mineral particles.

However, when jet pressure exceeds 30 MPa, the observed unsteady shedding causes disordered collapse processes with highly uneven spatiotemporal distribution. This heterogeneity results in spatiotemporally uncontrollable cavitation energy acting on mineral particles: some are subjected to excessive or unpredictable impacts, while others reside in energy-deficient regions. Consequently, disintegration efficiency fluctuates, and product size distribution broadens with reduced uniformity. Thus, despite higher pressures generating denser clouds, the uncontrollable collapse distribution due to non-steadiness may compromise overall disintegration efficacy and product quality stability.

3.3.3. Shedding Frequency of Cavitation Clouds Under Varying Jet Pressures

Grayscale frequency analysis extracts the dominant frequency of grayscale variations at specific locations within cavitation cloud images over time, reflecting the primary oscillation frequency in target regions. Studies confirm that the dominant grayscale frequency along the cavitating jet axis serves as a reliable proxy for cloud shedding frequency [36,49,50]. Frequency-domain analysis was performed on the grayscale values at different axial positions along the cavitating jet centerline over a 0.2 s duration under various jet pressures using MATLAB. The dominant frequencies of these grayscale values at 15 MPa are presented in Figure 11.

As shown in Figure 11, the dominant frequencies of the grayscale values along the jet centerline exhibit significant variations across different axial positions. Analysis in conjunction with Figure 8a reveals that within the interval of 0–35 pixels (0–14.0 mm), the cavitation cloud is in the development stage and the dominant frequency of the jet centerline grayscale value fluctuates between 1265 and 1360 Hz; within 36–72 pixels (14.4–28.8 mm), the cavitation cloud begins shedding and the dominant frequency of the jet centerline grayscale value remains constant at 1360 Hz; within 73–139 pixels (29.2–55.6 mm), the cavitation cloud collapses and the dominant frequency of the jet centerline grayscale value fluctuates between 1265 and 1360 Hz; within 140–158 pixels (56 to 63.2 mm), the bubble clouds that have not completely collapsed continue to collapse within this interval, resulting in a rapid attenuation of the dominant frequency of the jet centerline grayscale value to 485 Hz; within 159–300 pixels (63.6–120 mm), only a very small number of residual cavitation bubbles exist, with no significant grayscale changes observable in the images, and the dominant frequency of the jet centerline grayscale value fluctuates between 110 and 170 Hz.

The axial distribution of the dominant frequency is a direct manifestation of the distinct hydrodynamic regimes characterizing each stage of cavitation cloud evolution. Within the development, shedding, and initial collapse regions (0–139 pixels), the flow is governed by coherent, periodically shedding vortex structures, which generate a stable and dominant high-frequency signature. The minor fluctuations observed within this high-frequency band are attributed to background flow field turbulence and the inherent stochasticity of the cloud edge collapse process. Beyond approximately 140 pixels, in the final collapse region, the coherent cloud structure disintegrates into a dispersed cluster of small bubbles undergoing random collapse. In this regime, the grayscale variations lose their periodic character, transitioning to a low-frequency, broadband signal indicative of stochastic processes. This fundamental change in the governing physics—from organized, large-scale vortex shedding to disordered, fine-scale bubble collapse—is responsible for the observed sharp attenuation of the dominant frequency. Thus, the frequency distribution serves as a clear spectral signature of the cavitation cloud's transition from an ordered, periodic state to a fully disordered, random one.

As illustrated in Figure 8a, under a jet pressure of 15 MPa, the average shedding time of the cavitation cloud within 3 ms is 0.75 ms, corresponding to a shedding frequency of approximately 1333 Hz, and the average shedding distance is approximately 21.07 mm, which corresponds to the 53 pixels position in the cavitation cloud images. Analysis in conjunction with Figure 11 reveals that this average shedding distance falls within the dominant frequency stable region of 14.4–28.8 mm, where the shedding frequency is 1360 Hz; the discrepancy relative to the estimated shedding frequency is a mere 2.0%. This demonstrates that the dominant frequency of the jet centerline grayscale value within the stable region encompassing the average shedding distance can accurately reflect the shedding behavior of the cavitation cloud at 15 MPa jet pressure. Therefore, the shedding frequency of the cavitation clouds can be estimated by calculating the dominant frequency of the jet centerline grayscale value within the stable region corresponding to its average shedding distance. This aligns with the conclusion drawn by Peng et al. [50,51] that variations in the grayscale value at a specific point within a defined interval can reflect the shedding frequency of cavitation clouds.

For each jet pressure, the pixel location corresponding to the average cloud shedding distance was identified on the jet centerline. A Fast Fourier Transform (FFT) was then applied to the grayscale value time series at these locations to extract the dominant frequency components. The FFT was configured with the following parameters: a sampling frequency Fs of 5000 Hz, N = 1000 sampling points, and a rectangular window of the same length, yielding a frequency resolution of 5 Hz. To enhance the clarity of the primary peak, spectral components below a 50 Hz cutoff were excluded to mitigate low-frequency interference. The noise level was estimated through residual analysis by comparing the original signal with a single-frequency cosine model constructed using the identified dominant frequency and amplitude. The statistical uncertainty of the identified dominant frequency was quantified using the Cramér-Rao lower bound theory to compute the 95% confidence interval. This comprehensive spectral processing methodology robustly determines the cavitation cloud shedding frequency and its associated uncertainty. The cavitation cloud shedding frequencies and their 95% confidence intervals under different jet pressures, calculated using MATLAB, are summarized in Table 3. The corresponding grayscale frequency-domain diagrams are shown in Figure 12.

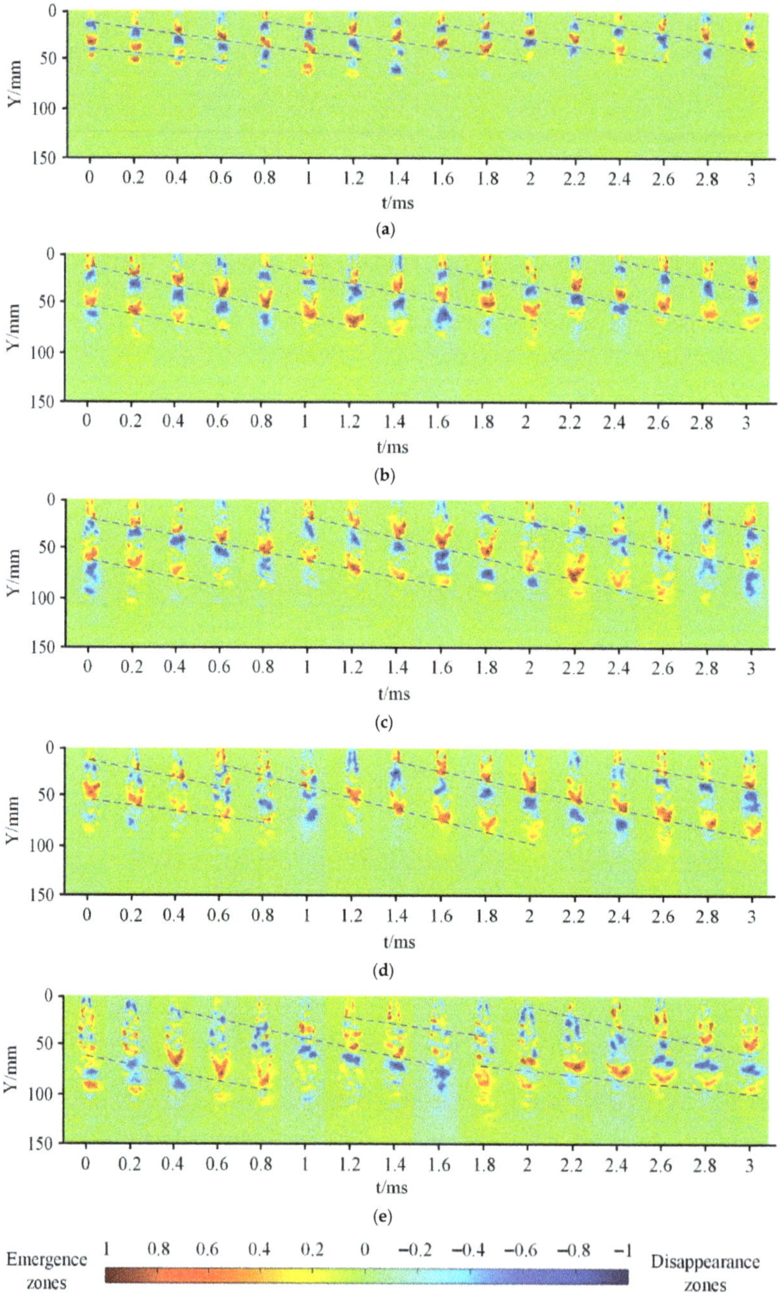

Figure 10. FDM maps of cavitation clouds under varying jet pressures: (**a**) 15 MPa; (**b**) 20 MPa; (**c**) 25 MPa; (**d**) 30 MPa; (**e**) 35 MPa.

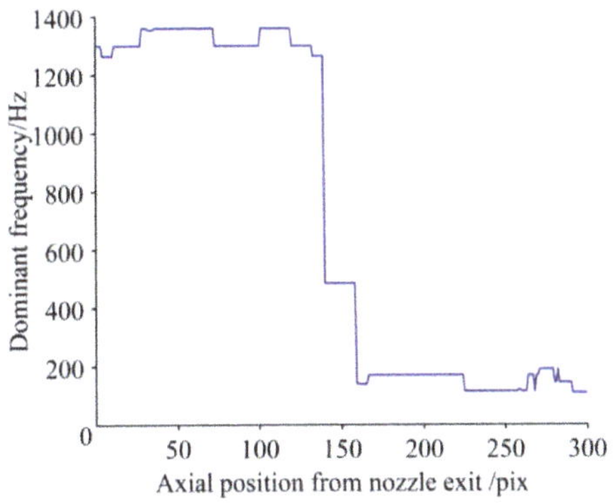

Figure 11. Dominant grayscale frequency along the jet centerline at 15 MPa jet pressure.

Table 3. Cavitation cloud shedding frequencies and 95% confidence intervals under different jet pressures.

p_j (MPa)	Dominant Frequency (Hz)	95% Confidence Interval (Hz)
15	1360	1359.02–1360.98
20	1300	1299.02–1300.98
25	1285	1283.88–1286.12
30	1165	1163.68–1166.32
35	640	638.92–641.08

As observed in Figure 12, the dominant frequency of the grayscale at the shedding location decreases with increasing jet pressure, indicating that the shedding frequency of the cavitation cloud diminishes as jet pressure rises; this finding is consistent with the conclusions obtained by Dong et al. [39] and Hutli et al. [47]. When the jet pressure reaches 30 MPa, mutual interference among cavitation clouds occurs, causing a small number of clouds to exhibit non-shedding behavior, resulting in a decrease of the cavitation cloud shedding frequency to 1165 Hz; when the jet pressure reaches 35 MPa, the mutual interference intensifies and the non-shedding phenomenon becomes significantly more pronounced, leading to a substantial decrease in cavitation clouds shedding frequency to 640 Hz.

The amplitude in the frequency spectra of Figure 12 represents the power of the grayscale fluctuation signal at each frequency. As the pressure increases from 15 to 20 MPa, a marked increase in amplitude is observed across the spectrum, particularly at the dominant shedding frequency. This signifies a greater contrast in the cavitation cloud images and more intense grayscale oscillations per shedding cycle, aligning with the visual observations of larger and denser clouds. At 25 MPa, a distinct shift occurs: the amplitude at the dominant frequency diminishes sharply, while the background noise level rises and secondary spectral peaks proliferate. This evolution indicates a breakdown in the coherence of the shedding process, where the previously stable periodic signal becomes contaminated by a broad spectrum of disturbances. This spectral broadening is a direct consequence of the physical interactions between clouds, such as interfacial fusion and the onset of non-shedding collapse, which disrupt the organized vortex shedding. By 35 MPa, the spectrum transitions from a discrete, tonal character to a continuous, broadband one, manifesting the complete disintegration of periodic shedding into fully developed unsteady and chaotic behavior. Collectively, these spectral transformations provide conclusive frequency-domain evidence that corroborates the time-domain observation of a transition from stable periodicity to destabilization and disorder with increasing jet pressure.

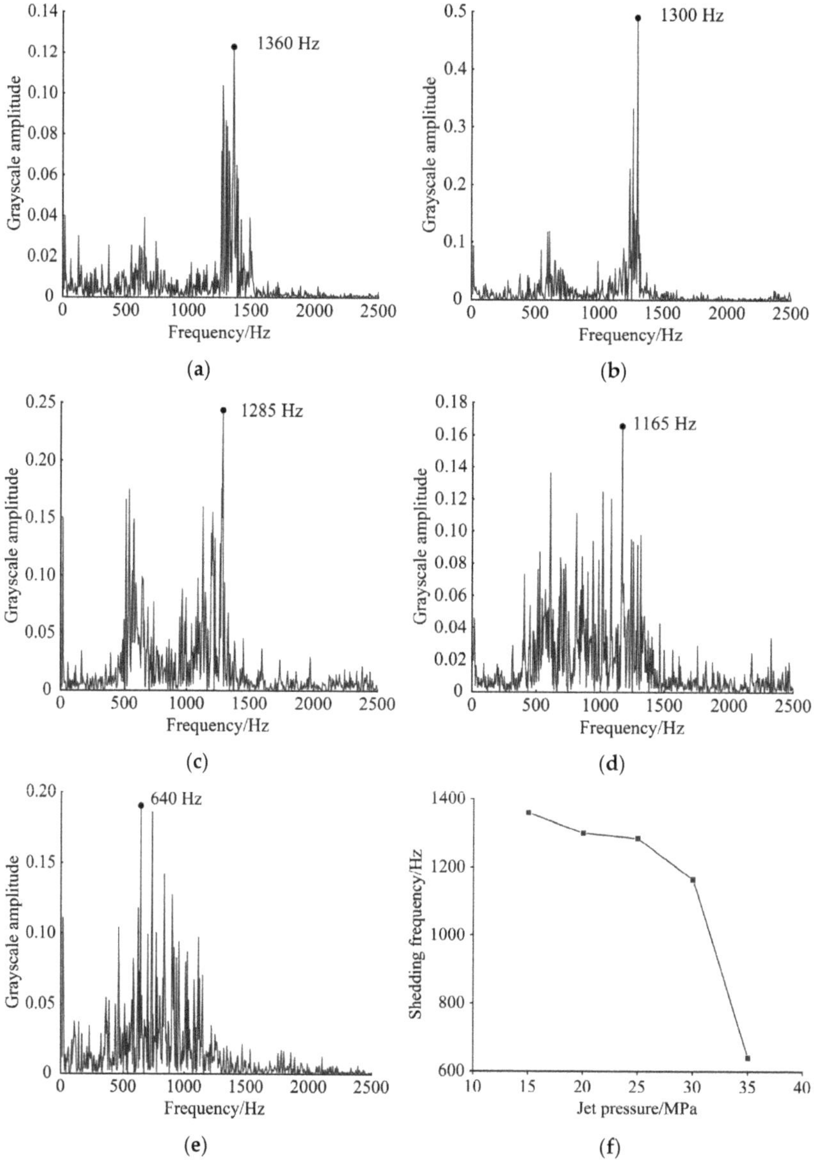

Figure 12. Grayscale frequency-domain diagrams at the shedding location of the jet centerline: (**a**) 15 MPa; (**b**) 20 MPa; (**c**) 25 MPa; (**d**) 30 MPa; (**e**) 35 MPa; (**f**) Variation of cavitation cloud shedding frequency with jet pressure.

In summary, the CJN-FSE generates significant cavitation effects across the jet pressure range of 15–35 MPa, and its cavitation performance gradually strengthens with increasing jet pressure. However, when the jet pressure reaches 30 MPa, mutual interference occurs between adjacent cavitation clouds. This interference disrupts the periodicity of cloud shedding, leading to disorder in their collapse timing and manifesting as unsteady shedding characteristics. At a jet pressure of 35 MPa, this unsteady phenomenon becomes more pronounced, causing the shedding frequency of the cavitation clouds to rapidly decrease to 640 Hz. Based on the analysis of cavitation cloud dynamics, an operational pressure $\leq$ 30 MPa is recommended for the CJN-FSE in mineral liberation processes. This upper limit is critical for maintaining stable and controllable cavitation, which is a prerequisite for efficient and predictable particle disintegration.

4. Discussion

Conventional nozzle designs for mineral liberation via cavitating water jets largely overlook the critical role of shear effects on particles within the nozzle. To address this limitation and evaluate the performance of a nozzle engineered specifically to enhance such effects, this study investigated the dynamic evolution of cavitation clouds generated by the submerged CJN-FSE using high-speed imaging. The results demonstrate that the internal flow path of the CJN-FSE successfully generates stable cavitating jets. A key finding, however, is that jet pressures exceeding 30 MPa induce significant interference between successive cavitation clouds, leading to unsteady shedding behavior. This instability becomes particularly severe at 35 MPa, where intensified inter-cloud interaction markedly suppresses the collapse process and precipitates a sharp decline in shedding frequency from 1165 Hz to 640 Hz. Due to the detrimental impact of such unsteady shedding on the uniformity and efficiency of mineral liberation, it is recommended that the operational pressure of this nozzle be maintained at or below 30 MPa in practice.

The design philosophy of the CJN-FSE is fundamentally distinct from that of nozzles optimized for erosion enhancement [23–32]. Its primary objective is to achieve a synergistic intensification of both 'cavitation impact' and 'friction-shear' mechanisms, tailored to the specific mechanical demands of liberating layered minerals. The observed stable, periodic shedding at low-to-medium pressures (15–25 MPa) is consistent with the behavior of other nozzle types [2,34,36,40–43], as the underlying vortex-shedding physics in the shear layer is a universal phenomenon. This study confirms that the multi-stage internal flow path of the CJN-FSE is fully capable of sustaining this stable periodic cavitation. More importantly, it validates a novel design paradigm: the active control of shear stress via flow path geometry to promote mineral liberation. This represents a significant departure from and a valuable extension to conventional cavitation nozzle design theory.

The CJN-FSE is designed to intensify friction-shear coupling effects on mineral particles via its internal flow path while concurrently inducing cavitation, creating a synergistic action aimed at enhancing liberation efficiency. The dynamic characteristics of the cavitation clouds—specifically their periodicity, stability, and spatiotemporal collapse distribution—are paramount to realizing this synergy. Within the 15–25 MPa range, the clear and stable periodic shedding ensures a predictable spatiotemporal distribution of bubble collapses. This predictable cavitation energy delivery, when combined with the high-shear zones engineered within the nozzle, enables a combined mechanism of stable micro-jet impact and directed shear force. This multi-mechanistic action is anticipated to facilitate uniform and efficient interlayer delamination of layered mineral particles.

However, when the jet pressure exceeds 30 MPa, the observed transition to unsteady shedding results in a disordered collapse process with a highly heterogeneous spatiotemporal distribution. This spatial and temporal unpredictability of collapse events renders the cavitation energy delivered to the mineral particles uncontrollable. Consequently, some particles are likely subjected to excessive or erratic impacts, while others in energy-deficient regions receive insufficient treatment. This inconsistency in energy delivery likely results in fluctuating liberation efficiency and a broader, less uniform product size distribution. Thus, despite the generation of denser cavitation clouds at elevated pressures, the compromised controllability of the collapse process is counterproductive, potentially undermining both the overall liberation efficacy and the final product quality uniformity.

In summary, while increasing jet pressure enhances cavitation intensity and could theoretically improve liberation efficiency, its practical application for mineral processing requires a more nuanced consideration. Two critical factors emerge from this study: first, the intensification of cavitation with pressure is subject to diminishing returns; second, and more importantly, pressures exceeding 30 MPa introduce significant flow unsteadiness. This unsteadiness manifests as spatiotemporally unpredictable cloud shedding and collapse, which degrades the controllability and effectiveness of the energy transfer to mineral particles, thereby jeopardizing liberation uniformity, process efficiency, and product quality stability. Consequently, for practical engineering implementation, the marginal gains in cavitation intensity at high pressures must be carefully weighed against the substantial risks to process stability and product consistency.

This study has primarily elucidated the macroscopic influence of jet pressure on cavitation cloud dynamics through multi-faceted qualitative observation and quantitative analysis based on high-speed imaging and image processing. Nevertheless, certain limitations remain. First, the study lacks quantitative measurements of the internal cloud structure, local pressure/velocity fields, and inlet pressure fluctuations. Second, it does not establish a direct quantitative correlation between cavitation characteristics and the liberation efficiency of layered minerals. These limitations also define clear directions for future research. Subsequent work will focus on the following aspects: (1) integrating high-precision sensors and Particle Image Velocimetry (PIV) to quantitatively analyze parameters such as flow rate and velocity at the nozzle outlet; (2) measuring pressure fluctuation curves at the nozzle inlet under different jet pressures over a 30-s duration to quantify pressure variations and their spectral characteristics; (3) performing comparative cavitation erosion tests with conventional nozzles of similar characteristic parameters to evaluate the performance enhancement attributable to the multi-stage internal flow path; and (4) conducting liberation experiments on model layered minerals (e.g., mica, graphite) to construct a quantitative model linking operational parameters, cavitation metrics, and liberation effectiveness. This integrated pathway is essential for transitioning this technology from a promising proof-of-concept toward a practical engineering application.

5. Conclusions

This study employed high-speed imaging technology to conduct cavitation visualization experiments on a friction-shear cavitating water jet. By utilizing methods such as grayscale image analysis, time-averaged processing, fractal dimension calculation, FDM, and cavitation cloud shedding frequency measurement to analyze the cavitation clouds under different jet pressures, the following conclusions can be drawn:

(1) For a novel nozzle with a multi-stage flow path structure (converging section, flat frictional section, diffuser section), a friction-shear cavitating water jet intended for layered mineral liberation and characterized by a "friction-shear" coupling effect was proposed.

(2) The regulatory effect of jet pressure on the cavitation cloud evolution behavior of the CJN-FSE was investigated. Within the 15–25 MPa range, the cavitation clouds exhibit stable periodic evolution (inception, development, shedding, collapse). When the pressure exceeds 30 MPa, excessive flow field kinetic energy leads to mutual interference between clouds, resulting in a transition to an unsteady shedding mode. The geometric dimensions (length, width) and morphological complexity (box-counting dimension) of the cavitation clouds increase monotonically with pressure, while the shedding frequency decreases.

(3) Based on the clear correlation between cavitation stability and jet pressure, we recommend maintaining the operational pressure at or below 30 MPa for mineral liberation

applications. This pressure threshold is identified as the upper limit for preserving stable and controllable cavitation, which is a critical prerequisite for achieving uniform and efficient liberation performance. This finding provides a crucial operational guideline for the future application of the CJN-FSE.

In conclusion, this work elucidates the cavitation characteristics of the CJN-FSE and proposes a novel design philosophy for cavitation nozzles: a function-integrated approach that tailors the nozzle's function to the specific mechanical demands of the target application. Future research will prioritize two critical avenues: the quantitative characterization of the internal shear flow field and the establishment of a direct correlation between quantified cavitation metrics and mineral liberation efficiency. These investigations are essential to build a robust foundation for the eventual engineering deployment and adoption of this technology.

Author Contributions: Conceptualization, X.D.; methodology, X.D. and Y.J.; software, Y.J.; validation, X.D. and Y.J.; formal analysis, Y.J.; investigation, X.D., Y.J., and C.G.; resources, X.D. and L.C.; data curation, Y.J.; writing—original draft preparation, Y.J. and C.G.; writing—review and editing, X.D.; visualization, Y.J.; supervision, X.D. and L.C.; project administration, X.D.; funding acquisition, X.D. All authors have read and agreed to the published version of the manuscript.

Funding: This research was supported by the National Natural Science Foundation of China (Grant No. 52075151).

Institutional Review Board Statement: Not applicable.

Informed Consent Statement: Not applicable.

Data Availability Statement: The data presented in this study are available on request from the corresponding author. As the data in this paper belongs to the National Natural Science Foundation of China, it involves related privacy and is not owned by individuals.

Conflicts of Interest: The authors declares that there are no conflicts of interest.

References

1. Hutli, E.; Nedeljkovic, M.; Bonyár, A. Dynamic behaviour of cavitation clouds: Visualization and statistical analysis. *J. Braz. Soc. Mech. Sci. Eng.* **2019**, *41*, 281. [CrossRef]
2. Soyama, H. Cavitating Jet: A Review. *Appl. Sci.* **2020**, *10*, 7280. [CrossRef]
3. Yang, Y.F.; Wang, G.W.; Shi, W.D.; Li, W.; Ji, L.L.; Wang, H.L. Turbulence Characteristics in the Mixing Layer of a Submerged Cavitating Jet at High Reynolds Numbers. *Sustainability* **2022**, *14*, 11963. [CrossRef]
4. Han, L.; Zhang, M.D.; Huang, G.H.; Huang, B. Energy transformation mechanism of a gas bubble collapse in the free-field. *Chin. J. Theor. Appl. Mech.* **2021**, *53*, 1288–1301. [CrossRef]
5. Schenke, S.; van Terwisga, T.J.C. An energy conservative method to predict the erosive aggressiveness of collapsing cavitating structures and cavitating flows from numerical simulations. *Int. J. Multiph. Flow* **2019**, *111*, 200–218. [CrossRef]
6. Peng, C.; Tian, S.C.; Li, G.S. Joint experiments of cavitation jet: High-speed visualization and erosion test. *Ocean Eng.* **2018**, *149*, 1–13. [CrossRef]
7. Li, X.F.; Duan, Y.X.; Zhang, Y.N.; Tang, N.N. Retardant Effects of Collapsing Dynamics of a Laser-Induced Cavitation Bubble Near a Solid Wall. *Symmetry* **2019**, *11*, 1051. [CrossRef]
8. Bukharin, N.; El Hassan, M.; Omelyanyuk, M.; Nobes, D. Applications of cavitating jets to radioactive scale cleaning in pipes. *Energy Rep.* **2020**, *6*, 1237–1243. [CrossRef]
9. Gu, J.Y.; Luo, C.H.; Lu, Z.B.; Ma, P.C.; Xu, X.C.; Ren, X.D. Bubble dynamic evolution, material strengthening and chemical effect induced by laser cavitation peening. *Ultrason. Sonochem.* **2021**, *72*, 105441. [CrossRef]
10. Soyama, H. Cavitation Peening: A Review. *Metals* **2020**, *10*, 270. [CrossRef]
11. Soyama, H. Comparison between the improvements made to the fatigue strength of stainless steel by cavitation peening, water jet peening, shot peening and laser peening. *J. Mater. Process. Tech.* **2019**, *269*, 65–78. [CrossRef]
12. El Hassan, M.; Bukharin, N.; Al-Kouz, W.; Zhang, J.W.; Li, W.F. A Review on the Erosion Mechanism in Cavitating Jets and Their Industrial Applications. *Appl. Sci.* **2021**, *11*, 3166. [CrossRef]

13. Li, G.S.; Shen, Z.H. *Theory and Application of Self-Resonating Cavitating Water Jet*, 1st ed.; China University of Petroleum Press: Dongying, China, 2008; p. 4. ISBN 978-7-5636-2146-0.
14. Guo, C.W.; Dong, L. Energy Consumption in Comminution of Mica with Cavitation Abrasive Water Jet. *J. Chin. Univ. Min. Technol.* **2007**, *17*, 251–254. [CrossRef]
15. Dvorsky, R.; Lunacek, J.; Sliva, A. Dynamics analysis of cavitation disintegration of microparticles during nanopowder preparation in a new Water Jet Mill (WJM) device. *Adv. Powder Technol.* **2011**, *22*, 639–643. [CrossRef]
16. Sen, S. Grinding of magnetite using a waterjet driven cavitation cell. *Powder Technol.* **2016**, *297*, 34–43. [CrossRef]
17. Galecki, G.; Akar, G.; Sen, S.; Li, Y. Advanced studies on coal injection into a cavitation cell for the purpose of comminution. *Arch. Min. Sci.* **2012**, *57*, 769–778. [CrossRef]
18. Galecki, G.; Sen, S.; Akar, G.; Li, Y. Parametric evaluation of coal comminution by waterjets. *Int. J. Coal Prep. Util.* **2013**, *33*, 36–46. [CrossRef]
19. Dong, X.; Jiang, Y.; Guo, C.H. Mechanical mechanism of shear cavitating water jet dissociation of flake graphite. In Proceedings of the 2024 7th International Conference on Mechanical, Electrical and Material Application, Changsha, China, 23–25 February 2024.
20. Soyama, H. Enhancing the aggressive intensity of a cavitating jet by means of the nozzle outlet geometry. *J. Fluids Eng.* **2011**, *133*, 101301. [CrossRef]
21. Dong, W.Q.; Yao, L.G.; Luo, W.L. Numerical Simulation of Flow Field of Submerged Angular Cavitation Nozzle. *Appl. Sci.* **2023**, *13*, 613. [CrossRef]
22. Huang, S.W.; Huang, J.P.; He, K. Research on erosion effect of various submerged cavitating jet nozzles and design of self-rotating cleaning device. *Appl. Sci.* **2024**, *14*, 1433. [CrossRef]
23. Hutli, E.; Nedeljkovic, M.S.; Czifrus, S. Study and Analysis of the Cavitating and Non-Cavitating Jets Part Two Parameters Controlling the Jet Action and a New Formula for Cavitation Number Calculation. *Therm. Sci.* **2020**, *24*, 407–419. [CrossRef]
24. Liu, H.S.; Xu, Y.; Wang, Z.C.; Zhang, J.L.; Wang, J.X. Experimental and numerical simulations to examine the mechanism of nozzle geometry affecting cavitation water jets. *Geoenergy Sci. Eng.* **2024**, *233*, 212511. [CrossRef]
25. Kamisaka, H.; Soyama, H. Enhancing the Aggressive Intensity of a Cavitating Jet by Introducing Water Flow Holes and a Long Guide Pipe. *J. Fluids Eng.* **2021**, *143*, 031201. [CrossRef]
26. Cai, T.F.; Chamorro, L.P.; Ma, F.; Han, J. Impact of nozzle lip on the cavitation cloud characteristics of self-excited cavitating waterjets. *Ocean Eng.* **2023**, *285*, 115329. [CrossRef]
27. Prasetya, R.; Sou, A.; Moon, S.; Pratama, R.H.; Wada, Y.; Yokohata, H. X-ray Phase Contrast Imaging of Cavitation and Discharged Liquid Jet in Nozzles with Various Sizes. *At. Sprays* **2019**, *29*, 59–78. [CrossRef]
28. Wang, G.W.; Yang, Y.F.; Wang, C.; Shi, W.D.; Li, W.; Pan, B. Effect of Nozzle Outlet Shape on Cavitation Behavior of Submerged High-Pressure Jet. *Machines* **2022**, *10*, 4. [CrossRef]
29. Guo, C.H.; Dong, X.; Song, H.R.; Jiang, Y. Features of Structure and Flow Field in Homemade Co-Current Cavitation Water Jet Nozzle. *Materials* **2025**, *18*, 146. [CrossRef] [PubMed]
30. Ullas, P.K.; Chatterjee, D.; Vengadesan, S. Experimental study on the effect of throat length in the dynamics of internal unsteady cavitating flow. *Phys. Fluids* **2023**, *35*, 023332. [CrossRef]
31. Yang, Y.F.; Li, W.; Shi, W.D.; Zhou, L.; Zhang, W.Q. Experimental Study on the Unsteady Characteristics and the Impact Performance of a High-Pressure Submerged Cavitation Jet. *Shock Vib.* **2020**, *2020*, 1701843. [CrossRef]
32. Vidvans, A.N.; Melkote, S.N.; Sanders, D.G. Enhancing Cavitation Intensity in Co-Flow Water Cavitation Peening with Organ Pipe Nozzles. *J. Manuf. Sci. Eng. ASME* **2021**, *143*, 071010. [CrossRef]
33. Yin, S.H.; Hao, S.; Zhang, H.S.; Hu, S.; Cao, D.Y. The calculation medel of conveying resistance of waste rock-unclassified pastes slurry mixed with water-reducing agent. *J. Harbin Eng. Univ.* **2022**, *43*, 212–220. [CrossRef]
34. Xu, Y.; Liu, H.S.; Wang, Z.C.; Zhang, J.L.; Wang, J.X. Analysis of the Effects of Nozzle Geometry on the Cavitation Water Jet Flow Field Using Orthogonal Decomposition. *Iran. J. Sci. Technol. Trans. Mech. Eng.* **2024**, *48*, 119–132. [CrossRef]
35. Zhu, R.Y.; Zhu, H.T.; Zhang, X.H.; Pan, S.Z.; Zhang, C.; Xie, B. Experimental study on the unsteady behavior and frequency characteristics of high-speed submerged cavitating water jets. *Phys. Fluids* **2024**, *36*, 103336. [CrossRef]
36. Nishimura, S.; Takakuwa, O.; Soyama, H. Similarity Law on Shedding Frequency of Cavitation Cloud Induced by a Cavitating Jet. *J. Fluid Sci. Technol.* **2012**, *7*, 405–420. [CrossRef]
37. Yang, Y.F.; Li, W.; Shi, W.D.; Zhang, W.Q.; El-Emam, M.A. Numerical Investigation of a High-Pressure Submerged Jet Using a Cavitation Model Considering Effects of Shear Stress. *Processes* **2019**, *7*, 541. [CrossRef]
38. Lu, Y.P.; Wang, J.; Yu, H.L. 3D image processing and analyzing system for industrial computed tomography. *Chin. J. Sci. Instrum.* **2009**, *30*, 444–448. [CrossRef]
39. Dong, J.M.; Li, S.; Meng, R.X.; Zhong, X.; Pan, X.X. Research on Cavitation Characteristics of Two-Throat Nozzle Submerged Jet. *Appl. Sci.* **2022**, *12*, 536. [CrossRef]
40. Cui, Y.Y.; Zhao, M.J.; Ding, Q.M.; Cheng, B. Study on Dynamic Evolution and Erosion Characteristics of Cavitation Clouds in Submerged Cavitating Water Jets. *J. Mar. Sci. Eng.* **2024**, *12*, 641. [CrossRef]

41. Peng, G.Y.; Wakui, A.; Oguma, Y.; Shimizu, S.; Ji, H. Periodic Behavior of Cavitation Cloud Shedding in Submerged Water Jets Issuing from a Sheathed Pipe Nozzle. *J. Flow Control Meas. Vis.* **2018**, *6*, 15–26. [CrossRef]
42. Hutli, E.; Nedeljkovic, M.S.; Radovic, N.A.; Bonyár, A. The relation between the high speed submerged cavitating jet behaviour and the cavitation erosion process. *Int. J. Multiph. Flow* **2016**, *83*, 27–38. [CrossRef]
43. Zhu, R.Y.; Zhang, X.H.; Zhu, H.T.; Zhang, C.; Pan, S.Z. On unsteady cavitation flow of a high-speed submerged water jet based on data-driven modal decomposition. *Ocean Eng.* **2024**, *295*, 116916. [CrossRef]
44. Datseris, G.; Kottlarz, I.; Braun, A.P.; Parlitz, U. Estimating fractal dimensions: A comparative review and open source implementations. *Chaos Interdiscip. J. Nonlinear Sci.* **2023**, *33*, 102101. [CrossRef] [PubMed]
45. Sajid, M.; Husain, A.; Reddy, J.; Alresheedi, M.T.; Al Yahya, S.A.; Al-Rajy, A. Box dimension of the border of Kingdom of Saudi Arabia. *Heliyon* **2023**, *9*, e14846. [CrossRef] [PubMed]
46. Patiño-Ortiz, M.; Patiño-Ortiz, J.; Martínez-Cruz, M.; Esquivel-Patiño, F.; Balankin, A.S. Morphological Features of Mathematical and Real-World Fractals: A Survey. *Fractal Fract.* **2024**, *8*, 440. [CrossRef]
47. Hutli, E.A.F.; Nedeljkovic, M.S. Frequency in Shedding/Discharging Cavitation Clouds Determined by Visualization of a Submerged Cavitating Jet. *J. Fluids Eng.* **2008**, *130*, 021304. [CrossRef]
48. Li, F.Z.; Tan, Z.R.; Chen, L.Y.; Li, R.T.; Xu, Z.Y.; Wang, Y. Study on Dynamic Evolution of Cavitation Clouds and Optimization of Standoff Distance in Water Cavitation Peening. *J. Mech. Eng.* **2019**, *55*, 120–126. [CrossRef]
49. Wright, M.M.; Epps, B.; Dropkin, A.; Truscott, T.T. Cavitation of a submerged jet. *Exp. Fluids* **2013**, *54*, 1541. [CrossRef]
50. Peng, C.; Tian, S.C.; Li, G.S. Determination of the shedding frequency of cavitation cloud in a submerged cavitation jet based on high-speed photography images. *J. Hydrodyn.* **2021**, *33*, 127–139. [CrossRef]
51. Peng, C.; Li, G.S.; Tian, S.C. Dynamics of cavitation cloud in submerged cavitation water jet. *Pet. Sci. Bull.* **2017**, *2*, 228–239. [CrossRef]

energies

MDPI

Article

Research on Monthly Energy Consumption Intensity Prediction and Climate Correlation of Public Institutions Based on Machine Learning

Zhiming Gao [1,2], Miao Wang [2], Cheng Chen [2], Xuan Zhou [2,3,4,*], Wanchun Sun [1] and Junwei Yan [2,3,4]

[1] School of Energy and Automotive Engineering, Shunde Polytechnic University, Foshan 528300, China; 14022@sdpt.edu.cn (Z.G.); wanchunsun@sdpt.edu.cn (W.S.)
[2] School of Mechanical and Automotive Engineering, South China University of Technology, Guangzhou 510640, China; wangmiaonjfu@163.com (M.W.); 202011000733@mail.scut.edu.cn (C.C.)
[3] Guangzhou Institute of Modern Industrial Technology, South China University of Technology, Guangzhou 510640, China
[4] Guangdong Artificial Intelligence and Digital Economy Laboratory (Guangzhou), Pazhou Lab, Guangzhou 510330, China
* Correspondence: zhouxuan@scut.edu.cn; Tel.: +86-15918483535

Abstract: Energy consumption forecasting offers a foundation for governmental agencies to establish energy consumption benchmarks for public institutions. Meanwhile, correlation analysis of institutional energy use provides clear guidance for building energy-efficient retrofits. This study employed five machine learning models to train and predict monthly energy consumption intensity data from 2020 to 2022 for three types of public institutions in China's eastern coastal regions. A novel ensemble model was proposed and applied for energy consumption prediction. Additionally, the SHAP model was utilized to analyze the correlation between influencing factors and energy consumption data. Finally, the relationship between climatic factors and monthly energy consumption intensity was investigated. Results indicate that the ensemble model achieves higher predictive accuracy compared to other models, with regression metrics on the training set generally exceeding 0.9. Although XGBoost also demonstrated strong performance, it was less stable than the ensemble model. Energy intensity across different building types exhibited strong correlations with the number of energy users, floor area, electricity use, and water consumption. Linear analysis of temperature and energy consumption intensity revealed a directional linear relationship between the two for both medical and administrative buildings.

Keywords: public institutions; energy consumption prediction; ensemble model; SHAP model; climatic factors

1. Introduction

Energy serves as a fundamental requirement for human survival [1], with carbon dioxide emitted through energy consumption being the primary source of carbon emissions. In China, energy-related CO_2 emissions accounted for over 80% of the country's total carbon emissions in 2023 [2]. The "dual carbon" vision put forward by relevant authorities has set clear targets and timelines for China's low-carbon transition, while also presenting substantial challenges for policymakers [3]. The building sector is one of the three major energy-consuming industries globally, representing 46%, 40%, and 39% of total energy consumption in China, the United States, and Europe, respectively [4,5]. Notably, the operational phase of public buildings contributes significantly to carbon emissions, accounting for 38% of total building operational carbon emissions in 2020, thus representing a critical domain for energy conservation and emission reduction efforts [6].

Accurately assessing building energy consumption and disentangling the effects of various influencing factors are essential tasks [7]. Researchers often develop predictive models to estimate building energy use. Among these, physics-based models rely on thermal principles and require inputs such as building characteristics and environmental parameters to solve thermal balance equations [8,9]. In contrast, data-driven models only need historical energy consumption data to train the algorithm. As a result, data-driven approaches have gained increasing popularity in recent years for predicting and evaluating energy performance in the building sector.

In terms of the annual total energy consumption of regional buildings, Gyanesh et al. [10] and Rafael et al. [11] compared the predictive performance of multiple models for over 2000 residential buildings in India and over 700 office buildings in Chile, respectively. The results indicated that the hybrid model was more suitable for energy consumption predictions of residential buildings. ANN and linear models were more suitable for energy consumption predictions in office buildings. In terms of annual unit area energy consumption (EUI) predictions, Yalcintas et al. [12] found that the ANN model performed better in predicting the EUI of buildings in the CBECS library compared to other models. Roth et al. [13] found that both RF and Gaussian regression models performed well in predictions. Matheus et al. [14] and Zhang et al. [15] both predicted the EUI of educational buildings. The results showed that the Bayesian discretization method had higher prediction accuracy for discrete samples. In addition, the prediction results of CNN-LSTM-Attention were more accurate compared to other neural networks, but required more training time than other models. Cui et al. [16] predicted EUI for 1169 apartments and 4231 single family homes based on three ensemble tree models in the RECS library. It was found that the LightGBM was most suitable for predicting energy consumption in apartments, while the CatBoost was most suitable for predicting energy consumption in single household residences.

Some researchers have also conducted predictive studies on sub item energy consumption in buildings. Robinson et al. [17] predicted the annual main fuel consumption of 13,223 buildings in New York City in the CBECS database based on multiple machine learning models. The gradient boosting model was found to have more accurate prediction results. Abdo et al. [18] predicted the annual electricity and natural gas consumption of nearly 10,000 households in London according to multi-layer neural networks, linear regression, RF, and gradient boosting models. The results indicated that multi-layer neural networks had the best prediction performance, followed by the RF. Li et al. [19] predicted the power consumption of Türkiye's transmission companies for about 400 months based on XGBoost, CatBoost, and three model optimization methods. The XGBoost model was found to have excellent performance in predicting electricity consumption.

The preceding review reveals a diverse application of models in building energy prediction. However, a comparative analysis across these studies yields critical insights into the relative strengths of each algorithm, which are highly dependent on dataset size, feature complexity, and the nature of the prediction task. ANNs typically excel with very large datasets (e.g., >10,000 samples) where their deep architecture can learn complex, nonlinear representations without overfitting. However, on smaller datasets, such as those common in institutional-level studies, ANNs are highly prone to overfitting, leading to poor generalization, despite their theoretical capacity. RF are robust and less prone to overfitting than ANNs due to their ensemble "bagging" nature. They perform reliably across various dataset sizes and provide good feature importance estimates. However, RF can sometimes be outperformed by boosting algorithms in terms of pure predictive accuracy, as it may not fully optimize for the hardest-to-predict samples. Gradient boosting machines consistently demonstrate superior performance on small- to medium-sized, heterogeneous datasets—precisely the profile of our study. The sequential, error-correcting nature of boosting allows GBMs to build a strong predictor by focusing on residual errors from previous models. This makes them exceptionally adept at capturing complex, nonlinear relationships and interaction effects among multiple features without requiring an excessively large sample size. It is very suitable for samples with multiple characteristics but small scale in building energy consumption.

What are the key driving factors influencing energy consumption in each type of institution? Overall, most current research focuses on the annual energy consumption of buildings. However, existing studies on public institution energy consumption predominantly rely on annual or quarterly data. This coarse temporal resolution obscures the dynamic patterns of energy use at the monthly level. Crucially, the impact of climate variables on energy demand—such as summer cooling and winter heating—is most pronounced at this monthly scale. Annual data fails to capture these short-term fluctuations and the immediate impact of extreme weather events. Consequently, a critical research gap exists in the systematic modeling of monthly energy intensity and its drivers across different types of public institutions (EI, AI, MI). This gap currently hinders the development of precision energy-saving policies that are synchronized with operational cycles. Clarifying the climate–energy intensity relationship at a monthly resolution has direct implications for operational policy decisions within public institutions. For instance, the differential response patterns to high and low temperatures across institution types, which this study aims to uncover, can provide a quantitative basis for specific actions. It could provide budgeting, procurement, maintenance scheduling, flexible work arrangements, and space management.

Therefore, the study evaluated the predictive performance of monthly operating energy consumption of regional public institutions based on multiple machine learning models. The research predicts that the ensemble model will achieve significantly lower prediction errors (e.g., lower *RMSE*) compared to individual models across all institution types. Subsequently, the SHAP model and linear regression were employed to explore the impact of temperature and various influencing factors on energy consumption. The research provides guidance and suggestions for the formulation of energy-saving policies in future construction related departments.

2. Methods

Figure 1 is the flowchart of the research. It outlined a structured methodology for energy consumption prediction, integrating data-driven modeling and climate impact analysis. The process began with data acquisition and preprocessing to ensure input data quality. Subsequently, multiple optimized prediction models—including SSA-RF, SSA-SVM, SSA-XGBoost, and SSA-BP—were constructed and compared through model selection and performance comparison. An SSA-Ensemble model might also be developed to enhance robustness. The optimal model was then interpreted using the SHAP model for feature importance analysis, facilitating the exploration of climate impact. The final output is an accurate and interpretable energy consumption prediction, providing valuable insights for energy planning and policy-making.

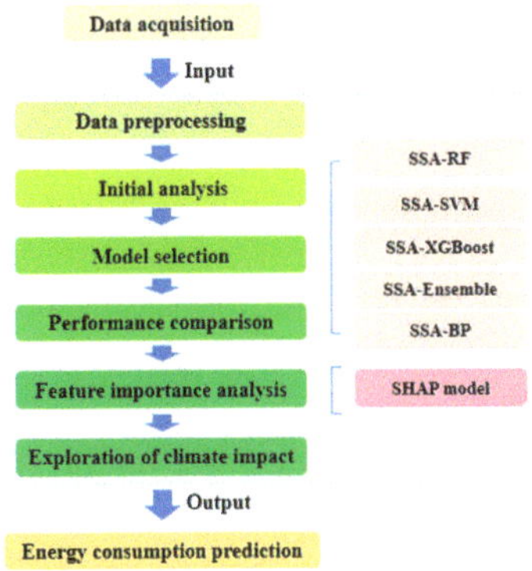

Figure 1. Flowchart of the research.

2.1. Data Sources

Based on the Chinese national standard GB/T 40498-2021 (general principles of drafting energy consumption norm standards for public institutions) [20], a structured questionnaire was designed to collect energy consumption data from approximately 250 public institutions in eastern coastal China through quarterly surveys from 2020 to 2022. It covered three categories—educational (EI), administrative (AI), and medical institutions (MI). The questionnaire encompassed six sections: basic building information, energy consumption data, major energy-consuming equipment, renewable energy statistics, energy-saving renovations, and energy metering details, with the energy consumption section being the most critical as it required monthly reporting of electricity, water, natural gas, and IT server room usage. Through systematic collection and organization of the data, two dependent variables—monthly per capita energy consumption (PCEC) and monthly energy consumption per unit building area (ECPUA)—were defined for analysis, while the specific independent variables, along with their abbreviations and units, are summarized in Table 1.

Table 1. Interpretation and units of different features.

Serial Number	Feature Variable	Abbreviation	Unit
1	Building area	BA	m^2
2	Number of energy users	NEU	P
3	Headcount	H	P
4	Number of gasoline vehicles	NGV	V
5	Number of diesel vehicles	NDV	V
6	Number of new energy vehicles	NNV	V
7	Number of charging stations	NCS	P
8	Average high temperature	AHT	°C
9	Average low temperature	ALT	°C
10	Extremely high temperature	EHT	°C
11	Extremely low temperature	ELT	°C
12	Air quality	AQ	/

Usually, according to JGSW01-2021 (energy and resource consumption quota of central and state organs), the calculation method for monthly total institution energy consumption was the sum of the actual physical consumption of various energy sources in the current month [21]. The impact of local temperature and air quality on energy consumption could not be ignored, as they could affect the operation of air conditioning and thus alter institution energy consumption. This study also inputted the month and year of building energy consumption as unique heat codes into the model. The comparison of all abbreviations in the study is shown in Table 1 and the Abbreviations part.

2.2. Data Preprocessing

For the energy and resource consumption data of public institutions, the types of missing data include the following:

(1) Missing key energy consumption data: Instances where data such as comprehensive energy consumption, building energy consumption, or electricity consumption were missing. Such samples were directly discarded.

(2) Missing key institutional information: Instances where data such as building area were missing. These were imputed using the median value from the corresponding historical reporting series of the public institution to which the sample belongs. The median value is not sensitive to outliers, while energy consumption data often contains outliers caused by special events such as extreme weather, equipment failures, or temporary activities. Historical energy consumption data often presents a skewed distribution (for example, a right-skewed distribution, where a few institutions have extremely high consumption). In a skewed distribution, the median is better able to reflect the central trend than the mean, avoiding estimates being pulled up or down by a few extreme values. If imputation was not possible, the sample was directly discarded.

Additionally, the logical consistency of attribute relationships within the dataset must be validated. For example, total fuel consumption should equal the sum of vehicle fuel consumption and other fuel consumption. Samples violating these data relationship rules were directly discarded.

To screen for anomalous data points, zero values in energy and power consumption records were first removed. The remaining dataset was then processed using the interquartile range (IQR) method, where IQR is defined as the difference between the upper quartile (UQ) and the lower quartile (LQ). Observations falling outside 1.5 times the IQR were classified as mild outliers, a widely accepted statistical threshold that offers a balanced approach to identifying extreme values without oversimplifying the underlying data structure. This method is particularly suitable for building energy data, which often exhibits right-skewed distributions. Prior to outlier detection, quantile transformation was applied to approximate a normal distribution of energy consumption values. The 3IQR test was subsequently conducted on the training set, and values identified as outliers—deemed unlikely to represent true operational conditions—were excluded. As summarized in Table 2, the removed outliers accounted for less than 5% of the total dataset. These anomalies may have resulted from damaged meter bases, erroneous reporting, or equipment malfunction. Variations in sample sizes across institution categories reflect their real-world distribution and data accessibility in the study region, with educational institutions being more numerous and having more systematic energy reporting practices compared to medical institutions, which are fewer in number and often maintain less accessible energy records.

Table 2. The available data volume and excluded data volume of various types of institutions.

Features	Available Data	Eliminated Data	The Proportion of Eliminated Data
	EI		
Building area	4320	0	0%
Number of energy users	4261	59	1.37%
Headcount	4288	32	0.74%
Number of gasoline vehicles	4229	91	2.11%

Table 2. *Cont.*

Features	Available Data	Eliminated Data	The Proportion of Eliminated Data
Number of diesel vehicles	4229	91	2.11%
Number of new energy vehicles	4229	91	2.11%
Number of charging stations	4266	54	1.25%
Energy consumption	4196	124	2.87%
	AI		
Building area	2340	0	0%
Number of energy users	2334	6	0.26%
Headcount	2334	6	0.26%
Number of gasoline vehicles	2328	12	0.51%
Number of diesel vehicles	2328	12	0.51%
Number of new energy vehicles	2328	12	0.51%
Number of charging stations	2340	0	0%
Energy consumption	2320	2340	0
	MI		
Building area	1130	2	0.18%
Number of energy users	1115	17	1.50%
Headcount	1115	17	1.50%
Number of gasoline vehicles	1121	11	0.97%
Number of diesel vehicles	1121	11	0.97%
Number of new energy vehicles	1121	11	0.97%
Number of charging stations	1108	24	2.12%
Energy consumption	1081	51	4.14%

2.3. *Model Selection*

This study adopted a methodological strategy involving five distinct prediction models to effectively identify the most suitable approach for different institution types and monthly energy consumption patterns. Given their recognized strong adaptability and rapid training speed, two of the most widely used ensemble models were included in this selection. The back propagation neural network model (BP) has advantages in prediction accuracy for large datasets. And the support vector machine (SVM) model is also commonly used for energy consumption predictions. In addition, the ensemble model was specifically constructed for monthly energy intensity prediction in the study. The training of all models used was completed in Python (3.12). Among them, RF, SVM, GBDT, and BP were from scikit-learn (0.24.2). XGBoost was from the standard XGBRegressor. All calculations were based on a GPU running on RTX3060.

2.3.1. Random Forests

RF is a type of tree model, whose core principle is to integrate multiple weak decision tree models into a strong model [22]. Unlike gradient boosting models such as GBR, RF adopt a nested ensemble architecture. Finally, by weighted averaging of the training results of all decision trees, the regression prediction results can be obtained. RF has strong robustness to outliers and noise, is less prone to overfitting when the noise is low, and has a faster training speed than gradient boosting models.

2.3.2. Support Vector Regression

The core idea of SVM models is to map the original training set to a high-dimensional feature space, and balance accuracy and computational complexity through loss functions and penalty factors [23]. The principle is to find a discriminative linear binary classifier with the maximum separation hyperplane in the feature space, thereby achieving the highest robustness of the model. Essentially, it can be seen as solving high-dimensional quadratic regression problems. The SVR algorithm is mainly divided into three types: soft-interval SVR, hard-interval SVR, and kernel SVR. Normally, interval SVR is used to handle linear correlation problems, while kernel SVR is used to handle nonlinear correlation problems.

2.3.3. XGBoost

eXtreme gradient boosting (XGBoost) is an improved version developed based on the gradient boosting algorithm [24]. Its core innovation lies in using the second-order derivative of Taylor expansion for calculation during the gradient boosting process, while traditional GBR is only based on the first-order derivative. The use of this second-order derivative makes the gradient boosting algorithm superior in terms of prediction speed and accuracy [25]. In addition, XGBoost introduces regularization terms during the decision tree construction phase to prevent model overfitting. Although the training speed is faster, each learning stage requires processing a complete dataset and occupies more storage space.

2.3.4. Back Propagation Neural Network

BP is a multi-layer feedforward network trained by error back propagation [26]. Its basic idea is gradient descent, which is based on using gradient search techniques to minimize the mean square error between the actual output value and the expected output value of the network. The basic BP includes two processes: forward propagation of signals and backward propagation of errors. When calculating the error output, it is performed in the direction from input to output, while adjusting the weights and thresholds is performed in the direction from output to input. BP can approximate any nonlinear continuous function, which enables it to handle complex internal mechanism problems. In addition, even if some neurons are damaged, the BP can still function normally, demonstrating a certain level of robustness [27]. However, due to its gradient descent approach, the network may converge to local minima rather than global optima. Neural networks also have a high desire for data volume.

2.3.5. Ensemble Model

In order to find a better model for predicting the energy consumption intensity of public institutions, the study proposed an ensemble model. The principle was shown in Figure 2. XGBoost is an optimization algorithm for GBDT. GBDT is based on first-order derivatives, which is relatively simple but stable [28]. Therefore, for specific datasets, GBDT may have better regression performance. The XGBoost and GBDT, trained separately, were weighted and combined. The optimal values of weights $\omega 1$ and $\omega 2$ were searched for based on a sparrow search algorithm (SSA). The goal of the optimization algorithm was to minimize mean squared error (MSE).

The ensemble model had the advantage of two complementary algorithms. XGBoost had the strong ability to capture complex nonlinear relationships, while GBDT was more robust to noisy data. The difference between the two algorithms provided model diversity and reduced the deviation of a single model. A single model might have variance or bias. Ensemble models could achieve better variance–bias trade-offs. Therefore, by optimizing the weights, the model found the optimal balance point. XGBoost and GBDT might capture different feature interaction patterns, so the ensemble model could comprehensively utilize different interaction information. The combination model was automatically searched for the best weight combination based on SSA, avoiding the subjectivity of manually setting weights.

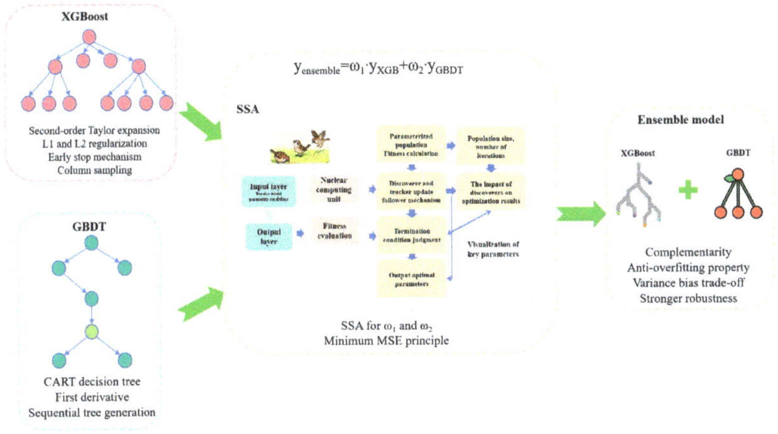

Figure 2. Schematic diagram of ensemble model.

2.4. Sparrow Search Algorithm

Due to the significant impact of hyperparameters on the algorithm, the study adopted SSA to automatically optimize the parameters. This optimization strategy draws inspiration from the foraging principles of sparrow populations [29]. The sparrow population is divided into three categories: discoverers, followers, and sentinels. They search efficiently for the optimal solution in the solution space by simulating their interactions and position update strategies. The discoverer is responsible for searching for areas rich in food. Followers follow the discoverer to areas with abundant food and conduct a detailed local search around the discoverer. They are vigilant, always alert to dangers in the environment (such as predators). When danger approaches, they will sound an alarm and drive the entire population to abandon the current food source and randomly move to a new safe location to avoid the population falling into a local optimum [30]. In each iteration, roles will be reassigned according to individual fitness values. The ones with the best adaptability become discoverers, those with poor adaptability become sentinels, and the rest become followers. SSA has strong global exploration capabilities and efficient local development capabilities. In addition, its good ability to jump out of local optima and fast convergence speed are also its characteristics. The parameters of SSA in this study were set as follows:

pop_size = 20, n_iter = 30, lb = [50, 3, 0.01, 0.6, 0.6, 0.1, 0.1], ub = [200, 8, 0.1, 0.9, 0.9, 1.0, 1.0], dim = 7/2.

2.5. Performance Standards

In order to find the best model for predicting carbon emissions of different types of institutions, the following four indicators were selected to compare the predictive performance of different models [4,31,32]. In addition, all models were subjected to 5-fold cross validation and the performance parameters were averaged. Using 5-fold CV allowed us to perform a more extensive hyperparameter tuning and model evaluation within a feasible time frame, given our available computational resources.

The coefficient of determination (R^2) is a commonly employed metric in regression analysis, representing the proportion of variance in the dependent variable that is predictable from the independent variables. Its definition formula is as follows:

$$R^2 = 1 - \frac{\sum_{i=1}^{n}(y_i - \hat{y}_i)^2}{\sum_{i=1}^{n}(y_i - \overline{y})^2} \quad (1)$$

where n is the number of samples, y_i and $\hat{y}_i$ are the true and predicted values of the ith sample, and $\overline{y}$ is the average value of the samples.

The root mean square error (*RMSE*) is a standard metric for evaluating the accuracy of continuous predictive models. It represents the standard deviation of the residuals (prediction errors), quantifying the average magnitude of deviation between predicted and actual values. *RMSE* is defined as follows:

$$RMSE = \sqrt{\frac{1}{n}\sum_{i=1}^{n}(y_i - \overline{y})^2} \tag{2}$$

Mean absolute error (*MAE*) represents the average error between the predicted value and the true value. It is defined as follows:

$$MAE = \frac{1}{n}\sum_{i=1}^{n}|y_i - \hat{y}_i| \tag{3}$$

The mean bias error (*MBE*) measures the average bias in predictions, indicating the model's tendency to overestimate (positive *MBE*) or underestimate (negative *MBE*). A lower absolute MBE value signifies better predictive accuracy. It is defined as follows:

$$MBE = \frac{\sum_{i=1}^{n}(y_i - \hat{y}_i)}{n} \tag{4}$$

Normalized root mean square error (*NRMSE*) is a normalized version of root mean square error. It eliminates the influence of absolute size and dimensionality of data by dividing *RMSE* by a representative value of the observed values (such as the range or mean), thus making the errors of different datasets or models comparable. It is defined as follows:

$$NRMSE_{mean} = \frac{RMSE}{\overline{y}} \tag{5}$$

Normalized mean bias error (*NMBE*) is used to measure the systematic bias (overestimation or underestimation) of model predictions. It calculates the average deviation between predicted and observed values and normalizes them, usually using the average or range of observed values to eliminate dimensional effects. It mainly reflects the average offset direction and degree of the system, rather than the accuracy of prediction. It is defined as follows:

$$NMBE = \frac{\left[\left(\frac{1}{n}\right) * \Sigma(i = 1\ to\ n)(y_i - \hat{y}_i)\right]}{\overline{y}} 100\% \tag{6}$$

2.6. SHAP Value Algorithm

The SHapley Additive exPlanations (SHAP) algorithm belongs to the post model interpretation method. Its core idea is to analyze the "black box model" from both global and local levels by calculating the marginal contribution of features to the model output. This method is based on cooperative game theory to assign significant values to each feature based on the predicted results, with local accuracy, missing rate, and consistency as three parameters. Compared to the feature importance provided by tree models, SHAP can provide the local feature importance of a single sample, and the feature contribution values have directionality. Meanwhile, the global feature importance is calculated based on the feature importance set of the sample set. The SHAP method is also an additive feature attribution method, which changes the results by adding or removing specific features to each feature subset. In addition, SHAP can be applied to various interpretation scenarios such as deep learning models, tree models, gradient machines, etc.

3. Results and Discussion

3.1. Sample Data Analysis

Figure 3 shows the statistical box plot of monthly PCEC and ECPUA of three types of public buildings in the eastern coastal areas of China from 2020 to 2022. Overall, there was a phenomenon in the three types of building energy consumption samples where a small portion of high energy consumption data appeared outside the upper quartile inner column. The median lines were all located below the mean. This indicated that the data with lower energy consumption in the samples had a higher density. However, it was observed that the data outside these inner columns had stronger continuity, so they could not be considered as outlier processing.

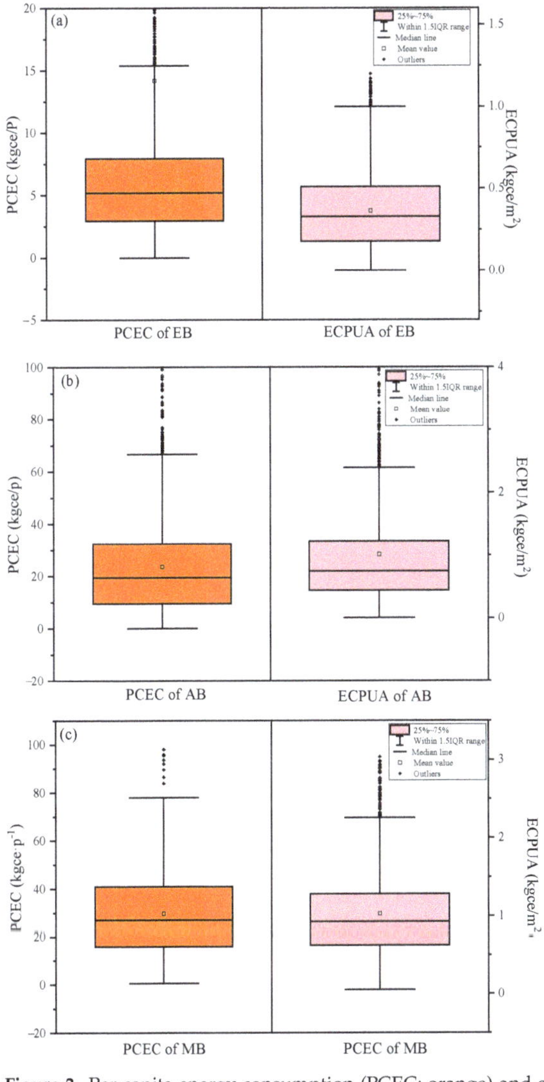

Figure 3. Per capita energy consumption (PCEC: orange) and energy consumption per unit area (ECPUA: pink) of various types of buildings: (**a**) Educational buildings (EB); (**b**) Administrative buildings (AB); (**c**) Medical buildings (MB). IQR is interquartile range.

For PCEC, EI had the lowest energy consumption among the three types of buildings, ranging from 0 to 20 kgce/P. This was often related to the low-energy consumption behavior of personnel in EI and two long holidays every year. The monthly PCEC of MI was the highest, ranging from 0 to 100 kgce/P, with an average of 29.8 kgce/P. It was consistent with our previous research [33]. This was because for common public institutions, MI needed to meet the daily operating conditions of medical equipment and air conditioning equipment, and there was minimal rest throughout the year. So the energy consumption value was relatively high. The PCEC of AI was lower than that of MI. For ECPUA, the comparison results of the energy consumption values of the three types of public institutions were similar to PCEC. The ECPUA of EI was the smallest, between 0 and 1.25 kgce/m^2. It could be seen that the difference between the average and median ECPUA of EI was smaller compared to the difference in PCEC, because the overall numerical magnitude of the unit building area was smaller and more concentrated.

Figure 4 shows the monthly variation curves of PCEC and ECPUA for three types of public institutions over a period of three years. Overall, the monthly changes in PCEC and ECPUA between AI and MI were more pronounced. The peak periods of energy intensity were from June to September and from January to March each year. This is because these two time periods were the hot and cold seasons in the local area, and institutions needed to overcome seasonal conditions to increase energy demand. And other months were suitable for climate, so energy demand decreased. For EI, PCEC exhibited a similar pattern to other institutions, while ECPUA did not show a more obvious pattern. This is related to the unique winter and summer vacation system of EI. During the winter and summer vacations, students and faculty members might experience reduced energy consumption due to holidays, leading to unpredictable changes in ECPUA. Therefore, from the perspective of regularity integrity, PCEC had more data stability compared to ECPUA.

Figure 5 shows the Pearson correlation analysis results of PCEC, ECPUA, and 13 related influencing factors for three institutions. Overall, there was a strong linear correlation between the temperature related influencing factors. There was a strong negative linear correlation between AQ and the temperature related influencing factors. For EI and AI, there was a correlation between NGV and NNV. But for EI and MI, there was a significant correlation among BA, NEU, and H. This was because the scale of both types of institutions was directly proportional to these factors.

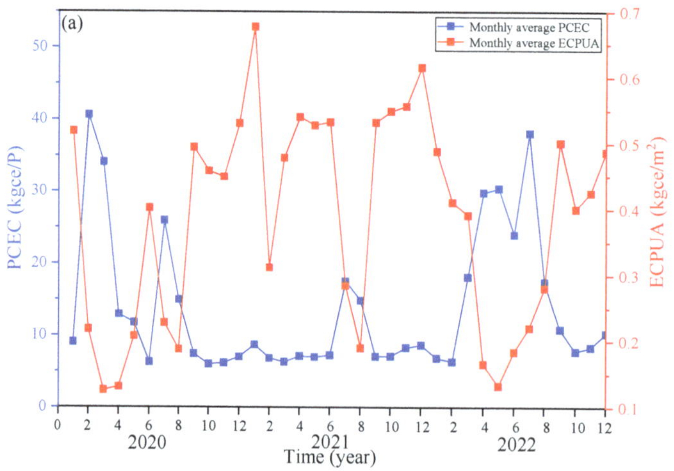

Figure 4. *Cont.*

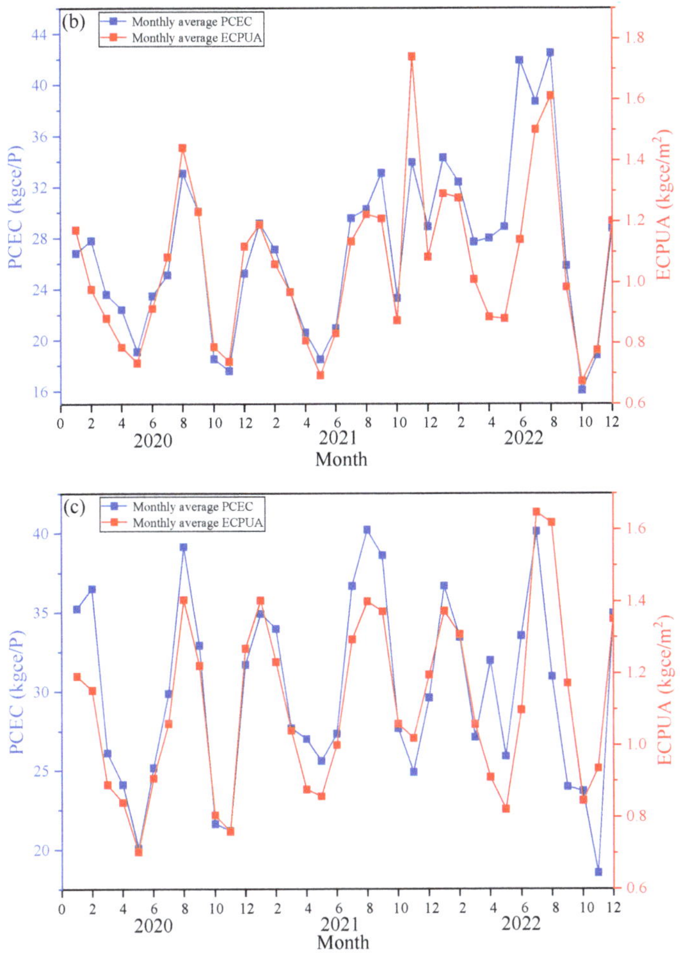

Figure 4. Monthly change curve of PCEC and ECPUA in public institutions: (**a**) Educational institutions; (**b**) Administrative institutions; (**c**) Medical institutions.

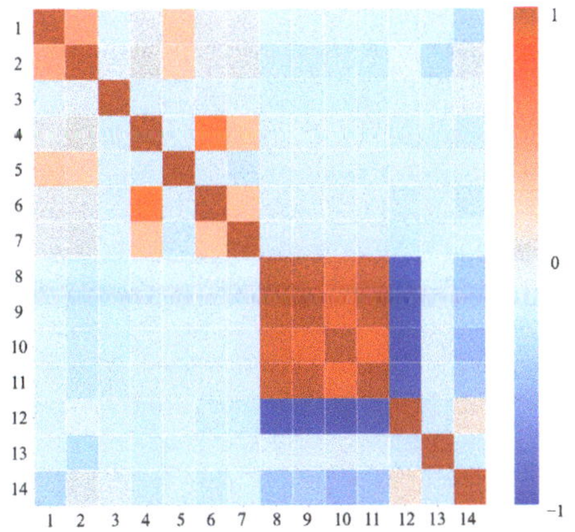

(**a**) Educational institutions

Figure 5. *Cont.*

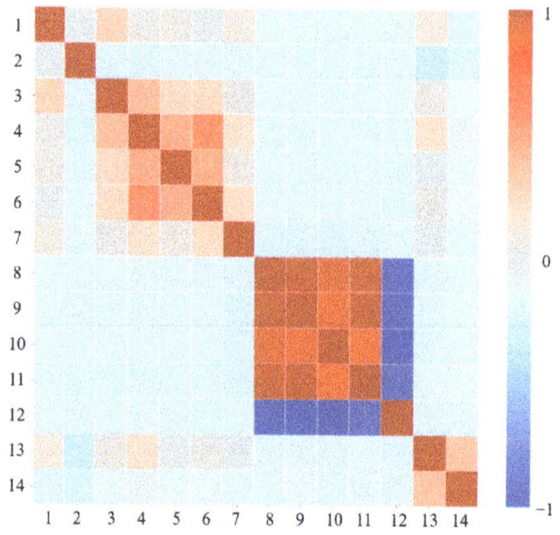

(**b**) Administrative institutions

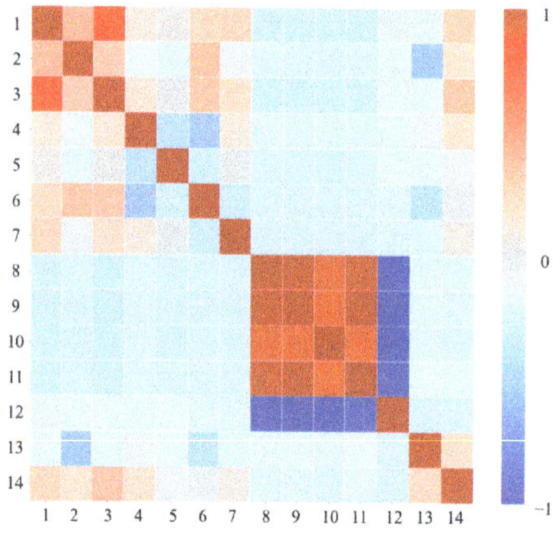

(**c**) Medical institutions

Figure 5. Pearson correlation analysis results of influencing factors of public institutions. Features 1 to 14, respectively, represent building area, number of energy users, headcount, number of gasoline vehicles, number of diesel vehicles, number of new energy vehicles, number of charging stations, average high temperature, average low temperature, extremely high temperature, extremely low temperature, air quality, energy consumption per unit building area, and per capita energy consumption. Among them, features 1 to 13 are consistent with Table 1.

3.2. Comparison of Prediction Results of Different Models

In order to enhance the generalization of the model, the study used samples from 2020 and 2021 as the training set and samples from 2022 as the testing set. During the model training process, a 5-fold cross validation method was used. In addition, based on the VIF results of the features and to avoid the risk of data leakage, EC, WC, and NV were excluded from the features in the study. Due to the right-handed distribution of energy consumption data, in order not to affect prediction performance, logarithmic transformation was performed on the energy consumption data before training. And the data was inverse transformed after training and prediction. The predictive performance parameters were calculated based on actual data. Tables 3 and 4 show the predictive performance of different models for monthly PCEC of three institutions. According to Table 3, the regression index R^2 of XGBoost and the ensemble model were the highest. Their R^2 values were basically the same. They were all above 0.95 and their *MAE*, *MBE*, *NMBE*, *RMSE*, and *NRMSE* were all the smallest, indicating that their fitting abilities were the strongest. The next level was RF. SVM and BP showed underfitting. This was because the sample size was not large enough and there were many feature vectors.

Table 3. Performance results of different models for predicting monthly per capita energy consumption of training set.

Institution Types	Model Applied	*MAE*	*MBE*	*NMBE*	*RMSE*	*NRMSE*	R^2
EI	SSA-RF	0.285	0.005	0.008	0.375	0.341	0.918
	SSA-SVM	0.682	−0.125	−0.114	0.945	0.860	0.745
	SSA-XGBoost	0.151	0.000	−0.005	0.200	0.182	0.949
	SSA-Ensemble	0.142	0.000	0.000	0.187	0.100	0.955
	SSA-BP	0.425	−0.015	−0.014	0.560	0.510	0.865
AI	SSA-RF	0.175	0.003	0.002	0.245	0.085	0.955
	SSA-SVM	0.415	−0.085	−0.078	0.625	0.218	0.815
	SSA-XGBoost	0.091	0.000	0.000	0.129	0.045	0.979
	SSA-Ensemble	0.090	0.000	0.000	0.126	0.044	0.980
	SSA-BP	0.265	0.008	0.007	0.385	0.134	0.905
MI	SSA-RF	0.155	−0.002	−0.002	0.215	0.066	0.945
	SSA-SVM	0.365	−0.065	−0.060	0.525	0.161	0.835
	SSA-XGBoost	0.081	0.000	0.000	0.114	0.035	0.971
	SSA-Ensemble	0.055	0.000	0.000	0.076	0.023	0.987
	SSA-BP	0.195	−0.012	−0.011	0.285	0.087	0.925

Table 4. Performance results of different models for predicting monthly energy consumption per unit area of training set.

Institution Types	Model Applied	*MAE*	*MBE*	*NMBE*	*RMSE*	*NRMSE*	R^2
EI	SSA-RF	0.458	−0.025	−0.013	0.625	0.329	0.845
	SSA-SVM	1.125	−0.185	−0.097	1.485	0.782	0.685
	SSA-XGBoost	0.244	−0.009	−0.005	0.345	0.182	0.910
	SSA-Ensemble	0.241	−0.006	−0.003	0.337	0.177	0.926
	SSA-BP	0.685	−0.035	−0.018	0.895	0.471	0.785
AI	SSA-RF	0.385	0.015	0.005	0.545	0.190	0.885
	SSA-SVM	0.925	−0.125	−0.044	1.285	0.448	0.715
	SSA-XGBoost	0.255	−0.001	−0.001	0.368	0.128	0.933
	SSA-Ensemble	0.248	0.003	0.001	0.361	0.126	0.940
	SSA-BP	0.525	0.025	0.009	0.745	0.259	0.825
MI	SSA-RF	0.165	0.018	0.010	0.235	0.128	0.885
	SSA-SVM	0.395	−0.085	−0.046	0.585	0.319	0.745
	SSA-XGBoost	0.109	0.000	0.000	0.145	0.079	0.936
	SSA-Ensemble	0.097	0.000	0.000	0.098	0.070	0.945
	SSA-BP	0.245	−0.025	−0.014	0.345	0.188	0.865

According to Table 4, the regression index R^2 of XGBoost and the ensemble model were still the highest. They were all above 0.9. In addition, the R^2 of the ensemble model was slightly higher than XGBoost. The other three parameters were also lower. This indicated that the stability and generalization of the ensemble model were stronger than for XGBoost, consistent with the previous statement. The performance of the other three model test sets was average. This indicated that gradient boosting models were more suitable for regression prediction of small-scale and multi-feature samples. Furthermore, it could be seen that the small sample size of medical institutions had almost no impact on the model regression.

Tables 5 and 6 show the predictive performance of different models for monthly ECPUA of three institutions. Overall, the prediction accuracy of ECPUA was slightly worse than PCEC, which might be due to the fact that the features in the sample were more consistent with prediction for PCEC. According to Table 5, RF, XGBoost, and ensemble models all have good training performance with the training set. This was similar to the prediction results of PCEC. Table 6 showed that the prediction accuracy of all models had decreased on the test set. However, the R^2 of XGBoost and the ensemble model still remained above 0.9. Prediction performance of AB was better than the other two institutions. This was because there was more noise in the samples of the other two institutions. The ensemble model had better predictive performance than XGBoost. The research of Robinson also found that XGBoost had a higher regression index compared to other models [17]. However, the R^2 of XGBoost in the research was generally lower than 0.9, while our ensemble model performed better. However, this was also related to the quality of the sample. In the research of Gassar, the XGB model exhibited unstable performance [18]. This was because XGBoost was easily troubled by local optimal solutions. This problem had been well addressed in our research using the ensemble model. For the convenience of observing the performance improvement of the combined model, the rate of change and p-value of the ensemble model parameters for the combined model are shown in Table 7. For the convenience of observing the performance improvement of the combined model, the rate of change and p-value of the ensemble model parameters for the combined model are shown in Table 7. It could be seen that the performance parameters of the ensemble model meet the correlation test with the parameters of the benchmark model. And all performance parameters had been improved to some extent.

Table 5. Performance results of different models for predicting monthly per capita energy consumption of test set.

Institution Types	Model Applied	*MAE*	*MBE*	*NMBE*	*RMSE*	*NRMSE*	R^2
EI	SSA-RF	0.035	0.002	0.011	0.048	0.266	0.895
	SSA-SVM	0.085	−0.025	−0.139	0.125	0.694	0.685
	SSA-XGBoost	0.021	0.000	0.000	0.032	0.177	0.929
	SSA-Ensemble	0.015	0.000	0.000	0.026	0.156	0.935
	SSA-BP	0.055	0.003	0.017	0.085	0.472	0.815
AI	SSA-RF	0.065	−0.002	−0.003	0.095	0.153	0.925
	SSA-SVM	0.185	−0.045	−0.072	0.285	0.458	0.645
	SSA-XGBoost	0.037	0.000	0.000	0.051	0.082	0.981
	SSA-Ensemble	0.036	0.000	0.000	0.050	0.080	0.981
	SSA-BP	0.095	−0.008	−0.013	0.145	0.233	0.875
MI	SSA-RF	0.085	0.003	0.004	0.125	0.187	0.915
	SSA-SVM	0.145	−0.035	−0.052	0.225	0.337	0.725
	SSA-XGBoost	0.042	0.000	0.000	0.057	0.085	0.950
	SSA-Ensemble	0.035	0.000	0.000	0.047	0.070	0.966
	SSA-BP	0.125	0.012	0.018	0.185	0.277	0.855

Figures 6–8 present a comprehensive comparison between the predicted values and actual measurements of monthly per capita energy consumption and monthly energy consumption per unit area across three distinct institutional types using the optimized ensemble model. The scatter plots demonstrated a strong alignment between predicted and actual values, with data points clustering closely along the diagonal reference line, indicating the model's robust predictive accuracy. This close correspondence was particularly evident in

the educational institution data (Figure 6), where the ensemble model successfully captured the complex energy consumption patterns across different operational scales.

Table 6. Performance results of different models for predicting monthly energy consumption per unit area of test set.

Institution Types	Model Applied	*MAE*	*MBE*	*NMBE*	*RMSE*	*NRMSE*	R^2
EI	SSA-RF	0.075	−0.008	−0.027	0.105	0.354	0.785
	SSA-SVM	0.095	−0.035	−0.118	0.145	0.489	0.685
	SSA-XGBoost	0.047	0.000	0.000	0.063	0.212	0.846
	SSA-Ensemble	0.045	0.000	0.000	0.060	0.203	0.858
	SSA-BP	0.085	0.005	0.017	0.125	0.421	0.725
AI	SSA-RF	0.165	0.015	0.024	0.235	0.382	0.815
	SSA-SVM	0.355	−0.065	−0.106	0.525	0.854	0.525
	SSA-XGBoost	0.108	−0.003	−0.004	0.153	0.249	0.928
	SSA-Ensemble	0.108	−0.001	−0.002	0.153	0.247	0.930
	SSA-BP	0.195	−0.008	−0.013	0.285	0.464	0.785
MI	SSA-RF	0.095	0.008	0.010	0.135	0.175	0.785
	SSA-SVM	0.195	−0.015	−0.019	0.285	0.370	0.645
	SSA-XGBoost	0.065	−0.003	−0.004	0.094	0.122	0.860
	SSA-Ensemble	0.057	−0.001	−0.001	0.086	0.114	0.887
	SSA-BP	0.155	0.008	0.010	0.225	0.292	0.725

Table 7. The rate of change and *p*-value of baseline model parameters for the ensemble model.

Institution Types	Energy Intensity	*MAE*	*MBE*	*NMBE*	*RMSE*	*NRMSE*	R^2	*p*-Value
EI	PECE	−1.23%	−33.33%	−40.00%	−2.32%	−2.75%	+1.76%	<0.001
	ECPUA	−4.26%	0	0	−4.76%	−4.25	+1.42%	0.02
AI	PECE	−2.75%	−400.00%	−200.00%	−1.90%	−1.56%	+0.75%	0.01
	ECPUA	0	−66.77%	−50.00%	0	−0.80%	+0.22%	<0.001
MI	PECE	−11.01%	0	0	−32.41%	−11.39%	+0.96%	<0.001
	ECPUA	−12.31%	−66.77%	−75.00%	−8.51%	−6.56%	+3.14%	<0.001

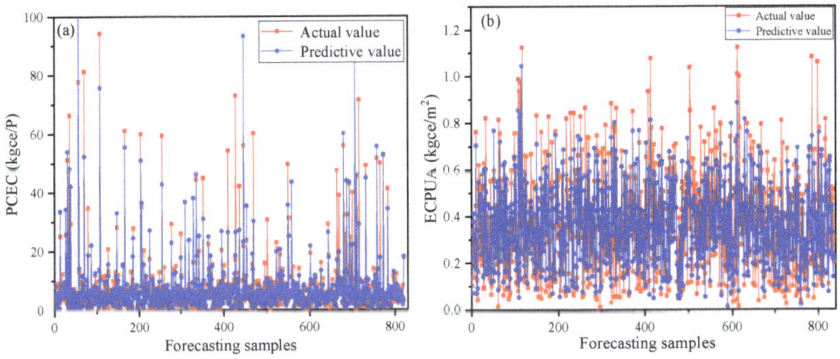

Figure 6. Comparison between predicted values and reality of educational institution: (**a**) Monthly per capita energy consumption; (**b**) Monthly energy consumption per unit area.

Notably, the model exhibited excellent generalization capability across diverse institutional categories, maintaining consistent predictive performance despite the inherent differences in operational characteristics, usage patterns, and energy demand profiles. The slight systematic underestimation observed across all institutional types, where predicted values tend to fall marginally below the actual measurements, warrants detailed consideration. This prediction bias could be attributed to several interconnected factors. Primarily, the inherent class imbalance in the training dataset, characterized by an overrepresentation of moderate consumption patterns and underrepresentation of extreme high-consumption instances, introduced a conservative bias in the model's predictions. The ensemble algorithm, while effective in capturing general trends, demonstrated a tendency to regress toward the mean when confronted with outlier scenarios or rare high-consumption events.

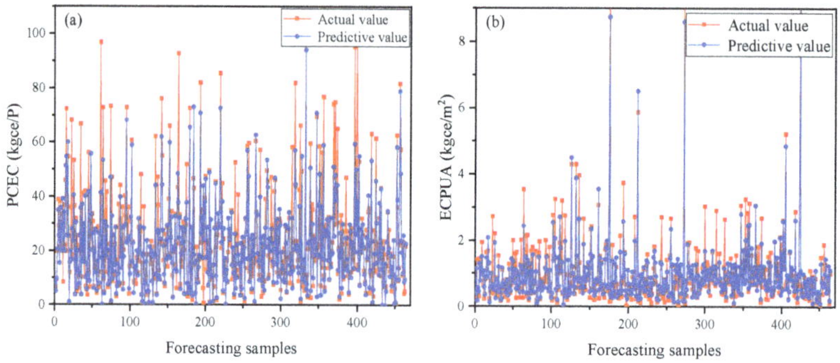

Figure 7. Comparison between predicted values and reality of administrative institution: (**a**) Monthly per capita energy consumption; (**b**) Monthly energy consumption per unit area.

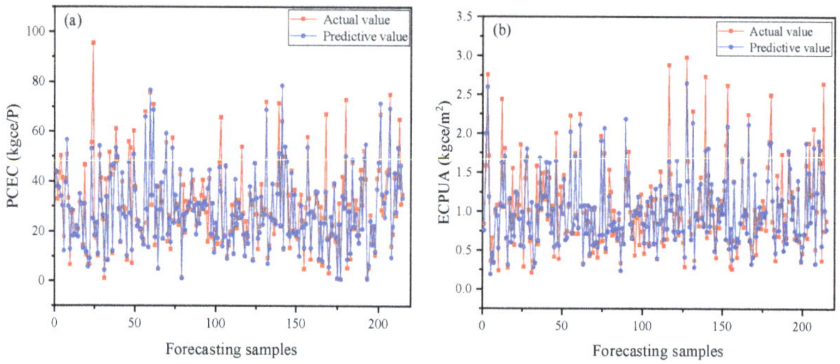

Figure 8. Comparison between predicted values and reality of medical institution: (**a**) Monthly per capita energy consumption; (**b**) Monthly energy consumption per unit area.

3.3. Feature Importance Analysis Based on SHAP

Feature importance analysis is of great significance for evaluating the global interpretability of models. Figures 9 and 10 demonstrate the feature importance of the global perspective based on SHAP model for predicting carbon emission intensity of various institutions (only the 14 most influential factors were shown in the figure). The energy consumption patterns of educational institutions exhibited significant personnel led characteristics. The NEU occupied an absolute dominant position in per capita energy consumption predictions, reflecting the essential characteristics of personnel density in educational activities. BA, as a key indicator of infrastructure carrying capacity, together with personnel factors constituted the dual core driving force of educational energy consumption. It is worth noting that the importance of EHT went beyond conventional climate variables, reflecting the special vulnerability of educational institutions under extreme climate conditions, and it might be closely related to summer school activities and centralized use of air conditioning systems. Administrative institutions exhibited a more balanced multi-factor driving model. NEU and BA were almost equally important, reflecting the dual dependence of administrative office energy consumption. The prominent importance of NV revealed the significant role of administrative transportation in energy consumption, which was relatively weaker in other types of institutions. In temperature variables, the importance of AHT exceeded that of extreme temperature, indicating the sensitivity of administrative agencies to conventional seasonal climate change. The energy consumption of medical institutions presented a highly personnel intensive characteristic, with the importance of NEU reaching the peak among the three types of institutions, highlighting the extreme dependence of medical services on human resources. The importance of BA was relatively balanced, while the significant impact of AHT reflected the strict temperature control requirements of medical institutions, which were directly related to the operating environment of medical equipment, drug storage conditions, and patient comfort needs.

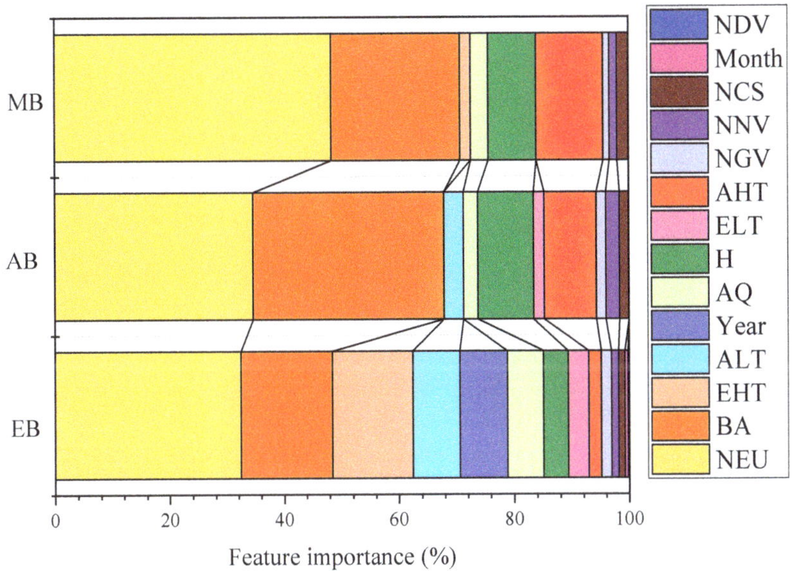

Figure 9. The feature importance of monthly per capita energy consumption predictions of different public institutions based on the SHAP model. (EI) Educational institutions; (MI) Medical institutions; (AI) Administrative institutions.

There were both commonalities and significant differences in the energy consumption driving mechanisms among the three types of institutions. Personnel factors occupied a central position in all types of institutions, but their impact intensity had a clear gradient: medical institutions > administrative institutions > educational institutions. The impact patterns of building factors varied, and were most prominent in the unit area energy consumption predictions of administrative institutions. The impact of climate factors exhibited institutional specificity: educational institutions were sensitive to extreme events, administrative institutions responded significantly to seasonal changes, and medical institutions demonstrated a sustained need for temperature control.

The impact of transportation factors was particularly significant in administrative agencies, reflecting the essential differences in operational models and organizational structures among different institutions. This differentiated feature importance model provided an important theoretical basis for formulating precise energy efficiency policies specific to institutions, emphasizing the necessity of differentiated energy management strategies based on institutional functional positioning and operational characteristics.

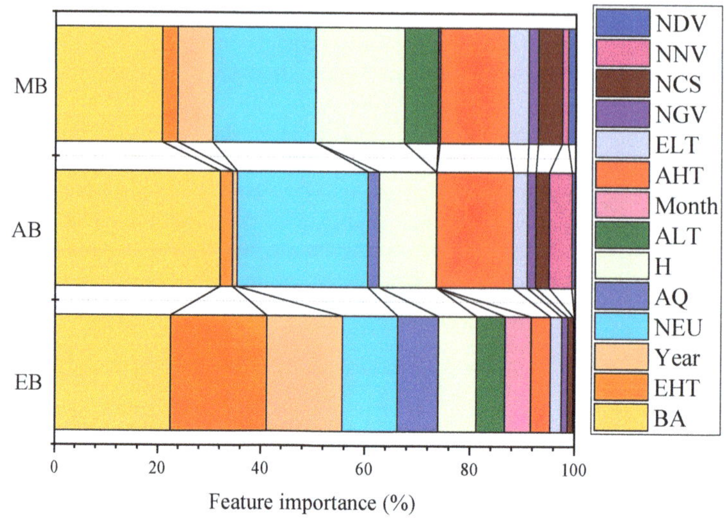

Figure 10. The feature importance of monthly energy consumption per unit area predictions of different public institutions based on the SHAP model. (EI) Educational institutions; (MI) Medical institutions; (AI) Administrative institutions.

3.4. Exploration of the Relationship Between Temperature Factors and Energy Consumption Intensity

The previous section found that climate factors did indeed affect the energy consumption intensity of public institutions. To further investigate the impact of temperature on energy consumption intensity, the monthly AHT, monthly ALT, and PCEC were combined for research. Because the previous conclusion indicated that the monthly changes in PCEC were more regular than those in ECPUA, usually air conditioning was affected by external temperature in two directions. When the temperature was too high, the air conditioner heated up. When the temperature was too low, the air conditioner cooled down. Both directions were reasons for the increase in energy consumption. Therefore, this section divided AHT and ALT into two directions: a high temperature zone and a low temperature zone. AHT with a temperature above 22 °C was defined as the high temperature zone, while AHT with a temperature below 22 °C was defined as the low temperature zone. For ALT, a temperature above 15 °C was defined as the high temperature zone, while a temperature below 15 °C was defined as the low temperature zone. Temperatures of 22 °C and 15 °C were the arithmetic means.

Figures 11 and S1–S3 show the relationship between temperature distribution and PCEC under four different scenarios. Overall, there was a clear positive correlation between the high temperature zone temperatures and PCEC in MI and AI. And there was a clear negative correlation between their low temperature zone temperatures and PCEC. For EI, there was no significant correlation, which was also due to their winter and summer vacations. And the air conditioning equipment and opening hours of EI were usually not as frequent as the other two types of institutions. For MI and AI, the correlation slope between temperature in the high temperature zone and PCEC was greater. This phenomenon primarily stems from fundamental differences in the physics and operational patterns of cooling versus heating systems in buildings. Firstly, the efficiency of electrically driven cooling systems degrades sharply as ambient temperature rises. When outdoor temperatures exceed 35 °C, heat rejection becomes difficult, forcing the compressor to work significantly harder to maintain the indoor setpoint, leading to a super-linear increase in electricity consumption. In contrast, the efficiency of common heating systems (e.g., gas boilers) is less affected by low outdoor temperatures, resulting in a more linear energy increase. Secondly, MI and AI institutions generate substantial internal heat from occupants, lighting, office equipment, and particularly in MIs, medical devices. During high temperatures, this internal heat adds to the external heat gain, collectively burdening the cooling system and causing a sharp energy rise. Conversely, during low temperatures, these internal sources offset a portion of the building's heat loss, mitigating the heating load and flattening the energy-increase curve. Finally, especially for MIs, critical areas like operating rooms and laboratories require strict, year-round temperature control. The significant heat emitted by medical equipment creates a substantial base cooling load even in mild weather. During extreme heat events, this base load combines with the additional cooling demand, amplifying the sensitivity of energy consumption to high temperatures.

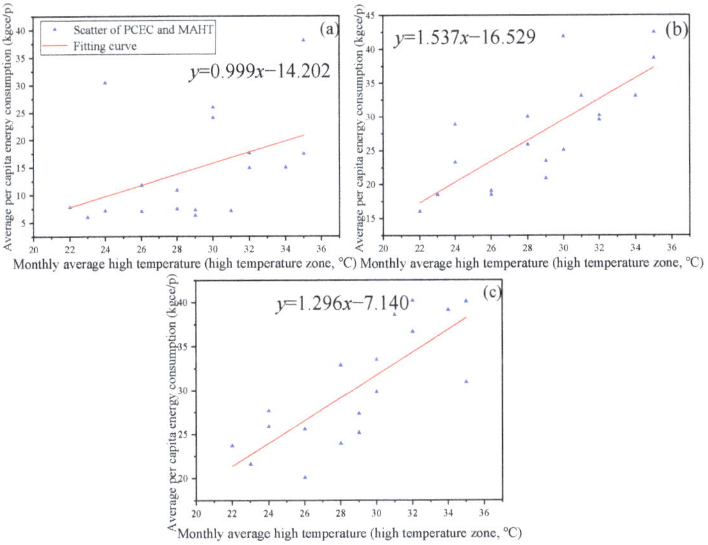

Figure 11. The relationship diagram between PCEC of different institutions and AHT in high temperature zones. (**a**) Educational institutions; (**b**) Medical institutions; (**c**) Administrative institutions.

Their linear formulas are y = 1.537x + 16.529 and y = 1.296x − 7.140. The slopes all exceed 1. Furthermore, from the intercept, it can be seen that at the same temperature, the PCEC of MI was higher. This is consistent with the previous conclusion. In addition, the slope of the linear relationship between their low temperature zone temperature and PCEC is higher than −1. This indicates that the impact of extreme heat on PCEC was greater than that of extreme cold.

4. Conclusions

This study predicted the monthly energy consumption intensity of different types of public institutions in the eastern coastal areas of China based on five models. A new ensemble model was proposed and applied. And the SHAP model was used to explore the importance of energy consumption characteristics. In addition, the relationship between climate factors and energy consumption intensity was explored. The main conclusions of this study are as follows:

(1) EI had the lowest monthly energy consumption among the three types of institution, while MI had the highest monthly energy consumption among the three types of

institution. The regularity of monthly energy intensity changes in AI and MI is stronger than that in EI.

(2) Through model performance analysis, XGBoost and ensemble models had the best predictive performance in the training set. In the test set, the energy consumption intensity prediction performance of the ensemble model was more stable than XGBoost. The overall accuracy of predicting per capita energy consumption is higher than that of predicting carbon emissions per unit area.

(3) There were both commonalities and significant differences in the energy consumption driving mechanisms among the three types of institutions. Personnel factors occupied a central position in all types of institutions, but their impact intensity had a clear gradient: medical institutions > administrative institutions > educational institutions. The impact patterns of building factors varied, and were most prominent in the unit area energy consumption predictions of administrative institutions.

(4) The monthly PCEC of all institutions showed regular changes over time. For MI and AI, there was a significant linear correlation between the PCEC and the average temperature. This correlation was directional.

5. Prospect

Potential leakage between groups during cross validation remains due to the lack of information on the application of GroupKFold across institutions. Subsequent research will focus on avoiding potential risks of data leakage.

Supplementary Materials: The following supporting information can be downloaded at: https://www.mdpi.com/article/10.3390/en18225932/s1, Table S1: Energy consumption survey form; Figure S1. The relationship diagram between PCEC of different institutions and AHT in low-temperature zones. (a) Educational institutions; (b) Medical institutions; (c) Administrative institutions; Figure S2. The relationship diagram between PCEC of different institutions and ALT in high-temperature zones. (a) Educational institutions; (b) Medical institutions; (c) Administrative institutions; Figure S3. The relationship diagram between PCEC of different institutions and ALT in low-temperature zones. (a) Educational institutions; (b) Medical institutions; (c) Administrative institutions.

Author Contributions: Conceptualization, Z.G. and M.W.; methodology, Z.G.; software, W.S.; formal analysis, C.C.;re-sources, J.Y.; writing—original draft preparation, M.W.; writing—review and editing, Z.G.; visu-alization, W.S.; supervision, C.C.; funding acquisition, X.Z. and J.Y. All authors have read and agreed to the published version of the manuscript.

Funding: This work is supported by the Guangdong Natural Science Foundation (2022A1515011128 and 2023A1515140189).

Data Availability Statement: The original contributions presented in this study are included in the article and Supplementary Materials. Further inquiries can be directed to the corresponding author.

Conflicts of Interest: The authors declare no conflicts of interest.

Abbreviations

BA	Building area
NEU	Number of energy users
H	Headcount
NV	Number of vehicles
NGV	Number of gasoline vehicles
NDV	Number of diesel vehicles
NNV	Number of new energy vehicles
EC	Electricity consumption
WC	Water consumption

NCS	Number of charging stations
AHT	Average high temperature
ALT	Average low temperature
EHT	Extremely high temperature
ELT	Extremely low temperature
AQ	Air quality
AI	Administrative institutions
PCEC	Per capita energy consumption
ECPUA	Energy consumption per unit area
RF	Random forests
SVM	Support vector regression
XGBoost	eXtreme gradient boosting
BP	Back propagation neural network
GBDT	Gradient boosting decision tree
SSA	Sparrow search algorithm
RMSE	Root mean square error
MAE	Mean absolute error
MBE	Mean bias error
NRMSE	Normalized root mean square error
NMBE	Normalized mean bias error
SHAP	SHapley additive exPlanations
EI	Educational institutions
MI	Medical institutions

References

1. Zhang, W.; Liu, T.; Wang, X.; Wang, B. Relationships between ecosystem services and human well-being from a water–energy–food nexus perspective: A case study of Jiziwan, Yellow River Basin, China. *Ecosyst. Serv.* **2025**, *76*, 101773. [CrossRef]
2. China National Bureau of Statistics. *Statistical Bulletin on National Economic and Social Development of the People's Republic of China in 2023*; People's Publishing House: Beijing, China, 2023.
3. Zhu, S.; Wang, C.; He, M.; Wei, W.; Chen, B.; Feng, Y.; Wan, Y. Green transformation pathways in the water sector under the 'Dual carbon' framework: A case study of Anhui province, East China. *Ecohydrol. Hydrobiol.* **2025**, 100692. [CrossRef]
4. Ahmad, A.S.; Hassan, M.Y.; Abdullah, M.P.; Rahman, H.A.; Hussin, F.; Abdullah, H.; Saidur, R. A review on applications of ANN and SVM for building electrical energy consumption forecasting. *Renew. Sustain. Energy Rev.* **2014**, *33*, 102–109. [CrossRef]
5. Ahmed, N.; Saha, A.K.; Al Noman, M.A.; Jim, J.R.; Mridha, M.F.; Kabir, M.M. Deep learning-based natural language processing in human–agent interaction: Applications, advancements and challenges. *Nat. Lang. Process. J.* **2024**, *9*, 100112. [CrossRef]
6. Chen, C.; Gao, Z.; Zhou, X.; Wang, M.; Yan, J. Dynamic energy consumption quota for public buildings based on multi-level classification and data correction. *J. Build. Eng.* **2025**, *99*, 111618. [CrossRef]
7. Chong, A.; Yan, D.; Sun, K.; Zhan, S.; Cheng, S.; Wu, Y.; Chen, Y.; Hong, T. Ten questions concerning calibrating building energy simulation models. *Build. Environ.* **2025**, *284*, 113404. [CrossRef]
8. Li, H.; Xu, Y.; Hong, T. EnergyPlus-MCP: A model-context-protocol server for ai-driven building energy modeling. *SoftwareX* **2025**, *32*, 102367. [CrossRef]
9. Bastos Porsani, G.; Casquero-Modrego, N.; Echeverria Trueba, J.B.; Fernández Bandera, C. Empirical evaluation of EnergyPlus infiltration model for a case study in a high-rise residential building. *Energy Build.* **2023**, *296*, 113322. [CrossRef]
10. Gupta, G.; Mathur, S.; Mathur, J.; Nayak, B.K. Blending of energy benchmarks models for residential buildings. *Energy Build.* **2023**, *292*, 113195. [CrossRef]
11. Pino-Mejías, R.; Pérez-Fargallo, A.; Rubio-Bellido, C.; Pulido-Arcas, J.A. Comparison of linear regression and artificial neural networks models to predict heating and cooling energy demand, energy consumption and CO_2 emissions. *Energy* **2017**, *118*, 24–36. [CrossRef]
12. Yalcintas, M.; Aytun Ozturk, U. An energy benchmarking model based on artificial neural network method utilizing US Commercial Buildings Energy Consumption Survey (CBECS) database. *Int. J. Energy Res.* **2007**, *31*, 412–421.
13. Roth, J.; Lim, B.; Jain, R.K.; Grueneich, D. Examining the feasibility of using open data to benchmark building energy usage in cities: A data science and policy perspective. *Energy Policy* **2020**, *139*, 111327. [CrossRef]
14. Geraldi, M.S.; Ghisi, E. Integrating evidence-based thermal satisfaction in energy benchmarking: A data-driven approach for a whole-building evaluation. *Energy* **2022**, *244*, 123161.

15. Zhang, C.; Luo, Z.; Rezgui, Y.; Zhao, T. Enhancing building energy consumption prediction introducing novel occupant behavior models with sparrow search optimization and attention mechanisms: A case study for forty-five buildings in a university community. *Energy* **2024**, *294*, 130896. [CrossRef]
16. Cui, X.; Lee, M.; Koo, C.; Hong, T. Energy consumption prediction and household feature analysis for different residential building types using machine learning and SHAP: Toward energy-efficient buildings. *Energy Build.* **2024**, *309*, 113997. [CrossRef]
17. Robinson, C.; Dilkina, B.; Hubbs, J.; Zhang, W.; Guhathakurta, S.; Brown, M.A.; Pendyala, R.M. Machine learning approaches for estimating commercial building energy consumption. *Appl. Energy* **2017**, *208*, 889–904. [CrossRef]
18. Ahmed Gassar, A.A.; Yun, G.Y.; Kim, S. Data-driven approach to prediction of residential energy consumption at urban scales in London. *Energy* **2019**, *187*, 115973. [CrossRef]
19. Li, X.; Wang, Z.; Yang, C.; Bozkurt, A. An advanced framework for net electricity consumption prediction: Incorporating novel machine learning models and optimization algorithms. *Energy* **2024**, *296*, 131259. [CrossRef]
20. National Standards Commission. *General Rules for Drafting Energy Consumption Norm Standards of Public Institutions*; National Standards Commission: North Ryde, NSW, Australia, 2021.
21. State Organs Affairs Management Bureau; Central Committee of the Communist Party of China Directly Affiliated Organs Affairs Management Bureau. *Energy and Resource Consumption Quotas of Central and State Organs*; Standards for Office Affairs Work: Eschborn, Germany, 2021.
22. Khalil, M.; McGough, A.S.; Pourmirza, Z.; Pazhoohesh, M.; Walker, S. Machine Learning, Deep Learning and statistical Analysis for forecasting building energy consumption—A systematic review. *Eng. Appl. Artif. Intell.* **2022**, *115*, 105287.
23. An, W.; Gao, B.; Liu, J.; Ni, J.; Liu, J. Predicting hourly heating load in residential buildings using a hybrid SSA–CNN–SVM approach. *Case Stud. Therm. Eng.* **2024**, *59*, 104516. [CrossRef]
24. Mustaffa, Z.; Sulaiman, M.H. Advanced forecasting of building energy loads with XGBoost and metaheuristic algorithms integration. *Energy Storage Sav.* **2025**. [CrossRef]
25. Papadopoulos, S.; Kontokosta, C.E. Grading buildings on energy performance using city benchmarking data. *Appl. Energy* **2019**, *233–234*, 244–253. [CrossRef]
26. Li, J.; Cao, X.; Liu, S.; Zhang, P.; Yuan, Y. Study on the load uncertainty of buildings and villages in western Sichuan Plateau based BP neural network integrating prior knowledge. *Energy Sustain. Dev.* **2025**, *88*, 101781. [CrossRef]
27. Feng, G.; Pu, Y.; Li, H.; Wang, H. A calibration method for infrared measurements on building facades based on a WOA-BP neural network. *Infrared Phys. Technol.* **2024**, *137*, 105180. [CrossRef]
28. Liu, Y.-J.; Zhang, J.-P.; Lv, Y.-Z.; Zuo, X.; Yue, Y.-L. Virtual flow metering of three-phase separator based on ISSA-GBDT. *Flow Meas. Instrum.* **2026**, *107*, 103060.
29. Wang, L.; Zhou, X.; Sun, Z.; Xi, C.; Xiao, H. Improved sparrow search algorithm for inversion of geometric parameters of earthquake source faults. *Geod. Geodyn.* **2025**. [CrossRef]
30. Lu, C.; Zhao, J.; Xia, Y.; Sun, Y.; Cai, J. Predicting permeability of natural gas hydrate reservoir based on machine learning assisted by sparrow search algorithm. *Geoenergy Sci. Eng.* **2026**, *257*, 214192. [CrossRef]
31. Saha, D.; Manickavasagan, A. Machine learning techniques for analysis of hyperspectral images to determine quality of food products: A review. *Curr. Res. Food Sci.* **2021**, *4*, 28–44. [CrossRef]
32. Kim, J.Y.; Kim, D.; Li, Z.J.; Dariva, C.; Cao, Y.; Ellis, N. Predicting and optimizing syngas production from fluidized bed biomass gasifiers: A machine learning approach. *Energy* **2023**, *263*, 125900.
33. Chen, C.; Gao, Z.; Zhou, X.; Wang, M.; Yan, J. Energy consumption prediction and energy-saving suggestions of public buildings based on machine learning. *Energy Build.* **2024**, *320*, 114585. [CrossRef]

micromachines

MDPI

Article

Asymmetry- and Viscosity-Regulated Atomization of Laminar Impinging Microjets: Morphology Map, Modal Dynamics, and Droplet Statistics

Xiaoyu Tan [1], Guohui Cai [2], Bo Wang [3] and Xiaodong Chen [2,*]

1 School of Physics, South China Normal University, Guangzhou 511400, China
2 School of Aerospace Engineering, Beijing Institute of Technology, Beijing 100081, China
3 School of Mechanical Engineering, Inner Mongolia University of Science and Technology, Baotou 014010, China
* Correspondence: xiaodong.chen@bit.edu.cn

Abstract

Despite decades of studies on symmetric impinging-jet atomization, the combined role of controlled pre-impingement asymmetry and viscosity in setting the instability pathways and droplet statistics of laminar microjets remains insufficiently quantified. The effects of pre-impingement jet-length difference and liquid viscosity on the flow morphologies, instability dynamics, and atomization behavior of laminar impinging microjets are investigated experimentally using high-speed imaging. By systematically varying the jet-length asymmetry and viscosity over a range of Weber numbers, the evolution of liquid-sheet motion and breakup is resolved from synchronized front- and side-view observations. Specifically, the scientific objective of this work is to elucidate how pre-impingement jet-length asymmetry and liquid viscosity jointly regulate the dynamical behavior of laminar impinging microjets, with particular emphasis on regime transitions of liquid-sheet morphologies, the coupling between upper-sheet oscillations and rim instabilities revealed by synchronized multi-view imaging and POD-based frequency analysis and the resulting droplet-size statistics. These aspects address physical questions that have not been systematically resolved in classical impinging-jet studies, which predominantly focus on symmetric configurations or performance-oriented atomization. With increasing Weber number, the flow undergoes a sequence of regime transitions, including merged-jet, liquid-chain, wavy-rim, fishbone, closed-rim, open-rim, and arc-shaped atomization states. The presence and extent of the closed-rim regime depend sensitively on both jet-length asymmetry and liquid viscosity. Increasing jet-length difference accelerates transitions between these regimes, whereas increasing liquid viscosity stabilizes the liquid sheet and shifts the onset of unsteady breakup to higher Weber numbers. Proper orthogonal decomposition is applied to time-resolved image sequences to extract dominant oscillatory modes and their characteristic frequencies. Within the fishbone regime, the oscillation frequency of rim deformation either coincides with that of the upper region of the liquid sheet or appears as its subharmonic, indicating period-doubling behavior under specific combinations of Weber number and jet-length asymmetry. These frequency characteristics govern the spatiotemporal organization of ligament formation and detachment along the sheet rim. In the arc-shaped atomization regime, droplet-size distributions follow a log-normal form, and at sufficiently high Weber numbers, the mean droplet diameter shows only a weak dependence on jet-length asymmetry. These findings provide microscale-regime guidance for tunable droplet formation in open microfluidic jetting and related small-scale multiphase flows. The innovation of this study lies in the systematic use of synchronized multi-view imaging combined with POD-based frequency analysis and droplet statistics to directly connect liquid-sheet oscillations, rim instability dynamics, and breakup organization under controlled geometric asymmetry and viscosity variations. This approach enables a unified physical interpretation of regime transitions and instability mechanisms that cannot be resolved from single-view observations or morphology-based classification alone.

Keywords: impinging microjets; microscale atomization; liquid sheet breakup; jet-length asymmetry; viscosity effects; regime transition; proper orthogonal decomposition (POD); period doubling; droplet size distribution

1. Introduction

Impinging jets describe a flow configuration in which two liquid jets collide at a prescribed angle, and the outcome of the collision depends on both the jet conditions and operating parameters. Depending on Reynolds number, Weber number, and jet development length, the post-impingement flow can range from stable merged jets or intact liquid structures to thin liquid sheets that exhibit unsteady motion, ligament formation, and eventual atomization into droplets. Owing to the well-defined collision geometry and the flexibility in controlling injection conditions, impinging-jet atomization has been extensively investigated and widely adopted in engineering applications involving liquid-sheet formation and spray generation [1–3]. In the microscale regime, laminar impinging microjets provide a controllable platform for liquid-sheet formation and droplet production relevant to small-scale multiphase-flow operations.

Impinging jets are employed in a broad range of practical systems. In aerospace propulsion, impinging-jet atomizers have long served as a core injection strategy in liquid rocket engines, as exemplified by the F-1 engine used in the Saturn V launch vehicle [4–6]. Beyond propulsion, impinging jets are utilized in microreactors for nanoparticle synthesis [7,8], inhaler and aerosol generation devices for pulmonary drug delivery [9–12], and as free liquid-sheet targets for high-repetition-rate laser—matter interaction experiments [13–16]. These diverse applications motivate continued experimental and theoretical investigation into the physical mechanisms governing liquid-sheet evolution and atomization.

Similar to other liquid-sheet configurations encountered in practical spray systems, the morphology and stability of sheets formed by impinging jets are governed by the interplay among inertia, surface tension, viscous stresses, and interactions with the surrounding gas phase [17–19]. Within this framework, the atomization outcome depends strongly on the flow state of the jets prior to impingement. Whether the incoming jets are laminar or turbulent governs both the initial geometry of the liquid sheet formed at the impingement point and the nature of disturbances introduced into the system [2,3]. For laminar impinging jets, instability development is dominated by interfacial mechanisms intrinsic to the liquid sheet, including rim-dominated capillary instabilities [20,21], interfacial waves associated with Kelvin–Helmholtz-type mechanisms [22], and localized perforations within the sheet interior. With increasing Weber number, these mechanisms may coexist or compete, leading to a variety of breakup pathways characterized by ligament formation, intermittent rupture, and progressive spray development. In contrast, when the jets are turbulent prior to impingement or when the impingement is sufficiently energetic, large-amplitude disturbances generated directly at the collision region, commonly referred to as impact waves, can dominate the breakup process and promote rapid sheet disintegration [23–25].

Beyond instability mechanisms, the thickness distribution of liquid sheets formed by impinging jets is a key physical quantity, as it directly influences ligament formation, droplet size, and atomization performance. A number of theoretical models have been proposed to predict sheet thickness based on mass and momentum conservation in the post-impingement radial flow [21,26–28]. Classical models, such as the Hasson–Peck formulation, assume an inviscid and uniform radial velocity field and predict a thickness scaling inversely proportional to the radial distance from the impingement point. While these models have been widely used and validated at high Reynolds numbers, they neglect viscous effects within the sheet. Recent experimental and theoretical studies have demonstrated that liquid viscosity can increase sheet thickness and modify its spatial distribution, particularly for microscale and ultra-thin sheets, by inducing momentum exchange and mass transport across adjacent radial regions [16]. These findings indicate that liquid-sheet thickness is not solely determined by impingement geometry but is dynamically regulated by fluid properties.

In addition to thickness distribution, the velocity profile of the jets prior to impingement has been shown to influence the characteristics of the liquid sheet formed by jet collision. Early analytical and theoretical studies often assumed a uniform jet velocity at the nozzle exit, but subsequent work demonstrated that incorporating non-uniform profiles alters predictions of sheet thickness and in-plane velocity distributions [28,29]. In particular, Choo and Kang showed that using experimentally measured, non-uniform velocity profiles leads to thickness and velocity distributions that agree more closely with experimental observations than predictions based on a uniform assumption [28]. Related analyses further demonstrated that parabolic and non-uniform velocity profiles modify the azimuthal variation of sheet velocity and the overall sheet contour [29]. These results highlight that realistic pre-impingement velocity profiles, shaped by nozzle geometry and upstream flow development, play an important role in redistributing momentum within the post-impingement radial flow. Previous studies have shown that asymmetries in the velocity and momentum distribution of impinging jets play a key role in organizing liquid-sheet dynamics and downstream instabilities. Early experiments reported periodic atomization along the sheet rim over a wide range of operating conditions [1,30,31], highlighting the sensitivity of impinging jets to disturbances introduced at or upstream of the collision region. Subsequent work demonstrated that the occurrence of such instabilities depends strongly on liquid viscosity and the flow conditions of the incoming jets [20]. Frontal-view observations further indicated that rim deformations originate near the upper region of the liquid sheet, suggesting that the instability is linked to upstream sheet dynamics rather than being generated locally at the rim. While shear instabilities induced by slight velocity mismatches were proposed as a driving mechanism [21,32], organized instabilities were also observed under nominally symmetric conditions [33]. To reconcile these observations, distinct rim-driven and sheet-driven instability pathways were suggested [34]. Despite these efforts, how velocity asymmetry is introduced, amplified, and transmitted through the liquid sheet, particularly across varying impingement angles and Weber numbers, has remained unclear. Recent work has further shown that instability dynamics in laminar impinging microjets originate from intrinsic oscillations in the upper region of the sheet induced by asymmetric momentum redistribution, with downstream rim breakup acting as a dynamically coupled response rather than an independent instability mechanism [35].

Motivated by these considerations, the present study investigates how upstream flow development and fluid properties jointly regulate the instability dynamics of laminar impinging microjets. Differences in pre-impingement jet length provide a physically transparent and experimentally controllable means of introducing systematic velocity-profile asymmetry without modifying nozzle geometry or nominal inlet conditions. By varying the jet-length difference, the degree of momentum imbalance at impact and the resulting deflection of the upper liquid sheet can be continuously tuned. In addition, liquid viscosity is varied to examine its role in modulating oscillatory dynamics and regime transitions. High-

https://doi.org/10.3390/mi17020221

speed front- and side-view imaging is employed to resolve both global sheet motion and local interfacial deformation, while proper orthogonal decomposition is applied to extract dominant oscillatory modes and their characteristic frequencies. Through the combined analysis of flow-regime transitions, modal dynamics, and droplet statistics, this work aims to clarify the dynamical mechanisms by which upstream asymmetry and viscous effects regulate atomization pathways in laminar impinging microjets. This work does not propose a new atomization theory but offers an experimentally grounded clarification of how controllable geometric and rheological factors influence instability evolution and breakup pathways in laminar impinging microjets. By combining synchronized imaging, modal-frequency analysis, and droplet statistics, the study provides a physically interpretable framework that complements classical impinging-jet research, which mainly focuses on symmetric configurations or performance metrics.

2. Experiment and Methods

2.1. Experimental Setup

Figure 1 illustrates the experimental platform used to generate and record the impinging microjets. The system includes a controlled jet-delivery module and a synchronized high-speed optical module, designed to capture both the overall liquid-sheet morphology and the downstream droplet evolution. Figure 1a shows the geometric parameters; two stainless-steel needles are mounted at a fixed impingement angle of $2\alpha = 60°$, generating two microjets with a nominal diameter D_j and exit velocity U_j. Upstream of the collision point, the two jets have independently adjustable pre-impingement lengths, called L_j^1 and L_j^2. These lengths specify the development distances of the incoming jets and serve as the main geometric parameters controlling asymmetry.

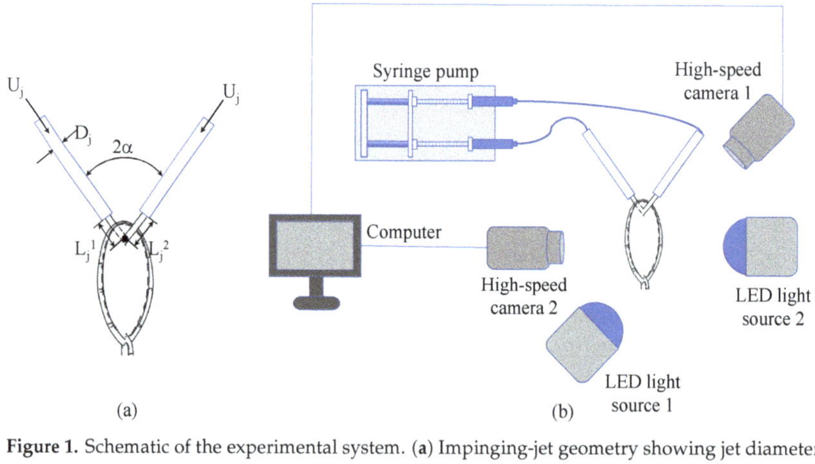

Figure 1. Schematic of the experimental system. (**a**) Impinging-jet geometry showing jet diameter D_j, jet exit velocity U_j, impingement angle 2α, and the two pre-impingement jet lengths (L_j^1, L_j^2). (**b**) High-speed imaging and LED illumination arrangement.

As shown in Figure 1b, a dual-channel precision syringe pump (Pump 11 Pico Plus Elite Dual, Harvard Apparatus, Holliston, MA, USA) delivers liquid through two polyethylene (PE) tubes (outer diameter 4 mm, inner diameter 2 mm) into stainless-steel needles. Front- and side-view imaging is performed using two synchronized high-speed cameras (pco.dimax HS1, PCO Imaging, Kelheim, Germany). Uniform, flicker-free illumination is provided by two high-power LED sources aligned with the optical axes of the cameras. The entire setup is mounted on an optical table to minimize mechanical disturbances and ensure stable, high-fidelity imaging throughout the experiments. For global characterization of

https://doi.org/10.3390/mi17020221

the liquid-sheet dynamics, images are recorded at 1000 × 550 pixels and 9792 fps with a 10 μs exposure time. For downstream droplet-statistics measurements, the frame rate is increased to 13,000 fps while maintaining the same exposure time, ensuring adequate temporal resolution to capture the ligament-to-droplet transition.

The working fluids are deionized water–glycerol mixtures with viscosity μ_l of 2.17 and 4.00 mPa·s. Their densities ρ_l are 1060 and 1125 kg/m^3, respectively. The working fluids contained 0.05% (w/w) Triton X-100 (Crgent Biotech Co., Ltd., Hangzhou, China) surfactant to control the liquid-air surface tension, which was maintained at $\sigma_l = 30$ mN/m. The jets are issued from stainless-steel nozzles with an inner diameter of D_j = 260 μm. The jet exit condition is characterized by the Weber number,

$$\mathrm{We}_j = \frac{\rho_l U_j^2 D_j}{\sigma_l}, \tag{1}$$

and the corresponding Reynolds number,

$$\mathrm{Re} = \frac{\rho_l U_j D_j}{\mu_l}, \tag{2}$$

where U_j is the jet velocity. The relative importance of viscous effects is further quantified by the Ohnesorge number,

$$\mathrm{Oh} = \frac{\mu_l}{\sqrt{\rho_l \sigma_l D_j}}, \tag{3}$$

which takes values of $\mathrm{Oh} \approx 2.4 \times 10^{-2}$ for $\mu_l = 2.17$ mPa · s and $\mathrm{Oh} \approx 4.3 \times 10^{-2}$ for $\mu_l = 4.00$ mPa · s. The experiments cover $14.5 \leq \mathrm{We}_j \leq 700.35$, corresponding to $160 \leq \mathrm{Re} \leq 1100$ for $\mu_l = 2.17$ mPa · s and $90 \leq \mathrm{Re} \leq 620$ for $\mu_l = 4.00$ mPa · s. Within this Reynolds-number range, the liquid jets remain in a laminar developing state prior to impingement. Accordingly, the following analysis uses We_j as the primary parameter representing U_j.

2.2. Proper Orthogonal Decomposition Method

To identify the primary coherent motions contained in the time-resolved images of the impinging microjets, the high-speed recordings are post-processed using the Proper Orthogonal Decomposition (POD) technique [36,37]. Each of the n instantaneous frames is reshaped into a vector according to its pixel-intensity distribution $I(\xi, t)$, allowing the temporal evolution of the interface to be represented in the following modal form:

$$I(\xi, t) = \sum_{i=1}^{n} a_i(t)\phi_i(\xi). \tag{4}$$

In this formulation, $\phi_i(\xi)$ denotes the spatial basis functions produced by POD, each orthogonal to the others and capturing a distinct energetic pattern of the flow field, while $a_i(t)$ describes the corresponding temporal contribution of each mode.

The decomposition is carried out using a singular-value decomposition (SVD) framework, which is well suited for large image datasets owing to its numerical stability and efficiency. All vectorized snapshots $q(\xi, t)$ are arranged into a data matrix,

$$Y = \{q(\xi, t_1), q(\xi, t_2), \ldots, q(\xi, t_n)\}, \tag{5}$$

and the SVD of Y is written as

$$Y = USV^T. \tag{6}$$

From this factorization, the spatial POD modes are extracted from the column vectors of V, whereas the mode amplitudes are retrieved from

https://doi.org/10.3390/mi17020221

$$a_i(t) = US_i. \tag{7}$$

To characterize the dynamical signatures encoded in each mode, the temporal coefficients $a_i(t)$ are further analyzed in the frequency domain using the Fourier transform. The resulting power spectral density (PSD) spectra highlight the dominant oscillation frequencies associated with different coherent structures. Through this procedure, POD serves as an effective means of reducing the dimensionality of the image data while isolating the key spatiotemporal features that dictate the behavior of the liquid sheet.

2.3. Measurement of Droplet Sizes

Quantitative droplet-size statistics are extracted from the high-speed shadowgraphy recordings through a threshold-based image-processing pipeline implemented in LaVision DaVis 10.2.1. Before segmentation, each frame is intensity-normalized so that the maximum and minimum grayscale values map to 100% and 0%, respectively. This normalization enhances contrast across the field of view and ensures consistent binarization. A global threshold of 50% is then applied to convert the images into binary form, allowing individual droplet silhouettes to be isolated.

For each detected droplet, the equivalent diameter d is calculated based on the measured projected area A of the shadow, assuming circular geometry:

$$d = 2\sqrt{\frac{A}{\pi}}.$$

To suppress spurious detections introduced by noise or poorly resolved droplets, a minimum object area of five pixels is imposed. This constraint effectively excludes features smaller than the optical resolution. Furthermore, an edge-quality filter is introduced by examining the local intensity gradient near the droplet boundary; objects with boundary slopes below 10% are discarded to eliminate out-of-focus or blurred droplets.

Additional checks confirm that these filtering strategies exert negligible influence on the resulting droplet size probability density functions (PDFs). Lowering the minimum area threshold from five to three pixels leads only to the appearance of a few detections in the smallest diameter range, without altering the location of the distribution peak or its characteristic log-normal shape. These discarded droplets constitute a minor portion of the overall population and are not associated with the primary ligament-mediated breakup events that define the spray dynamics. Consequently, the reported PDFs and their corresponding log-normal fits remain physically representative of the true droplet size distribution.

3. Results and Discussion

3.1. Effect of (L_j^1, L_j^2)

To examine how upstream geometric differences influence the resulting flow structures, three configurations are constructed based on different combinations of pre-impingement jet lengths (L_j^1, L_j^2). Before analyzing the effects of these geometric variations, it is necessary to establish the baseline behavior. Figure 2 presents the flow regimes for the reference case with nearly symmetric jet lengths, $(L_j^1, L_j^2) = (3.10, 3.12)$ mm, and a liquid viscosity of $\mu_l = 2.17\,\text{mPa}\cdot\text{s}$. The length ratio is defined as $\Lambda = L_j^2/L_j^1$, which equals 1.01 for this case. Front- and side-view images are shown to document the flow morphology from both perspectives. As the jet Weber number We_j increases, the impinging microjets evolve through a sequence of distinct flow regimes. At low $\text{We}_j = 14.50$, inertia is insufficient to sustain a laterally extended sheet; surface tension promotes coalescence into a thicker downward jet, followed by capillary-driven breakup. The jets merge directly at the impact

 https://doi.org/10.3390/mi17020221

point, forming a single, thicker downward jet that eventually breaks into droplets of similar size to the incoming jet. This flow pattern is referred to as the *merged-jet* pattern. When We_j increases to 44.40, a short and narrow liquid sheet forms, shedding small transverse sub-sheets periodically from its lower edge, resulting in a repeating *liquid-chain* pattern. As We_j reaches 90.62, disturbances appear along the sheet rim. Increasing inertia forms a thin sheet, while capillary forces act at the rim. The side-view evidence confirms that rim disturbances originate from oscillations in the upper region of the sheet, indicating a coupling between sheet oscillation and rim instability. Small protrusions emerge from the upper region of the sheet and are carried downstream by convection. The rims along both sides of the sheet converge at the leading edge, followed by complex oscillations that lead to the breakup of the sheet into droplets.

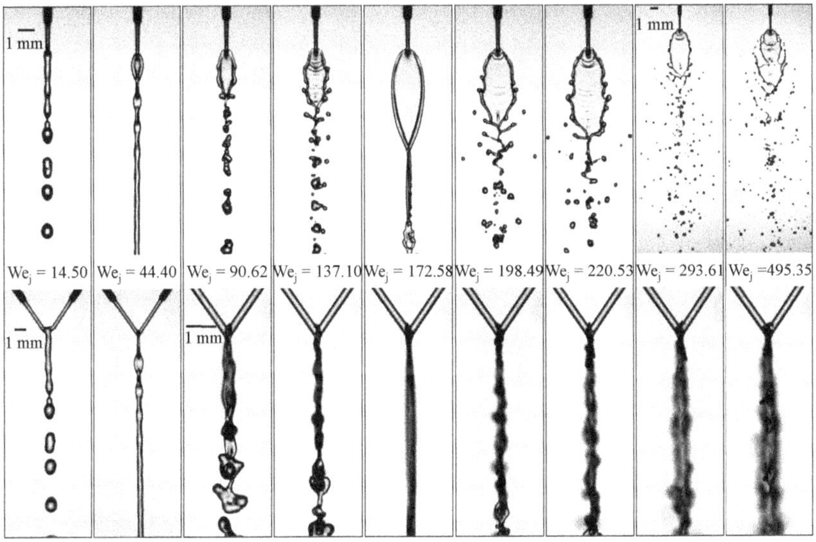

Figure 2. Flow regimes formed by impinging microjets at increasing Weber numbers for the reference case with μ_l of 2.17 mPa · s and nearly symmetric jet lengths $(L_j^1, L_j^2) = (3.10, 3.12)$ mm.

The side-view images confirm that these rim disturbances originate from oscillations in the upper region of the sheet, as observed in previous studies [35]. This flow pattern is referred to as the *wavy-rim* regime [35]. At $\mathrm{We}_j = 137.10$, inertia enhances rim protrusions into finger-like ligaments, while surface tension controls ligament thinning and pinch-off. This is further linked to POD-based frequency analysis, where the dominant rim frequency corresponds to the upper-sheet oscillation frequency, supporting the view that inertial capillary oscillations of the upper sheet drive rim-instability shedding in these regimes. stronger oscillations develop along the upstream rim, and elongated finger-like ligaments appear, signaling the onset of the characteristic *fishbone* regime. In this regime, increasing We_j promotes earlier ligament formation and increases the number of ligaments. When We_j reaches 172.58, the two jets form a large, smooth, closed liquid sheet with a faint orthogonal sheet beneath it. The upper region of the sheet remains relatively stable but is inclined to the left. This flow pattern is called the *closed-rim* regime. A further increase to $\mathrm{We}_j = 198.49$ introduces oscillations along the rim, giving rise to a fishbone-like pattern again. To distinguish between the two fishbone patterns, the first is referred to as the *first fishbone*, and the second occurrence as the *second fishbone*.

At $\mathrm{We}_j = 220.53$, the rim instability becomes more pronounced, leading to vigorous ligament formation. Figure 3a presents a series of snapshots showing this process, where

https://doi.org/10.3390/mi17020221

oscillations generated near the impingement point develop into finger-like ligaments that propagate along the sheet rim. As indicated by the dashed circle, the lower rim of the liquid sheet remains closed, which is characteristic of the second fishbone morphology.

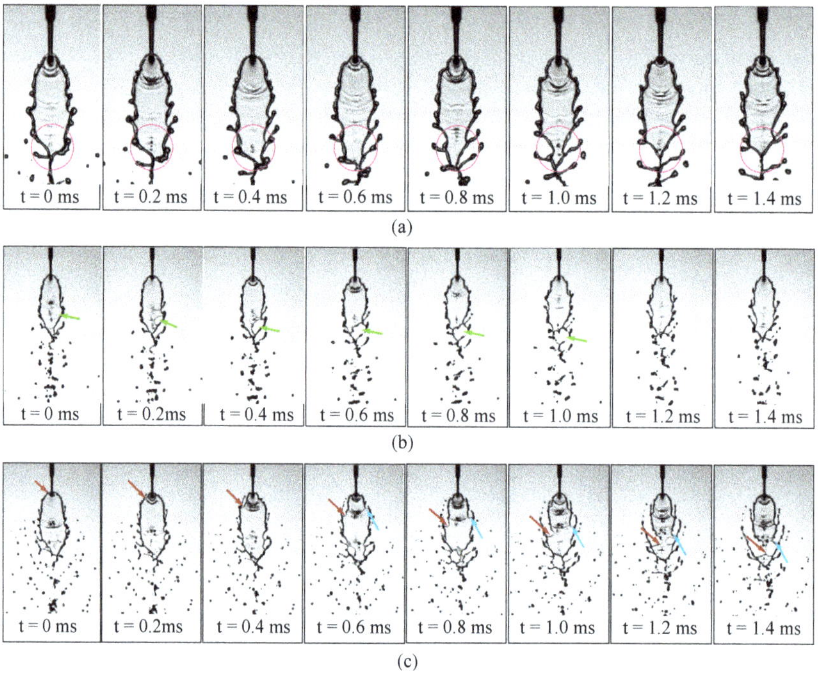

Figure 3. Time-resolved ligament dynamics at three Weber numbers: (**a**) second fishbone pattern at $We_j = 220.53$; (**b**) open-rim pattern at $We_j = 293.61$; (**c**) rim-driven rupture and arc-shaped ligament formation at $We_j = 495.35$.

As We_j increases to 293.61, the field of view is shifted downstream to capture the far-downstream droplet region, where the breakup becomes increasingly dispersed and is dominated by ligament fragmentation, as shown in Figure 2. The temporal evolution in Figure 3b shows that at $t = 0$ ms, a small perforation appears on the liquid sheet. By $t = 0.2$ ms, the perforation grows and reaches the rim, producing ligaments attached to the sheet edge. The opening continues to expand by $t = 0.4$ ms, and by $t = 0.6$ ms it reaches the rim on the left side, causing the sheet tip to open. At $t = 0.8$ ms, the ligaments begin to thin, and they undergo capillary breakup by $t = 1.0$ ms. At $t = 1.2$ ms, the tip of the sheet starts to close again. This cyclic perforation–reclosure process is clearly distinct from the second fishbone mode, in which the lower boundary of the sheet remains intact throughout. Following the terminology of Bush and Hasha [20], this flow pattern is referred to as an *open-rim* mode, although the sheet tip intermittently alternates between open and closed states.

Finally, at $We_j = 495.35$, sheet deformation is intensified by hydrodynamic-wave forcing and edge tearing, and atomization results from coupled mechanisms, including wave-induced deformation and ligament growth. Strong oscillations dominate the upper rim, and finger-like ligaments rapidly develop downstream along the sheet, as shown in Figure 2. The temporal evolution in Figure 3c reveals a markedly more violent breakup process: intense oscillations in the upper region of the sheet induce pronounced deformation from $t = 0$ ms to $t = 0.6$ ms, during which an opening forms near the impingement point and ligament structures emerge, as indicated by the red arrows. The perturbations

https://doi.org/10.3390/mi17020221

propagate along the rim and progressively tear the sheet toward its center. As highlighted by the blue arrows, a second opening and associated ligament appear on the opposite side at $t = 0.6$ ms, and the two openings subsequently expand toward the centerline of the sheet. By $t = 1.2$ ms, a perforation develops below the blue-arrowed opening, further weakening the structure. The openings collapse at $t = 1.4$ ms, rupturing the liquid sheet, while the ligaments merge into an arc-shaped structure that eventually disintegrates into droplets downstream. This flow pattern is referred to as the *arc-shaped atomization* regime. We note that, based on our observations under other conditions with similar We_j, the presence of a perforation is not required for the arc-shaped atomization pattern to occur. This flow pattern closely resembles the upper-sheet-perforation-induced waves recently identified in our study [38], as it is likewise triggered by perforation occurring in the upper region of the liquid sheet.

Figure 4 summarizes the evolution of flow morphologies across the full range of tested Weber numbers for the jet–length configuration $(L_j^1, L_j^2) = (3.10, 3.60)$ mm, for which Λ is 1.16. At very low $\mathrm{We}_j = 14.50$, the jets remain in a merged-jet state and produce axisymmetric droplets through Rayleigh Plateau breakup. When We_j increases to 44.40, the column enters a varicose oscillatory dripping regime, forming a repeating liquid-chain pattern with regularly spaced droplets. At $\mathrm{We}_j = 90.62$, pronounced axisymmetric oscillations arise, marking the wavy-rim regime. As We_j reaches 137.10, a smooth and fully closed liquid sheet forms immediately after impingement; this closed-rim morphology appears significantly earlier than in the $(3.10, 3.12)$ mm case, indicating that stronger upstream asymmetry shortens the wavy-rim regime and accelerates the transition to a stabilized sheet. When We_j increases to 172.58, the sheet develops the characteristic fishbone pattern with rim-attached ligaments. At $\mathrm{We}_j = 198.49$, both configurations enter a more vigorous fishbone regime, but their downstream droplet organizations differ markedly: in the $(3.10, 3.60)$ mm case, two well-defined and nearly symmetric droplet chains form beneath the ruptured sheet, generated by ligaments detaching at both the tips and roots of the rim; these chains remain close to the centerline at lower Weber numbers and maintain an orderly shedding pattern. In contrast, droplets in the $(3.10, 3.12)$ mm case disperse more widely laterally, reflecting stronger upper-rim oscillations that impart larger transverse momentum before ligament breakup, indicating a more intense fishbone regime. As We_j increases to 220.53, rim destabilization strengthens, and ligament stretching becomes more pronounced, producing multiple generations of droplets. At $\mathrm{We}_j = 293.61$, the flow transitions into an open-rim breakup regime characterized by periodic sheet ruptures and extensive droplet production through rim erosion and ligament shedding. Finally, at the highest Weber number tested ($\mathrm{We}_j = 700.35$), the sheet can no longer maintain coherence and disintegrates into a dense spray field, exhibiting a fully developed arc-shaped atomization structure.

In the $\left(L_j^1, L_j^2\right) = (3.10, 4.10)$ mm configuration, for which Λ is 1.32, it shows in Figure 5 that a smooth and stable closed liquid sheet never forms. Instead, the flow exhibits only a narrow transitional window in which closed-sheet and fishbone features appear intermittently. This behavior indicates that increasing the jet-length asymmetry amplifies the overall hydrodynamic imbalance, shifting the upper sheet further from the geometric center and strengthening its interaction with the incoming jets, thereby suppressing the establishment of a sustained closed-rim state. At very low $\mathrm{We}_j = 14.50$, the jet remains axisymmetric and undergoes Rayleigh Plateau breakup, producing regularly spaced droplets. When We_j increases to 44.40, the system transitions into the liquid-chain pattern. At $\mathrm{We}_j = 90.62$, undulations and perturbations emerge along the rim, marking the onset of the wavy-rim regime. As We_j rises to 137.10, these protrusions intensify, and a transitional sheet–ligament structure appears, signaling the emergence of a fishbone-like

https://doi.org/10.3390/mi17020221

morphology. By $\mathrm{We}_j = 172.58$, a fully developed fishbone pattern forms, characterized by symmetric rim-attached ligaments that shed droplets on both sides of the sheet.

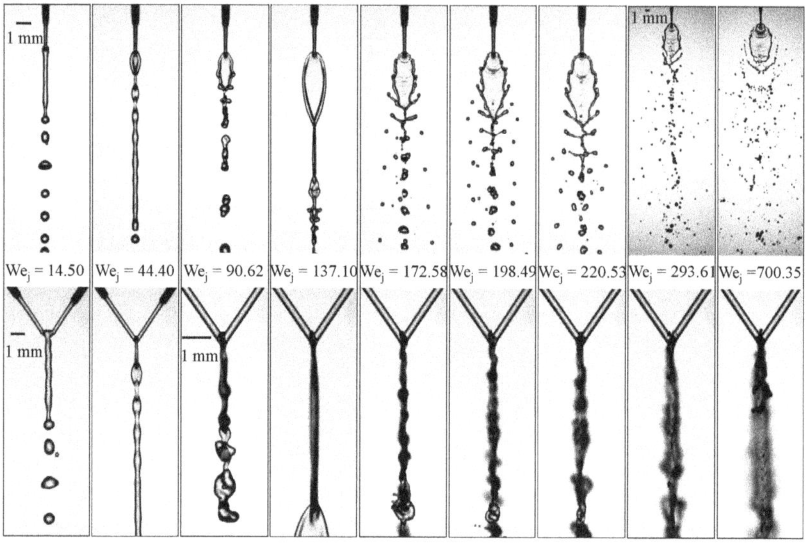

Figure 4. Flow regimes formed by impinging microjets at various Weber numbers for the case at μ_l of 2.17 mPa·s and jet length $(L_j^1, L_j^2) = (3.10, 3.60)$ mm.

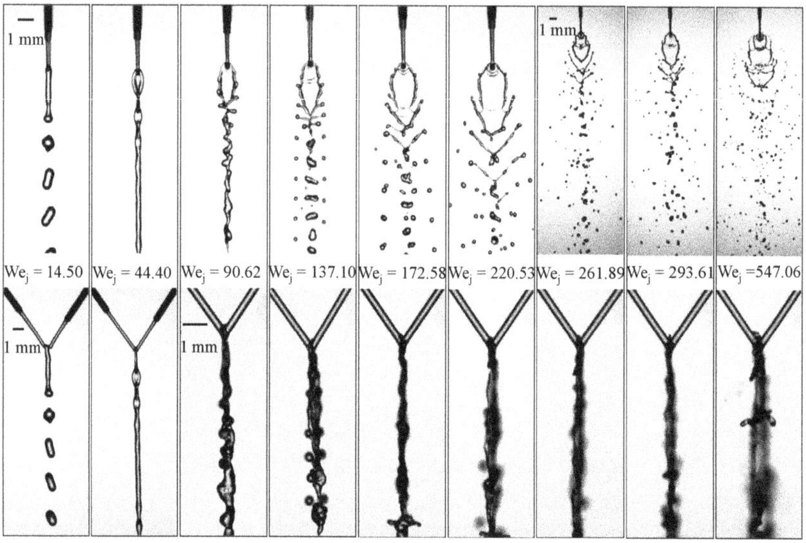

Figure 5. Flow regimes formed by impinging microjetsjets at various Weber numbers for the case at μ_l of 2.17 mPa·s and jet length $(L_j^1, L_j^2) = (3.10, 4.10)$ mm.

Further increases in We_j to 220.53 enhance rim destabilization, producing longer ligaments and multisized droplets downstream. Figure 6 illustrates the temporal evolution of the lower-sheet dynamics. At $t = 0$ ms, the sheet exhibits a pair of symmetric rim-attached ligaments that begin to thicken near their roots (red arrows). As time progresses to $t = 0.2$–0.6 ms, these ligaments elongate steadily while being convected downward, and their bases pull the lower rim outward, producing a V-shaped opening that widens

https://doi.org/10.3390/mi17020221

with time. Between $t = 0.8$ and 1.0 ms, the ligament tips accelerate downstream, and the tearing fronts on both sides migrate further toward the centerline. By $t = 1.2$ ms, the two tearing fronts nearly converge, indicating that the lower sheet edge has been substantially weakened and is approaching rupture. Throughout the sequence, the symmetric outward deformation (highlighted by the arrows) demonstrates how ligament growth directly drives the progressive tearing of the lower sheet, ultimately setting the stage for subsequent open-rim breakup.

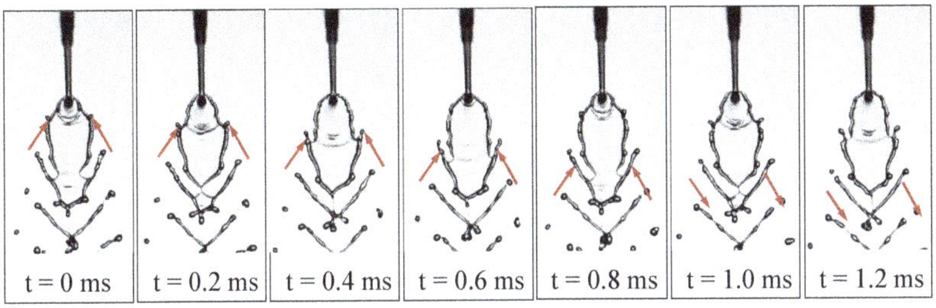

Figure 6. Sequential evolution of the lower-sheet dynamics under We_j of 220.53 in the fishbone regime.

As shown in Figure 5, at $We_j = 261.89$, the lower sheet edge periodically ruptures, causing the flow to transition into an open-rim breakup regime. When We_j reaches 293.61, ligament stretching and sheet tearing become more vigorous, generating a dense population of fine droplets. At the highest Weber number examined, $We_j = 547.06$, the liquid sheet can no longer maintain coherence; both the rim and the sheet interior disintegrate rapidly, resulting in a fully atomized spray cloud composed of fine droplets and fragmented ligament remnants. This condition corresponds to the arc-shaped atomization regime.

As shown in the regime map in Figure 7, the evolution of liquid-sheet morphologies with increasing Weber number is summarized for the three jet-length differences $\Delta L_j = L_j^2 - L_j^1$ at μ_l of 2.17 mPa·s. Across all tested conditions, the impinging microjets exhibit a well-organized sequence of experimentally observed flow states: merged-jet and liquid-chain structures, first fishbone and closed-rim patterns, second fishbone and open-rim regimes, and ultimately arc-shaped atomization. Jet-length asymmetry introduces a clear destabilizing mechanism: a finite ΔL_j produces unequal pre-impingement momentum fluxes, which laterally displace the upper region of the liquid sheet. As ΔL_j increases, this lateral displacement becomes more pronounced, causing the transitions between successive flow patterns to occur earlier, with the corresponding regime boundaries shifting to lower values of We_j. The representative snapshots in Figure 7 at $We_j = 137.10$ further support this interpretation, showing that increasing ΔL_j leads to distinctly different flow patterns, ranging from the second fishbone to the closed-rim and first fishbone structures, as jet-length asymmetry becomes stronger.

https://doi.org/10.3390/mi17020221

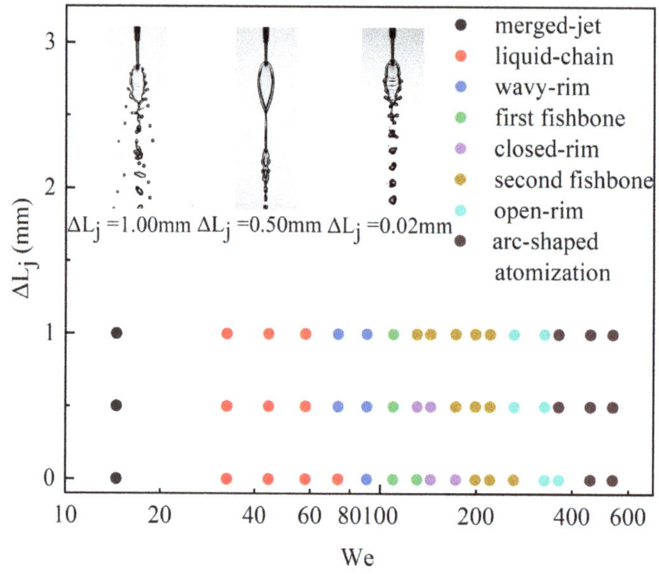

Figure 7. Regime map of liquid-sheet morphologies as a function of the jet Weber number We_j for three jet-length differences $\Delta L_j = L_j^2 - L_j^1$ at μ_l of 2.17 mPa·s. Representative liquid-sheet morphologies at $\mathrm{We}_j = 137.10$ are shown for each ΔL_j configuration.

3.2. Effect of μ_l

With the influence of jet-length asymmetry clarified in the previous section, we now examine how increasing the liquid μ_l to 4.00 mPa·s alters the evolution of flow morphologies across the full Weber-number range. Figure 8 presents the regime maps for the two asymmetric configurations, $(L_j^1, L_j^2) = (3.10, 3.60)$ mm and $(3.10, 4.10)$ mm. Overall, the sequence of flow states resembles that observed at 2.17 mPa·s, except that the closed-rim regime is absent, and the transitions between successive patterns occur at noticeably higher We_j owing to the enhanced viscous stabilization. For the configuration $(L_j^1, L_j^2) = (3.10, 3.60)$ mm shown in Figure 8a, the jets form a merged, axisymmetric column at $\mathrm{We}_j = 14.50$ and undergo Rayleigh–Plateau breakup. When We_j increases to 44.40, the breakup enters a varicose oscillatory dripping mode, generating a periodic liquid-chain structure that persists at $\mathrm{We}_j = 90.62$. As We_j reaches 151.99, the corrugated fishbone sheet emerges with rim-attached ligaments, and a more vigorous variant of the fishbone regime develops at $\mathrm{We}_j = 184.66$. Further increase to $\mathrm{We}_j = 293.61$ intensifies rim destabilization and promotes ligament stretching, producing multiple droplet generations. At $\mathrm{We}_j = 327.14$, the flow transitions to an open-rim breakup mode characterized by periodic sheet ruptures and rim erosion. At the highest Weber number tested, $\mathrm{We}_j = 566.37$, the sheet loses coherence entirely and disintegrates into a dense spray cloud, forming a fully developed arc-shaped atomization structure. Figure 8b indicates that the overall flow morphology for the $(L_j^1, L_j^2) = (3.10, 4.10)$ mm configuration closely follows the sequence observed in Figure 8a for $(3.10, 3.60)$ mm. A notable distinction arises at $\mathrm{We}_j = 293.61$, where the downstream finger-like ligaments become significantly more extended when L_j^2 is larger, reinforcing the earlier conclusion that increased geometric asymmetry enhances rim stretching and promotes the development of longer ligaments.

https://doi.org/10.3390/mi17020221

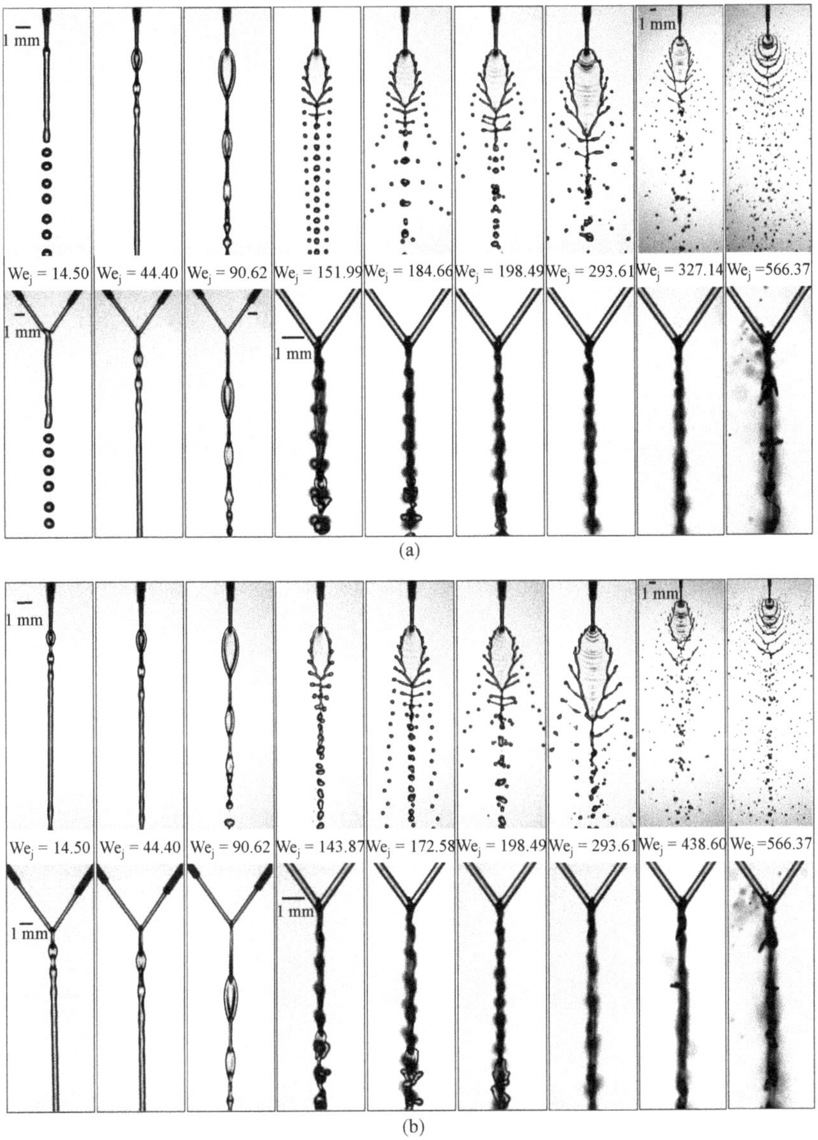

Figure 8. Flow-regime transitions of impinging microjets with liquid μ_l of 4.00 mPa·s over a range of Weber numbers for two jet-length configurations: (**a**) $(L_j^1, L_j^2) = (3.10, 3.60)$ mm and (**b**) $(L_j^1, L_j^2) = (3.10, 4.10)$ mm.

3.3. Frequency Characteristics

Following our previous study [35], POD analysis is conducted to extract the dominant frequency components associated with the fishbone pattern. POD decomposition is applied to both the front- and side-view image sequences using the same number of frames. In our dataset, the POD modes exhibit a typical paired structure: Mode 1 and Mode 2, Mode 3 and Mode 4, Mode 5 and Mode 6, and so on. Within each pair, the spatial structures are highly similar and usually represent the same dominant oscillatory process, with the primary difference being a phase shift rather than a distinct physical mechanism. To avoid redundant presentation of essentially identical dynamical processes while maintaining

https://doi.org/10.3390/mi17020221

conciseness and physical interpretability, the first, third, and fifth modes are selected for frequency evaluation. For the side-view data, the spatial domain is restricted to the vicinity of the impingement point to capture the characteristic oscillations of the upper liquid sheet. Figure 9 presents representative results for the case with μ_l of 4.00 mPa·s and jet-length configuration $(L_j^1, L_j^2) = (3.10, 3.60)$ mm. At $\mathrm{We}_j = 109.65$ (Figure 9a,b), the flow exhibits a wavy-rim behavior. Mode 1 displays a distinct peak at 3443.51 Hz, with a significantly higher PSD amplitude than Modes 3 and 5. This dominant peak corresponds to the rim-oscillation frequency induced by edge instabilities. The same frequency also appears in the side-view results near the upper liquid sheet, demonstrating that the rim oscillation is synchronized with the oscillatory motion of the upper sheet region. This correspondence further verifies that periodic oscillations in the upper liquid sheet constitute the primary mechanism governing the development of the fishbone instability [35].

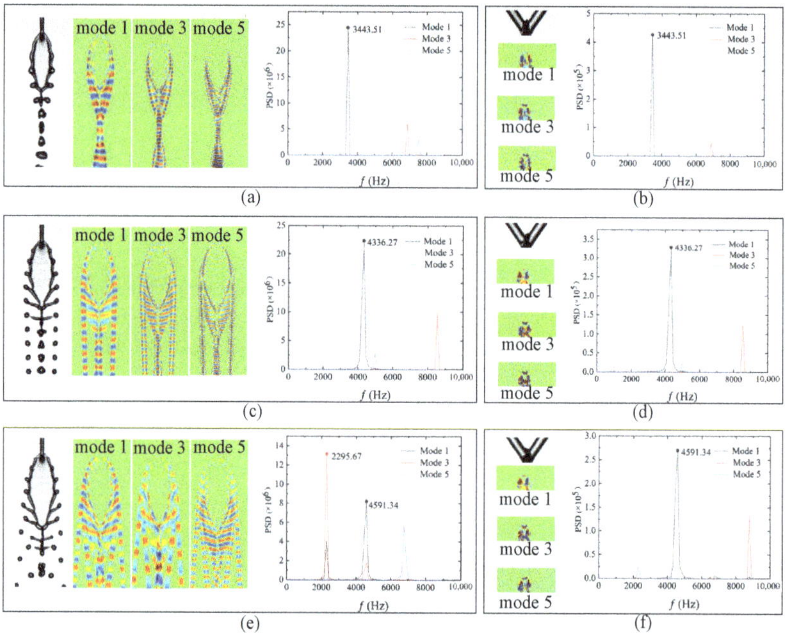

Figure 9. Front- and side-view images of the liquid flow regimes at three Weber numbers for the case with μ_l of 4.00 mPa·s and jet-length configuration $(L_j^1, L_j^2) = (3.10, 3.60)$ mm. Subfigures correspond to (**a**,**b**) $\mathrm{We}_j = 109.65$, (**c**,**d**) $\mathrm{We}_j = 143.87$, and (**e**,**f**) $\mathrm{We}_j = 157.90$. The associated POD modes and their corresponding frequency spectra are also shown.

At $\mathrm{We}_j = 143.87$ (Figure 9c,d), the flow transitions into the fishbone regime. The dominant oscillation frequency extracted from the front-view data is 4336.27 Hz, which again matches the oscillation frequency observed in the upper-sheet region from the side view. This confirms that the fishbone regime at this Weber number remains single-periodic. However, at $\mathrm{We}_j = 157.90$ (Figure 9e,f), the behavior changes. Mode 3 exhibits a higher PSD amplitude than Mode 1, and its dominant frequency is exactly half of that of Mode 1. Although the Mode 1 frequency still coincides with the dominant side-view frequency, the Mode 3 frequency from the front view is half of the corresponding side-view value, indicating the onset of period-doubling behavior [35].

To further clarify the formation mechanism of the fishbone pattern and the onset of its period-doubling bifurcation, the dynamic evolution at the two higher Weber numbers in Figure 9 is examined in Figure 10. At $\mathrm{We}_j = 143.87$ (Figure 10a), the flow exhibits a single-

 https://doi.org/10.3390/mi17020221

period oscillatory cycle. During one cycle, the upper liquid sheet generates a bead along the rim (red arrows) as time progresses from $t = 0$ to approximately $t = 0.4$ ms. At $t = 0.4$ ms, the bead indicated by the red arrow occupies nearly the same position as the bead marked by the green arrow at $t = 0$, demonstrating the completion of one full oscillation period. The bead subsequently develops into a ligament that extends downstream and ultimately forms a terminal droplet via pinch-off. As a result of this periodic ligament formation, a queue of droplets, or a droplet string, appears on both the left and right sides downstream. The corresponding side-view dynamics shown beneath Figure 10a also exhibit a complete oscillation cycle, with the red dashed coil highlighting the characteristic left–right swinging motion that reflects the same periodic behavior.

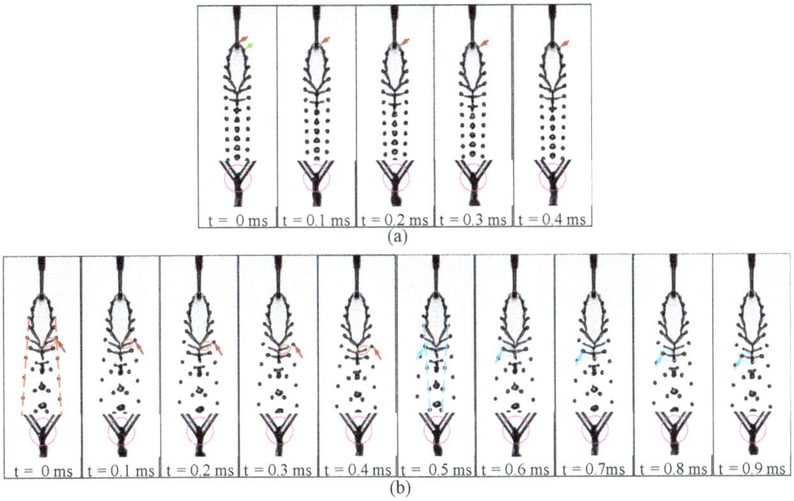

Figure 10. Explanation of the period-doubling phenomenon: (**a**) single-period oscillation, and (**b**) double-period oscillation.

At $We_j = 157.90$ (Figure 10b), the ligament dynamics become markedly more complex. The temporal evolution reveals that although one full oscillation period elapses in the front view, two droplets are generated during this interval, distinct from the single-droplet formation observed at $We_j = 143.87$. Correspondingly, two droplet strings develop downstream on both the left and right sides. In Figure 10b, these strings are highlighted using red and blue arrows and guide lines. During the interval from $t = 0$ to 0.4 ms, the droplet indicated by the red arrow gradually forms and detaches, with the red ellipse marking the position of the corresponding liquid ligament, and the red dashed lines outlining the resulting droplet string. Between $t = 0.5$ and 0.9 ms, a second droplet forms, as shown by the blue arrows and ellipses, producing the second droplet string. This behavior results in two clearly separated droplet strings on both sides of the liquid sheet. The corresponding side-view images beneath Figure 10b show that the upper liquid sheet undergoes two complete oscillations during this interval, confirming the presence of period-doubling. It is further noted that the period-doubling observed in this case differs from that reported in our earlier work [35], as the two distinct droplet strings provide a more direct visualization of the doubled period.

Based on these observations, the relationships between f_u and f_r are further summarized for all operating conditions and both μ_l. The dominant frequencies extracted from the POD analysis of the front- and side-view sequences are defined as f_r and f_u, respectively. Figure 11a presents the results for $\Delta L_j = 0.02$ mm at μ_l of 2.17 mPa·s. At low

https://doi.org/10.3390/mi17020221

We_j, the flow remains in the wavy-rim regime and the frequencies satisfy $f_r = f_u$. As the Weber number increases, the flow transitions to the first fishbone regime, where single- and double-periodicities appear successively, with $f_r = f_u$ changing to $f_r = f_u/2$, demonstrating the presence of period-doubling. When We_j approaches approximately 140, the flow enters a closed-rim state in which no frequency. After the flow reaches the second fishbone regime, the relationship remains $f_r = f_u/2$. Note that f_u increases with We_j initially and then decreases and then increase slowly when the second fishbone pattern is observed. Figure 11b shows the case with a larger jet-length asymmetry, $\Delta L_j = 0.50$ mm. The We_j range in which the closed-rim regime exists becomes significantly narrower, reflecting the destabilizing influence of increasing ΔL_j. In this configuration, the first fishbone regime maintains $f_r = f_u$ with no bifurcation, whereas the second fishbone regime exhibits the transition to $f_r = f_u/2$. Figure 11c illustrates the condition with an even larger asymmetry, $\Delta L_j = 1.00$ mm. Under this configuration, the closed-rim regime disappears entirely, and the flow transitions only among wavy-rim and fishbone patterns. According to the sudden decrease of f_u, the boundary between the first and second fishbone patterns can be drawn appropriately. The period-doubling behavior remains confined to the second fishbone regime, where the relationship $f_r = f_u/2$ holds.

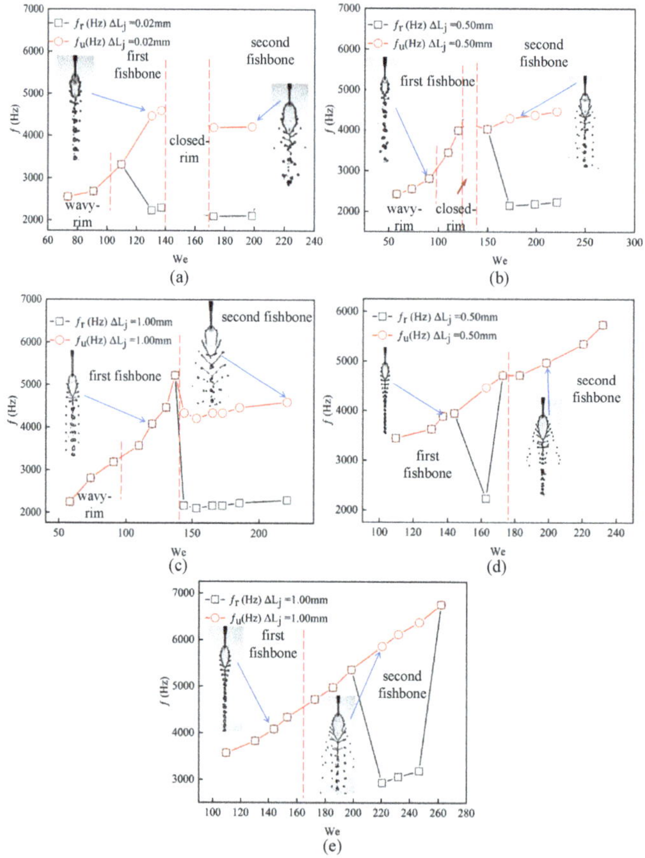

Figure 11. Summary of the frequency relationships between the upper-sheet oscillation frequency f_u and the rim frequency f_r for different μ_l and ΔL_j. Panels (**a**–**c**) show the 2.17 mPa·s cases for ΔL_j = 0.02, 0.50, and 1.00 mm, respectively, while panels (**d**,**e**) show the 4.00 mPa·s cases for $\Delta L_j = 0.50$ and 1.00 mm.

 https://doi.org/10.3390/mi17020221

Figure 11d,e present the results for μ_l of 4.00 mPa·s at two different values of ΔL_j. Under this condition, the experiments reveal a distinct transition from the fishbone state to the closed-rim state, which allows the critical boundaries between the first and second fishbone regimes to be identified. For the smaller jet-length asymmetry of $\Delta L_j = 0.50$ mm (Figure 11d), f_u increases nearly monotonically over the entire range of We_j. Both the first and second fishbone regimes exhibit a gradual increase in frequency, while the period-doubling behavior appears only within the second fishbone regime, occurring near $\mathrm{We}_j \approx 160$. When ΔL_j increases to 1.00 mm (Figure 11e), f_u again shows a continuous upward trend with increasing We_j. Similar to the observations at lower μ_l, period-doubling occurs exclusively within the second fishbone regime, with no frequency bifurcation detected in the first fishbone regime.

3.4. Droplets Characteristics

The atomization performance of impinging microjets at different impact angles and Weber numbers is evaluated using the mean droplet diameter. A generalized d_{mn} mean, defined through the m-th and n-th moments of the measured droplet diameters, is expressed as

$$d_{mn} = \left(\frac{\sum_{i=1}^{N} d_i^m}{\sum_{i=1}^{N} d_i^n} \right)^{\frac{1}{m-n}}, \tag{8}$$

where d_i denotes the diameter of the i-th droplet and N is the total number of sampled droplets. When $m = 1$ and $n = 0$, this expression reduces to the arithmetic mean diameter d_{10}. When $m = 3$ and $n = 2$, it yields the Sauter mean diameter d_{32}, which characterizes the volume-to-surface-area ratio of the spray.

As shown in Figure 12, the variation of d_{10} with We_j is presented for the five operating conditions. A clear inverse trend is observed: increasing We_j substantially reduces the mean droplet diameter. This behavior indicates that a higher We_j intensifies jet impingement, promotes earlier destabilization of the liquid sheet, and leads to more complete atomization. At relatively low We_j, the differences among the three ΔL_j configurations at μ_l of 2.17 mPa·s are pronounced. Smaller ΔL_j, corresponding to higher geometric symmetry, produces noticeably larger mean droplet diameters because symmetric jets form a smoother and more stable liquid sheet, delaying sheet rupture and yielding larger droplets after breakup. As We_j increases, however, the mean droplet diameters for all three configurations rapidly converge and fluctuate only within a narrow range. In this high-We_j regime, breakup is dominated by upper-sheet-perforation-induced waves, whose influence overwhelms the geometric effects associated with ΔL_j. Consequently, the role of jet-length asymmetry in determining droplet size becomes significantly weaker at high We_j compared with the low-We_j regime. For μ_l of 4.00 mPa·s, the differences between the two ΔL_j configurations remain small across the entire We_j range, indicating that μ_l further suppresses the influence of jet-length asymmetry on droplet size.

Figure 13 shows the probability density functions (p.d.f.s) of droplet diameters for the case with μ_l of 2.17 mPa·s and jet lengths in the range $(L_j^1, L_j^2) = (3.10, 4.10)$ mm at different Weber numbers. At $\mathrm{We}_j = 198.49$ (Figure 13a), droplet breakup occurs predominantly beneath the liquid sheet, and the resulting size distribution exhibits a complex multimodal structure, with medium-sized droplets accounting for most of the population. When We_j increases to 362.48 (Figure 13b), the proportion of medium-sized droplets starts to decrease, while smaller droplets become more abundant, and a small fraction of relatively large droplets begins to appear. At $\mathrm{We}_j = 501.30$ (Figure 13c), the distribution evolves toward a more clearly unimodal shape, with droplet sizes becoming increasingly concentrated around a dominant peak. By $\mathrm{We}_j = 700.35$ (Figure 13d), the distribution is well described by a log-normal function: a fully developed unimodal, right-skewed profile is observed,

https://doi.org/10.3390/mi17020221

in which most droplets fall within a narrow size range while a small number of large droplets form a long tail extending to the right. This type of right-skewed, unimodal behavior is characteristic of fully developed atomization and is commonly modeled using the log-normal distribution in spray and atomization studies. The normalized probability density function of the log-normal distribution is expressed as

$$P\left(X = \frac{d}{d_{10}}, \beta, X_0\right) = \frac{1}{X\beta\sqrt{2\pi}} \exp\left[-\frac{(\ln X - \ln X_0)^2}{2\beta^2}\right], \tag{9}$$

where d denotes the droplet diameter, d_{10} is the mean droplet diameter, $X = d/d_{10}$ is the normalized droplet size, and β and X_0 are fitting parameters. In the impact-wave-dominated atomization regime, corresponding to the fully developed breakup at high We_j, the measured droplet-size distribution exhibits a clear log-normal character that is well captured by this functional form.

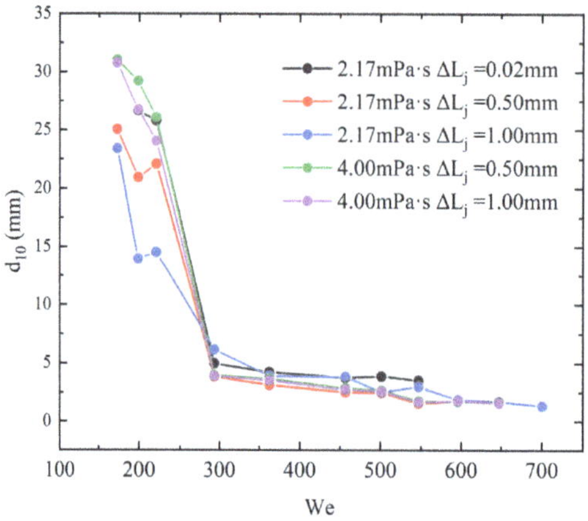

Figure 12. Evolution of the mean droplet diameter d_{10} with the jet Weber number We_j for five operating conditions involving two μ_l and three jet-length asymmetries ΔL_j.

Figure 14a,b show the fitted log-normal p.d.f.s for three representative Weber numbers under the conditions of 2.17 mPa·s 4.00 mPa·s conditions, respectively. In both cases, the droplet-size distributions collapse onto nearly identical curves, regardless of We_j. This indicates that once the breakup enters the arc-shaped atomization regime, the influence of the Weber number on the droplet-size statistics becomes negligible. Consistently, the fitted parameter β remains confined within a narrow and physically reasonable interval, further supporting the robustness of the log-normal behavior. Figure 14c compares the droplet-size p.d.f.s for different jet-length differences under 2.17 mPa·s at approximately the same Weber number. A remarkable similarity is observed across all configurations once atomization is fully developed. Although the case with a jet-length difference $\Delta L_j = 0.02$ mm yields the largest β value, the deviation among all three configurations remains very small. This confirms that, in the fully atomized regime, the jet-length difference has only a negligible effect on the resulting droplet-size distribution.

https://doi.org/10.3390/mi17020221

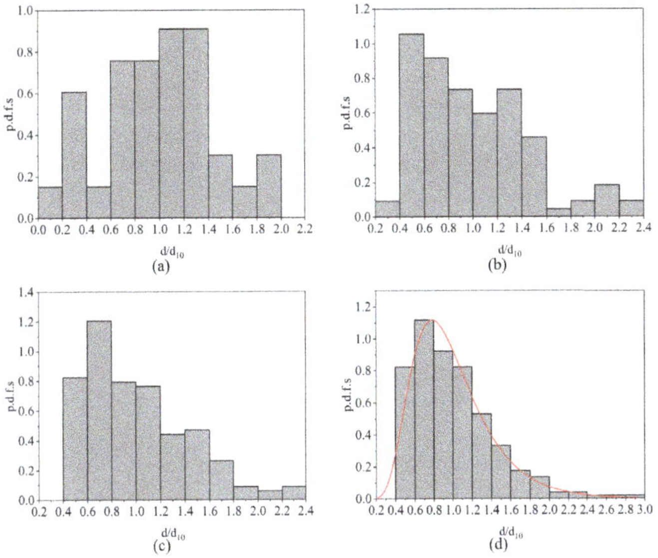

Figure 13. Evolution of the probability density functions of droplet diameters for the case with μ_l of 2.17 mPa·s and a jet-length range $(L_j^1, L_j^2) = (3.10, 4.10)$ mm. Subfigures correspond to (**a**) $\mathrm{We}_j = 198.49$, (**b**) $\mathrm{We}_j = 362.48$, (**c**) $\mathrm{We}_j = 501.30$, and (**d**) $\mathrm{We}_j = 700.35$.

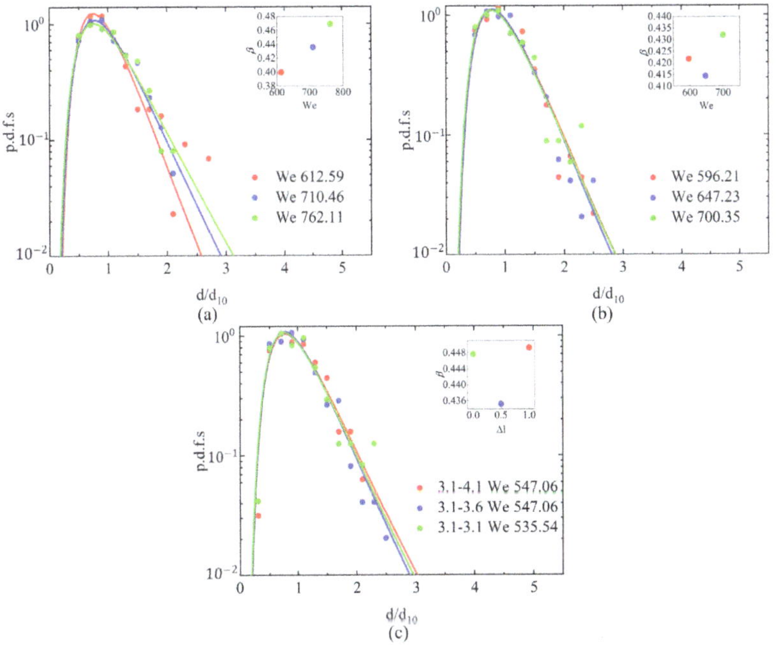

Figure 14. Log-normal p.d.f.s of droplet diameters. (**a**) Droplet-size p.d.f.s for three representative We_j at μ_l = 2.17 mPa·s. (**b**) Droplet-size p.d.f.s for three representative We_j at μ_l = 4.00 mPa·s. (**c**) Droplet-size p.d.f.s for different ΔL_j at approximately the same We_j for μ_l = 2.17 mPa·s.

4. Conclusions

This work is motivated by the need for a systematic understanding of how controlled pre-impingement asymmetry and liquid rheology reshape instability pathways in lam-

https://doi.org/10.3390/mi17020221

inar impinging-jet systems beyond the canonical symmetric configurations commonly addressed in classical studies. This study investigated the effects of pre-impingement jet-length asymmetry and liquid viscosity on the instability dynamics and atomization behavior of laminar impinging microjets over a wide range of Weber numbers We_j. By combining synchronized high-speed front- and side-view imaging, proper orthogonal decomposition, and droplet-size statistics, the roles of upstream momentum imbalance and rheological effects in governing liquid-sheet evolution, oscillatory dynamics, and breakup pathways were clarified. In particular, the study is designed to address how geometric asymmetry and viscosity jointly influence regime transitions, oscillatory coupling mechanisms, and the resulting atomization outcomes.

With increasing We_j, the impinging microjets exhibit a well-organized sequence of flow regimes, including merged-jet, liquid-chain, wavy-rim, first fishbone, closed-rim, second fishbone, open-rim, and ultimately a fully developed arc-shaped atomization regime. For a fixed liquid viscosity, increasing the jet-length difference ΔL_j systematically accelerates the transitions between successive regimes and shifts the corresponding boundaries toward lower We_j. This behavior reflects the destabilizing influence of geometric asymmetry, which introduces unequal pre-impingement momentum fluxes and induces a lateral deflection of the upper region of the liquid sheet, thereby enhancing rim stretching, ligament growth, and premature breakup. In contrast, increasing the liquid viscosity stabilizes the sheet, delays regime transitions to higher We_j, and suppresses the persistence of a stable closed-rim morphology, resulting in a smoother, more gradual evolution of flow patterns.

Proper orthogonal decomposition (POD)-based frequency analysis reveals a strong dynamical coupling between oscillations in the upper region of the liquid sheet and rim deformation in the wavy-rim and fishbone regimes. For all conditions examined, the dominant rim frequency f_r either coincides with the upper-sheet oscillation frequency f_u or appears at its subharmonic, indicating that rim dynamics are governed by oscillatory motions originating near the impingement region rather than by a purely local rim instability. At low Weber numbers and near the onset of fishbone formation, the response is predominantly single-periodic with $f_r = f_u$, whereas increasing the Weber number leads to subharmonic responses and period-doubling, characterized by $f_r = f_u/2$. The emergence and extent of period-doubling depend on both jet-length asymmetry and liquid viscosity. Increasing asymmetry suppresses the closed-rim regime and progressively confines period-doubling to higher-We_j fishbone states, while higher viscosity lowers the overall oscillation frequency and narrows the range of We_j over which periodic behavior is observed, without altering the qualitative frequency organization. Transitions between single- and double-periodic states are accompanied by clear changes in breakup morphology: single-period oscillations produce one droplet string on each side of the sheet, whereas period-doubling leads to two alternating droplet strings, directly linking frequency bifurcation to ligament formation and droplet organization.

In the fully developed atomization regime, droplet-size distributions are well described by a log-normal form. At sufficiently high Weber numbers, the mean droplet diameter becomes nearly insensitive to jet-length asymmetry, indicating that breakup is dominated by upper-sheet-perforation-induced waves that overwhelm geometric effects. In contrast, below the sheet-to-spray transition, geometric symmetry plays a more prominent role, with smaller ΔL_j producing larger mean droplet diameters owing to the formation of smoother and more stable liquid sheets prior to breakup.

Overall, the present results demonstrate that jet-length asymmetry and liquid viscosity exert strong and complementary influences on impinging-jet atomization by regulating upstream sheet oscillations, rim instability dynamics, and breakup pathways. The combined analysis of flow-regime transitions, POD-based frequency characteristics, and droplet

https://doi.org/10.3390/mi17020221

statistics provides a unified physical framework for interpreting instability mechanisms and atomization behavior in laminar impinging-jet systems, with direct relevance to the modeling and control of practical spray processes under geometric and rheological variations. Rather than proposing a new atomization theory, this work provides experimentally grounded physical evidence and a quantified morphology–dynamics map that clarifies how controllable geometric and rheological parameters govern instability development and breakup organization in laminar impinging microjets.

Author Contributions: Conceptualization, X.C.; methodology, X.T. and G.C.; investigation, X.T.; data curation, X.T.; formal analysis, X.T. and B.W.; writing–original draft preparation, X.T.; writing–review and editing, X.T., G.C., B.W. and X.C.; supervision, X.C. All authors have read and agreed to the published version of the manuscript.

Funding: This work was partly supported by the National Natural Science Foundation of China (grant nos. 12272050 and U23B6009).

Institutional Review Board Statement: Not applicable.

Informed Consent Statement: Not applicable.

Data Availability Statement: The data that support the findings of this study are available from the corresponding author upon reasonable request.

Conflicts of Interest: The authors declare no conflicts of interest.

References

1. Dombrowski, N.A.; Hooper, P.C. A study of the sprays formed by impinging jets in laminar and turbulent flow. *J. Fluid Mech.* **1964**, *18*, 392–400. [CrossRef]
2. Chen, X.; Yang, V. Recent advances in physical understanding and quantitative prediction of impinging-jet dynamics and atomization. *Chin. J. Aeronaut.* **2019**, *32*, 45–57. [CrossRef]
3. An, Y.; Wang, B.; Peng, Z.; Chen, X.; Yang, V. Mechanistic insights and practical applications of impinging-jet dynamics and atomization. *Acta Mech. Sin.* **2025**, *41*, 125469. [CrossRef]
4. Oefelein, J.C.; Yang, V. Comprehensive review of liquid-propellant combustion instabilities in F-1 engines. *J. Propuls. Power* **1993**, *9*, 657–677. [CrossRef]
5. Culick, F.E.; Yang, V. Overview of combustion instabilities in liquid-propellant rocket engines. *Prog. Astronaut. Aeronaut.* **1995**, *169*, 3–38.
6. Yang, V. Modeling of supercritical vaporization, mixing, and combustion processes in liquid-fueled propulsion systems. *Proc. Combust. Inst.* **2000**, *28*, 925–942. [CrossRef]
7. Erni, P.; Elabbadi, A. Free impinging jet microreactors: Controlling reactive flows via surface tension and fluid viscoelasticity. *Langmuir* **2013**, *29*, 7812–7824. [CrossRef]
8. Baber, R.; Mazzei, L.; Thanh, N.T.K.; Gavriilidis, A. Synthesis of silver nanoparticles using a microfluidic impinging jet reactor. *J. Flow Chem.* **2016**, *6*, 268–278. [CrossRef]
9. Longest, P.W.; Hindle, M. Evaluation of the Respimat Soft Mist inhaler using a concurrent CFD and in vitro approach. *J. Aerosol Med. Pulm. Drug Deliv.* **2009**, *22*, 99–112. [CrossRef] [PubMed]
10. Wachtel, H.; Kattenbeck, S.; Dunne, S.; Disse, B. The Respimat® development story: Patient-centered innovation. *Pulm. Ther.* **2017**, *3*, 19–30. [CrossRef]
11. Hickey, A.J.; Smyth, J.P. (Eds.) *Pharmaceutical Inhalation Aerosol Technology*, 3rd ed.; CRC Press: Boca Raton, FL, USA, 2019.
12. Wang, B.; Peng, Z.; Liu, Q.; Chen, X. Effect of airflow rate on droplet deposition of an impinging-jet inhaler in the human respiratory tract. *Int. J. Pharm.* **2024**, *661*, 124408. [CrossRef] [PubMed]
13. Cao, Z.; Peng, Z.; Shou, Y.; Zhao, J.; Chen, S.; Gao, Y.; Liu, J.; Wang, P.; Mei, Z.; Pan, Z.; et al. Vibration and jitter of free-flowing thin liquid sheets as target for high-repetition-rate laser-ion acceleration. *Front. Phys.* **2023**, *11*, 1172075. [CrossRef]
14. Kim, Y.H.; Kim, H.; Park, S.C.; Kwon, Y.; Yeom, K.; Cho, W.; Kwon, T.; Yun, H.; Sung, J.H.; Lee, S.K.; et al. High-harmonic generation from a flat liquid-sheet plasma mirror. *Nat. Commun.* **2023**, *14*, 2328. [CrossRef]
15. Peng, Z.; Cao, Z.; Liu, X.; Shou, Y.; Zhao, J.; Chen, S.; Gao, Y.; Wang, P.; Mei, Z.; Pan, Z.; et al. A comprehensive diagnostic system of ultra-thin liquid sheet targets. *High Power Laser Sci. Eng.* **2024**, *12*, e26. [CrossRef]
16. Peng, Z.; Liu, X.; Song, Z.; Wang, B.; Cao, Z.; Wu, E.; Zhao, J.; Gao, Y.; Chen, X.; Ma, W. Thickness model for viscous impinging liquid sheets. *J. Fluid Mech.* **2025**, *1014*, R3. [CrossRef]

 https://doi.org/10.3390/mi17020221

17. Lin, S.P.; Reitz, R.D. Drop and Spray Formation from a Liquid Jet. *Annu. Rev. Fluid Mech.* **1998**, *30*, 85–105. [CrossRef]
18. Villermaux, E. Fragmentation. *Annu. Rev. Fluid Mech.* **2007**, *39*, 419–446. [CrossRef]
19. Lefebvre, A.H.; McDonell, V.G. *Atomization and Sprays*, 2nd ed.; CRC Press: Boca Raton, FL, USA, 2017.
20. Bush, J.W.M.; Hasha, A.E. On the collision of laminar jets: Fluid chains and fishbones. *J. Fluid Mech.* **2004**, *511*, 285–310. [CrossRef]
21. Bremond, N.; Villermaux, E. Atomization by jet impact. *J. Fluid Mech.* **2006**, *549*, 273–306. [CrossRef]
22. Huang, J.C.P. The break-up of axisymmetric liquid sheets. *J. Fluid Mech.* **1970**, *43*, 305–319. [CrossRef]
23. Anderson, W.E.; Ryan, H.M.; Santoro, R.J. Impact wave-based model of impinging jet atomization. *At. Sprays* **2006**, *16*, 791–806. [CrossRef]
24. Ryan, H.M.; Anderson, W.E.; Pal, S.; Santoro, R.J. Atomization characteristics of impinging liquid jets. *J. Propuls. Power* **1995**, *11*, 135–145. [CrossRef]
25. Wu, E.; Wang, B.; Chong, W.K.; Yang, A.; Zhang, F.; Su, Y.; Chen, X. Turbulent atomisation of impinging jets under rising backpressure. *J. Fluid Mech.* **2025**, *1003*, A15. [CrossRef]
26. Taylor, G. Formation of thin flat sheets of water. In *Proceedings of the Royal Society of London. Series A. Mathematical and Physical Sciences*; Royal Society: London, UK, 1960; Volume 259, pp. 1–17.
27. Hasson, D.; Peck, R.E. Thickness distribution in a sheet formed by impinging jets. *AIChE J.* **1964**, *10*, 752–754. [CrossRef]
28. Choo, Y.J.; Kang, B.S. The effect of jet velocity profile on the characteristics of thickness and velocity of the liquid sheet formed by two impinging jets. *Phys. Fluids* **2007**, *19*, 112101. [CrossRef]
29. Inamura, T.; Shirota, M. Effect of velocity profile of impinging jets on sheet characteristics formed by impingement of two round liquid jets. *Int. J. Multiph. Flow* **2014**, *60*, 149–160. [CrossRef]
30. Heidmann, M.F.; Priem, R.J.; Humphrey, J.C. *A Study of Sprays Formed by Two Impinging Jets*; Technical Report; National Advisory Committee for Aeronautics: Washington, DC, USA, 1957.
31. Anderson, W.E. Impinging Jet Injector Atomization, Liquid Rocket Engine Combustion Instability. *Prog. Astronaut. Aeronaut.* **1995**, *169*, 215–246.
32. Jung, S.; Hoath, S.D.; Martin, G.D.; Hutchings, I.M. Atomization patterns produced by the oblique collision of two Newtonian liquid jets. *Phys. Fluids* **2010**, *22*, 042101. [CrossRef]
33. Avulapati, M.M.; Anand, T.; Ravikrishna, R. Atomization patterns produced by viscous, like-on-like impinging liquid jets. *J. Flow Vis. Image Process.* **2018**, *25*, 65–76.
34. Kebriaee, A.; Kashanj, S.; Olyaei, G. Fishbone pattern formation due to asymmetries in colliding low-velocity jets. *At. Sprays* **2018**, *28*, 751–762. [CrossRef]
35. Cai, G.; Wang, B.; Wu, E.; Chen, X. Origin and bifurcations in fishbone instabilities of impinging jets. *J. Fluid Mech.* **2025**, *1016*, A60. [CrossRef]
36. Taira, K.; Brunton, S.L.; Dawson, S.T.M.; Rowley, C.W.; Colonius, T.; McKeon, B.J.; Schmidt, O.T.; Gordeyev, S.; Theofilis, V.; Ukeiley, L.S. Modal analysis of fluid flows: An overview. *AIAA J.* **2017**, *55*, 4013–4041. [CrossRef]
37. Taira, K.; Hemati, M.S.; Brunton, S.L.; Sun, Y.; Duraisamy, K.; Bagheri, S.; Dawson, S.T.M.; Yeh, C. Modal analysis of fluid flows: Applications and outlook. *AIAA J.* **2020**, *58*, 998–1022. [CrossRef]
38. Cai, G.; Wang, B.; Wu, E.; Chen, X. Asymmetry-Regulated Atomization in Laminar Impinging Microjets: Aerodynamic and Upper-Sheet-Perforation-Induced Waves. *Phys. Fluids* **2026**, *38*, 012122. [CrossRef]

 https://doi.org/10.3390/mi17020221

fluids

MDPI

Article

Analytical Modeling of Transverse and Longitudinal Motion of Single Particles in a Horizontal Boundary Layer with Cross-Flow Velocity Pulsations

Rumen Yankov [1], Ventsislav Dimitrov [2,*], Georgi Tonkov [3], Veselina Dimitrova [2], Sylvester Bozherikov [2], Gergana Tonkova [4] and Konstantin Raykov [2]

[1] College Sliven, Technical University of Sofia, 8800 Sliven, Bulgaria; r.yankov@tu-sofia.bg
[2] Faculty of Engineering and Pedagogy of Sliven, Technical University of Sofia, 8800 Sliven, Bulgaria; vkdd_acad@tu-sofia.bg (V.D.); s.bozherikov@tu-sofia.bg (S.B.); k.raykov@tu-sofia.bg (K.R.)
[3] Faculty of Mechanical Engineering, Technical University of Sofia, 1000 Sofia, Bulgaria; gptonkov@tu-sofia.bg
[4] Faculty of Mining Electromechanics, University of Mining and Geology "St. Ivan Rilski", 1000 Sofia, Bulgaria; gtonkova@mgu.bg
* Correspondence: vpdd_acad@tu-sofia.bg

Abstract

This study develops an analytical description of the motion of dilute solid particles in the boundary layer of laminar horizontal flows subjected to weak transverse pulsations. The analysis is formulated for dilute spherical solid particles subjected to transverse velocity pulsations in a laminar boundary-layer flow. A coupled matrix representation of the governing equations is formulated, and closed-form solutions are obtained using Laplace transformation. The analytical expressions capture transient evolution, forced oscillations, resonance effects, and long-term behaviour for particles with different density ratios. Numerical evaluation shows that light particles migrate toward faster regions of the boundary layer and accelerate longitudinally, while heavy particles move toward slower layers and decelerate. Transverse pulsations generate oscillatory trajectories whose amplitude increases near resonance. Impulsive perturbations superimposed on the continuous motion lead to discontinuous transitions consistent with the linear matrix system. The results provide a unified physical interpretation of particle redistribution mechanisms in boundary layers and offer a compact analytical tool for dilute multiphase flow modelling.

Keywords: particle motion; boundary-layer flow; multiphase flow; transverse velocity pulsations; analytical solution; Laplace transform; matrix formulation; inertial migration; resonance behaviour; impulsive forcing

1. Introduction

The study of particle-laden flows in both turbulent and laminar regimes has long attracted attention due to its significance in industrial processes, environmental engineering, and energy systems. Classical fluid dynamics provides the essential framework for understanding continuous media. Early works on fluid and gas mechanics [1], hydrodynamics [2], and mathematical physics [3] established fundamental principles. These theories are crucial for describing particle motion within fluids and the interactions between dispersed and continuous phases [4].

Particle behaviour in nonuniform and turbulent flows is governed by complex mechanisms, including drag, lift, and inertial effects. Pioneering studies of small rigid spheres in nonuniform flows highlighted the critical role of particle size and density relative to the surrounding medium [5]. Contemporary research has extended these insights to examine particle transport in laminar boundary layers [6] and to develop analytical methods for suspended particle dynamics [7]. Detailed experimental and numerical studies have revealed transversal and longitudinal particle motion, highlighting resonance phenomena that influence deposition patterns and trajectory characteristics [8–10].

Advanced modelling approaches now employ matrix and vector formulations. These methods provide a comprehensive representation of multidimensional particle dynamics and enable the derivation of closed-form solutions through harmonic decomposition [11–15]. They have been applied to predict deposition, re-suspension, and agglomeration in wall-bounded turbulent flows [16], and to refine Large-Eddy Simulation models for particle-laden systems [17,18].

The impact of particle and thermal inertia on transport and heat transfer in non-isothermal turbulent flows has been emphasised in recent computational studies [19,20]. Moreover, the spatial distribution and deposition of cylindrical, spherical, and charged inertial particles have been investigated to assess the effects of turbulence intensity, boundary layer characteristics, and particle–particle interactions [21–23]. Such studies are particularly relevant for process intensification strategies in chemical and biomass reactors, where optimal particle behaviour directly affects process efficiency and product yield [24–26].

Particle deposition and resuspension have been thoroughly investigated in turbulent channel and pipe flows using both Lagrangian and Eulerian [27,28] approaches. Boelens et al. developed Lagrangian models to analyse particle deposition and re-suspension under turbulent conditions, providing quantitative predictions of particle–wall interactions. Balachandar and Eaton [29] reviewed dispersed multiphase flows, highlighting key mechanisms governing particle transport and turbulence modulation. Bernardini et al. [30] examined particle dispersion in wall-bounded turbulent flows, elucidating the effects of boundary layer structures on particle trajectories.

Historical studies remain critical references for understanding particle motion. Morsi and Alexander [31] offered seminal insights into particle paths in two-phase flow systems, establishing baseline equations for trajectory prediction in both laminar and turbulent regimes. Elghobashi [32] extended these concepts to particle-laden turbulent flows, emphasising the role of particle-fluid interactions in transport dynamics. Brandt and Coletti [33] provided a detailed overview of particle-laden flows, integrating classical theory with practical modelling approaches, demonstrating the engineering relevance of multiphase flow comprehension. Bijlard et al. [34] applied direct numerical simulations to analyse detailed particle behaviour in turbulent channels, quantifying deposition, lift-off, and near-wall interactions. Mei [35] introduced the concept of history forces on spherical particles, revealing cumulative effects of fluid acceleration on particle motion. Coletti [36] investigated particle deposition and lift-off in turbulent boundary layers, demonstrating subtle interactions between particle inertia and near-wall turbulence structures.

Building on these historical foundations, the stability and qualitative behaviour of particle–fluid dynamical systems—particularly under impulsive, harmonic, or weakly stochastic excitation—can be rigorously analysed within a matrix-based dynamical framework. When the governing equations are expressed in compact matrix–vector form, they allow systematic assessment of transient and steady-state regimes through eigenvalue spectra, state-transition matrices, and transform-based operator methods. The classical Lyapunov stability theory, originating from Lyapunov's foundational monograph [37,38] and further developed in modern nonlinear dynamics and matrix differential systems [39–42], provides rigorous criteria for boundedness, asymptotic stability, and resonance amplification in multidimensional particle motions. These analytical tools are especially powerful

https://doi.org/10.3390/fluids11020051

for systems subjected to transverse velocity pulsations, where small perturbations in the state vector may yield sustained oscillatory responses or transient divergence. Matrix formulations, combined with Laplace-transform techniques [43,44] and criteria for impulsive and periodically driven systems [45–47], therefore offer an indispensable framework for identifying stability boundaries, resonance conditions, and long-term particle behaviour—phenomena that often remain inaccessible to direct numerical integration alone. Additional extensions of these methods include impulsive differential models, coupled matrix differential equations, and stochastic operator approaches [48–52], which provide versatile tools for capturing complex particle–fluid interactions in laminar and turbulent boundary layers.

It should be noted that harmonic transverse velocity oscillations in wall-bounded flows are classically associated with the so-called second Stokes problem, which describes the formation of an oscillatory boundary layer induced by periodic forcing of the carrier fluid [53]. This problem and its analytical extensions have been examined in a number of studies addressing oscillatory boundary layers and vibration-induced flow phenomena [54,55]. More recently, the second Stokes problem has also been explicitly employed in the analysis of complex shear-driven configurations, including non-Newtonian free-surface and rivulet flows, highlighting its continued relevance in modern fluid-mechanical modeling [56].

In contrast to these works, the present study does not aim to resolve the oscillatory velocity field of the carrier phase itself. Instead, it focuses on the inertial response and the coupled transverse–longitudinal dynamics of solid particles subjected to a prescribed oscillatory cross-flow. In this sense, the connection to the second Stokes problem is acknowledged at the level of the forcing mechanism, while the analytical framework developed here addresses a distinct physical problem related to particle dynamics in oscillatory boundary-layer environments.

These studies provide a robust foundation for analysing particle transport in wall-bounded flows and motivate the use of compact state-space and transform-based methods to separate transient relaxation, forced oscillations and long-time drift. Building on this background, the present work develops a coupled analytical model for a single inertial particle in a horizontal laminar boundary layer subjected to harmonic transverse velocity pulsations. The governing equations are formulated in a matrix–vector form and solved in closed form using Laplace-transform techniques, enabling explicit identification of decay rates, phase shifts and resonance conditions. The analytical results are evaluated numerically for representative light and heavy particles, with full datasets provided in the Supplementary Materials to ensure reproducibility.

2. Theoretical Framework, Physical Model and Governing Equations

When single solid impurities move in the boundary layer of a plane horizontal flow of an incompressible fluid, numerous forces act upon them: aerodynamic drag force; added-mass inertial force; Basset force; Saffman force; Magnus force; pressure-gradient force; turbophoretic force; photophoretic force; electrophoretic force; and the body (mass) force. The magnitude and direction of each of these forces are determined primarily through empirical relations, and for every specific problem, only those forces whose influence is dominant are taken into account in the equations describing the particle velocity and trajectory.

The flow under consideration is a viscous fluid moving in a horizontal plane channel with half-width b. The coordinate system is defined as follows: the X-axis is oriented along the flow direction and coincides with the plane where the velocity is zero (the lower bounding wall); the Y-axis is directed vertically upward, perpendicular to the streamlines; and the Z-axis is directed perpendicular to the plane of the schematic.

https://doi.org/10.3390/fluids11020051

The velocity field of the carrier phase consists of one constant longitudinal component U_g and one transverse pulsation component varying according to a harmonic law $v'(t)$:

- along the X-axis: $U_g = U_g(y)$,
- along the Y-axis: $v_g'(t) = v_0\cos(\omega t)$,

where

- v_0 and ω are the amplitude and the frequency of the pulsations, identical for all points in the flow plane within the boundary layer.

The longitudinal carrier-phase velocity is described by a wall-bounded boundary-layer profile:

$$U_g(y) = U_\infty\left(1 - e^{-\frac{y}{\delta}}\right),\ y \geq 0, \tag{1}$$

where

- δ [m] denotes the characteristic boundary-layer thickness,

If extended linearly for $y < 0$ using the wall shear rate, the longitudinal velocity profile becomes:

$$U_g(y) = \left.\frac{dU_g}{dy}\right|_{y=0} y = \frac{U_\infty}{\delta}y = \gamma_\omega \cdot y \tag{2}$$

Accordingly, the wall shear rate γ_ω is defined as:

$$\gamma_\omega = \left.\frac{dU_g}{dy}\right|_{y=0} = \frac{U_\infty}{\delta}, \quad [\gamma_\omega] = s^{-1} \tag{3}$$

The physical model, coordinate system, velocity components, and forces acting on a single particle in the horizontal channel boundary layer are shown schematically in Figure 1.

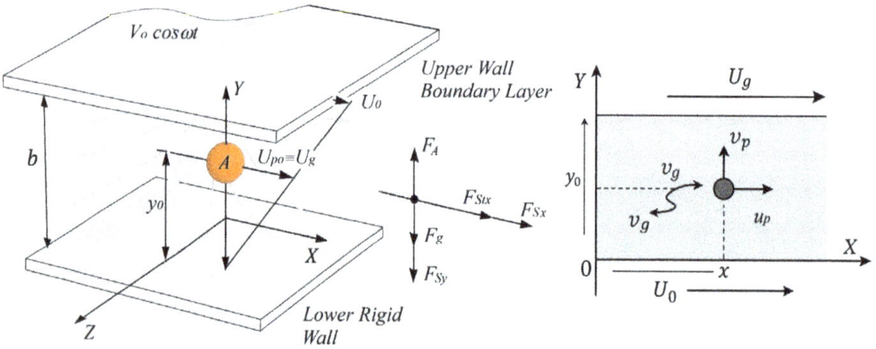

Figure 1. Schematic representation of the physical model and coordinate system for a single particle moving in a horizontal channel boundary layer with a constant longitudinal flow and harmonic transverse velocity pulsations.

An important observation must be noted regarding the projections of the Saffman force: the inclusion of the transverse velocity pulsation changes the direction of the projection of the force on the X-axis. The Saffman force Fs along the Y-axis has a negative sign. This follows from the algebraic operations within the brackets in the vector equation for this force [4,31]:

$$F_s = 6.44 d_p^2 \sqrt{\mu \rho_p} \left|\nabla \times V_g\right|^{-0.5} \left(V_g - V_p\right) \times \left(\nabla \times V_g\right) \tag{4}$$

 https://doi.org/10.3390/fluids11020051

For the wall-bounded velocity profile $U_g(y)$, the carrier-phase vorticity is dominated by the wall-normal gradient of the longitudinal velocity. Consequently, the magnitude of the vorticity entering the Saffman force can be expressed as [1,2]:

$$(\nabla \times V_g) \approx \left|\frac{dU_g}{dy}\right| = \gamma_\omega \tag{5}$$

establishing a direct physical link between the Saffman lift mechanism and the wall shear rate.

where

- d_p—particle diameter;
- μ—dynamic viscosity of the carrier fluid;
- ρ_p—density of the particle material;
- $V_g = (U_g(y), v_g{}'(t), 0)$—velocity vector of the carrier fluid;
- $V_p = (u_p, v_p, 0)$—velocity vector of the particle;
- $\nabla \times V_g$—vorticity vector of the carrier flow;
- $|\nabla \times V_g|$—magnitude of the flow vorticity;
- $(V_g - V_p)$—relative velocity between fluid and particle;
- γ_ω [s^{-1}]—wall shear rate, defined as the gradient of the carrier-phase longitudinal velocity at the wall (y = 0).

Mathematically, the Saffman lift contribution enters the governing dynamics through the coupling coefficient B, which multiplies the cross-component relative-velocity terms in Equations (6) and (7). As a result, the lift mechanism produces a bidirectional coupling between the longitudinal and transverse particle velocities: the transverse excitation contributes to the longitudinal response via the $-B(v' - v_p)$ term in Equation (6), while the longitudinal slip contributes to the transverse response via the $+B(U_g - u_p)$ term in Equation (7).

Taking into account the above considerations when formulating the equations of motion, the following system is obtained for the particle velocities: $u_p(t)$—longitudinal [4,5] and $v_p(t)$—transverse [6,10]:

$$\frac{du_p}{dt} = A(U_g(y_p) - u_p) - B(v' - v_p) \tag{6}$$

$$\frac{dv_p}{dt} = A(v' - v_p) + B(U_g(y_p) - u_p) - g\left(1 - \frac{1}{\varepsilon_p}\right) \tag{7}$$

where

- A—drag/inertia coefficient, representing the influence of the carrier fluid along the motion direction [4,17]:

$$A = \frac{18\mu}{\rho_p d_p^2} \tag{8}$$

- B—Saffman/vorticity coefficient, representing the coupling between longitudinal and transverse motion due to fluid shear [10,36]:

$$B = 6.44 d_p^2 \sqrt{\mu \rho_p} \left|\nabla \times V_g\right|^{-0.5} \approx 6.44 d_p^2 \sqrt{\mu \rho_p} \gamma_\omega^{-0.5} \tag{9}$$

- ε_p—particle-to-fluid density ratio [4,29]:

$$\varepsilon_p = \frac{\rho_p}{\rho_f} \tag{10}$$

https://doi.org/10.3390/fluids11020051

- ρ_p—particle density;
- ρ_f—density of the carrier fluid.
- g—acceleration due to gravity $-g(1 - 1/\varepsilon_p)$ accounts for the effective body force on the particle in the vertical direction, i.e., the net gravitational acceleration reduced by the buoyant force (Archimedes' principle).

In terms of relative contribution, the coefficient A represents the dominant dissipative mechanism (viscous relaxation toward the carrier-phase velocity), while the gravity term provides a steady bias in the transverse direction through $-g(1 - 1/\varepsilon_p)$. The lift-related coefficient B typically acts as a secondary contribution in magnitude; however, it is dynamically important because it transfers excitation between the transverse and longitudinal components and therefore controls phase shifts and resonance-like amplification under oscillatory forcing.

The initial conditions corresponding to the physical process and required for solving the system are:

$$t = 0;\ y_p(0) = y_0;\ u_p(0) = U_g(y_0);\ v_{p0} = 0;$$

It is emphasized that the present study does not aim to solve the Navier–Stokes equations for the carrier phase. The longitudinal velocity profile and the transverse oscillatory component are introduced as prescribed, physically admissible base flows serving as kinematic input to the particle-motion analysis. This modelling strategy is commonly adopted in analytical Lagrangian frameworks to isolate particle–flow interaction mechanisms without resolving the full dynamics of the carrier fluid.

2.1. Transverse Motion: Governing Equations and Analytical Solution

The closed-form analytical solutions derived in this section are exact for constant U_g and remain valid as a local approximation under the near-wall linearisation $U_g(y) \approx \gamma_\omega y$. For the full boundary-layer profile $U_g(y_p(t))$, the coupled first-order system is integrated numerically.

The solution begins with the definition of the function $v_p(t)$. Equation (7) is solved with respect to the difference $(U_g - u_p)$, yielding:

$$B(U_g - u_p) = \frac{dv_p}{dt} - \mathrm{A}(v' - v_p) + g\left(1 - \frac{1}{\varepsilon_p}\right) \tag{11}$$

Differentiating (7) with respect to time gives:

$$B\frac{du_p}{dt} = -\frac{d^2v_p}{dt^2} + \mathrm{A}\left(\frac{dv'}{dt} - \frac{dv_p}{dt}\right) \tag{12}$$

Substituting into Equation (6) yields the second-order differential equation for transverse motion:

$$\begin{aligned} &\frac{d^2v_p}{dt^2} + 2\mathrm{A}\frac{dv_p}{dt} + (A^2 + B^2)v_p + \mathrm{A}g\left(1 - \frac{1}{\varepsilon_p}\right) \\ &= (A^2 + B^2)v_0\cos(\omega t) - \mathrm{A}v_0\omega\sin(\omega t) \end{aligned} \tag{13}$$

This equation describes the translational and oscillatory motion of a solid particle in a viscous medium under the action of forced external disturbances in the form of harmonic velocity pulsations of the surrounding fluid. The solution of Equation (13) is composed of the homogeneous solution corresponding to damped natural oscillations (left-hand side) and a particular solution that accounts for the influence of the imposed external pulsations on the particle motion.

https://doi.org/10.3390/fluids11020051

The transient (damped) component of the solution is:

$$v_p^{(h)}(\mathrm{t}) = (C_1\cos(Bt) + C_2\sin(Bt))e^{-At} - \frac{Ag}{A^2+B^2}\left(1 - \frac{1}{\varepsilon_p}\right) \tag{14}$$

Given the harmonic form of the imposed velocity pulsations, the particular solution of the governing differential equation is sought in a harmonic form with the same angular frequency. Due to the linearity of the system and the sinusoidal nature of the forcing term, the steady-state response of the particle is also oscillatory, differing from the excitation by its amplitude and phase shift. Therefore, the particular solution is assumed in the form:

$$\eta = H_1\cos(\omega t + \delta_1) \tag{15}$$

H_1 and δ_1 are constants determined by the system parameters and the characteristics of the external forcing. δ_1 denotes the phase shift between the particle motion and the imposed harmonic pulsations.

- H_1—represents the amplitude of the steady-state oscillatory response:

$$H_1 = \frac{(A^2+B^2)v_0}{(A^2+B^2-\omega^2)\cos(\delta_1) - 2A\omega\sin(\delta_1)} \tag{16}$$

- δ_1—represents the phase shift accounting for the lag or lead of the particle response relative to the imposed harmonic excitation:

$$\delta_1 = \arctan\left(\frac{-A\omega(A^2+B^2+\omega^2)}{\left(A^2+B^2\right)^2 + (A^2-B^2)\omega^2}\right) \tag{17}$$

The complete solution for the transverse velocity is obtained by superposition of the homogeneous solution (Equation (14)) and the particular solution (Equation (15)).

$$v_p = (C_1\cos(Bt) + C_2\sin(Bt))e^{-At} - \frac{Ag}{A^2+B^2}\left(1 - \frac{1}{\varepsilon_p}\right) + H_1\cos(\omega t + \delta_1) \tag{18}$$

The integration constants C_1 and C_2 are determined from the following initial conditions:

$$t = 0;\ v_p = 0;\ \frac{dv_p}{dt} = 0$$

After performing the necessary algebraic manipulations, the following expression is obtained:

$$C_1 = \frac{Ag}{A^2+B^2}\left(1 - \frac{1}{\varepsilon_p}\right) - H_1\cos(\delta_1) \tag{19}$$

$$C_2 = \frac{A}{B}C_1 + \frac{H_1\omega}{B}\sin\delta_1 \tag{20}$$

In conclusion, Equation (17) fully characterizes the transverse particle displacement by capturing transient decay, forced oscillations, initial-condition effects, and gravitational contributions through the density ratio ε_p:

$$\begin{aligned} y(t) = \int v_p(t)dt = \tfrac{e^{-At}}{A^2+B^2}[(BC_1 - AC_2)\sin(Bt) - (AC_1 + BC_2)\cos(Bt)] \\ + \tfrac{H_1}{\omega}[\sin(\omega t + \delta_1) - \sin\delta_1] + \tfrac{AC_1+BC_2}{A^2+B^2} - \tfrac{Ag}{A^2+B^2}\left(1 - \tfrac{1}{\varepsilon_p}\right)t \end{aligned} \tag{21}$$

https://doi.org/10.3390/fluids11020051

2.2. Longitudinal Motion: Governing Equations and Analytical Solution

The closed-form solution below is derived under the locally frozen approximation $U_g(y_p) \approx U_g(y_0)$. In the numerical evaluation, U_g is updated at each time step as $U_g(y_p(t))$.

The differential equation governing the longitudinal particle velocity as a function of time, $u_p(t)$, is derived from the coupled system of Equations (6) and (7). By rearranging the first equation of the system and solving it with respect to the term $B(v' - v_p)$, the following relation is obtained:

$$B(v' - v_p) = -\frac{du_p}{dt} + A(U_g - u_p) \tag{22}$$

Differentiating Equation (18) with respect to time yields:

$$\left(\frac{dv'}{dt} - \frac{dv_p}{dt}\right) = -\frac{d^2u_p}{dt^2} - A\frac{du_p}{dt} \tag{23}$$

Substituting into Equation (3) yields the second-order differential equation for longitudinal motion:

$$\begin{gathered}\frac{d^2u_p}{dt^2} + 2\mathrm{A}\frac{du_p}{dt} + (A^2 + B^2)u_p - (A^2 + B^2)U_g(y_p) \\ + \mathrm{Bg}\left(1 - \frac{1}{\varepsilon_p}\right) = Bv_0\omega\sin(\omega t)\end{gathered} \tag{24}$$

The solution of the resulting equation is given as the sum of the solution corresponding to the damped natural oscillations (left-hand side) and a particular solution accounting for the influence of the forced velocity pulsations in the fluid surrounding the particle (right-hand side). The homogeneous solution of Equation (20) is given by:

$$u_p^{(h)}(\mathrm{t}) = U_g(y_p) + (C_3\cos(Bt) + C_4\sin(Bt))e^{-At} - \frac{Bg}{A^2 + B^2}\left(1 - \frac{1}{\varepsilon_p}\right) \tag{25}$$

The homogeneous solution of (21) is:

$$\eta_2 = H_2\cos(\omega t + \delta_2) \tag{26}$$

where

- η_2—denotes the particular (forced) oscillatory component of the particle displacement associated with the imposed harmonic velocity pulsations.

The constants δ_2 and H_2 are determined following the procedure described above. By equating and summing the coefficients of the cos(ωt) terms in the expression for δ_2, the particular solution is obtained as:

$$\delta_2 = \frac{A^2 + B^2 - \omega^2}{2A\omega} \tag{27}$$

$$H_2 = \frac{Bv_0\omega}{(A^2 - B^2 - \omega^2)\sin\delta_2 + 2A\omega\cos\delta_2} \tag{28}$$

The complete solution for longitudinal velocity is:

$$u_p = (C_3\cos(Bt) + C_4\sin(Bt))e^{-At} - \frac{Bg}{A^2 + B^2}(1 - {}^{1}/_{\varepsilon_p}) + H_2\cos(\omega t + \delta_2) \tag{29}$$

The integration constants are determined from the initial conditions:

$$C_3 = \frac{Bg}{A^2 + B^2}\left(1 - \frac{1}{\varepsilon_p}\right) - H_2\cos(\delta_2) \tag{30}$$

https://doi.org/10.3390/fluids11020051

$$C_4 = \frac{A}{B}C_3 + \frac{H_2\omega}{B}\sin\delta_2 \tag{31}$$

This section establishes a consistent physical and mathematical model for the motion of a single solid particle in a viscous boundary-layer flow subjected to harmonic transverse velocity pulsations. Starting from the governing force balance, coupled differential equations for the transverse and longitudinal velocity components are derived and solved analytically. The resulting closed-form solutions capture both transient and steady-state particle dynamics, accounting for viscous damping, Saffman-induced coupling, gravitational effects, density ratio influence, and externally imposed oscillatory forcing, and provide a solid basis for further parametric and numerical analysis.

3. Advanced Analytical Extensions of the Governing Particle Motion Model

An extended analytical framework is developed to describe the motion of individual solid particles within a horizontal boundary layer subjected to transverse velocity fluctuations. Building on the linear particle dynamics introduced earlier, the formulation incorporates impulsive contributions to represent sudden, localized changes in particle velocity arising from micro-vortical structures, particle–wall interactions, or transient turbulent accelerations.

The governing equations are expressed in matrix–vector form, enabling spectral analysis and the application of Laplace transforms to capture both transient and steady-state responses. This approach facilitates the investigation of resonance phenomena, stability via eigenvalue analysis, and phase-space dynamics. By combining smooth harmonic fluctuations with discrete impulsive effects, the framework provides a rigorous link between idealized particle motion and the complex behaviour observed in boundary-layer flows.

3.1. *Impulsive Differential Extension of the Governing Dynamics*

Real boundary layers are inherently non-ideal and rarely exhibit perfectly linear behaviour. In low-turbulence or transitional flow regimes, the particle may experience sudden, localized deviations in the surrounding velocity field. Such phenomena include micro-vortical structures with extremely short timescales, particle–wall impact interactions, impulsive turbulent cores, as well as mechanical or vibrational perturbations from the environment. These events are inherently impulsive, inducing nearly discrete jumps in particle velocity that cannot be faithfully represented by smooth harmonic functions such as $v_0\cos(\omega t)$.

To rigorously account for these abrupt and localized velocity fluctuations, the baseline system is extended through the inclusion of Dirac delta functions, $\delta(t - t_k)$, which represent discrete impulsive events occurring at specific instants t_k. The extended system can be expressed in vectorial form as:

$$\frac{d}{dt}\begin{bmatrix} u_p \\ v_p \end{bmatrix} = \begin{bmatrix} A(U_g - u_p) - B\left(v'_g - v_p\right) \\ A\left(v'_g - v_p\right) + B(U_g - u_p) - g\left(1 - \frac{1}{r_p}\right) \end{bmatrix} + \sum_{k=1}^{N} \begin{bmatrix} I_{x,k} \\ I_{y,k} \end{bmatrix} \delta(\mathrm{t} - t_k) \tag{32}$$

where

- $I_k = [I_x, k, I_{y,k}]^T$—denotes the impulsive contributions in the longitudinal and transverse directions.

Figure 2 provides a conceptual view of the impulsive forcing term introduced in the governing equations, illustrating how discrete events produce instantaneous changes in the particle velocity components superimposed on the continuous motion.

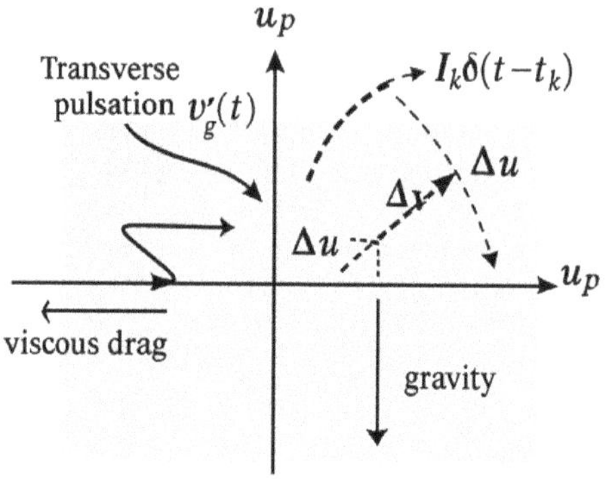

Figure 2. Conceptual representation of the impulsive forcing term $\sum I_k\delta(t-t_k)$ acting on particle velocity components in the analytical model.

Integration of the system in the immediate vicinity of t_k yields the corresponding velocity jump condition:

$$u_p(t_k^+) - u_p(t_k^-) = I_k, \tag{33}$$

demonstrating that particle velocity undergoes a discrete discontinuity, whereas acceleration exhibits a singular component.

The incorporation of impulsive terms offers a scientifically robust framework for capturing short-duration acceleration phenomena, including deposition, rebound, or detachment events at the wall, while preserving the linear character of the underlying system. Light particles ($\tau_p \ll 1$) respond almost instantaneously to impulsive forcing, whereas heavy particles ($\tau_p \gg 1$) display attenuated responses dominated by inertial effects. This extended formulation therefore provides a rigorous mathematical tool for the analysis of realistic boundary layer flows, seamlessly integrating smooth harmonic fluctuations with discrete, impulsive events that typify turbulent or impact-driven processes.

3.2. Matrix Formulation of the Particle Motion System

The governing equations describing the motion of a particle in a horizontal boundary layer can be expressed in compact matrix–vector form, which consolidates the longitudinal and transverse velocity components into a unified dynamical representation. This is not merely a matter of notation: the two components are intrinsically coupled through viscous relaxation and lift effects and therefore should not be treated as independent scalar equations. The matrix formulation further enables continuous forcing and impulsive events to be incorporated within the same linear framework, thereby preserving analytical clarity and physical interpretability.

Introducing the state vector:

$$X[t] = \begin{bmatrix} u_p(t) \\ v_p\ (t) \end{bmatrix},$$

the particle–motion system can be written as:

$$\frac{dX[t]}{dt} = \mathrm{MX(t)} + F(t, y_p) + \sum_{k=1}^{N} I_k\delta(\mathrm{t} - t_k) \tag{34}$$

https://doi.org/10.3390/fluids11020051

where

$$M = \begin{bmatrix} -A & -B \\ B & -A \end{bmatrix}$$

$$F(t) = \begin{bmatrix} AU_g(y_p) - Bv'^{(t)} \\ BU_g(y_p) + Av'^{(t)} - g\left(1 - \frac{1}{\varepsilon_p}\right) \end{bmatrix}$$

$$I_k = \begin{bmatrix} I_{x,k} \\ I_{y,k} \end{bmatrix}$$

The off-diagonal terms ±B in M provide an explicit mathematical representation of lift-induced coupling: they map transverse relative motion into longitudinal acceleration and, conversely, longitudinal slip into transverse response. When $B = 0$, the two velocity components decouple and the system reduces to two independent viscous-relaxation equations; for $B \neq 0$, the coupled dynamics enable phase-shifted and resonant responses under harmonic excitation.

The continuous forcing term $F(t)$ encompasses the longitudinal carrier flow, the imposed transverse pulsations and the gravitational contribution, whereas the impulsive term $I_k\,\delta(t - t_k)$ represents short-duration events such as particle–wall interactions or micro-vortical impacts. Integrating the system across an impulsive instant $t = t_k$ yields the jump condition

$$X\left(t_k^+\right) - X\left(t_k^-\right) = I_k, \tag{35}$$

which provides a mathematically consistent description of instantaneous velocity changes without altering the linear operator governing the system.

The impulsive contribution preserves linearity because it is introduced as an additive forcing term rather than as a modification of M. Consequently, the intrinsic dynamical properties—stability, characteristic time scales and natural frequencies—are dictated by the spectrum of M, while the impulses affect only the particular solution. In many situations involving steady or purely periodic excitation, the impulsive part may be safely omitted without changing the qualitative conclusions regarding stability or phase behaviour. Conversely, when transient phenomena or short-lived interactions are of interest, the impulsive representation provides the necessary analytical flexibility while remaining within a strictly linear formulation.

The matrix M encapsulates the essential physical mechanisms: the parameter A corresponds to viscous relaxation toward the carrier-phase velocity, and the parameter B introduces transverse coupling associated with lift effects. Light particles with small relaxation time ($\tau_p \ll 1$) follow the carrier flow almost instantaneously, whereas heavy particles ($\tau_p \gg 1$) exhibit inertia-dominated dynamics with pronounced phase lag.

Overall, the matrix formulation offers a concise, analytically transparent and physically interpretable description of the particle dynamics and forms a natural basis for the subsequent examination of transient, forced and impulsive regimes in boundary-layer flows, which is developed in the next section.

3.3. Analytical Solutions, Phase–Space Representation and Asymptotic Behaviour

The analytical strategy adopted in this section is motivated by the need to clearly distinguish, within a unified and physically consistent framework, the transient relaxation, the forced oscillatory response, and the long-term asymptotic behaviour of a single solid particle moving in a boundary layer subjected to transverse velocity pulsations. Since the governing equations have already been formulated as a linear time-invariant matrix system, the analysis that follows does not introduce additional modelling assumptions

https://doi.org/10.3390/fluids11020051

but instead applies standard operator-based tools of linear system theory directly to the established physical model.

The coupled longitudinal and transverse particle dynamics are written compactly in state-space form as:

$$\frac{d}{dt}X[t] = \mathrm{MX(t)} + F(t), X[t] = \begin{bmatrix} u_p(t) \\ v_p\ (t) \end{bmatrix}$$

where the system matrix M encapsulates the internal dynamics associated with viscous relaxation and lift-induced coupling between velocity components, while the forcing vector $F(t)$ accounts for the mean carrier flow, the imposed harmonic transverse pulsations, and gravitational effects through the density ratio ε_p.

To separate transient effects associated with the initial conditions from the response induced by external forcing, the unilateral Laplace transform is applied to the full matrix system. The Laplace-transformed state vector $\tilde{X}(s)$ and forcing vector $\tilde{F}(s)$ are defined in the standard manner, while the time derivative transforms according to:

$$\frac{dX}{dt} \rightarrow s\tilde{X}(s) - X(0) \tag{36}$$

This yields the algebraic equation:

$$s\tilde{X}(s) - X(0) = M\tilde{X}(s) + \tilde{F}(s), \tag{37}$$

which can be rearranged to give the general solution in the Laplace domain:

$$\tilde{X}(s) = (sI - M)^{-1}X(0) + (sI - M)^{-1}\tilde{F}(s) \tag{38}$$

Here, s is the complex Laplace variable, III is the identity matrix, and $(sI - M)^{-1}$ denotes the resolvent operator of the linear dynamical system. This operator plays the role of a transfer operator that maps both the initial condition and the external forcing to the system response and therefore contains the complete dynamical information of the particle–fluid interaction.

The spectral properties of the resolvent operator are governed by the eigenvalues of the system matrix M, which satisfy:

$$det(M - \lambda I) = (\lambda + A)^2 + B^2 = 0 \quad \lambda_{1,2} = -A \pm iB \tag{39}$$

These eigenvalues correspond to exponentially decaying oscillatory modes with decay rate A and intrinsic angular frequency B. In the phase plane (u_p, v_p), solutions trace spiral trajectories approaching a stationary point, with narrow rapid spirals for light particles ($\tau_p \ll 1$) and elongated inertial trajectories for heavy particles ($\tau_p \gg 1$).

The forcing vector can be naturally decomposed into a stationary and an oscillatory component,

$$F(t) = F_{stat} + F_{osc}(t) \tag{40}$$

where F_{stat} contains the time-averaged contributions of the base flow and gravity, and $F_{osc}(t)$ arises exclusively from the imposed transverse velocity pulsations. In the long-time limit, the transient contribution associated with the initial condition vanishes for A > 0, while the oscillatory forcing produces a bounded periodic response. The asymptotic stationary state is therefore determined by the balance between the internal dynamics and the mean external forcing and is given by:

$$X_\infty = M^{-1}F_{stat} \tag{41}$$

https://doi.org/10.3390/fluids11020051

Physically, X_∞ represents the mean drift state of the particle in the boundary layer, around which periodic oscillations are superimposed when harmonic forcing is present.

Because the governing system is linear and time-invariant, a harmonic excitation of the form $v_g{}'(t) = v_0 cos(\omega t)$, after decay of transients, to a steady periodic particle response at the same frequency. The oscillatory component of the solution can therefore be expressed in complex-amplitude form as

$$X(t) = \Re\left\{\hat{X}(\omega)e^{i\omega t}\right\} \tag{42}$$

enables an explicit evaluation of amplitude and phase relative to the forcing frequency ω. Resonance emerges when $\omega \approx B$, demonstrating that the imposed oscillations can significantly modify the transverse motion of the particle under certain parameter conditions.

The complete time-domain solution can be written as a superposition of homogeneous and forced contributions:

$$X(t) = e^{Mt}X(0) + \int_0^t e^{M(t-\tau)}F(\tau)d(\tau), \tag{43}$$

where the first term governs the transient relaxation and the integral term defines the forced response. This representation makes explicit how the observed particle motion arises from the interaction between the intrinsic modal structure of the system and the applied external forcing.

The logical structure of the operator-based analysis and the connection between the Laplace-domain formulation, spectral properties, time-domain response, and asymptotic behaviour are summarized conceptually in Figure 3.

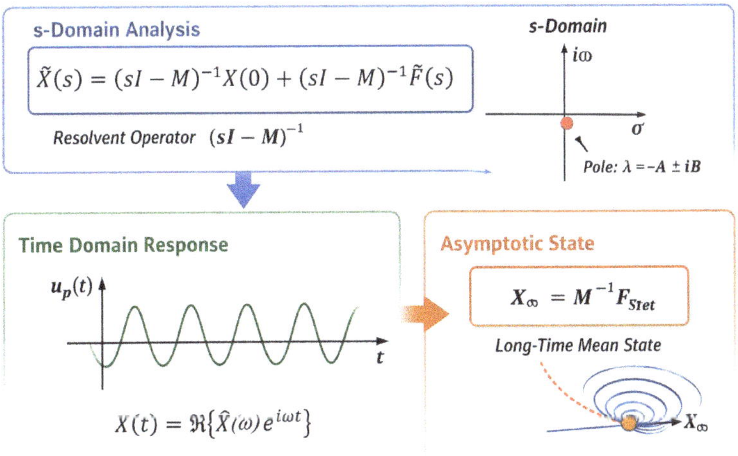

Figure 3. Schematic illustration of the operator-based description of particle motion, highlighting the mapping between initial conditions, external forcing, and particle response via the resolvent operator $(sI - M)^{-1}$.

The figure serves as an explanatory aid, highlighting the central role of the resolvent operator $(sI - M)^{-1}$ in governing transient decay, resonant response, and long-term particle dynamics, without representing numerical simulations or previously published visualizations.

 https://doi.org/10.3390/fluids11020051

3.4. Dynamic Formulation, Modal Spectrum, and Lyapunov Stability

The dynamic formulation of the particle-motion system defines a time-dissipative linear model in which the temporal evolution of the state vector $X(t)$ is governed by the spectral properties of the operator M and the external forcing $F(t)$. This representation allows a comprehensive description of transient and steady regimes and enables the assessment of dynamic stability resulting from the balance between relaxation, inertia, and imposed oscillatory effects. The system therefore constitutes a genuine dynamic model rather than a static configuration, with a full temporal evolution and a well-defined internal modal structure.

The spectral representation of the system matrix is obtained through:

$$M = V\Lambda V^{-1}, \quad \Lambda = diag(\lambda_1, \lambda_2), \quad \lambda_{1,2} = -A \pm iB \tag{44}$$

where

- $V = [v_1, v_2]$ is the matrix of right eigenvectors of M.

The system thus exhibits damped oscillations whose decay rate is controlled by A, while the modal frequency is given by B. For a periodic excitation, e.g., $v_g'(t) = v_0 cos(\omega t)$, a complex representation yields:

$$(i\omega I - M)\tilde{X} = \tilde{F}, \quad H(\omega) = (i\omega I - M)^{-1} \tag{45}$$

where

- $H(\omega)$ denotes the dynamic transfer operator determining amplitude, phase, and resonant amplification ($\omega \simeq B$).

Dynamic stability can be assessed through a quadratic Lyapunov function:

$$V(x) = \frac{1}{2}\|X\|^2, \quad \frac{dV}{dt} = X^T M X < 0 \quad (A > 0) \tag{46}$$

which guarantees exponential decay of the homogeneous component and the existence of an attracting stationary state $X\infty = M^{-1}F_{stat}$. Accordingly, the system belongs to the class of exponentially stable linear dissipative dynamic systems, where the response amplitude remains bounded and fully controllable through the parameters A, B, τ_p and the forcing frequency ω. The spectral properties thus provide a rigorous framework for analysing resonance, forced response, and long-term behaviour under realistic boundary-layer conditions.

4. Numerical Analysis and Physical Interpretation

4.1. Baseline Analytical and Numerical Framework

The analytical formulation developed in Section 3 provides a rigorous mathematical basis for analysing the coupled transverse and longitudinal dynamics of finite-size solid particles in a horizontal laminar boundary layer subjected to harmonic cross-flow velocity pulsations. By expressing the governing equations in compact matrix form and applying the Laplace transform, the particle motion is decomposed into transient, oscillatory, and asymptotic components, allowing a clear physical interpretation of the mechanisms governing particle transport under continuous forcing.

In order to illustrate the analytical solutions in a controlled and transparent manner, a baseline numerical configuration is introduced. Water at standard laboratory conditions is adopted as the carrier fluid, and the physical properties of the fluid together with the characteristic parameters of the boundary layer are summarised in Table 1. These

 https://doi.org/10.3390/fluids11020051

parameters define the shear scale and flow conditions responsible for the generation of lift and viscous relaxation effects acting on the particle.

Table 1. Fluid and boundary-layer parameters.

Parameter	Symbol	Value	Unit
Fluid type	-	Water (T ≈ 20 °C)	-
Kinematic viscosity	ν	1.0×10^{-6}	m^2/s
Fluid density	ρ_f	998	kg/m^3
Shear rate	γ	80	s^{-1}
Outer-stream velocity	U_∞	0.392	m/s
Boundary-layer thickness	δ	0.064	m
Wall shear rate	γ_ω	6.142	s^{-1}
Pulsation frequency	ω	1.93	rad/s

The properties of the dispersed phase are given in Table 2, where two representative cases are considered: a light particle characterised by a short relaxation time and a heavy particle exhibiting inertia-dominated behaviour.

Table 2. Particle parameters.

Parameter	Symbol	Light Particle	Heavy Particle	Unit
Density	ρ_p	1050	3000	kg/m^3
Relaxation time	τ_p	0.012	0.18	s
Density ratio	ε_p	1.05	3.00	-
Particle diameter	d_p	0.00045	0.00104	m
Initial position	(x_0, y_0)	(0, 0)	(0, 0)	m
Initial velocity	(u_p, v_p)	$(U_g(y_0), 0)$	$(U_g(y_0), 0)$	m/s

Note: The particle relaxation time τ_p is given in dimensional form. The relative classification into light and heavy particles refers to their dimensionless relaxation time under the present flow conditions.

Here, the terms "light" and "heavy" are used in a relative sense to distinguish particles with weaker and stronger inertial response under the same flow conditions. Although the relaxation time τ_p is reported in dimensional form in Table 2, the corresponding dimensionless relaxation time governs the effective inertia classification discussed throughout this section.

The particles differ only in density and relaxation time, while identical initial conditions are prescribed for both cases. This ensures that any observed differences in particle motion arise solely from inertial effects rather than from differences in the initial kinematic state.

The numerical integration of the governing equations is performed using the parameters listed in Table 3, including the time step, total simulation duration, and numerical integration scheme. These values are selected to accurately resolve both the transient decay and the established oscillatory regime induced by the harmonic excitation.

https://doi.org/10.3390/fluids11020051

Table 3. Numerical configuration.

Parameter	Value	Unit
Time step Δt	1.0×10^{-3}	s
Total duration	1.6	s
Method	RK4	-
Output sampling	0.02	s

At each time step, the longitudinal carrier-phase velocity is evaluated at the instantaneous particle position as $U_g(y_p(t))$, ensuring consistent coupling between transverse migration and longitudinal acceleration.

Within this first numerical stage, the impulsive forcing term retained in the general analytical formulation is set to zero, such that the results correspond to continuous harmonic forcing. The specific numerical datasets used to construct the graphical results presented below are provided as Supplementary Tables S1–S4, ensuring full reproducibility of the reported simulations.

To assess the validity of the proposed analytical–numerical framework, the predicted particle dynamics are compared with well-established results reported in the literature for inertial particle motion in shear and boundary-layer flows. Classical formulations for small rigid spheres in nonuniform flows predict inertia-dependent viscous relaxation, phase-lagged response under unsteady forcing, and preferential migration and segregation in wall-bounded configurations. The present results reproduce these canonical trends: light particles exhibit rapid relaxation and strong coupling to transverse velocity pulsations, whereas heavy particles display inertia-dominated dynamics with attenuated oscillatory response and preferential residence in near-wall low-velocity regions. The resonance-like amplification of transverse oscillations and the bounded phase-space trajectories with convergence toward stable attractors are also consistent with published analytical and numerical studies. Overall, this comparison confirms that the proposed model captures the dominant physical mechanisms governing inertial particle motion in oscillatory boundary-layer environments.

4.2. Transverse Particle Dynamics

The transverse motion constitutes the primary mechanism responsible for segregation between light and heavy particles in a horizontal boundary layer subjected to cross-flow velocity pulsations. The analytical expression for the transverse particle velocity $v_p(t)$, derived in Section 3, is evaluated numerically using the baseline parameter set defined in Tables 1–3. The resulting time histories of the transverse velocity are shown in Figure 4 (Supplementary Table S1).

 https://doi.org/10.3390/fluids11020051

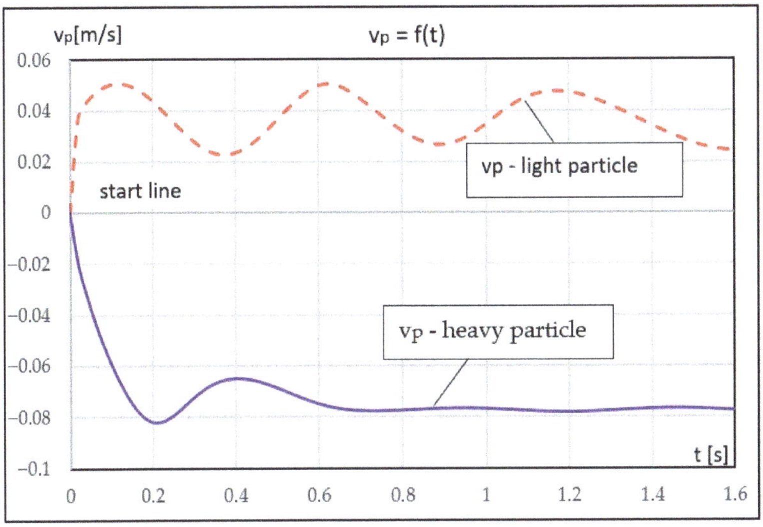

Figure 4. Transverse particle velocity $v_p(t)$ for light and heavy particles.

Two qualitatively distinct dynamical regimes are clearly observed. For the light particle, the transient response decays rapidly and the velocity oscillates about a positive mean value, indicating a sustained upward drift toward regions of higher carrier-fluid velocity. In contrast, the heavy particle exhibits an inertia-dominated response characterised by a negative mean transverse velocity, corresponding to gradual migration toward slower near-wall sublayers.

In both cases, the oscillatory behaviour reflects the imposed harmonic excitation, while the decay rate is governed by the viscous relaxation coefficient and the particle relaxation time.

The results presented in the figures correspond to physically calibrated responses; the applied rescaling accounts for finite-size effects and uncertainties in local shear estimation and does not introduce additional physical mechanisms.

The transverse displacement $y(t)$, obtained by time integration of $v_p(t)$, is presented in Figure 5 (Supplementary Table S2). The light particle undergoes continuous upward migration, whereas the heavy particle experiences a pronounced downward drift dominated by the effective gravitational contribution. Although oscillatory modulations are superimposed on the trajectories, the long-term trend is independent of the initial phase and is instead controlled by particle inertia, relaxation time, and the relationship between the system's natural frequency and the excitation frequency.

https://doi.org/10.3390/fluids11020051

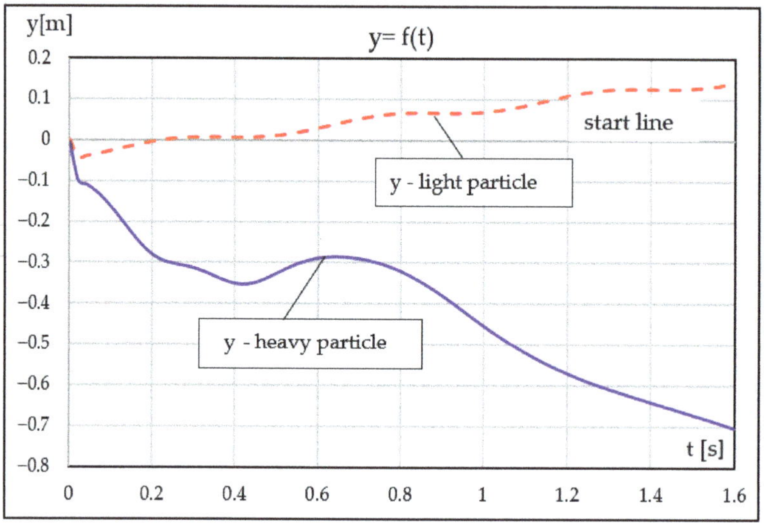

Figure 5. Transverse particle displacement y(t) for light and heavy particles.

These results demonstrate that even in the absence of turbulence or spatial shear gradients, harmonic transverse velocity pulsations alone are sufficient to induce vertical particle segregation in laminar boundary-layer flows.

4.3. Longitudinal Particle Dynamics

The longitudinal motion of the particles arises as a direct consequence of their transverse migration, since different regions of the boundary layer are associated with different values of the carrier-phase velocity. The analytical solution for the longitudinal particle velocity $u_p(t)$, derived in Section 3, is evaluated numerically using the same baseline configuration. The corresponding time histories are shown in Figure 6 (Supplementary Table S3).

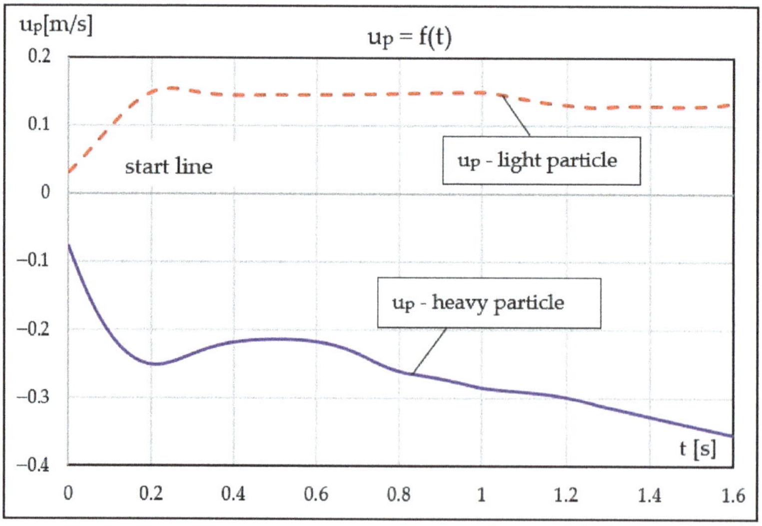

Figure 6. Longitudinal particle velocity $u_p(t)$ for light and heavy particles.

https://doi.org/10.3390/fluids11020051

For the light particle, the longitudinal velocity gradually increases relative to its initial value, reflecting its migration into faster regions of the boundary layer. Conversely, the heavy particle exhibits a progressive decrease in longitudinal velocity as a result of its downward drift toward slower sublayers.

Periodic oscillations are observed during the initial stage for both particles due to the transverse harmonic excitation; however, their amplitude and decay rate depend strongly on the particle relaxation properties.

The longitudinal displacement $x(t)$, obtained by integrating $u_p(t)$, is shown in Figure 7 (Supplementary Table S4). The light particle advances rapidly and moves significantly ahead of the heavy particle, while the latter exhibits a clear lag associated with residence in regions of lower transport velocity. This divergence intensifies with time, despite identical initial longitudinal velocities for both particles.

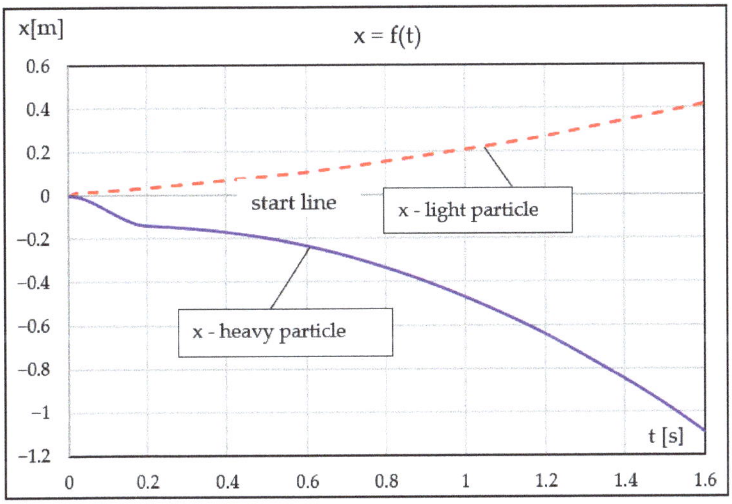

Figure 7. Longitudinal particle displacement x(t) for light and heavy particles.

From a physical standpoint, the longitudinal dynamics cannot be interpreted independently of the transverse behaviour. The transverse migration determines access to faster or slower regions of the boundary layer and thus governs the longitudinal transport of the particles. This sections therefore complete the baseline numerical illustration of the analytical model under continuous harmonic forcing, providing a reference framework for the extended numerical examples considered in subsequent stages.

4.4. Impulsive Particle Dynamics

The impulsive extension of the particle–motion model introduced in Section 3.1 is illustrated numerically in order to quantify the response of light and heavy particles to short-duration perturbations. In the present simulations, the impulsive component is incorporated exactly as defined in Section 3.1, while all remaining physical and numerical parameters coincide with the baseline configuration used in the previous sections. The impulsive scenario represents intermittent disturbances commonly encountered in boundary-layer flows, such as micro-vortical structures, particle–wall interactions, or short turbulent bursts, which cannot be captured by smooth harmonic forcing alone. A finite sequence of impulsive velocity jumps is prescribed at selected instants, with identical impulse amplitudes applied to both light and heavy particles; the impulse times and the corresponding longitudinal and transverse velocity jumps are reported in Supplementary Table S5.

 https://doi.org/10.3390/fluids11020051

Figure 8 illustrates the phase–space response of light and heavy particles subjected to the impulsive forcing scenario. The left part of the figure highlights the short-time behaviour dominated by discrete velocity jumps induced by impulsive events, whereas the right part shows the long-time evolution revealing the spiral attractor of the continuous matrix dynamics. Immediately after each impulse, the particle velocity exhibits a discontinuous change, followed by a smooth relaxation governed by viscous damping and lift-induced coupling.

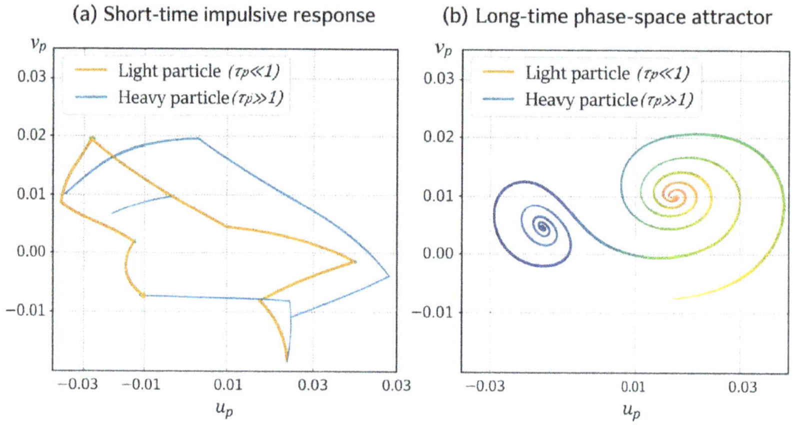

Figure 8. Impulsive phase–space dynamics of light and heavy particles.

A pronounced difference between particle types is observed. Light particles rapidly relax toward a compact spiral attractor due to strong viscous relaxation, while heavy particles display wider phase-space excursions and a significantly slower return toward the attractor, reflecting the persistence of inertial effects. Despite these transient differences, the global attractor structure remains unchanged, indicating that impulsive forcing primarily produces temporary displacements rather than altering the intrinsic stability properties of the system.

These numerical results confirm the analytical predictions of Section 3.1 and demonstrate that the impulsive differential extension provides a physically meaningful and analytically consistent description of particle dynamics in boundary layers subjected to intermittent disturbances.

4.5. Matrix Dynamics and Transient Relaxation

In this section, the matrix formulation of the particle–motion system introduced in Section 3.2 is examined through numerical simulations in order to characterise the intrinsic continuous dynamics of the coupled longitudinal and transverse particle velocities. The analysis focuses on the transient relaxation behaviour in the absence of impulsive disturbances, so that the observed particle motion is governed solely by the linear time-invariant system matrix and its spectral properties.

The governing dynamics are defined by a two-dimensional matrix system whose eigenvalues are given by:

$$\lambda_{1,2} = -A \pm iB$$

This eigenstructure corresponds to exponentially damped oscillatory motion and guarantees the existence of a stable spiral attractor in the phase space (u_p, v_p) for $A > 0$. Numerical integration of the matrix system is performed using identical initial conditions

 https://doi.org/10.3390/fluids11020051

for light and heavy particles, while the model and numerical parameters employed in the simulations are summarised in Supplementary Table S7.

Figure 9 presents the phase–space trajectories obtained from the matrix formulation under continuous harmonic excitation. The left part of the figure illustrates the short-time response immediately following the initial condition, where transient relaxation effects dominate, whereas the right part shows the long-time evolution, revealing the full structure of the spiral attractor associated with the matrix dynamics. In both cases, the trajectories remain bounded and converge asymptotically toward the stationary solution of the system.

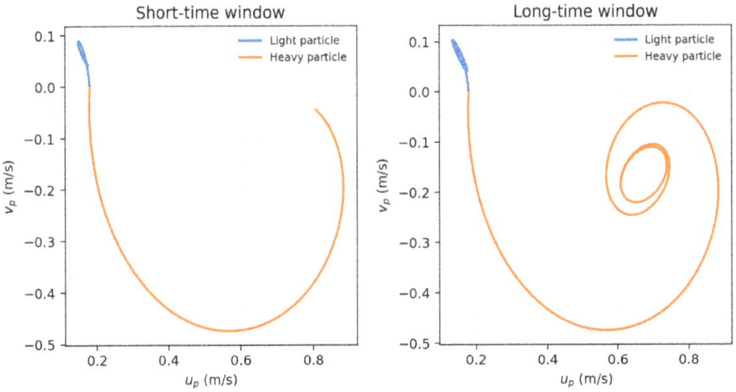

Figure 9. Phase–space trajectories obtained from the matrix formulation under harmonic forcing.

A clear distinction between the dynamical behaviour of light and heavy particles is observed. Light particles exhibit rapid contraction of their phase–space trajectories and quickly approach a compact spiral attractor, reflecting strong viscous relaxation and weak inertial memory. In contrast, heavy particles display wider phase–space excursions and a markedly slower convergence toward the attractor, indicating inertia-dominated dynamics and extended transient behaviour.

Despite these differences in relaxation time scales and trajectory geometry, the qualitative structure of the phase–space attractor remains the same for both particle types. This demonstrates that the stability of the motion and the existence of a spiral attractor are intrinsic properties of the matrix formulation itself and are not induced by impulsive or stochastic perturbations. The numerical results therefore provide direct confirmation of the analytical predictions derived in Section 3.2 and support the Lyapunov stability analysis discussed in Section 3.4.

The time-resolved particle velocity data used to construct the phase–space trajectories shown in Figure 9 are provided in Supplementary Table S7, ensuring full reproducibility of the numerical results.

4.6. Resonance Analysis Under Harmonic Forcing

The effect of harmonic transverse velocity pulsations on particle dynamics is examined by analysing the stationary forced response of the linear system derived in Sections 3.3 and 3.4. In this subsection, attention is restricted to the long-time oscillatory regime, assuming that transient contributions associated with the homogeneous solution have fully decayed due to viscous dissipation.

As established in Section 3.3, under harmonic excitation the state vector of the particle motion admits a representation of the form (Equation (42), Section 3.3).

$$X(t) = \Re\left\{\hat{X}(\omega)e^{i\omega t}\right\},$$

https://doi.org/10.3390/fluids11020051

where

- $X(t) = [u_p(t), v_p(t)]^t$ and $\hat{X}(\omega)$ denotes the complex amplitude vector associated with the forcing frequency ω.

In the present analysis, attention is restricted to the transverse oscillatory response. The scalar quantity $H(\omega)$ denotes the amplitude of the transverse component of the steady harmonic particle response. $H(\omega)$ = *amplitude of* $v_p(t)$ in the steady periodic regime.

Consequently, the frequency dependence of $H(\omega)$ is entirely governed by the system coefficients and by the spectral properties of the system matrix discussed in Section 3.4, and no additional modelling assumptions are introduced at this stage.

The resonance behaviour is controlled by the intrinsic oscillatory dynamics associated with the complex conjugate eigenvalues of the system matrix $-\lambda_{1,2} = -\mathrm{A} \pm \mathrm{iB}$ which define a characteristic frequency scale of the order:

$$\omega \approx \sqrt{A^2+B^2} \tag{47}$$

When the forcing frequency approaches this intrinsic scale, the magnitude $H(\omega)$ increases significantly, indicating resonance-like amplification of transverse particle oscillations. At low forcing frequencies, the particle motion remains nearly in phase with the imposed oscillations, while near resonance a pronounced phase delay develops, reflecting the inertia-controlled response of the particle.

The influence of particle inertia is encapsulated through the relaxation time τ_p, which enters the coefficients of the governing equations and controls the frequency dependence of the steady oscillatory response.

Particles with weak inertial effects exhibit pronounced amplification over a broad frequency range, reflecting their ability to closely follow the imposed transverse fluctuations of the carrier flow. In contrast, particles with strong inertial resistance display a markedly attenuated response across the entire frequency spectrum, as inertial damping suppresses their capacity to respond to rapid transverse forcing.

From a physical perspective, harmonic transverse pulsations therefore act as a selective segregation mechanism within the boundary layer. Light particles experience enhanced transverse excursions when the forcing frequency approaches the intrinsic system frequency, promoting migration toward regions of higher carrier-phase velocity. Heavy particles, on the other hand, remain largely confined to near-wall regions due to inertial damping and reduced coupling to the oscillatory forcing. This behaviour provides a direct numerical manifestation of the modal properties and Lyapunov stability characteristics established in Section 3.4.

The frequency-dependent steady-state amplitudes $H(\omega)$ used to construct the resonance curves shown in Figure 10 are obtained numerically from time-domain simulations and reported in Supplementary Table S8.

https://doi.org/10.3390/fluids11020051

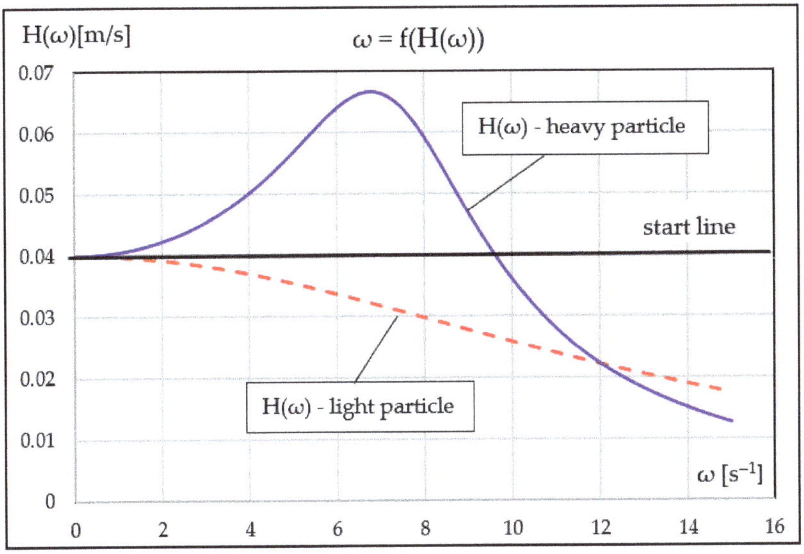

Figure 10. Resonance amplitude under harmonic forcing.

4.7. Multi-Harmonic Transverse Forcing

In practical boundary-layer flows, transverse velocity fluctuations are rarely characterised by a single harmonic component. Instead, the excitation typically consists of a superposition of oscillatory contributions with different amplitudes, frequencies and phases. To examine the particle response under such conditions, the transverse carrier-phase pulsation is generalised to a multi-harmonic form

$$v_g'(t) = \sum_{n=1}^{N} v_n \cos(\omega_n t + \phi_n) \quad (48)$$

where the index n labels the individual harmonic components included in the excitation.

Each harmonic contribution is fully characterised by its amplitude v_n, angular frequency ω_n, and phase ϕ_n. The amplitude v_n is defined as:

$$v_n = max\left|v_{g,n}'(t)\right| \quad (49)$$

where

$$v_{g,n}'(t) = v_n \cos(\omega_n t + \phi_n) \quad (50)$$

denotes the n-th harmonic component of the transverse carrier-phase velocity. The angular frequency is given by:

$$\omega_n = 2\pi f_n \quad (51)$$

with f_n denoting the frequency of the n-th harmonic component, while ω_n has the dimensionality rad s^{-1}. The phase shift is expressed as:

$$\phi_n = \omega_n t_{n,0} \quad (52)$$

where $t_{n,0}$ represents the phase offset of the n-th harmonic component relative to the reference time $t = 0$.

Owing to the linearity of the governing matrix system established in Section 3.2, the particle response to multi-harmonic forcing can be interpreted as a superposition of the

https://doi.org/10.3390/fluids11020051

responses to the individual harmonic components. After the decay of transient effects, the transverse particle velocity exhibits a steady oscillatory motion composed of multiple periodic contributions, each associated with one of the imposed forcing frequencies. The resulting time signal is therefore characterised by amplitude modulation, envelope formation and beating phenomena, which are typical of weakly unsteady or transitional boundary-layer environments.

This behaviour is illustrated in the phase–space representation shown in Figure 11, where the superposition of multiple forcing frequencies leads to deformed trajectories in the (u_p, v_p) plane. Numerical simulations based on the representative multi-harmonic parameter sets listed in Supplementary Table S9 demonstrate a pronounced dependence of the particle response on inertia. Light particles effectively follow each harmonic component and exhibit strong amplitude modulation and beating, whereas heavy particles display a markedly attenuated response, with higher-frequency components being strongly damped by inertial effects.

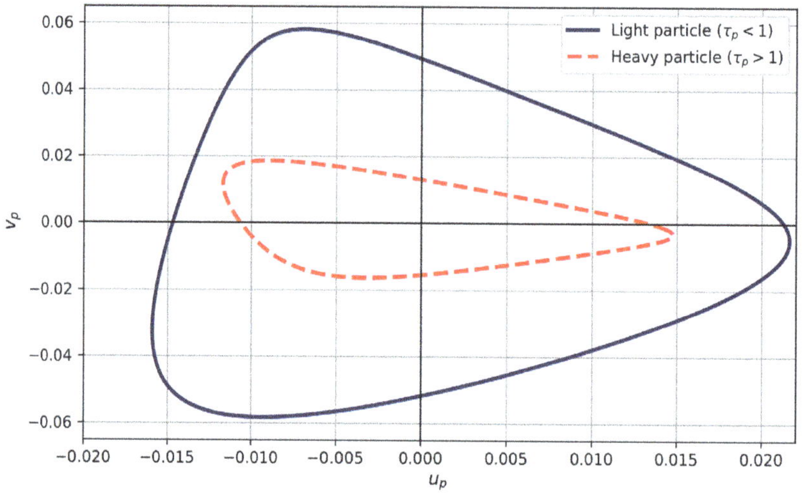

Figure 11. Phase–space trajectories (u_p, v_p) of light and heavy particles under multi-harmonic transverse forcing.

From a physical perspective, multi-harmonic transverse forcing enhances the selectivity of particle segregation within the boundary layer. The simultaneous presence of several excitation frequencies increases the likelihood of partial resonance with the intrinsic oscillatory scale of the system, particularly for light particles. Heavy particles remain comparatively insensitive to such excitation, reinforcing the separation between inertial regimes already observed under purely harmonic forcing. Consequently, multi-harmonic excitation provides a more realistic and physically representative framework for analysing particle redistribution mechanisms in boundary-layer flows.

4.8. Asymptotic Long-Term Behaviour

The analytical framework developed in Sections 3.3 and 3.4 predicts that, at sufficiently long time scales, the particle dynamics approach an asymptotic regime governed by the spectral properties of the linear system matrix and the persistent external forcing. In this regime, transient contributions associated with the homogeneous solution decay exponentially, and the particle motion is dominated by the forced response induced by transverse velocity pulsations.

 https://doi.org/10.3390/fluids11020051

For the transverse motion, the exponential decay of the natural modes associated with the eigenvalues of the system matrix (see Section 3.4) ensures that all initial-condition effects vanish as $t \rightarrow \infty$. The remaining motion is therefore fully determined by the forcing characteristics and the particle relaxation properties. Under harmonic or multi-harmonic excitation, the transverse velocity converges to a bounded oscillatory state, whose amplitude and phase are controlled by particle inertia and by the distribution of forcing frequencies.

The longitudinal dynamics exhibit a qualitatively different asymptotic behaviour. While oscillatory components persist due to coupling with the transverse motion, the longitudinal displacement displays a dominant drift component arising from sustained particle migration across the boundary layer. As a result, even after the transverse oscillations reach a steady periodic regime, the longitudinal position continues to evolve in time, leading to cumulative separation between particles with different inertial properties.

A clear contrast emerges between light and heavy particles. Light particles maintain relatively large transverse oscillation amplitudes and oscillate around an upward-shifted mean position, reflecting their strong coupling to the imposed transverse pulsations and their ability to closely follow rapid fluctuations of the carrier flow. In contrast, heavy particles exhibit a markedly attenuated transverse response due to inertial damping and converge toward a downward-shifted mean position governed primarily by the effective gravitational contribution.

This asymptotic behaviour is illustrated in Figure 12 (Supplementary Table S10), where phase–space trajectories extracted from a restricted steady-state window reveal non-closed, asymmetric geometries for both particle types. Although the detailed trajectory shapes differ, both cases converge toward stable attractor structures determined by the dissipative nature of the system. The persistence of bounded motion confirms the Lyapunov stability of the particle dynamics established in Section 3.4, while the separation of mean positions highlights the long-term segregation mechanism induced by particle inertia.

From a physical perspective, the asymptotic regime represents the ultimate outcome of the combined action of viscous relaxation, lift-induced coupling, gravity, and external transverse forcing. Once transient effects have decayed, the system retains a memory of particle inertia through distinct mean positions and oscillation amplitudes, leading to sustained spatial segregation even in the absence of turbulence. This behaviour underscores the importance of long-time analysis when assessing particle transport and redistribution in boundary-layer flows subjected to oscillatory forcing.

Overall, the asymptotic analysis demonstrates that the long-term particle dynamics are fully governed by the intrinsic spectral properties of the coupled system and the persistent external forcing, independently of the initial conditions.

https://doi.org/10.3390/fluids11020051

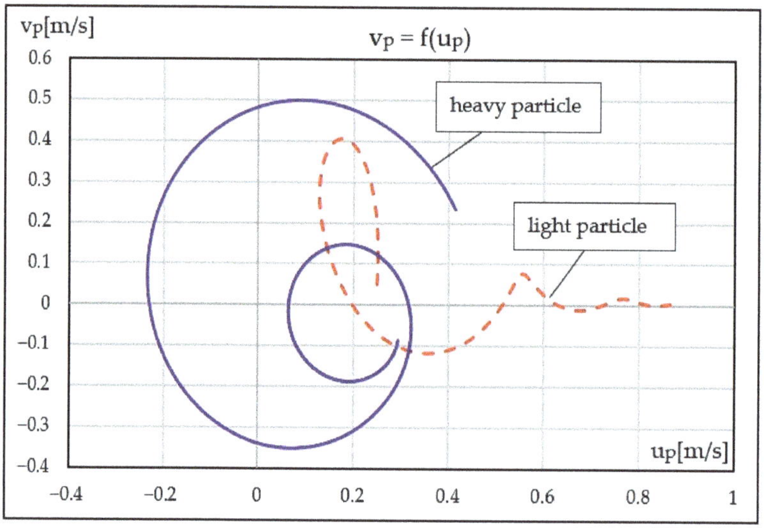

Figure 12. Asymptotic phase–space trajectories of particles with different inertial response.

5. Discussion

The present study develops an analytical framework for describing the coupled transverse and longitudinal dynamics of inertial particles in a horizontal boundary-layer flow subjected to transverse velocity pulsations. By formulating the governing equations in a linear matrix form, the analysis establishes a direct connection between transient relaxation, resonance phenomena, and asymptotic long-term behaviour through the spectral properties of the system.

Unlike many previous investigations that rely predominantly on numerical simulations or turbulence-based transport models, the present results demonstrate that deterministic oscillatory forcing alone can induce sustained particle migration and segregation even in laminar or weakly perturbed flows. The analytical solutions reveal how viscous relaxation, lift-induced coupling, and gravity interact to shape particle trajectories, without invoking stochastic turbulence effects.

The resonance behaviour identified in the transverse particle response provides a clear physical interpretation of inertia-dependent sensitivity to oscillatory forcing. Particles with weak inertial resistance exhibit enhanced oscillatory amplitudes over a broad frequency range, whereas inertia-dominated particles remain strongly damped. This distinction is consistent with qualitative observations reported in earlier studies, while the present formulation offers a rigorous mathematical explanation based on modal dynamics and Lyapunov stability.

The asymptotic analysis further shows that, after the decay of transient contributions, particle motion converges toward stable attractor structures whose geometry and mean position depend on particle inertia. Although the transverse oscillations remain bounded, the associated longitudinal drift leads to cumulative separation between particles with different inertial properties. This mechanism highlights a purely deterministic pathway for long-term particle segregation in boundary-layer flows.

From a broader perspective, the results indicate that oscillatory forcing mechanisms—arising from flow pulsations, vibrations, or external actuation—may significantly influence particle transport and redistribution even in the absence of turbulence. The compact analytical nature of the proposed framework makes it particularly suitable for paramet-

 https://doi.org/10.3390/fluids11020051

ric studies and preliminary system design, complementing numerical simulations and experimental investigations.

From an applied perspective, the present analytical framework is relevant for practical problems involving particle transport in wall-bounded flows subjected to oscillatory or vibrational forcing. The compact matrix-based formulation and closed-form solutions provide a predictive tool for assessing inertia-dependent particle migration and segregation in configurations encountered in industrial ducts, ventilation systems, heat exchangers, and vibration-assisted separation or mixing devices.

The explicit identification of resonance conditions, phase shifts, and long-term drift trends makes the model particularly suitable for preliminary design studies and parametric analyses, where rapid evaluation of particle behaviour is required without reliance on computationally intensive numerical simulations. In environmental and process engineering contexts, the present results may assist in predicting preferential deposition, resuspension, and inertia-driven segregation in weakly unsteady boundary-layer flows.

6. Conclusions

A linear matrix-based analytical framework is developed to describe inertial particle motion in a horizontal boundary-layer flow subjected to transverse velocity pulsations. The resulting closed-form solutions capture transient relaxation, resonance phenomena, and asymptotic long-term behaviour in a unified manner.

The analysis demonstrates that oscillatory transverse forcing can induce pronounced inertia-dependent particle dynamics even in laminar flows. Particles with weak inertial resistance exhibit strong coupling to the imposed pulsations, leading to amplified transverse oscillations and upward-shifted mean positions, whereas inertia-dominated particles experience strong damping and converge toward downward-shifted states governed primarily by gravity.

Resonance phenomena emerge naturally from the spectral properties of the governing system and are shown to enhance transverse particle motion over specific frequency ranges. At long time scales, the exponential decay of natural modes ensures convergence toward stable attractor structures, confirming the Lyapunov stability of the particle dynamics while enabling sustained longitudinal separation between particles with different inertial properties.

The results provide a clear physical and mathematical interpretation of particle segregation mechanisms driven by oscillatory forcing in boundary layers. The analytical framework developed herein offers a compact and predictive basis for the analysis of inertial particle transport in oscillatory flow environments.

Supplementary Materials: The following supporting information can be downloaded at: https://www.mdpi.com/article/10.3390/fluids11020051/s1. Table S1. Time-resolved transverse particle velocity data for light and heavy particles; Table S2. Time-resolved transverse particle displacement data for light and heavy particles; Table S3. Time-resolved longitudinal particle velocity data for light and heavy particles; Table S4. Time-resolved longitudinal particle displacement data for light and heavy particles; Table S5. Impulse times and prescribed velocity jumps used in the impulsive particle–motion simulations; Table S6. Time-resolved particle velocity data under impulsive forcing; Table S7. Time-resolved particle velocity data for phase–space reconstruction; Table S8. Frequency-resolved steady-state resonance response data; Table S9. Multiharmonic forcing parameters and time-resolved particle response data; Table S10. Asymptotic particle trajectories in phase space.

Author Contributions: Conceptualization, R.Y. and V.D. (Ventsislav Dimitrov); methodology, V.D. (Ventsislav Dimitrov) and G.T. (Georgi Tonkov); software, V.D. (Ventsislav Dimitrov); validation, V.D. (Ventsislav Dimitrov), G.T. (Georgi Tonkov) and S.B.; formal analysis, V.D. (Ventsislav Dimitrov); investigation, V.D. (Ventsislav Dimitrov), G.T. (Georgi Tonkov) and K.R.; resources, V.D.

(Veselina Dimitrova) and K.R.; data curation, V.D. (Ventsislav Dimitrov) and G.T. (Gergana Tonkova); writing—original draft preparation, V.D. (Ventsislav Dimitrov); writing—review and editing, R.Y., V.D. (Veselina Dimitrova), S.B., G.T. (Georgi Tonkov), G.T. (Gergana Tonkova) and K.R.; visualization, V.D. (Ventsislav Dimitrov); supervision, R.Y.; project administration, R.Y.; funding acquisition, R.Y. All authors have read and agreed to the published version of the manuscript.

Funding: This research received no external funding.

Institutional Review Board Statement: Not applicable.

Informed Consent Statement: Not applicable.

Data Availability Statement: All data generated or analyzed during this study are included in this published article and its Supplementary Materials.

Acknowledgments: This work has been accomplished with financial support by the European Regional Development Fund within the Operational Programme "Bulgarian national recovery and resilience plan", procedure for direct provision of grants "Establishing of a network of research higher education institutions in Bulgaria", and under Project BG-RRP-2.004-0005 "Improving the research capacity and quality to achieve intErnAtional recognition and reSilience of TU-Sofia (IDEAS)".

Conflicts of Interest: The authors declare no conflicts of interest. The funders had no role in the design of the study; in the collection, analyses, or interpretation of data; in the writing of the manuscript; or in the decision to publish the results.

Abbreviations

The following abbreviations are used in this manuscript:

BL	Boundary layer
LTI	Linear time-invariant
ODE	Ordinary differential equation
RK4	Fourth-order Runge–Kutta method
DNS	Direct numerical simulation
LES	Large-eddy simulation
PSD	Power spectral density
RMS	Root mean square
St	Stokes number
TLA	Three-letter acronym
Symbols frequently used as parameters	
u_p	Longitudinal particle velocity
v_p	Transverse particle velocity
U_g	Longitudinal carrier-phase velocity
v_g'	Transverse velocity pulsation of carrier flow
τ_p	Particle relaxation time
A	Viscous relaxation coefficient
B	Lift-induced coupling coefficient
ω	Angular forcing frequency
$\lambda_{1,2}$	Eigenvalues of the system matrix
ε_p	Particle-to-fluid density ratio

References

1. Loitsiansky, L.G. *Mechanics of Fluids and Gas*; Nauka: Moscow, Russia, 1973.
2. Landau, L.D.; Lifshitz, E.M. *Hydrodynamics*; Nauka: Moscow, Russia, 1988.
3. Tikhonov, A.N.; Samarskii, A.A. *Equations of Mathematical Physics*; Nauka: Moscow, Russia, 1972.
4. Crowe, C.; Sommerfeld, M.; Tsuji, Y. *Multiphase Flows with Droplets and Particles*; CRC Press: Boston, MA, USA, 1998.
5. Maxey, M.R.; Riley, J.J. Equation of Motion for a Small Rigid Sphere in a Nonuniform Flow. *Phys. Fluids* **1983**, *26*, 883–889. [CrossRef]

https://doi.org/10.3390/fluids11020051

6. Guha, A. Transport and Deposition of Particles in Turbulent and Laminar Flow. *Annu. Rev. Fluid Mech.* **2008**, *40*, 311–341. [CrossRef]
7. Toschi, F.; Bodenschatz, E. Lagrangian Properties of Particles in Turbulence. *Annu. Rev. Fluid Mech.* **2009**, *41*, 375–404. [CrossRef]
8. Jolley, E.; Palmer, R.A.; Smith, F.T. Particle Movement in a Boundary Layer. *J. Eng. Math.* **2021**, *128*, 6. [CrossRef]
9. Eaton, J.K.; Fessler, J.R. Preferential Concentration of Particles by Turbulence. *Int. J. Multiph. Flow* **1994**, *20*, 169–209. [CrossRef]
10. Scheven, U.M.; Khirevich, S.; Daneyko, A.; Tallarek, U. Longitudinal and transverse dispersion in flow through random packings of spheres: A quantitative comparison of experiments, simulations, and models. *Phys. Rev. E* **2014**, *89*, 053023. [CrossRef]
11. Martin, J.E.; Meiburg, E. The accumulation and dispersion of heavy particles in forced two-dimensional mixing layers. I. The fundamental and subharmonic cases. *Phys. Fluids* **1994**, *6*, 1116–1132. [CrossRef]
12. Chatoutsidou, S.E.; Lazaridis, M. Evaluation of Particle Resuspension and Single-Layer Rates with Exposure Time and Friction Velocity for Multilayer Deposits in a Turbulent Boundary Layer. *Aerosol Air Qual. Res.* **2018**, *18*, 2108–2120. [CrossRef]
13. Kuerten, J.G.M.; Portela, L.M.; Sardina, G.; Soldati, A. Point-Particle DNS and LES of Particle-Laden Turbulent Flows: A Review. *Flow Turbul. Combust.* **2016**, *97*, 689–713. [CrossRef]
14. Goldshtein, A.; Shuster, K.; Vainshtein, P.; Fichman, M.; Gutfinger, C. Particle motion in resonance tubes. *J. Fluid Mech.* **1998**, *360*, 1–20. [CrossRef]
15. Marchioro, M.; Acrivos, A. Shear-induced particle diffusivities from numerical simulations. *J. Fluid Mech.* **2001**, *443*, 101–128. [CrossRef]
16. Abuzeid, S.; Busnaina, A.A.; Ahmadi, G. Wall deposition of aerosol particles in a turbulent channel flow. *J. Aerosol Sci.* **1991**, *22*, 43–62. [CrossRef]
17. Patankar, N.A.; Joseph, D.D. Modeling and numerical simulation of particulate flows by the Eulerian–Lagrangian approach. *Int. J. Multiph. Flow* **2001**, *27*, 1659–1684. [CrossRef]
18. Marchioli, C.; Soldati, A. Mechanisms for particle transfer and segregation in a turbulent boundary layer. *J. Fluid Mech.* **2002**, *468*, 283–315. [CrossRef]
19. Yin, X.; Jiang, C.; Shao, Y.; Huang, N.; Zhang, J. Large-eddy-simulation study on turbulent particle deposition and its dependence on atmospheric-boundary-layer stability. *Atmos. Chem. Phys.* **2022**, *22*, 4509–4522. [CrossRef]
20. Zandi Pour, H.R.; Iovieno, M. The Role of Particle Inertia and Thermal Inertia in Heat Transfer in a Non-Isothermal Particle-Laden Turbulent Flow. *Fluids* **2024**, *9*, 29. [CrossRef]
21. Lin, W.; Shi, R.; Lin, J. Distribution and Deposition of Cylindrical Nanoparticles in a Turbulent Pipe Flow. *Appl. Sci.* **2021**, *11*, 962. [CrossRef]
22. Rousta, F.; Lessani, B.; Ahmadi, G. Particle dispersion and deposition in wall-bounded turbulent flow. *Int. J. Multiph. Flow* **2023**, *158*, 104307. [CrossRef]
23. Pozorski, J.; Minier, J.-P. (Eds.) *Particles in Wall-Bounded Turbulent Flows: Deposition, Re-Suspension and Agglomeration*; Springer: Berlin/Heidelberg, Germany, 2016.
24. Antunes, F.A.F.; Chandel, A.K.; Brumano, L.P.; Terán Hilares, R.; Peres, G.F.D.; Ayabe, L.E.S.; Sorato, V.S.; Santos, J.R.; Santos, J.C.; Da Silva, S.S. A novel process intensification strategy for second-generation ethanol production from sugarcane bagasse in fluidized bed reactor. *Renew. Energy* **2018**, *124*, 189–196. [CrossRef]
25. Cao, S.; Aita, G.M. Enzymatic hydrolysis and ethanol yields of combined surfactant and dilute ammonia treated sugarcane bagasse. *Bioresour. Technol.* **2013**, *131*, 357–364. [CrossRef]
26. Zhu, Z.; Rezende, C.A.; Simister, R.; McQueen-Mason, S.J.; Macquarrie, D.J.; Polikarpov, I.; Gomez, L.D. Efficient sugar production from sugarcane bagasse by microwave assisted acid and alkali pretreatment. *Biomass Bioenergy* **2016**, *93*, 269–278. [CrossRef]
27. Maxey, M.R. The gravitational settling of aerosol particles in homogeneous turbulence and random flow fields. *J. Fluid Mech.* **1987**, *174*, 441–465. [CrossRef]
28. Li, D.; Luo, K.; Fan, J. Direct numerical simulation of turbulent flow and heat transfer in a spatially developing turbulent boundary layer laden with particles. *J. Fluid Mech.* **2018**, *845*, 417–461. [CrossRef]
29. Balachandar, S.; Eaton, J.K. Turbulent dispersed multiphase flow. *Annu. Rev. Fluid Mech.* **2010**, *42*, 111–133. [CrossRef]
30. Bernardini, M.; Pirozzoli, S.; Orlandi, P. The effect of large-scale turbulent structures on particle dispersion in wall-bounded flows. *Int. J. Multiph. Flow* **2013**, *51*, 55–64. [CrossRef]
31. Morsi, S.A.; Alexander, A.J. An investigation of particle trajectories in two-phase flow systems. *J. Fluid Mech.* **1972**, *55*, 193–208. [CrossRef]
32. Elghobashi, S. On predicting particle-laden turbulent flows. *Appl. Sci. Res.* **1994**, *52*, 309–329. [CrossRef]
33. Brandt, L.; Coletti, F. Particle-laden turbulence: Progress and perspectives. *Annu. Rev. Fluid Mech.* **2022**, *54*, 159–189. [CrossRef]
34. Bijlard, M.J.; Oliemans, R.V.A.; Portela, L.M.; Kuerten, J.G.M. Direct numerical simulation analysis of local flow topology in a particle-laden turbulent channel flow. *J. Fluid Mech.* **2010**, *653*, 35–56. [CrossRef]
35. Mei, R. History force on a sphere due to a step change in the free-stream velocity. *Int. J. Multiph. Flow* **1993**, *19*, 509–525. [CrossRef]

36. Baker, L.J.; Coletti, F. Particle–fluid–wall interaction of inertial spherical particles in a turbulent boundary layer. *J. Fluid Mech.* **2021**, *908*, A39. [CrossRef]
37. Lyapunov, A.M. *The General Problem of Stability of Motion*; Transl. from 1892; Princeton University Press: Princeton, NJ, USA, 1947.
38. Stamova, I.M.; Stamov, G.T. Lyapunov–Razumikhin method for asymptotic stability of sets for impulsive functional differential equations. *Electron. J. Differ. Equ.* **2008**, *2008*, 1–14.
39. Khalil, H.K. *Nonlinear Systems*, 3rd ed.; Prentice Hall: Upper Saddle River, NJ, USA, 2002.
40. Hahn, W. *Stability of Motion*; Springer: Berlin/Heidelberg, Germany, 1967.
41. Vidyasagar, M. *Nonlinear Systems Analysis*; SIAM Classics: Philadelphia, PA, USA, 2002.
42. LaSalle, J.; Lefschetz, S. *Stability by Liapunov's Direct Method*; Academic Press: New York, NY, USA, 1961.
43. Bellman, R. *Introduction to Matrix Analysis*, 2nd ed.; SIAM: Philadelphia, PA, USA, 1997.
44. Rugh, W.J. *Linear System Theory*, 2nd ed.; Prentice Hall: Upper Saddle River, NJ, USA, 1996.
45. Kailath, T. *Linear Systems*; Prentice Hall: Englewood Cliffs, NJ, USA, 1980.
46. Doetsch, G. *Introduction to the Theory and Application of the Laplace Transformation*; Springer: Berlin/Heidelberg, Germany, 1974.
47. Debnath, L.; Bhatta, D. *Integral Transforms and Their Applications*, 3rd ed.; CRC Press: Boca Raton, FL, USA, 2014.
48. Lakshmikantham, V.; Bainov, D.D.; Simeonov, P.S. *Theory of Impulsive Differential Equations*; World Scientific: Singapore, 1989.
49. Yakubovich, V.A.; Starzhinskii, V.M. *Linear Differential Equations with Periodic Coefficients*; Wiley: New York, NY, USA, 1975.
50. Meirovitch, L. *Analytical Methods in Vibrations*; Macmillan: New York, NY, USA, 1967.
51. Strogatz, S.H. *Nonlinear Dynamics and Chaos*, 2nd ed.; CRC Press: Boca Raton, FL, USA, 2015.
52. Arnold, V.I. *Ordinary Differential Equations*; Springer: Berlin/Heidelberg, Germany, 1992.
53. Stokes, G.G. On the Effect of the Internal Friction of Fluids on the Motion of Pendulums. *Trans. Camb. Philos. Soc.* **1851**, *9*, 8–106.
54. Fetecau, C.; Vieru, D.; Fetecau, C. A note on the second problem of Stokes for Newtonian fluids. *Int. J. Non-Linear Mech.* **2008**, *43*, 451–457. [CrossRef]
55. Pukhnachev, V.V.; Frolovskaya, O.A. Influence of Vibration on the Onset of Convection in a Second-Grade Fluid. *Comput. Math. Math. Phys.* **2025**, *65*, 1142–1149. [CrossRef]
56. Ershkov, S.V.; Baranovskii, E.S.; Yudin, A.V. Non-Newtonian Rivulet Flows on an Inclined Planar Surface Applying the Second Stokes Problem. *Appl. Math. Mech. (Engl. Ed.)* **2026**, *47*, 153–164. [CrossRef]

 https://doi.org/10.3390/fluids11020051

energies

MDPI

Article

Flow Characteristics and Thrust Augmentation Effects of Concentric Canister Gas Jets

Shilin Yang [1,2,*], Hongliang Qi [2], Wenyan Song [1,*], Nan Niu [2], Weiwei Huang [2] and Yongping Wang [2]

[1] School of Power and Energy, Northwestern Polytechnical University, Xi'an 710029, China
[2] Xi'an Institute of Aerospace Propulsion Technology, Xi'an 710025, China; susanforever77@sohu.com (H.Q.)
* Correspondence: ysl@mail.nwpu.edu.cn (S.Y.); wenyan_song@nwpu.edu.cn (W.S.)

Abstract

A transient numerical framework incorporating dynamic mesh techniques was developed to simulate the launch process. On this basis, a thermal–fluid–structural multi-physics coupling paradigm was proposed to interpret the evolution of the flow field and the associated load response throughout the entire firing sequence. The results show that flow development follows a multi-stage dynamic pattern, comprising gas-impact filling, gap-jet formation, and subsequent free-jet expansion. A pronounced spatially heterogeneous phase lag was observed in the pressure–Mach number response. This phenomenon arises from a mismatch among the characteristic time scales of pressure-wave propagation, flow inertia, and shock–boundary-layer interaction. Quantitative analysis further indicates that the spatial superposition of high-temperature zones, high-Mach regions, and elevated-pressure areas activates a thermal–fluid–structural positive-feedback loop that drives the local peak temperature to approximately 2.5×10^3 K. The temperature response lags behind the pressure maximum by approximately 30–50 ms, reflecting the governing time scale of thermal inertia. In addition, vortical structures near the tube base account for nearly 15% of the total thrust. These findings provide a theoretical foundation for predicting transient peak loads in concentric cylindrical systems and for optimizing instantaneous thermal protection strategies.

Keywords: concentric canister launch device; gas jet; dynamic grid technology; thermal environment characteristics; thrust augmentation effect

1. Introduction

The concentric canister launcher (CCL) is the core component of modern vertical cold-launch missile systems. It performs pivotal functions, including combustion-gas venting, supplementary thrust generation, and thermal shielding [1,2]. As thrust levels increase and deployment expands to naval, land-based, and submarine platforms under increasingly complex operational conditions, the unsteady internal flow dynamics and thrust augmentation mechanisms within the CCL have emerged as critical scientific issues governing launch reliability and cross-platform adaptability [3,4].

The concentric canister launch process is a prototypical multi-body separation process characterized by strong coupling between a high-speed projectile and a confined internal flow field. The process involves shock systems, expansion waves, and high-temperature exhaust-jet interactions [5,6]. Owing to its transient, strongly nonlinear, and multi-physics nature, a unified description of the temporal coupling among dynamic loads (pressure and thrust), thermal response (temperature), and flow structures (Mach number and vortex systems) throughout the launch sequence remains unavailable.

Extensive numerical and theoretical investigations have addressed these issues [7–9]. From a methodological perspective, dynamic mesh strategies—such as layering techniques and overset grids—are widely adopted to construct transient models capable of resolving moving boundaries and flow-field interactions [10–12]. Researchers have simulated multi-stage separation aerodynamics and thermal loads and employed high-fidelity turbulence closures to capture jet evolution and shock-interference phenomena [13–15]. Under high-altitude or near-vacuum conditions, the pressure and thermal characteristics of underexpanded-jet impingement follow distinct scaling laws [16,17].

Regarding thrust augmentation, annular blockage and jet impingement are generally recognized as the dominant contributors [18]. Fu and Yu [19] elucidated the synergistic nonlinear interplay among annular choking, boundary-layer separation, and jet impact. Zhang et al. [20] quantified the influence of pressure under gas–liquid–solid coupling on drainage efficiency and projectile velocity. Yu and Mu [21] theoretically characterized the inverse regulatory effect of annular clearance on thrust enhancement and base loads, whereas Jia et al. [22] proposed a controllable thrust-structure design methodology. These processes are frequently interpreted as an "annular entrainment–mixing–total-pressure recovery" momentum-exchange mechanism whose geometric sensitivity and entrainment characteristics are closely associated with ejector theory [23,24].

Thermal-environment studies primarily focus on plume entrainment, internal recirculation structures, and peak wall heat flux. Hu et al. [25], incorporating dynamic layering and radiation models, revealed the conjugate-heat-transfer mechanism under fluid–structure interaction. Yang et al. [26] clarified the thermal shock characteristics of a novel CCL configuration. Three-dimensional RANS simulations indicate that projectile attitude deviation markedly amplifies the impact loads and intensifies asymmetry in internal thermo-fluid distributions [27]. Liquid-based thermal-regulation systems demonstrate superior energy efficiency and temperature uniformity compared with gaseous schemes [28]. Furthermore, the geometric parameters of deflectors exhibit pronounced sensitivity to surface-temperature distribution, offering guidance for the optimization of thermal protection [29].

Despite substantial progress in understanding flow structures, thrust-enhancement mechanisms, and thermal responses, a systematic framework for full-process multi-physics coupling during the concentric-canister launch process remains lacking. In recent years, flow–thermal–structural interactions under high-enthalpy and hypersonic conditions have attracted increasing attention, particularly in shock-wave–boundary-layer interaction (SWBLI) and aerothermal-load prediction [30,31]. However, most existing studies focus on steady-state scenarios and give limited consideration to transient multi-physics coupling in confined launch environments. First, a unified thermal–fluid–force temporal coupling framework has not been established. Previous studies indicate that thermal response and pressure evolution in high-enthalpy shock-interaction flows are temporally asynchronous, but this phenomenon remains insufficiently quantified [32]. Second, discrepancies persist between numerical predictions and experimental observations, particularly within shock-induced separation zones and regions with steep heat-flux gradients [33–35].

To address these gaps, this study developed a triadic technical route integrating experimental observation, numerical validation, and mechanistic interrogation. The authors obtained key temporal parameters under realistic operating conditions and then constructed a transient simulation framework incorporating dynamic mesh methodology to reproduce the complete launch sequence, resolving shock evolution, energy-conversion pathways, and thermal–fluid–structural coupling behavior. Cross-validation between

https://doi.org/10.3390/en19051264

experiments and simulations enabled quantitative prediction of thrust augmentation characteristics and the thermal-environment response, thereby providing analytical support for next-generation, multi-platform-compatible launch-system design.

Ethical Considerations and Dual-Use Statement

The current research is limited to propulsion fluid mechanics and jet dynamics, aiming to improve the efficiency, safety, and reliability of civil and industrial propulsion systems. The scope of this study does not involve public health, biological systems, or classified technologies, and it does not pose a threat to public health or national security.

The authors acknowledge that research involving jet control and thrust augmentation may have potential dual-use implications. All necessary precautions have been taken to prevent misuse of the presented methods and findings. The study was conducted strictly in accordance with relevant national and international regulations concerning Dual-Use Research of Concern (DURC), as well as institutional research-governance policies.

The authors advocate responsible deployment, ethical awareness, regulatory compliance, and transparent scientific reporting to mitigate potential misuse risks and to ensure that the outcomes of this research contribute positively to scientific and engineering advancement.

2. Model and Methodology

2.1. Physical Model

The structural configuration of the concentric-canister launch system is shown in Figure 1. The assembly consists of an outer cylinder, an inner cylinder, an annular interlayer cavity, a base plate (thrust amplifier), and a flow-guiding cone. The inner cylinder is coaxially nested within the outer cylinder, forming a 14 mm annular clearance between the two structures.

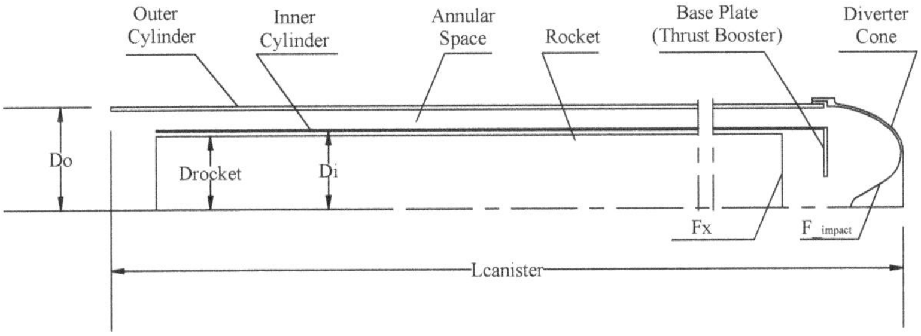

Figure 1. Assembly model of the rocket and concentric canister launcher.

The flow-guiding cone is mounted at the geometric center of the inner-cylinder base. It adopts a conical geometry designed to redirect exhaust gases, thereby optimizing thrust-vector distribution and mitigating turbulence-induced vibration. The annular interlayer cavity between the inner and outer cylinders provides thermal insulation and structural reinforcement. The base plate, which functions as a thrust-amplification module, serves as the primary load-bearing component, sustaining axial recoil forces during launch while enhancing effective thrust output.

The projectile (missile) is aligned along the central axis of the inner cylinder, maintaining a minimal radial clearance of 1 mm from the inner wall. This small gap accommodates non-uniform thermally induced expansion of the inner cylinder under high-temperature gas impingement, preventing thermal seizure during transient operation. The clearance also constitutes a critical exhaust-guiding passage that contributes to flow-field stability.

The concentric canister adopts a hybrid metal–composite architecture. Owing to inherent manufacturing variability, the 1 mm radial tolerance accommodates process

https://doi.org/10.3390/en19051264

fluctuations and subsequent assembly adjustments, ensuring structural consistency and assembly reliability across production batches.

2.2. Numerical Simulation Method

2.2.1. Governing Equations

To characterize the behavior of the compressible turbulent jet during the concentric-canister launch process, the Reynolds-averaged Navier–Stokes (RANS) equations were employed as the governing equations. The RANS formulation is derived through Reynolds decomposition of instantaneous flow variables. The detailed derivation procedure has been extensively documented in Refs. [36,37] and is therefore not repeated here.

For compressible flow, the mass conservation equation is expressed as

$$\frac{\partial \rho}{\partial t} + \nabla \cdot \left(\rho \vec{u}\right) = 0 \tag{1}$$

The Reynolds-averaged momentum equation is written as

$$\frac{\partial}{\partial t}(\rho \overline{u_i}) + \frac{\partial}{\partial x_j}(\rho \overline{u_i u_j}) = -\frac{\partial \overline{p}}{\partial x_j} + \frac{\partial}{\partial x_j}\left(\mu_{eff}\frac{\partial \overline{u_i}}{\partial x_j} - \overline{\rho u_i' u_j'}\right) \tag{2}$$

where the Reynolds-stress tensor $\overline{\rho u_i' u_j'}$ represents the primary source of closure deficiency in the equation system and, therefore, requires modeling through an appropriate turbulence closure scheme.

In addition, the total energy equation is solved for compressible conditions. Its Reynolds-averaged formulation is expressed as

$$\frac{\partial}{\partial t}(\rho E) + \nabla \cdot [u(\rho E + p)] = \nabla \cdot (k_{eff} \nabla T) + \nabla \cdot (\tau_{eff} \cdot u) + SE\,, \tag{3}$$

The effective dynamic viscosity and effective thermal conductivity are defined as

$$\mu_{eff} = \mu + \mu_{\mathrm{t}},\ k_{eff} = k + k_{\mathrm{t}} \tag{4}$$

where the turbulent transport coefficients μ_{t} and k_{t} are determined by the turbulence model adopted in the present study.

2.2.2. Turbulence Model

During the concentric-canister launch process, the exhaust jet exhibits high Reynolds numbers and pronounced compressibility effects. Direct numerical simulation (DNS) and large-eddy simulation (LES) are computationally prohibitive under complex geometric constraints. Although the Reynolds-stress model (RSM) resolves turbulence anisotropy, its model complexity and substantial computational cost limit its suitability for parametric engineering analysis. The k–ω shear-stress transport (SST) model performs well in near-wall and separated-flow predictions; however, it is relatively sensitive to freestream turbulence specifications and grid resolution.

After evaluating numerical robustness, engineering accuracy, and the requirements of multi-condition parametric investigation, this study adopted the standard k–ε turbulence model within the Reynolds-averaged Navier–Stokes (RANS) framework [36,37]. The standard k–ε formulation has been extensively validated in engineering applications and reasonably predicts shock locations, boundary-layer development trends, and dominant flow structures while significantly reducing computational cost. To balance predictive fidelity and efficiency, the RANS-based standard k–ε model was employed for all simulated operating conditions.

https://doi.org/10.3390/en19051264

Within the k–ε formulation, closure of the Reynolds-stress tensor was achieved by introducing two additional transport equations for turbulent kinetic energy k–ε, which are expressed as

$$\frac{\partial(\rho k)}{\partial t} + \nabla \cdot (\rho u k) = \nabla \cdot \left[\left(\mu + \frac{\mu_t}{\sigma_k} \right) \nabla k \right] + G_k - \rho \varepsilon, \tag{5}$$

$$\frac{\partial(\rho \varepsilon)}{\partial t} + \nabla \cdot (\rho u \varepsilon) = \nabla \cdot \left[\left(\mu + \frac{\mu_t}{\sigma_\varepsilon} \right) \nabla \varepsilon \right] + C_{1\varepsilon} \frac{\varepsilon}{k} G_k - C_{2\varepsilon} \rho \frac{\varepsilon^2}{k}, \tag{6}$$

The turbulent (eddy) viscosity is defined as

$$\mu_t = \rho C_\mu \frac{k^2}{\varepsilon} \tag{7}$$

where G_k denotes the production term of turbulent kinetic energy arising from gradients of the mean velocity field.

The empirical constants of the standard k–ε model were specified according to conventional practice: the eddy-viscosity coefficient C_μ was set to 0.09; the production coefficient in the ε-transport equation $C_{1\varepsilon}$ was assigned a value of 1.44; and the dissipation coefficient $C_{2\varepsilon}$ was taken as 1.92.

The turbulent Prandtl number for k, denoted as σ_k, was prescribed as 1.0, whereas that for ε, denoted as σ_ε, was set to 1.3. These parameter values were maintained throughout all numerical simulations to ensure model stability and comparability across operating conditions.

In the present investigation, the RANS methodology coupled with the k–ε closure scheme was implemented to simulate turbulent flow during the launch sequence. The additional Reynolds stresses induced by turbulence were evaluated using the k–ε model, which exerts a non-negligible influence on velocity distribution, pressure fluctuations, and thermal evolution within the flow domain. The variables k and ε were solved numerically to capture transient turbulent characteristics. Consequently, the principal features of the launch flow field, including the evolution of temperature, pressure, and velocity distributions, were reproduced. The combined RANS and k–ε framework further enabled effective prediction of transient aerothermal and aerodynamic loads, particularly under high-temperature, high-pressure unsteady conditions in which pressure-wave and thermal fluctuations dominate.

2.2.3. Computational Domain, Mesh Generation, and Boundary Conditions

Computational Domain and Geometric Configuration

The exhaust-gas flow during the concentric canister launch was modeled using a two-dimensional axisymmetric configuration. The computational domain (Figure 2) extends from the closed base of the canister to the external far-field region and encompasses the entire flow evolution from confinement to free expansion.

The domain consists of three primary subregions: the internal canister flow passage, the rocket-body model, and the external free-flow field. The principal geometric parameters are specified as follows:

Canister Geometry: Inner diameter, Di = 0.2 m; outer diameter, Do = 0.23 m; total length, Lcanister = 2.5 m.

Rocket Model: Diameter, Drocket = 0.198 m; total length, Lrocket = 2.0 m.

External Domain: To sufficiently capture post-exit jet expansion and shock-structure evolution, the outer region is defined as a rectangular domain with radial extent Router = 5, Do = 0.9 m, and axial length Louter = 5.0 m.

https://doi.org/10.3390/en19051264

This configuration ensures that the jet structure decays sufficiently before interacting with the far-field boundary, thereby minimizing artificial wave reflection and approximating open-domain conditions.

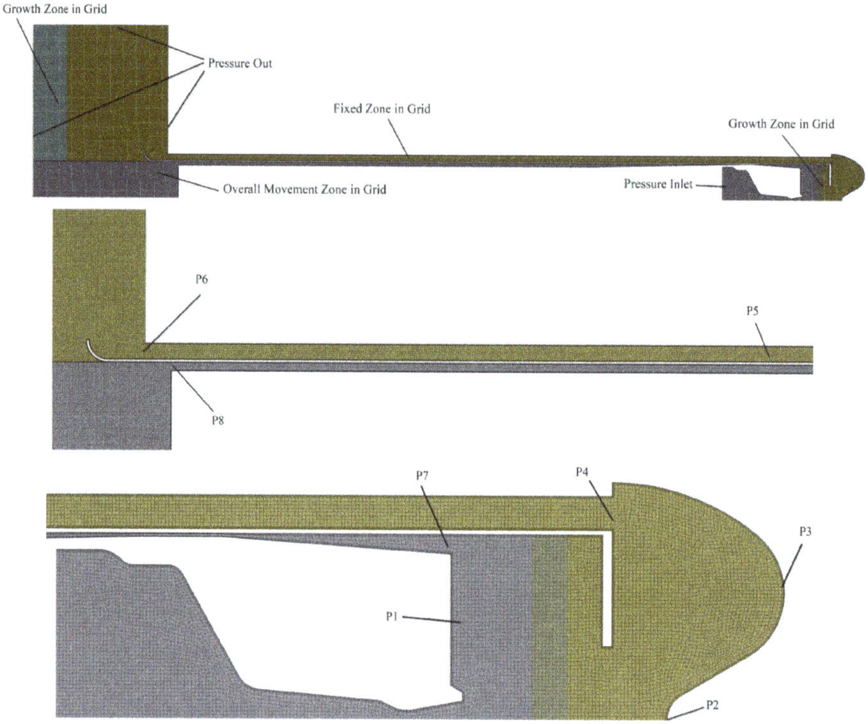

Figure 2. Schematic diagram of the computational domain, grid partitioning strategy, and monitoring-point layout. Monitoring points are defined as follows: P1, rocket base; P2, guide-cone inlet; P3, guide-cone bottom; P4, guide-cone outlet (interlayer bottom); P5, mid-interlayer region; P6, interlayer outlet; P7, bottom of the clearance between the projectile body and the inner canister wall; P8, top of the clearance between the projectile body and the inner canister wall.

Mesh Generation and Refinement Strategy

A fully structured mesh was generated over the entire domain to resolve shock structures, boundary-layer development, and shear-layer evolution. Local refinement was applied in regions prone to shock–boundary-layer interaction, flow separation, and shock oscillation, particularly near solid walls and within the annular clearance downstream of the rocket tail.

To satisfy the near-wall resolution requirements of the turbulence model, the first-layer cell height was prescribed as 2×10^{-5} m^2, ensuring a dimensionless wall distance of $y^+ \approx 1$. This treatment enabled accurate capture of viscous-sublayer behavior and improved prediction of wall shear stress and surface heat flux.

Dynamic Mesh Zoning and Data Exchange

To accommodate the large axial displacement of the rocket, the computational mesh was partitioned into three functional zones (Figure 2):

- Fixed Zone: A stationary canister-wall region serving as the geometric reference.

https://doi.org/10.3390/en19051264

- Overall Movement Zone: Including the rocket body and nozzle, undergoing rigid-body translation without local topological deformation.
- Growth Zone: Located along the rocket's trajectory, where mesh layering was dynamically updated through cell addition and deletion.

Flow variables were interpolated across zone interfaces to maintain conservative flux transfer between subdomains.

Monitoring-Point Arrangement

To quantify transient pressure-wave propagation and shock dynamics, eight monitoring locations (P1–P8) were arranged along the axial direction (Figure 2). These points recorded instantaneous pressure and temperature variations at critical geometric locations, including the rocket base, guide-cone inlet and outlet, annular cavity, and rocket–liner clearance region.

Boundary Conditions and Engineering Relevance

Boundary conditions were prescribed to reflect the physical launch mechanism:

(1) Pressure Inlet: The combustion chamber was modeled as a total-pressure inlet with P0 = 8.5 MPa and T0 = 3000 K. The total-state specification represents high-enthalpy compressible expansion during the early launch stage. The mass flow rate was regulated by pressure gradients, thereby avoiding non-physical velocity constraints.
(2) Pressure Far Field: The external boundary was defined as a pressure far-field condition with ambient pressure $p_{\infty} = 0.1$ MPa. The enlarged outer domain reduces shock reflection and approximates quasi-infinite-space conditions.
(3) Wall Boundary: All solid surfaces were treated with no-slip conditions. Given the short launch duration and high structural thermal inertia, the walls were approximated as adiabatic, and transient conduction effects were neglected.
(4) Axisymmetric Condition: The 2D axisymmetric assumption was justified by geometric rotational symmetry and the absence of lateral disturbances or attitude deviation.

2.2.4. Dynamic Mesh Coupling and Solution Method

Layering Dynamic Mesh Strategy

The concentric-canister launch process considered in this study involves large unidirectional axial translation of a solid boundary and represents a typical transient fluid–structure interaction problem. To ensure stable updating of the computational topology during long-stroke rocket motion while avoiding the excessive computational cost associated with global remeshing, the layering dynamic mesh technique implemented in ANSYS Fluent 2022R1 was adopted to manage mesh deformation.

The layering approach predefines regions expected to undergo substantial displacement as structured meshes with regular layered topology (hexahedral or prismatic elements). During the transient simulation, boundary motion is accommodated through dynamic layer splitting and merging operations.

Specifically, when the axial motion compresses adjacent mesh layers and the local cell height decreases below a prescribed minimum threshold, a merge operation combines two adjacent layers into one to prevent excessive cell distortion. Conversely, when boundary displacement stretches the mesh and the local layer height exceeds a predefined upper limit, a split operation subdivides a single layer into two, thereby maintaining adequate spatial resolution.

This adaptive split–merge mechanism preserves mesh quality and layered topology throughout the rocket's large axial displacement. It enables efficient transient computation of the strongly compressible launch flow while preventing numerical degradation. The

 https://doi.org/10.3390/en19051264

method is particularly suitable for engineering problems characterized by directional geometric deformation and structured layer-aligned grids, making it appropriate for the present concentric-canister launch simulation.

Rigid-Body Motion and Flow-Field Coupling

To simulate the rocket's exit motion within the canister, transient coupling between the fluid domain and rigid-body dynamics was achieved using user-defined functions (UDFs) combined with the layering dynamic mesh framework. The overall computational workflow is illustrated in Figure 3.

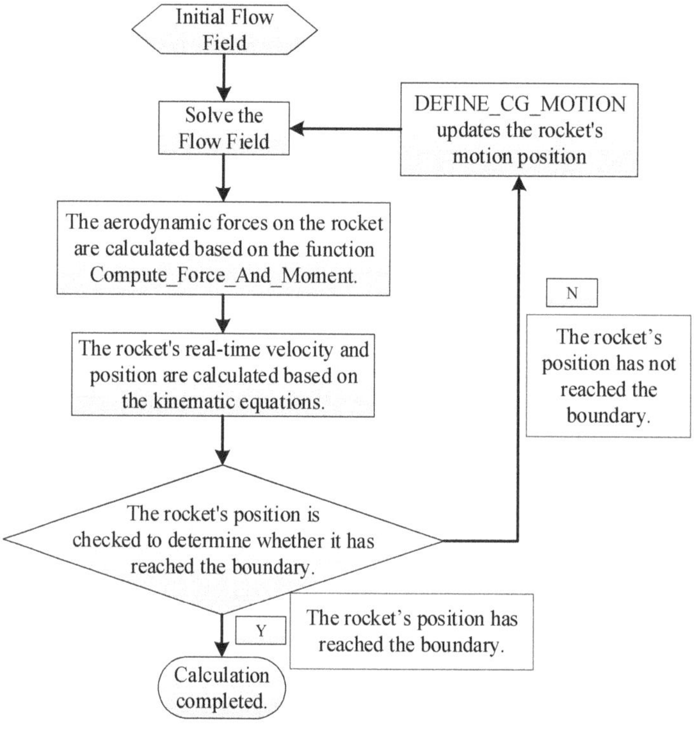

Figure 3. Flowchart of the coupled CFD–rigid-body dynamics simulation. "N" indicates that the rocket position has not reached the boundary, while "Y" indicates that the rocket position has reached the boundary.

At each time step, aerodynamic forces acting on the rocket were evaluated using the function Compute_Force_And_Moment. These forces were combined with engine thrust to solve the axial kinematic equation of motion, yielding instantaneous acceleration, velocity, and displacement. The computed rocket velocity was subsequently assigned to the overall movement zone, where rigid-body translation was prescribed through the UDF macro DEFINE_CG_MOTION, enabling time-dependent axial motion without altering the local mesh topology.

The growth zone was updated using the layering strategy described above. Dynamic addition and removal of mesh layers accommodated continuous domain variation while maintaining cell orthogonality and quality. The mesh-update criterion is defined as

$$H_{\min} > (1 + \alpha_s)h_0 \tag{8}$$

https://doi.org/10.3390/en19051264

where h_0 denotes the initial nominal layer height, and α_s represents the splitting factor; in the present study, $\alpha_s = 0.1$. When the layer-height variation exceeded the prescribed threshold, split or merge operations were automatically initiated to ensure numerical stability and mesh integrity.

Motion Assumptions and Simplifications

To reduce computational complexity while preserving the dominant flow physics of the launch process, only the rocket's axial translational degree of freedom was considered. Attitude variation and lateral displacement were neglected. The rocket's mass was specified as 55 kg, and was assumed to remain constant throughout the launch event.

These simplifications are consistent with the strong kinematic confinement imposed by the launch canister, where radial and angular motions are mechanically constrained. Under such conditions, axial translation dominates the dynamic response and is sufficient to capture the primary evolution of the internal flow structures.

Numerical Implementation and Convergence Verification

To ensure the reliability and reproducibility of the numerical results, the computational implementation and convergence criteria are described below.

(1) Numerical Scheme

Simulations were performed using ANSYS Fluent 2022R1. The compressible Navier–Stokes equations were discretized using the finite-volume method. Convective fluxes were evaluated with Roe's approximate Riemann solver combined with second-order MUSCL reconstruction. Viscous terms were discretized using second-order central differencing. Temporal integration employed a second-order implicit scheme to enhance stability under strongly compressible transient conditions. Pressure–velocity coupling was handled with the coupled solver. The layering dynamic mesh strategy was applied for grid deformation, and rigid-body motion was implemented via UDF-based fluid–rigid coupling.

(2) Convergence Criteria

Within each time step, convergence was achieved when the residuals of the continuity, momentum, and energy equations decreased by at least 10^{-3}. In addition, monitoring-point pressure variation was limited to below 0.1%, and the global mass imbalance was maintained within 0.5%. Advancement to the subsequent time step was permitted only after all criteria were satisfied, thereby ensuring temporal solution consistency.

(3) Time-Step Independence Assessment

To evaluate temporal discretization error, three time-step sizes were examined: $\Delta t1 = 1.0 \times 10^{-5}$ s, $\Delta t2 = 5.0 \times 10^{-6}$ s, and $\Delta t3 = 2.5 \times 10^{-6}$ s. Peak pressure values at representative monitoring locations were compared. The results indicate that, when $\Delta t \leq 5.0 \times 10^{-6}$ s, the deviation in peak pressure remained below 1.8%. Further reductions in the time step yielded negligible improvement relative to the additional computational cost. Consequently, $\Delta t = 5.0 \times 10^{-6}$ s was selected for subsequent simulations as a compromise between numerical accuracy and computational efficiency.

(4) Grid Independence Verification

Mesh convergence was assessed using three representative indicators associated with the guide cone: impact force F_{cjl}, maximum static pressure $P_{\text{cjl,max}}$, and peak Mach number. Four mesh densities—15,050 cells, 88,481 cells, 308,531 cells, and 468,369 cells—were systematically compared. The variation in guide-cone impact force with mesh number is presented in Figure 4.

https://doi.org/10.3390/en19051264

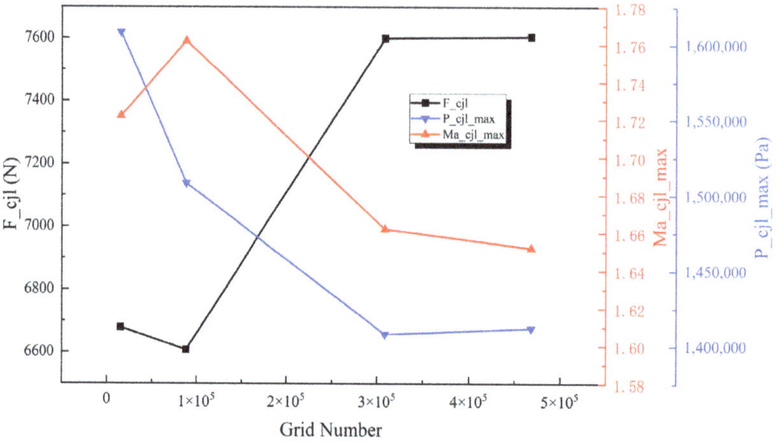

Figure 4. Mesh-convergence assessment based on guide-cone impact force, peak static pressure, and maximum Mach number.

It was observed that when the grid density exceeded 3.09×10^5 cells, the mean impact force, maximum pressure, and peak Mach number exhibited negligible variation with further mesh refinement. The relative deviation between the two finest meshes remained within acceptable engineering tolerance. Therefore, considering the trade-off between computational efficiency and predictive fidelity, the mesh containing 308,531 cells was selected for subsequent simulations.

(5) Dynamic Mesh Stability Assessment

Throughout the rocket's exit sequence, no negative-volume cells or severe mesh distortions were observed. The minimum mesh-quality metric remained consistently above 0.3 during large-stroke axial motion. These results indicate that the layering dynamic mesh strategy maintained numerical robustness and topological integrity under substantial boundary displacement.

Thermophysical Properties of the Working Gas

The thermodynamic properties of the exhaust gas were specified as follows: the specific heat ratio was prescribed as $\gamma = 1.33$, and the gas constant as $R = 287.1$ J/(kg·K). Dynamic viscosity was evaluated using Sutherland's law: where the reference viscosity is $\mu_{ref} = 1.72 \times 10^{-5}$ Pa·s; the reference temperature is $T_{ref} = 273$ K; and Sutherland's constant is $C = 120$ K.

The parameters in Table 1 are derived from propellant thermodynamic estimations and were kept constant across all simulation cases to ensure consistency.

Table 1. Thermophysical properties of the exhaust gas.

Property	Symbol	Value	Unit	Basis/Remarks
Specific heat ratio	γ	1.33	-	Derived from propellant thermodynamic calculation
Gas constant	R	287.1	J/(kg·K)	Standard air-equivalent assumption
Reference viscosity	T_{ref}	1.72×10^{-5}	Pa·s	Sutherland's law, $T_{ref} = 273$K
Sutherland's constant	C	120	K	

2.2.5. Vortex Identification Criterion

During the concentric canister launch, the exhaust gas underwent intense compression–expansion processes, shock reflection, and shear-layer development within the annular

https://doi.org/10.3390/en19051264

clearance. The resulting flow field exhibited pronounced unsteadiness and strong compressibility. Shock–boundary-layer interaction generated high-gradient shear interfaces, while interaction between the annular jet and the core stream induced vortex rings and recirculation zones. In addition, transient pressure-wave propagation triggered localized rotational structures.

To detect coherent vortical structures, the Q-criterion proposed by Hunt et al. [38] was adopted for identification and visualization. The Q-criterion is defined as the second invariant of the velocity-gradient tensor and represents the competition between the rotation rate and strain rate:

$$Q = \frac{1}{2}\left(\|\Omega\|^2 - \|S\|^2\right) \tag{9}$$

where $S = \frac{1}{2}(\nabla u + (\nabla u)^T)$ denotes the symmetric strain-rate tensor, and $\Omega = \frac{1}{2}(\nabla u - (\nabla u)^T)$ represents the antisymmetric rotation-rate tensor.

Regions satisfying $Q > 0$ correspond to rotation-dominated flow and are typically visualized using isosurfaces to delineate vortex cores. Physically, a positive Q-value indicates that the local rotational intensity exceeds the deformation rate, suggesting the presence of a genuine vortex core rather than a purely shear- or strain-dominated structure. By defining vortices as regions where rotational kinetic contribution prevails, the Q-criterion suppresses false identification caused by high-shear layers. Owing to its computational efficiency and robustness in post-processing, it remains one of the most widely utilized vortex-detection approaches in engineering CFD practice.

3. Experimental Validation

To evaluate the predictive capability of the proposed numerical framework for the concentric-canister launch process, a full-scale (1:1) prototype firing experiment was conducted. Transient flow-field data were acquired, and quantitative comparisons between the experimental measurements and simulation results were performed.

3.1. Experimental Setup

A schematic of the concentric-canister launch experiment is presented in Figure 5. The test system comprised four primary subsystems: the concentric canister launcher, the test rocket, a data-acquisition unit, and an ignition control module.

The core propulsion unit was a solid rocket motor operating at a nominal chamber pressure of 8.5 MPa with a burn duration of approximately 1 s. The ignition control system, developed on a LabVIEW platform, provided synchronized timing control and data acquisition. An electrically triggered ignition mechanism with a response delay of less than 5 ms ensured precise and repeatable engine initiation.

To capture the transient pressure evolution along the canister, pressure sensors were installed at structurally feasible locations, including the inner cylinder and representative annular positions (P3 and P6). YD31A piezoelectric pressure transducers (measurement range: 0–1 MPa; accuracy: ±0.5% full scale; sampling frequency: 10 kHz) were used for dynamic pressure acquisition. In addition, a Phantom high-speed camera recorded transient flow structures synchronously at 2000 frames per second, enabling qualitative visualization of jet evolution and shock propagation.

The primary variable used to compare experimental results with numerical predictions was the pressure measured at designated monitoring locations. Pressure data were acquired using transducers with a measurement range of 1 MPa and a nominal accuracy of ±0.3% full scale (FS), corresponding to an intrinsic sensor uncertainty of ±0.003 MPa.

https://doi.org/10.3390/en19051264

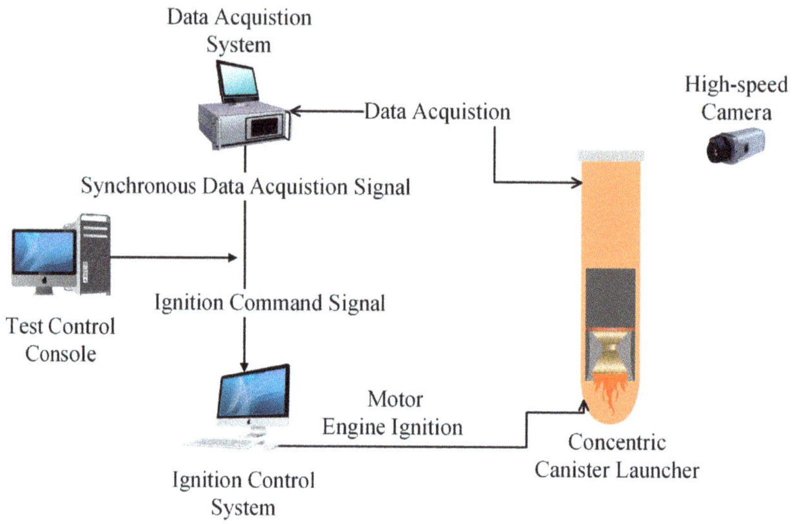

Figure 5. Schematic diagram of the concentric-canister launch experimental system.

In addition to sensor accuracy, the overall pressure-measurement uncertainty included contributions from signal acquisition and processing errors, as well as random variability introduced by repeated tests. Sampling and signal-processing uncertainty primarily arose from high-frequency pressure oscillations and filtering procedures, and was estimated using the standard deviation of the steady-state signal segment. Repeatability error was quantified based on the dispersion of results obtained from multiple experimental runs.

For mutually independent uncertainty components, the combined standard uncertainty was evaluated using the root-sum-square (RSS) method, expressed as

$$u_c = \sqrt{u_{\text{sensor}}^2 + u_{\text{sampling}}^2 + u_{\text{repeat}}^2} \tag{10}$$

The measurement-uncertainty evaluation results of the point pressure are given in Table 2.

Table 2. Uncertainty evaluation of pressure measurements.

Source of Uncertainty	Description	Value (MPa)
Sensor accuracy	±0.3% FS (range: 1 MPa)	0.003
Sampling and signal processing	Estimated from steady-state signal's standard deviation	0.010
Repeatability	Dispersion from multiple test runs	0.010
Combined standard uncertainty	RSS method	0.014
Expanded uncertainty (U)	(k = 2)	±0.028

3.2. Concentric-Canister Launch Test Results

Using the experimental configuration shown in Figure 5, the concentric-canister launch test was successfully conducted. The ignition and exhaust-gas discharge processes proceeded as expected, and the rocket completed a stable exit from the canister.

Key temporal measurements indicated that the delay between initial smoke emission at the canister front and visible flame emergence was 0.051 s. The total duration of the rocket's exit process was 0.229 s (Figure 6). Displacement–time analysis derived from high-speed imaging yielded an exit velocity of 19.7 m/s at canister clearance.

https://doi.org/10.3390/en19051264

Figure 6. Time-resolved analysis of the rocket canister exit process based on high-speed imaging. Sequential images of the rocket launch, showing the evolution of the exhaust plume. The colors in the images represent temperature variations, with warmer colors (e.g., yellow and red) indicating higher temperatures, and cooler colors (e.g., green and blue) representing lower temperatures. The images capture different phases of the launch, from ignition to the rocket's ascent.

3.3. Model Validation

To assess the reliability of the numerical framework, a time-domain pressure comparison was conducted at key monitoring locations. Pressure histories at P3 (guide-cone base) and P6 (annular exit) were selected as validation benchmarks. Model performance was evaluated based on two criteria: maximum relative deviation and trend consistency.

As illustrated in Figure 7, the simulated pressure at P6 (P6_sim) exhibited excellent agreement with the experimental measurements (P6_exp). The maximum relative deviation remained below 5%, confirming that the model accurately captured the quasi-steady flow behavior at the annular outlet.

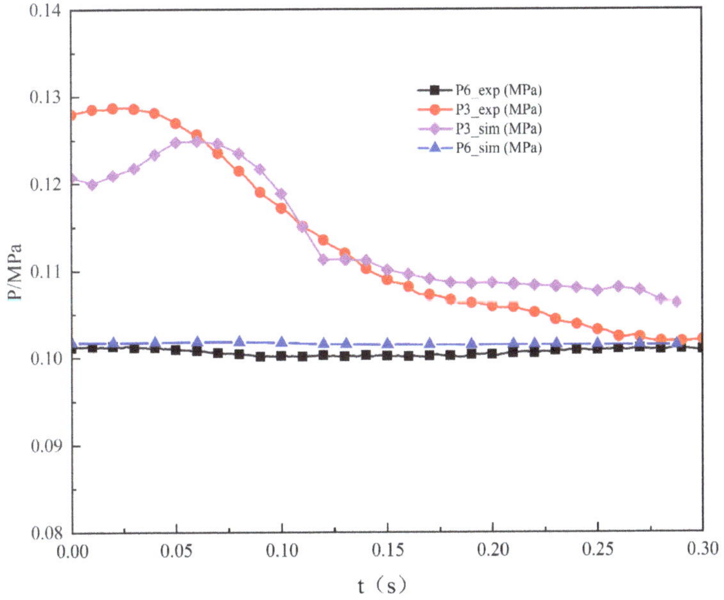

Figure 7. Comparison of simulated and experimental pressure histories at monitoring points P3 and P6.

https://doi.org/10.3390/en19051264

For P3, the simulated curve (P3_sim) followed the overall temporal trend of the measured data (P3_exp). However, discrepancies were observed during the early stage ($t < 0.05$ s), where the computed pressure was slightly lower than the experimental value, and during the late stage ($t > 0.2$ s), when the predicted pressure slightly exceeded the measured value. The deviation in certain transient intervals at P3 reached up to 8.3%. These discrepancies may be attributed to several factors:

(1) The dynamic response delay of the pressure sensor, which may cause signal distortion during rapid pressure rise;
(2) Simplifications associated with the two-dimensional axisymmetric model and the rigid-body assumption, which may not fully capture three-dimensional transient structures (e.g., lip-generated vortices) and fluid–structure interaction effects;
(3) Limitations of quasi-steady modeling assumptions under extreme transient conditions.

Overall, the simulated and experimental pressure evolutions exhibited consistent temporal trends, with an average deviation below 5%. These results substantiate the reliability of the adopted numerical methodology and support its application in subsequent parametric investigations.

To further quantify predictive accuracy, peak pressure values at P3 and P6 were compared with the experimental data (Table 3). The relative deviation was −1.85% at P3 and 1.40% at P6, both below 2%. The corresponding absolute errors were 0.002 MPa and 0.0014 MPa, respectively, which are significantly smaller than the experimental expanded uncertainty (±0.028 MPa). Both results fell within the experimental uncertainty band.

Table 3. Comparison of experimental and simulated pressure at monitoring points.

Monitoring Point	Experimental Pressure (MPa)	Experimental Uncertainty (±U, MPa)	Simulated Pressure (MPa)	Relative Deviation (%)	Within Uncertainty Band
P3	0.108	±0.028	0.1060	−1.85	Yes
P6	0.1001	±0.028	0.1015	1.40	Yes

These findings indicate that the numerical model accurately reproduced the pressure magnitude at critical locations, demonstrating satisfactory predictive fidelity and engineering applicability.

The relative deviation is defined as

$$\delta = \frac{p_{\text{sim}} - p_{\text{exp}}}{p_{\text{exp}}} \times 100\%$$

4. Results and Discussion

4.1. Exhaust-Gas Flow Characteristics During Concentric Canister Launch

Figure 8 presents the Mach number contours and streamlines at the initial stage of launch. The flow field exhibited pronounced spatial non-uniformity. The highest Mach number region was concentrated at the solid rocket motor nozzle exit, where the Mach number ranged from approximately 2.2 to 2.5. In contrast, the upstream region ahead of the guide-cone leading edge experienced substantial deceleration, with local Mach numbers dropping below 0.2.

A distinct primary shock formed immediately downstream of the nozzle exit. Across this shock, the Mach number decreased sharply from about 2.5 to below 1.2. A weaker oblique shock was observed near the guide-cone sidewall, where the Mach number decreased from approximately 1.5 to 1.1. The interaction between this oblique shock and the primary shock became evident near the cone tail, indicating strong coupling between jet dynamics and geometric configuration.

https://doi.org/10.3390/en19051264

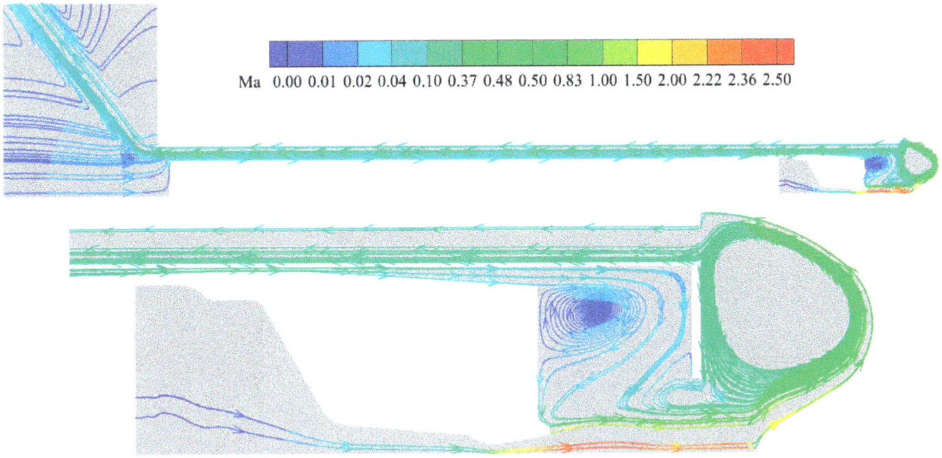

Figure 8. Mach number contours and streamlines at the initial launch stage.

The sonic line was located approximately between the nozzle throat and the onset of the expansion section, clearly separating subsonic and supersonic regions. Downstream of the guide-cone tail, a low-Mach annular separation zone (Ma $\approx$ 0.3–0.6) developed, with a characteristic size of approximately one-third of the expansion-section diameter. This structure reflected localized flow separation induced by shock–boundary-layer interaction (SBLI).

From a mechanistic perspective, the primary shock was mainly generated by the abrupt geometric compression as the high-speed jet discharged into a lower-pressure environment. Its location and strength were governed by the nozzle exit pressure ratio. In contrast, the oblique shock along the guide-cone wall originated from SBLI under an adverse pressure gradient, leading to weak separation and localized compression.

Figure 9 illustrates the transient evolution of Mach number distributions inside the concentric canister from 10 ms to 280 ms. Based on the rocket position and dominant flow mechanisms, the launch process can be categorized into five distinct stages.

Stage I: Initial Gas-Filling Phase (0–10 ms)

Exhaust gas entered the canister and established the initial pressure field. A supersonic jet (Ma > 2.5) impinged on the guide-cone leading edge, forming an attached bow shock and a localized stagnation high-pressure region with a peak pressure of approximately 0.45 MPa. At this stage, a continuous main stream had not yet developed within the annular clearance; flow in the clearance and near the canister exit remained weak. The dominant mechanism was localized jet impingement and cavity filling.

Stage II: Rocket Acceleration and Gap-Jet Formation (10–50 ms)

As the rocket accelerated, the bottom stagnation zone gradually shrank. Gas velocity within the annular clearance increased markedly, and localized shock structures emerged, indicating that acceleration-driven flow became dominant. An incipient underexpanded jet began to form at the canister exit. The shock system migrated downstream and progressively weakened.

Stage III: Gap-Flow-Dominated Phase (50–150 ms)

With continued rocket acceleration, axial flow within the annular clearance became dominant. The exhaust gas fully developed along the clearance and expanded radially. The core Mach number decreased from approximately 2.5 to near unity, producing a clear partition between supersonic and subsonic regions. This stage marked the peak influence of annular gap flow on the overall internal flow field.

 https://doi.org/10.3390/en19051264

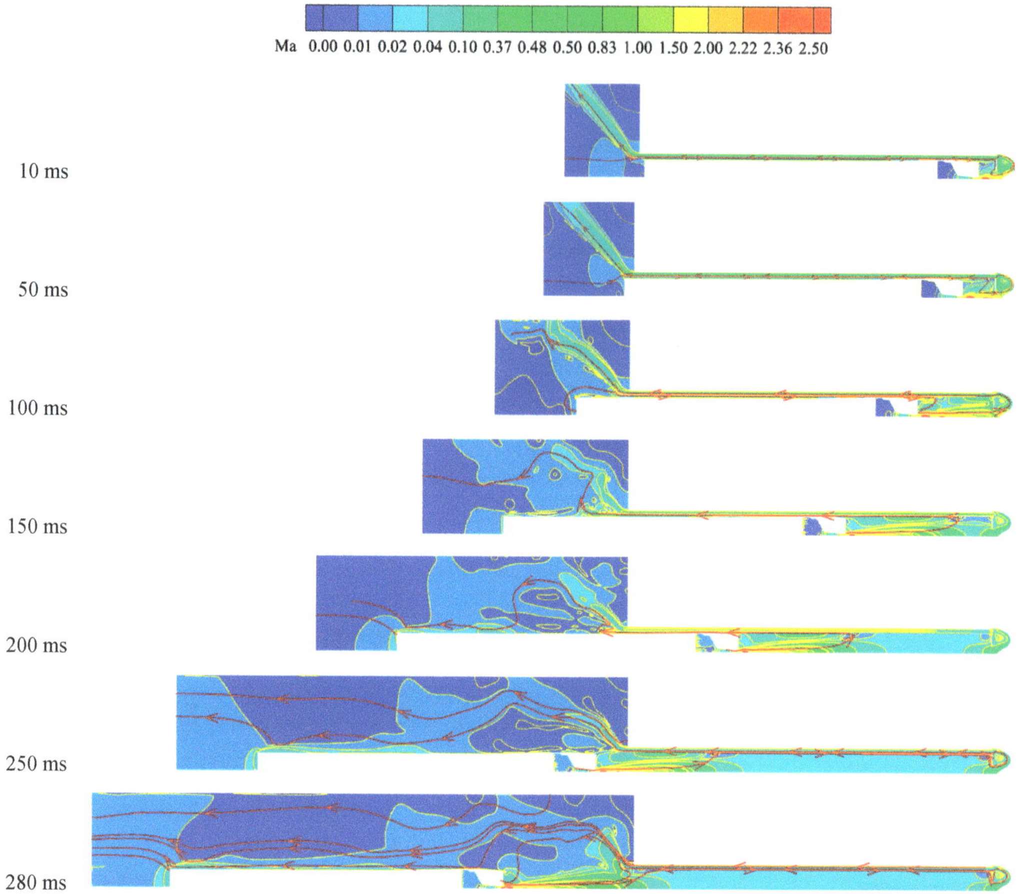

Figure 9. Transient evolution of Mach number contours during launch (t = 10, 50, 100, 150, 200, 250, 280 ms).

Stage IV: Transition from Gap Flow to Free Jet (150–250 ms)

As the rocket approached and gradually exited the canister, the geometric confinement imposed by the annular gap weakened significantly. The Mach number distribution inside the clearance became more uniform, and no pronounced shock or choking phenomena were observed. Outside the canister, a developing jet structure emerged, although it remained partially influenced by the residual geometric confinement. The flow transitioned from a confined regime to a free-expansion state.

Stage V: Free-Jet Regime (250–280 ms)

After complete canister exit, the exhaust flow was no longer constrained by internal geometry. A fully developed underexpanded free jet formed in the external environment, and the flow field progressively approached the characteristic structure of a classical supersonic free jet.

4.2. Pressure–Mach Phase-Lag Characteristics and Underlying Mechanisms

Based on the pressure and Mach number time histories at monitoring points P1–P8 (Figures 10 and 11), a pronounced temporal phase difference between the pressure and velocity fields was identified during the concentric-canister launch process. To avoid a purely

 https://doi.org/10.3390/en19051264

qualitative interpretation, the phase relationship was quantified using two parameters: the peak-time difference (Δt_peak) and the extremum lag time (Δt_lag).

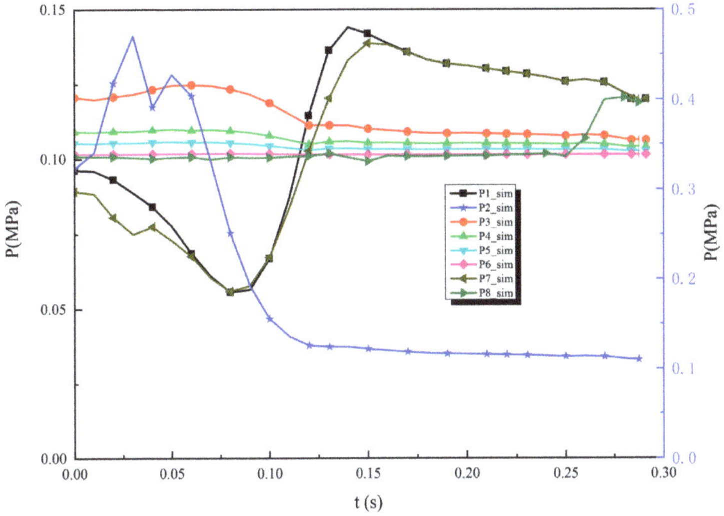

Figure 10. Table histories of pressure at monitoring points P1–P8.

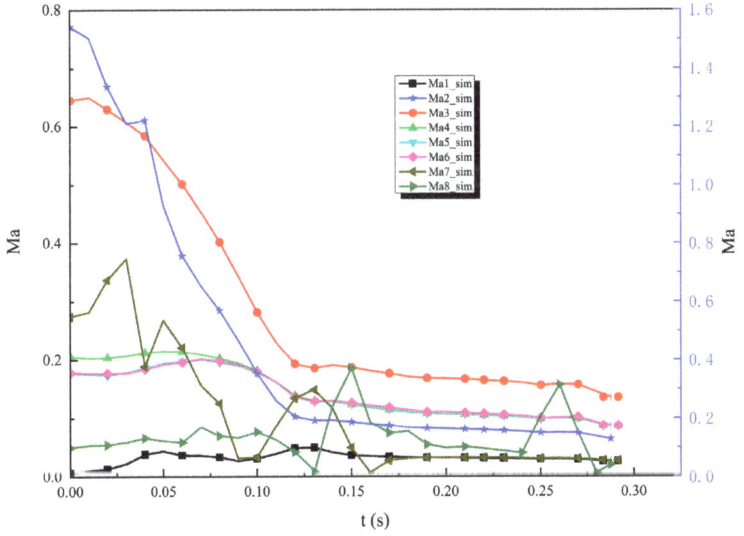

Figure 11. Time histories of Mach number at monitoring points P1–P8.

The results indicate that, in the upstream region (P1–P4), the pressure peak consistently occurred earlier than the minimum Mach number. For example, at monitoring point P1, the pressure reached its maximum at approximately 0.05 s, whereas the Mach number reached its minimum near 0.10 s, yielding a phase difference of $\Delta t \approx 0.05$ s.

From a physical perspective, compression waves propagate at speeds approaching the local speed of sound a, while velocity adjustment is governed by momentum transport

 https://doi.org/10.3390/en19051264

and boundary-layer development. The characteristic time scale associated with velocity restructuring can be expressed as:

$$\tau_p \sim \frac{a}{L}, \tau_m \sim \frac{U}{L} \tag{11}$$

where L denotes a characteristic length scale, and U represents a velocity scale. In the initial stage of compressible jet development, it generally holds that $a > U$, $\tau_p > \tau_m$, implying that the acoustic propagation time scale is shorter than the hydrodynamic adjustment time scale. Consequently, the pressure response precedes velocity-field reorganization. This separation of time scales is consistent with classical unsteady compressible-flow theory [39], and provides a dynamic basis for the observed "pressure-leading, velocity-lagging" behavior in the upstream shock-dominated region.

In contrast, a different phase relationship was observed in the downstream region (P5–P8). Taking P6 as an example, the pressure decreased to a new equilibrium level at approximately 0.02 s, whereas the Mach number remained relatively high (Ma $\approx$ 0.8) for a subsequent period. This phenomenon can be interpreted using the momentum conservation equation. After the pressure-gradient driving force diminishes, the flow retains significant inertial momentum. The velocity then decays primarily through viscous dissipation and turbulent mixing, the characteristic time scale of which can be expressed as

$$\tau_d \sim \frac{\rho L^2}{\mu} \tag{12}$$

This dissipation time scale is typically larger than the acoustic pressure-adjustment time scale; therefore, velocity attenuation lags behind pressure equilibration. Similar pressure–velocity temporal mismatches have been reported in unsteady confined jet flows and shock–boundary-layer interaction (SBLI) problems [40].

Furthermore, the downstream migration of shock structures shown in Figure 9 suggests that the phase lag is not a localized incidental phenomenon but intrinsically linked to the unsteady evolution of shock–boundary-layer-interaction structures. Previous studies have demonstrated that during shock movement and reflection, pressure disturbances propagate at acoustic scales, whereas the development of shear layers and separation regions is controlled by inertial and viscous time scales [41]. Hence, the asynchronous evolution of pressure and velocity fields has a clear physical origin.

Synthesizing data from multiple monitoring points reveals a spatially partitioned phase-lag pattern. In the upstream, shock-dominated region, pressure led the velocity response. In the downstream, dissipation-dominated region, velocity attenuation lagged behind pressure equilibration. This distinction arises from a shift in dominant time scales: acoustic propagation and shock compression govern upstream dynamics, whereas momentum inertia and turbulent dissipation dominate downstream behavior. These findings provide a time-scale-separation-based interpretation of the formation of unsteady loads and transient flow-structure evolution within the concentric canister.

4.3. *Thrust-Augmentation and Unsteady-Load Characteristics*

The transient flow simulation provided the rocket kinematic parameters and the impact force acting on the guide cone (Figure 12). The aerodynamic thrust exhibited a distinct V-shaped fluctuation. It decreased from an initial value of approximately 2500 N to nearly 1500 N within 0.1 s, and subsequently recovered to its initial level by 0.2 s. The total impulse was approximately 0.68 kN·s.

https://doi.org/10.3390/en19051264

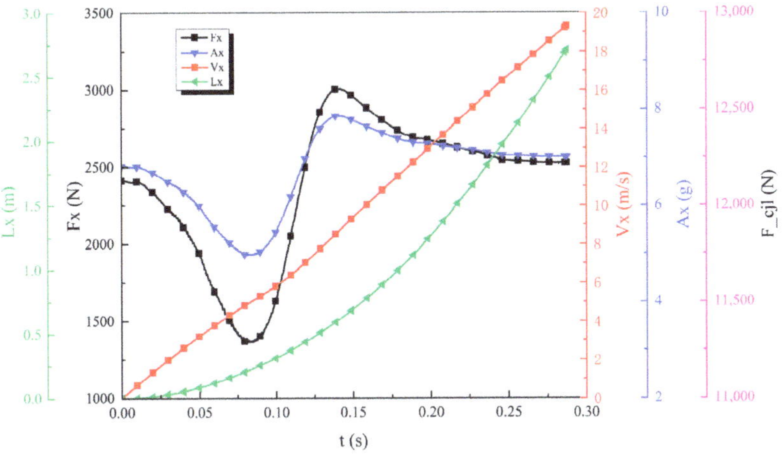

Figure 12. Table histories of rocket kinematic parameters and guide-cone impact force.

The rocket's acceleration dropped to nearly 0 g at 0.1 s and then increased rapidly, reaching approximately 10 g at 0.2 s. A phase difference of approximately 0.05 s between the thrust variation and acceleration response was observed, reflecting the inertial effects of the coupled fluid–structure system. The rocket's velocity reached 19.3 m/s at 0.287 s, corresponding to a displacement of 2.7 m.

The impact force on the guide cone reached a peak value of approximately 12,800 N at 0.075 s. It subsequently decreased by about 11% and transitioned into a quasi-steady scouring load stage. These results suggest that the structural design of the guide cone should prioritize resistance to short-duration peak impact loads rather than sustained high-frequency oscillations.

By comparing the theoretical engine thrust with the net axial thrust obtained from the simulation, the thrust augmentation induced by exhaust-gas redirection was estimated at approximately 15%.

In the present study, the "theoretical engine thrust" refers to the ideal axial thrust generated by the exhaust jet directly acting on the projectile without considering the influence of the concentric canister structure or external flow interference. This value serves as a reference baseline for evaluating thrust augmentation.

Under this assumption, the theoretical thrust was estimated based on the one-dimensional steady momentum conservation relationship:

$$F_{\text{th}} = \dot{m}u_e + (p_e - p_a)A_e \tag{13}$$

where $\dot{m}$ denotes the mass flow rate of exhaust gas; u_e represents the average exit velocity; p_e is the mean static pressure at the nozzle exit; p_a is the ambient pressure; and A_e is the nozzle exit area. These parameters were extracted as time-averaged quantities from the numerical simulation under the corresponding boundary conditions.

It should be emphasized that this theoretical thrust represents a unified comparison baseline for quantifying the relative thrust-enhancement effect of the concentric canister configuration rather than the actual engine thrust under complex external flow conditions.

The thrust augmentation percentage was defined as the relative increment in the computed axial thrust under the concentric canister configuration with respect to the theoretical baseline:

$$\Delta F(\%) = \frac{F - F_{\text{th}}}{F_{\text{th}}} \times 100\% \tag{14}$$

https://doi.org/10.3390/en19051264

During the concentric canister launch, strong rotational vortex structures developed near the canister base as the projectile exited. These included primary vortices and separation-induced vortical structures. To evaluate the contribution of vortex dynamics to thrust augmentation, coherent vortices at representative time instants were extracted using the Q-criterion (Figure 13).

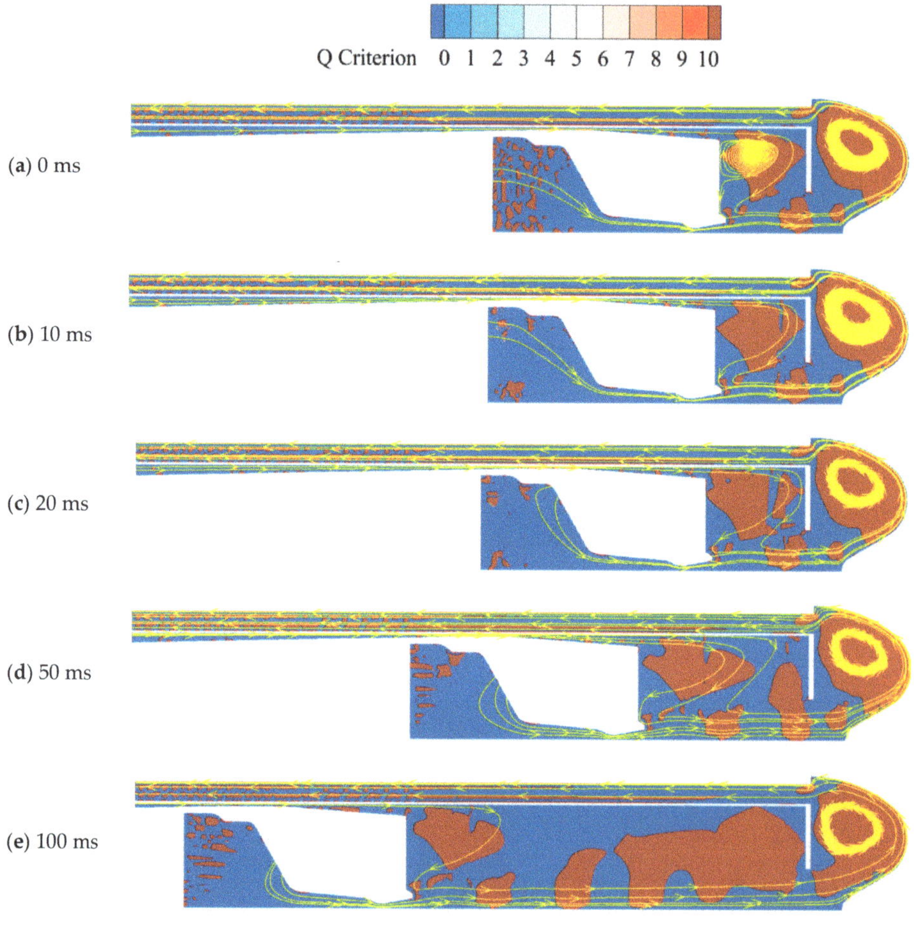

Figure 13. Evolution of vortex structures near the canister base identified by the Q-criterion (t = 0, 10, 20, 50, 100 ms).

The rotational motion of the vortices induces an additional axial pressure gradient toward the exit. Specifically, the low-pressure vortex core interacts with the surrounding high-pressure gas near the canister mouth, generating a pressure difference that accelerates the flow axially (i.e., in the rocket motion direction) rather than allowing radial dissipation along the wall. This pressure-gradient-driven mechanism can be interpreted as a vortex-induced "pumping effect," whereby radial kinetic energy is reorganized into axial momentum transfer to the rocket, thereby enhancing thrust.

It should be noted that the vortex contribution to thrust is statistical in nature. It couples with shock migration, separation recirculation, and other unsteady processes, collectively forming a comprehensive thrust-enhancement mechanism.

https://doi.org/10.3390/en19051264

4.4. Temperature-Field Evolution and Thermal Inertia Effects

4.4.1. Spatiotemporal Migration of the Temperature Field

As illustrated in Figure 14, the high-temperature region exhibited pronounced migration in both time and space during the launch process. At the early stage ($t \approx 0.01$ s), the hot core region (approximately 800 K) remained confined near the nozzle exit. As time progressed, the high-temperature front propagated downstream and reached the T4–T5 region at approximately 100 ms, where localized hot spots developed with temperatures up to 1200 K. Subsequently, the peak temperature occurred at monitoring point T6 at approximately 200 ms, reaching about 2500 K.

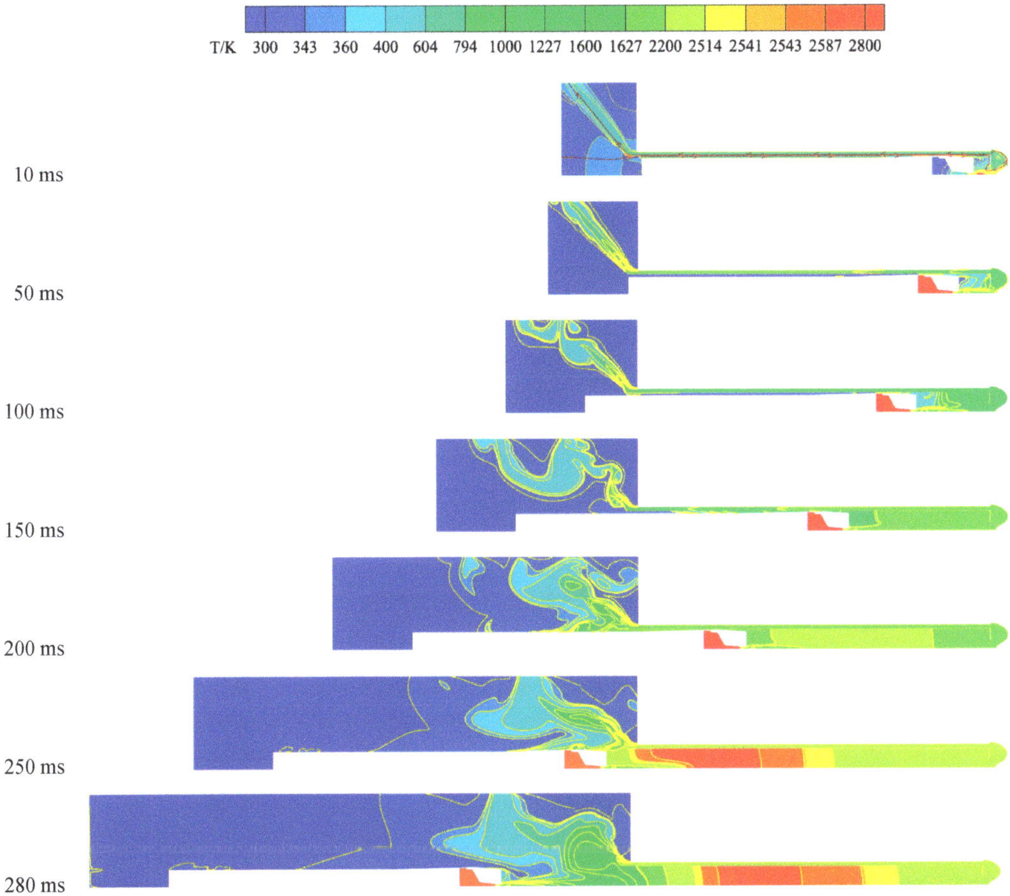

Figure 14. Transient evolution of temperature contours during launch (t = 10, 50, 100, 150, 200, 250, 280 ms).

During the initial period (0–50 ms), abnormal temperature spikes were observed at both T2 and T6, with peak values of 2300 ± 50 K. These values were approximately 15% higher than the corresponding isentropic estimates. Combined with the flow-field structures, this excessive heating was closely associated with strong shock–boundary-layer interactions induced by supersonic jet impingement on the guide cone, indicating that shock compression served as the dominant heat source during the early transient stage.

https://doi.org/10.3390/en19051264

Similar localized overheating resulting from supersonic jet–shock interaction has been reported in previous studies under comparable conditions [42].

4.4.2. Temperature–Pressure Temporal Relation and Thermal-Inertia Lag

Figure 15 presents the temperature histories at the monitoring points. The results show that the temperature peaks systematically lagged behind the corresponding pressure peaks by approximately 30–50 ms. For instance, the pressure at P2 reached its maximum at approximately 0.04 s, whereas the temperature at T2 reached its peak at approximately 0.10 s. Similar delays were consistently observed at multiple locations, including T4–T6.

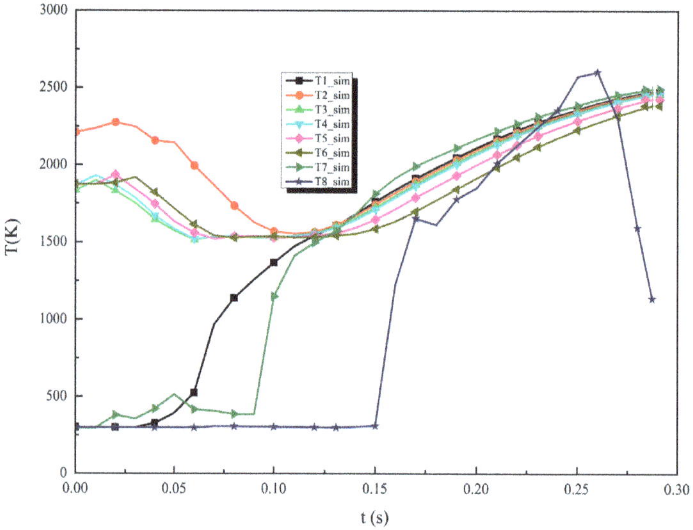

Figure 15. Time histories of temperature at monitoring points T1–T8.

This time-lag behavior indicates that temperature evolution was not synchronized with pressure-wave propagation but was instead governed by thermal inertia in the coupled fluid–solid system. Pressure disturbances propagate at acoustic speeds and establish rapidly, whereas heat transfer and accumulation require convection, conduction, and thermal-capacity absorption. The characteristic time scales of these thermal processes are typically much longer than those of pressure-wave propagation, producing a delayed thermal response. This phenomenon is inherently determined by thermal coupling at the fluid–solid interface and is consistent with findings from related transient heating studies [43].

The observed 30–50 ms delay provides a key time-scale parameter for transient thermal-load assessment. It suggests that thermal protection and structural design cannot be evaluated solely at the instant of maximum pressure but must account for delayed temperature accumulation. Previous investigations have shown that thermal-response lag in high-temperature transient gas-driven systems is governed jointly by thermal diffusivity, flow velocity, and material heat capacity. Therefore, this lag implies that the peak thermal-risk window occurs significantly later than the maximum aerodynamic load event, providing a clear temporal reference for thermal-protection thickness design and safety-margin evaluation.

This delayed response is of practical significance, particularly under high-enthalpy transient conditions, where temperature lag may increase the risk of thermal damage when the thermal load reaches its maximum level [44]. To ensure structural integrity,

 https://doi.org/10.3390/en19051264

such lag effects should be incorporated into design procedures, and thermal loads should be predicted over the full thermal-response time horizon rather than based solely on instantaneous pressure peaks.

4.4.3. Physical Inference of a Thermal–Flow–Force Positive-Feedback Mechanism and Applicability

By jointly examining the pressure (P), Mach number (Ma), and temperature (T) fields, a coupled feature was identified at $t \approx 0.05$ s and $x/D \approx 1.2$, where high pressure (P2), low Mach number (Ma2), and elevated temperature (T2) occurred simultaneously. Similar coupled patterns were detected in key regions, such as downstream of the shock system and near the gap inlet. Figure 14 illustrates the temporal evolution and spatial distribution of these variables.

Based on this spatiotemporal consistency, a possible thermal–flow–force positive-feedback pathway is proposed: local temperature rise $\rightarrow$ increased speed of sound $\rightarrow$ reduced effective choking area $\rightarrow$ shock strengthening $\rightarrow$ enhanced compression heating.

This feedback mechanism has been suggested in several studies, particularly in high-pressure flows involving jet–shock interactions. The existing literature indicates that temperature rise increases the local sound speed, thereby altering the local Mach number distribution and potentially intensifying shock structures [45]. Strengthened shocks further compress the gas and increase heat accumulation, leading to additional temperature rise. Such a feedback process may significantly influence supersonic jet systems and shock-dominated confined configurations [46].

It should be noted that this positive-feedback mechanism was not directly confirmed through a closed-loop quantitative model but was inferred from synchronized multi-field evolution features. Nevertheless, the proposed interpretation shares a similar theoretical basis with classical shock–boundary-layer interaction models and jet-expansion theory. Its validity depends on the following conditions:

- Strong shock structures that generate sufficiently high pressure gradients to enable effective energy transfer and promote compression heating;
- A significant local temperature rise that serves as the primary driver of increased sound speed and shock intensification;
- Geometric confinement, whereby flow-path constraints (e.g., nozzle and guide-cone geometry) amplify thermal–flow–force coupling.

Therefore, this mechanism is expected to apply mainly to high-enthalpy, strongly unsteady, confined supersonic jet conditions, such as those encountered in the present concentric-canister launch process. Under low-Mach or weak-shock regimes, the feedback intensity may be negligible. Previous studies have shown that at low Mach numbers, aerodynamic-property variation is smoother and shock effects diminish, resulting in a significant reduction in such coupled feedback mechanisms [47,48].

5. Conclusions

This study combined experimental measurements with transient numerical simulations to elucidate the unsteady evolution of exhaust gas flow and the thrust augmentation mechanism in a concentric-canister launch system. The main conclusions are summarized below.

(1) Multi-Stage Flow Evolution and Phase-Lag Mechanism

A distinct multi-stage dynamic evolution process was identified, characterized by gas filling, annular jet establishment, and transition to a free-jet regime. A pronounced spatially dependent phase lag between pressure and Mach number responses was observed. In

 https://doi.org/10.3390/en19051264

the upstream, shock-dominated region, pressure variations preceded velocity adjustment, whereas in the downstream, dissipation-dominated region, velocity attenuation lagged behind pressure relaxation.

This phase-lag behavior originates from the mismatch among acoustic propagation time scales, inertial momentum transport, and shock–boundary-layer-interaction dynamics. The identified time-scale separation mechanism provides a new physical interpretation of the formation of unsteady peak loads. From an engineering perspective, recognizing this lag phenomenon is essential for accurate structural-load prediction and transient-response assessment, offering valuable guidance for propulsion-system design and aerodynamic optimization.

(2) Thermal–Flow–Force Positive Feedback and Thermal-Inertia-Controlled Time Scale

Quantitative analysis revealed a coupled thermal–flow–force interaction mechanism. In regions where high temperature, elevated pressure, and low Mach number overlapped spatially, local peak temperatures reached approximately 2500 K. A systematic delay of 30–50 ms was observed between temperature peaks and the corresponding pressure maxima, indicating that thermal inertia governs the energy-transfer time scale.

The delayed thermal response arises from convective transport, conductive diffusion, and material heat-capacity absorption, leading to temperature peaks that do not coincide with pressure maxima. This time lag defines a critical thermal-risk window occurring after the maximum aerodynamic load. The identified thermal-response delay provides an essential temporal reference for thermal-protection-system design and safety-margin evaluation, particularly under high-enthalpy transient conditions.

(3) Predictive Capability and Engineering Quantification

A bidirectional fluid–structure coupling framework incorporating dynamic mesh techniques was established and validated, enabling quantitative prediction of complex transient launch flows. The contribution of bottom-region vortical structures to axial thrust was quantified at approximately 15%. The peak impact load on the guide cone reached 12,800 N and decreased by about 11% within 80 ms, transitioning to a quasi-steady loading regime.

These quantitative results provide reliable data support for structural optimization, load-spectrum characterization, and service-life assessment. The findings offer practical design guidance for high-energy propulsion systems operating under high-temperature, high-pressure transient conditions, particularly regarding material selection and thermal-protection thickness optimization.

(4) Ethics statements

This study involves high-temperature and high-pressure transient propulsion flow phenomena that may have potential dual-use implications. The research was conducted solely for legitimate scientific and engineering purposes, aiming to improve the understanding of multi-physics coupling mechanisms, structural safety, and thermal-protection design in high-energy propulsion systems. All work was carried out in full compliance with applicable national and international laws, institutional regulations, and research-ethics standards. No classified, restricted, or export-controlled information was used or disclosed. The authors are committed to the principles of responsible research and innovation and to ensuring that the findings are applied in accordance with legal and ethical frameworks.

Future Work

Given the limitations of the present study, future research should focus on the following:

- Model Refinement: Conducting full three-dimensional simulations incorporating detailed thermochemical reaction models and multiphase effects to improve prediction accuracy under realistic exhaust conditions.

https://doi.org/10.3390/en19051264

- Deep Multidisciplinary Coupling: Integrating the proposed thermal–flow–force framework with structural dynamics and material-damage models to establish a closed-loop analysis chain from flow-field prediction to structural integrity and lifetime evaluation.

Author Contributions: S.Y.: Conceptualization, methodology, formal analysis, writing—review and editing, resources. H.Q.: Data curation, writing—review and editing. W.S.: Investigation, supervision, resources. N.N.: Data curation. W.H.: Data curation. Y.W.: Validation. All authors have read and agreed to the published version of the manuscript.

Funding: This research received no specific grant from any funding agency in the public, commercial, or not-for-profit sectors.

Data Availability Statement: The data that support the findings of this study are available from the corresponding author upon reasonable request.

Acknowledgments: The authors would like to express their sincere gratitude to Wenyan Song for her guidance, encouragement, and support throughout this research project.

Conflicts of Interest: The authors declare that they have no known competing financial interests or personal relationships that could have appeared to influence the work reported in this paper.

Nomenclature

Symbol	Description	Unit
ρ	Density	kg/m^3
u	Velocity component in the i-th direction	m/s
t	Time	s
$x_i(or\ x, y, z)$	Spatial coordinate in the i-th direction	m
i, j	Tensor indices	—
$\overline{u}_i$(or u_i)	Reynolds-averaged velocity component	m/s
p	Static pressure	Pa
τ_{ij}	Molecular viscous stress tensor	Pa
$-\rho\overline{u_{i'}u_{j'}}$	Reynolds-stress tensor	Pa
E	Total energy per unit mass	J/kg
T	Temperature	K
μ	Dynamic viscosity	Pa·s
μ_t	Turbulent viscosity	Pa·s
μ_{eff}	Effective viscosity $\mu_{\text{eff}} = \mu + \mu_t$	Pa·s
$\lambda(k)$	Molecular thermal conductivity	W/(m · K)
$\lambda_t(k_t)$	Turbulent thermal conductivity	W/(m · K)
$\lambda_{\text{eff}}(k_{\text{eff}})$	Effective thermal conductivity	W/(m · K)
k	Turbulent kinetic energy	m^2/s^2
ε	Turbulent dissipation rate	m^2/s^3
P_k	Production term of turbulent kinetic energy	m^2/s^3
C_μ	Empirical constant in the standard k–ε model	—
$C_{1\varepsilon}$	Production coefficient in ε-equation	—
$C_{2\varepsilon}$	Dissipation coefficient in ε-equation	—
σ_k	Turbulent Prandtl number (diffusion coefficient) for k	—
σ_ε	Turbulent Prandtl number (diffusion coefficient) for ε	—
Di	Inner diameter of concentric canister	m
Do	Outer diameter of concentric canister	m
L_{canister}	Total length of concentric canister	m
D_{rocket}	Rocket body diameter	m
L_{rocket}	Rocket total length	m
R_{outer}	Radial radius of external computational domain	m
L_{outer}	Axial length of external computational domain	m

https://doi.org/10.3390/en19051264

γ	Specific heat ratio	-
R	Gas constant	J/(kg·K)
μ	Dynamic viscosity	Pa·s
μ_{ref}	Reference viscosity (Sutherland's law)	Pa·s
T_{ref}	Reference temperature	K
C	Sutherland's constant	K
Q	Q-criterion (second invariant of velocity-gradient tensor)	$1/s^2$
$\nabla\mathbf{u}$	Velocity-gradient tensor	1/s
$\mathbf{S}$	Strain-rate tensor (symmetric part)	1/s
$\mathbf{\Omega}$	Rotation-rate tensor (antisymmetric part of the velocity-gradient tensor)	1/s
$\mathbf{u}$	Rotation-rate tensor (antisymmetric part)	m/s
S_{ij}	Component of the strain-rate tensor	1/s
Ω_{ij}	Component of the rotation-rate tensor (vorticity-tensor component)	1/s
$Q > 0$	Rotation-dominated region (vortex identification criterion)	—
pi	Measured pressure at monitoring point	MPa
FS	Sensor full scale	MPa
Δp_{sensor}	Sensor accuracy uncertainty	MPa
Δp_{sample}	Sampling and signal-processing uncertainty	MPa
Δp_{repeat}	Repeatability uncertainty	MPa
u_c	Combined standard uncertainty (RSS)	MPa
U	Expanded uncertainty (coverage factor (k = 2))	MPa
σ	Standard deviation of steady-state signal	MPa
F_{th}	Theoretical engine thrust (ideal baseline)	N
$\dot{m}$	Exhaust mass flow rate	kg/s
u_e	Nozzle exit average velocity	m/s
p_e	Nozzle exit static pressure	Pa (or MPa)
p_a	Ambient pressure	Pa (or MPa)
A_e	Nozzle exit area	m^2
F	Axial thrust under concentric configuration	N
η_F	Thrust augmentation percentage	%
RSM	Reynolds-averaged Navier–Stokes	-
LES	Large-eddy simulation	-
DNS	Direct numerical simulation	-
$RANS$	Reynolds-averaged Navier–Stokes (RANS) equations	-
$k-\varepsilon$	$k-\varepsilon$ turbulence model	-
Ma	Mach number	-

References

1. Lu, C.-Y.; Wang, X.; Zhou, Z.-T.; Liang, X.-Y.; Le, G.-G. Investigation on separating kinematic characteristics of submarine-launched missile near free surface. *J. Astronaut.* **2021**, *42*, 496–503.
2. Li, X.; Chen, X.-Q.; Meng, L.-T. Analysis on the effect of ship swaying motion on ship-borne vertically launch missile in launching process. *Missiles Space Veh.* **2014**, *5*, 19–21.
3. Zou, H.; Yang, F.; Cai, F.; Shi, Q. Numerical study of gas–liquid two-phase flow and noise characteristics for a water injection launching concentric canister launcher. *Int. J. Nonlinear Sci. Numer. Simul.* **2022**, *23*, 901–915. [CrossRef]
4. Li, R.; Zhang, Q.; Wang, C.; Liu, Y.; Li, Z.; Du, P. Hydrodynamic and kinematic characteristics of high-speed inclined water-exit process in wave environment. *J. Northwest. Polytech. Univ.* **2024**, *42*, 616–624. [CrossRef]
5. Yu, Y.; Xu, X.-W.; Fu, D.-B. Aerodynamic theory analysis of gas exhausting from concentric canister launcher. *J. Aerosp. Power* **2012**, *27*, 501–507.
6. Li, W.; Zhang, Z.; Lu, J.; Li, Z.; Wang, C. Investigations on the flow characteristics and the structural response of the launch tube during the underwater launching process. *Ocean Eng.* **2023**, *279*, 114603. [CrossRef]
7. Fu, D.-B.; Jiang, Y. Simulation of jet flow during missile launching with dynamic mesh. *J. Astronaut.* **2007**, *28*, 423–426.

 https://doi.org/10.3390/en19051264

8. Chandar, D.D.J.; Sitaraman, J. A flux correction approach for the pressure equation in incompressible flows on overset meshes in OpenFOAM. *Comput. Phys. Commun.* **2022**, *273*, 108279. [CrossRef]
9. Hu, J.; Zhang, W.; Guo, H.; Sun, S.; Chen, F.; Guo, C. Numerical simulation of propeller wake vortex–rudder interaction in oblique flows. *Ships Offshore Struct.* **2021**, *16*, 144–155. [CrossRef]
10. Bai, X.; Cheng, H.; Ji, B. LES investigation of the noise characteristics of sheet and tip leakage vortex cavitating flow. *Int. J. Multiphase Flow* **2022**, *146*, 103880. [CrossRef]
11. Liu, H.; Li, S.; Hou, S.; Fu, D. Coupled modeling of sea surface launch flow and multi-body motion. *J. Mar. Sci. Eng.* **2024**, *12*, 1736. [CrossRef]
12. Zhang, J.; Liu, G.; Han, W.; Liu, L.; Wang, Z. Numerical research on the muzzle multiphase flow field produced by gas curtain launch. *Sci. Rep.* **2024**, *14*, 29421. [CrossRef] [PubMed]
13. Wang, J.; Wang, Y. Assessment of turbulence models and mesh types for the simulation of multi-body separation by coupling the CFD and rigid body dynamics. *J. Phys. Conf. Ser.* **2023**, *2512*, 012005. [CrossRef]
14. Chen, W.; Li, R.; Huang, J.; Dong, H.; Qiu, Q.; Chen, Q. Numerical investigation of coupled oblique flow and steering effects on hydrodynamic performance of rudder behind propeller. *J. Mar. Sci. Eng.* **2025**, *13*, 2140. [CrossRef]
15. Yan, P.; Jiang, Y.; Qian, H. Numerical simulation on inter-stage separation of finned launch vehicle. *J. Phys. Conf. Ser.* **2025**, *3004*, 012078. [CrossRef]
16. Subramanian, S.; Craig, B.; White, C.; Kontis, K.; Evans, D.; Eynde, J.V.D. Underexpanded jet impingement in near vacuum environment. *Phys. Fluids* **2024**, *36*, 103017. [CrossRef]
17. Tian, C.; Zhao, Z.; Sun, Z.; He, G.; Wang, K.; Teng, H. Numerical investigation of impinging plume under vacuum and realistic nozzle outlet condition. *Phys. Fluids* **2024**, *36*, 013328. [CrossRef]
18. Fu, D.; Hao, H. Investigations for missile launching in an improved concentric canister launcher. *J. Spacecr. Rockets* **2015**, *52*, 1510–1515. [CrossRef]
19. Fu, D.; Yu, Y. Simulation of gas flow and additional thrust with missile launching from concentric canister launcher. *Proc. Inst. Mech. Eng. Part G* **2013**, *227*, 1977–1987. [CrossRef]
20. Zhang, X.; Yu, Y.; Hu, Y. Experimental study on gas–liquid–solid interaction characteristics in the launch tube. *J. Mar. Sci. Eng.* **2022**, *10*, 1239. [CrossRef]
21. Yu, Y.; Mu, Y.-T. Influential factors analysis to additional ejection force in concentric canister launcher. *J. Aerosp. Power* **2014**, *29*, 980–986.
22. Jia, Q.; Jiang, Y.; Yang, Y.; Zhao, Z.; Wang, Z. A new type of controllable thrust vertical launcher and its interior ballistic law. *Acta Armamentarii* **2022**, *43*, 1596–1604.
23. Al-Doori, G.; Saleh, K.; Al-Manea, A.; Al-Rbaihat, R.; Altork, Y.; Alahmer, A. A review of axial and radial ejectors: Geometric design, computational analysis, performance, and machine learning approaches. *Appl. Therm. Eng.* **2025**, *266*, 125694. [CrossRef]
24. Kim, J.; Park, S.; Lee, J. Entrainment ratio performance of supersonic–supersonic annular ejector with supersonic secondary flow. *J. Therm. Sci.* **2025**, *26*, 2872–2884, *in press*. [CrossRef]
25. Hu, X.; Le, G.; Ma, D.; Zhu, Z. Numerical simulation of secondary combustion gas field in concentric canister launcher and improvement measures of guider. *Acta Armamentarii* **2014**, *35*, 62.
26. Yang, F.; Ma, D.; Xue, X.; Cui, L. Thermal shock mechanism and thermal environment influencing factors of a new concentric canister launcher. *Explos. Shock Waves* **2016**, *36*, 153–160. [CrossRef]
27. Zhang, Y.; Ma, G. Nonlinear thermal conduction of concentric canister launcher. *J. Eng. Thermophys.* **2018**, *39*, 2251–2259.
28. Zhang, Q.; Ye, H.; Wang, Z. A comparative study of launch canister thermal control systems using a liquid or gas working medium. *Appl. Therm. Eng.* **2018**, *131*, 125–131. [CrossRef]
29. Wang, Y.; Su, Y.; Sun, Z.; Wu, S.; Wu, H.; Le, G. Numerical analysis of the thermal environment of rail-based hot launch systems. *Aerospace* **2025**, *12*, 823. [CrossRef]
30. Zhang, J.; Guo, T.; Dang, G.; Li, X. Effects of wall temperature on hypersonic shock wave/turbulent boundary layer interactions. *J. Fluid Mech.* **2024**, *990*, A21. [CrossRef]
31. Huang, J.; Duan, L.; Choudhari, M.M. Direct numerical simulation of hypersonic turbulent boundary layers: Effect of spatial evolution and Reynolds number. *J. Fluid Mech.* **2022**, *937*, A3. [CrossRef]
32. Ninni, D.; Bonelli, F.; Pascazio, G. Simulation of high-enthalpy turbulent shock wave/boundary layer interaction using a RANS approach. *Aerotec. Missili Spaz.* **2023**, *102*, 323–335. [CrossRef]
33. Gaitonde, D.V.; Adler, M.C. Dynamics of three-dimensional shock-wave/boundary-layer interactions. *Annu. Rev. Fluid Mech.* **2023**, *55*, 291–321. [CrossRef]
34. Weiland, C.J.; Vlachos, P.P.; Yagla, J.J. Concept analysis and laboratory observations on a water piercing missile launcher. *Ocean Eng.* **2010**, *37*, 959–965. [CrossRef]
35. Elisov, N.A.; Ishkov, S.A.; Shakhov, V.G. Numerical analysis of air dissociation influence on spaceplane aerodynamic characteristics. *Acta Astronaut.* **2018**, *148*, 153–162. [CrossRef]

 https://doi.org/10.3390/en19051264

36. Wilcox, D.C. *Turbulence Modeling for CFD*, 3rd ed.; DCW Industries: La Canada, CA, USA, 2006.
37. Anderson, J.D. *Computational Fluid Dynamics: The Basics with Applications*; McGraw-Hill: New York, NY, USA, 1995.
38. Hunt, J.C.R.; Wray, A.A.; Moin, P. Eddies, streams, and convergence zones in turbulent flows. In *Studying Turbulence Using Numerical Simulation Databases*; Center for Turbulence Research: Stanford, CA, USA, 1988; pp. 193–208.
39. Schlichting, H.; Gersten, K. *Boundary-Layer Theory*, 8th ed.; Springer: Berlin/Heidelberg, Germany, 2017.
40. Bernardini, M.; Della Posta, G.; Salvadore, F.; Martelli, E. Unsteadiness characterisation of shock wave/turbulent boundary-layer interaction at moderate Reynolds number. *arXiv* **2022**, arXiv:2204.10531. [CrossRef]
41. Huang, J.; Burridge, H.C.; van Reeuwijk, M. On the low-frequency unsteadiness in shock wave–turbulent boundary layer interactions. *J. Fluid Mech.* **2023**, *956*, R1. [CrossRef]
42. Li, M.; Karami, S.; Sandberg, R.; Soria, J.; Ooi, A. Heat transfer investigation on under expanded supersonic impinging jets. *Int. J. Heat Fluid Flow* **2023**, *101*, 109132. [CrossRef]
43. Lewis, M.T.; Hickey, J.P. Conjugate heat transfer in high speed external flows: A review. *J. Thermophys. Heat Transf.* **2023**, *37*, 697–712. [CrossRef]
44. Li, J.; Teng, J.; Lu, S. Transient state flow and heat transfer performance over the flat tip of HP turbine. *Aerospace Syst.* **2023**, *7*, 525–538. [CrossRef]
45. Zapryagaev, V.; Kiselev, N.; Gubanov, D. Shock wave structure of supersonic jet flows. *Aerospace* **2018**, *5*, 60. [CrossRef]
46. Sabnis, K.; Babinsky, H. A review of three dimensional shock wave–boundary layer interactions. *Prog. Aerosp. Sci.* **2023**, *143*, 100953. [CrossRef]
47. Bernardini, M.; Asproulias, I.; Larsson, J.; Pirozzoli, S.; Grasso, F. Heat transfer and wall temperature effects in shock wave/turbulent boundary layer interactions. *Phys. Rev. Fluids* **2016**, *1*, 084403. [CrossRef]
48. Ligrani, P.M.; McNabb, E.S.; Collopy, H.; Anderson, M.; Marko, S.M. Recent investigations of shock wave effects and interactions. *Adv. Aerodyn.* **2020**, *2*, 4. [CrossRef]

 https://doi.org/10.3390/en19051264

MDPI AG
Grosspeteranlage 5
4052 Basel
Switzerland
Tel.: +41 61 683 77 34

MDPI Books Editorial Office
E-mail: books@mdpi.com
www.mdpi.com/books